ROBOTICS:
Control, Sensing, Vision, and Intelligence

CAD/CAM, Robotics, and Computer Vision

Consulting Editor
Herbert Freeman, Rutgers University

Fu, Gonzalez, and Lee: *Robotics: Control, Sensing, Vision, and Intelligence*
Groover, Weiss, Nagel, and Odrey: *Industrial Robotics: Technology, Programming, and Applications*
Levine: *Vision in Man and Machine*
Parsons: *Voice and Speech Processing*

ROBOTICS:
Control, Sensing, Vision, and Intelligence

K. S. Fu
School of Electrical Engineering
Purdue University

R. C. Gonzalez
Department of Electrical Engineering
University of Tennessee
and
Perceptics Corporation
Knoxville, Tennessee

C. S. G. Lee
School of Electrical Engineering
Purdue University

McGraw-Hill Book Company

New York   St. Louis   San Francisco   Auckland   Bogotá
Hamburg   London   Madrid   Mexico   Milan   Montreal   New Delhi
Panama   Paris   São Paulo   Singapore   Sydney   Tokyo   Toronto

To
Viola,
Connie, and
Pei-Ling

Library of Congress Cataloging-in-Publication Data

Fu, K. S. (King Sun),
    Robotics : control, sensing, vision, and intelligence.

    (McGraw-Hill series in CAD/CAM robotics and computer
vision)
    Bibliography: p.
    Includes index.
    1. Robotics.   I. Gonzalez, Rafael C.   II. Lee,
C. S. G. (C. S. George)   III. Title.
TJ211.F82   1987   629.8′ 92   86-7156
ISBN 0-07-022625-3
ISBN 0-07-022626-1 (solutions manual)

This book was set in Times Roman by House of Equations Inc.
The editor was Sanjeev Rao;
the production supervisor was Diane Renda;
the cover was designed by Laura Stover.
Project supervision was done by Lynn Contrucci.
R. R. Donnelley & Sons Company was printer and binder.

ROBOTICS: CONTROL, SENSING, VISION, AND INTELLIGENCE

    2 3 4 5 6 7 8 9 0 DOCDOC 8 9 8 7

ISBN 0-07-022625-3

ABOUT THE AUTHORS

K. S. Fu was the W. M. Goss Distinguished Professor of Electrical Engineering at Purdue University. He received his bachelor, master, and Ph.D. degrees from the National Taiwan University, the University of Toronto, and the University of Illinois, respectively. Professor Fu was internationally recognized in the engineering disciplines of pattern recognition, image processing, and artificial intelligence. He made milestone contributions in both basic and applied research. Often termed the "father of automatic pattern recognition," Dr. Fu authored four books and more than 400 scholarly papers. He taught and inspired 75 Ph.D.s. Among his many honors, he was elected a member of the National Academy of Engineering in 1976, received the Senior Research Award of the American Society for Engineering Education in 1981, and was awarded the IEEE's Education Medal in 1982. He was a Fellow of the IEEE, a 1971 Guggenheim Fellow, and a member of Sigma Xi, Eta Kappa Nu, and Tau Beta Pi honorary societies. He was the founding president of the International Association for Pattern Recognition, the founding editor in chief of the IEEE *Transactions of Pattern Analysis and Machine Intelligence*, and the editor in chief or editor for seven leading scholarly journals. Professor Fu died of a heart attack on April 29, 1985 in Washington, D.C.

R. C. Gonzalez is IBM Professor of Electrical Engineering at the University of Tennessee, Knoxville, and founder and president of Perceptics Corporation, a high-technology firm that specializes in image processing, pattern recognition, computer vision, and machine intelligence. He received his B.S. degree from the University of Miami, and his M.E. and Ph.D. degrees from the University of Florida, Gainesville, all in electrical engineering. Dr. Gonzalez is internationally known in his field, having authored or coauthored over 100 articles and 4 books dealing with image processing, pattern recognition, and computer vision. He received the 1978 UTK Chancellor's Research Scholar Award, the 1980 Magnavox Engineering Professor Award, and the 1980 M.E. Brooks Distinguished Professor Award for his work in these fields. In 1984 he was named Alumni Distinguished Service Professor

v

at the University of Tennessee. In 1985 he was named a distinguished alumnus by the University of Miami. Dr. Gonzalez is a frequent consultant to industry and government and is a member of numerous engineering professional and honorary societies, including Tau Beta Pi, Phi Kappa Phi, Eta Kappa Nu, and Sigma Xi. He is a Fellow of the IEEE.

C. S. G. Lee is an associate professor of Electrical Engineering at Purdue University. He received his B.S.E.E. and M.S.E.E. degrees from Washington State University, and a Ph.D. degree from Purdue in 1978. From 1978 to 1985, he was a faculty member at Purdue and the University of Michigan, Ann Arbor. Dr. Lee has authored or coauthored more than 40 technical papers and taught robotics short courses at various conferences. His current interests include robotics and automation, and computer-integrated manufacturing systems. Dr. Lee has been doing extensive consulting work for automotive and aerospace industries in robotics. He is a Distinguished Visitor of the IEEE Computer Society's Distinguished Visitor Program since 1983, a technical area editor of the IEEE *Journal of Robotics and Automation*, and a member of technical committees for various robotics conferences. He is a coeditor of *Tutorial on Robotics*, 2nd edition, published by the IEEE Computer Society Press and a member of Sigma Xi, Tau Beta Pi, the IEEE, and the SME/RI.

CONTENTS

PREFACE

This textbook was written to provide engineers, scientists, and students involved in robotics and automation with a comprehensive, well-organized, and up-to-date account of the basic principles underlying the design, analysis, and synthesis of robotic systems.

The study and development of robot mechanisms can be traced to the mid-1940s when master-slave manipulators were designed and fabricated at the Oak Ridge and Argonne National Laboratories for handling radioactive materials. The first commercial computer-controlled robot was introduced in the late 1950s by Unimation, Inc., and a number of industrial and experimental devices followed suit during the next 15 years. In spite of the availability of this technology, however, widespread interest in robotics as a formal discipline of study and research is rather recent, being motivated by a significant lag in productivity in most nations of the industrial world.

Robotics is an interdisciplinary field that ranges in scope from the design of mechanical and electrical components to sensor technology, computer systems, and artificial intelligence. The bulk of material dealing with robot theory, design, and applications has been widely scattered in numerous technical journals, conference proceedings, research monographs, and some textbooks that either focus attention on some specialized area of robotics or give a "broadbrush" look of this field. Consequently, it is a rather difficult task, particularly for a newcomer, to learn the range of principles underlying this subject matter. This text attempts to put between the covers of one book the basic analytical techniques and fundamental principles of robotics, and to organize them in a unified and coherent manner. Thus, the present volume is intended to be of use both as a textbook and as a reference work. To the student, it presents in a logical manner a discussion of basic theoretical concepts and important techniques. For the practicing engineer or scientist, it provides a ready source of reference in systematic form.

The mathematical level in all chapters is well within the grasp of seniors and first-year graduate students in a technical discipline such as engineering and computer science, which require introductory preparation in matrix theory, probability, computer programming, and mathematical analysis. In presenting the material, emphasis is placed on the development of fundamental results from basic concepts. Numerous examples are worked out in the text to illustrate the discussion, and exercises of various types and complexity are included at the end of each chapter. Some of these problems allow the reader to gain further insight into the points discussed in the text through practice in problem solution. Others serve as supplements and extensions of the material in the book. For the instructor, a complete solutions manual is available from the publisher.

This book is the outgrowth of lecture notes for courses taught by the authors at Purdue University, the University of Tennessee, and the University of Michigan. The material has been tested extensively in the classroom as well as through numerous short courses presented by all three authors over a 5-year period. The suggestions and criticisms of students in these courses had a significant influence in the way the material is presented in this book.

We are indebted to a number of individuals who, directly or indirectly, assisted in the preparation of the text. In particular, we wish to extend our appreciation to Professors W. L. Green, G. N. Saridis, R. B. Kelley, J. Y. S. Luh, N. K. Loh, W. T. Snyder, D. Brzakovic, E. G. Burdette, M. J. Chung, B. H. Lee, and to Dr. R. E. Woods, Dr. Spivey Douglass, Dr. A. K. Bejczy, Dr. C. Day, Dr. F. King, and Dr. L-W. Tsai. As is true with most projects carried out in a university environment, our students over the past few years have influenced not only our thinking, but also the topics covered in this book. The following individuals have worked with us in the course of their advanced undergraduate or graduate programs: J. A. Herrera, M. A. Abidi, R. O. Eason, R. Safabakhsh, A. P. Perez, C. H. Hayden, D. R. Cate, K. A. Rinehart, N. Alvertos, E. R. Meyer, P. R. Chang, C. L. Chen, S. H. Hou, G. H. Lee, R. Jungclas, Huarg, and D. Huang. Thanks are also due to Ms. Susan Merrell, Ms. Denise Smiddy, Ms. Mary Bearden, Ms. Frances Bourdas, and Ms. Mary Ann Pruder for typing numerous versions of the manuscript. In addition, we express our appreciation to the National Science Foundation, the Air Force Office of Scientific Research, the Office of Naval Research, the Army Research Office, Westinghouse, Martin Marietta Aerospace, Martin Marietta Energy Systems, Union Carbide, Lockheed Missiles and Space Co., The Oak Ridge National Laboratory, and the University of Tennessee Measurement and Control Center for their sponsorship of our research activities in robotics, computer vision, machine intelligence, and related areas.

K. S. Fu
R. C. Gonzalez
C. S. G. Lee

Professor King-Sun Fu died of a heart attack on April 29, 1985, in Washington, D.C., shortly after completing his contributions to this book. He will be missed by all those who were fortunate to know him and to work with him during a productive and distinguished career.

R. C. G.
C. S. G. L.

ONE

INTRODUCTION

> One machine can do the work of a
> hundred ordinary men, but no machine
> can do the work of one extraordinary man.
> *Elbert Hubbard*

1.1 BACKGROUND

With a pressing need for increased productivity and the delivery of end products of uniform quality, industry is turning more and more toward computer-based automation. At the present time, most automated manufacturing tasks are carried out by special-purpose machines designed to perform predetermined functions in a manufacturing process. The inflexibility and generally high cost of these machines, often called *hard automation systems,* have led to a broad-based interest in the use of robots capable of performing a variety of manufacturing functions in a more flexible working environment and at lower production costs.

The word *robot* originated from the Czech word *robota,* meaning work. Webster's dictionary defines robot as "an automatic device that performs functions ordinarily ascribed to human beings." With this definition, washing machines may be considered robots. A definition used by the Robot Institute of America gives a more precise description of industrial robots: "A robot is a *reprogrammable multi-functional* manipulator designed to move materials, parts, tools, or specialized devices, through variable programmed motions for the performance of a variety of tasks." In short, a robot is a reprogrammable general-purpose manipulator with external sensors that can perform various assembly tasks. With this definition, a robot must possess *intelligence,* which is normally due to computer algorithms associated with its control and sensing systems.

An industrial robot is a general-purpose, computer-controlled manipulator consisting of several rigid links connected in series by revolute or prismatic joints. One end of the chain is attached to a supporting base, while the other end is free and equipped with a tool to manipulate objects or perform assembly tasks. The motion of the joints results in relative motion of the links. Mechanically, a robot is composed of an arm (or mainframe) and a wrist subassembly plus a tool. It is designed to reach a workpiece located within its work volume. The work volume is the sphere of influence of a robot whose arm can deliver the wrist subassembly unit to any point within the sphere. The arm subassembly generally can move with three degrees of freedom. The combination of the movements positions the

1

wrist unit at the workpiece. The wrist subassembly unit usually consists of three rotary motions. The combination of these motions orients the tool according to the configuration of the object for ease in pickup. These last three motions are often called *pitch, yaw,* and *roll.* Hence, for a six-jointed robot, the arm subassembly is the positioning mechanism, while the wrist subassembly is the orientation mechanism. These concepts are illustrated by the Cincinnati Milacron T^3 robot and the Unimation PUMA robot arm shown in Fig. 1.1.

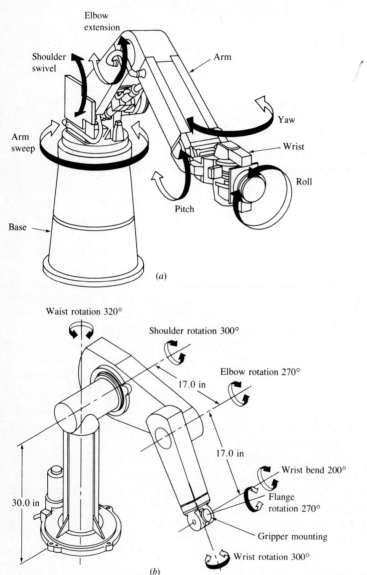

Figure 1.1 (*a*) Cincinnati Milacron T^3 robot arm. (*b*) PUMA 560 series robot arm.

Many commercially available industrial robots are widely used in manufacturing and assembly tasks, such as material handling, spot/arc welding, parts assembly, paint spraying, loading and unloading numerically controlled machines, space and undersea exploration, prosthetic arm research, and in handling hazardous materials. These robots fall into one of the four basic motion-defining categories (Fig. 1.2):

Cartesian coordinates (three linear axes) (e.g., IBM's RS-1 robot and the Sigma robot from Olivetti)

Cylindrical coordinates (two linear and one rotary axes) (e.g., Versatran 600 robot from Prab)

Spherical coordinates (one linear and two rotary axes) (e.g., Unimate 2000B from Unimation Inc.)

Revolute or articulated coordinates (three rotary axes) (e.g., T^3 from Cincinnati Milacron and PUMA from Unimation Inc.)

Most of today's industrial robots, though controlled by mini- and microcomputers, are basically simple positional machines. They execute a given task by

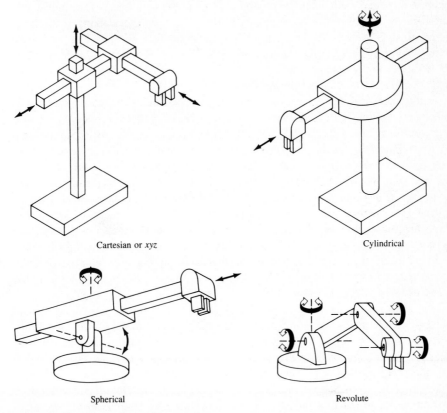

Cartesian or *xyz*

Cylindrical

Spherical

Revolute

Figure 1.2 Various robot arm categories.

playing back prerecorded or preprogrammed sequences of motions that have been previously guided or taught by a user with a hand-held control-teach box. Moreover, these robots are equipped with little or no external sensors for obtaining the information vital to its working environment. As a result, robots are used mainly in relatively simple, repetitive tasks. More research effort is being directed toward improving the overall performance of the manipulator systems, and one way is through the study of the various important areas covered in this book.

1.2 HISTORICAL DEVELOPMENT

The word *robot* was introduced into the English language in 1921 by the playwright Karel Capek in his satirical drama, *R.U.R.* (Rossum's Universal Robots). In this work, robots are machines that resemble people, but work tirelessly. Initially, the robots were manufactured for profit to replace human workers but, toward the end, the robots turned against their creators, annihilating the entire human race. Capek's play is largely responsible for some of the views popularly held about robots to this day, including the perception of robots as humanlike machines endowed with intelligence and individual personalities. This image was reinforced by the 1926 German robot film *Metropolis,* by the walking robot Electro and his dog Sparko, displayed in 1939 at the New York World's Fair, and more recently by the robot C3PO featured in the 1977 film *Star Wars.* Modern industrial robots certainly appear primitive when compared with the expectations created by the communications media during the past six decades.

Early work leading to today's industrial robots can be traced to the period immediately following World War II. During the late 1940s research programs were started at the Oak Ridge and Argonne National Laboratories to develop remotely controlled mechanical manipulators for handling radioactive materials. These systems were of the "master-slave" type, designed to reproduce faithfully hand and arm motions made by a human operator. The master manipulator was guided by the user through a sequence of motions, while the slave manipulator duplicated the master unit as closely as possible. Later, force feedback was added by mechanically coupling the motion of the master and slave units so that the operator could feel the forces as they developed between the slave manipulator and its environment. In the mid-1950s the mechanical coupling was replaced by electric and hydraulic power in manipulators such as General Electric's Handyman and the Minotaur I built by General Mills.

The work on master-slave manipulators was quickly followed by more sophisticated systems capable of autonomous, repetitive operations. In the mid-1950s George C. Devol developed a device he called a "programmed articulated transfer device," a manipulator whose operation could be programmed (and thus changed) and which could follow a sequence of motion steps determined by the instructions in the program. Further development of this concept by Devol and Joseph F. Engelberger led to the first industrial robot, introduced by Unimation Inc. in 1959. The key to this device was the use of a computer in conjunction with a manipulator to produce a machine that could be "taught" to carry out a variety of tasks automatically. Unlike hard automation machines, these robots could be reprogrammed and retooled at relative low cost to perform other jobs as manufacturing requirements changed.

While programmed robots offered a novel and powerful manufacturing tool, it became evident in the 1960s that the flexibility of these machines could be enhanced significantly by the use of sensory feedback. Early in that decade, H. A. Ernst [1962] reported the development of a computer-controlled mechanical hand with tactile sensors. This device, called the MH-1, could "feel" blocks and use this information to control the hand so that it stacked the blocks without operator assistance. This work is one of the first examples of a robot capable of adaptive behavior in a reasonably unstructured environment. The manipulative system consisted of an ANL Model-8 manipulator with 6 degrees of freedom controlled by a TX-O computer through an interfacing device. This research program later evolved as part of project MAC, and a television camera was added to the manipulator to begin machine perception research. During the same period, Tomovic and Boni [1962] developed a prototype hand equipped with a pressure sensor which sensed the object and supplied an input feedback signal to a motor to initiate one of two grasp patterns. Once the hand was in contact with the object, information proportional to object size and weight was sent to a computer by these pressure-sensitive elements. In 1963, the American Machine and Foundry Company (AMF) introduced the VERSATRAN commercial robot. Starting in this same year, various arm designs for manipulators were developed, such as the Roehampton arm and the Edinburgh arm.

In the late 1960s, McCarthy [1968] and his colleagues at the Stanford Artificial Intelligence Laboratory reported development of a computer with hands, eyes, and ears (i.e., manipulators, TV cameras, and microphones). They demonstrated a system that recognized spoken messages, "saw" blocks scattered on a table, and manipulated them in accordance with instructions. During this period, Pieper [1968] studied the kinematic problem of a computer-controlled manipulator while Kahn and Roth [1971] analyzed the dynamics and control of a restricted arm using bang-bang (near minimum time) control.

Meanwhile, other countries (Japan in particular) began to see the potential of industrial robots. As early as 1968, the Japanese company Kawasaki Heavy Industries negotiated a license with Unimation for its robots. One of the more unusual developments in robots occurred in 1969, when an experimental walking truck was developed by the General Electric Company for the U.S. Army. In the same year, the Boston arm was developed, and in the following year the Stanford arm was developed, which was equipped with a camera and computer controller. Some of the most serious work in robotics began as these arms were used as robot manipulators. One experiment with the Stanford arm consisted of automatically stacking blocks according to various strategies. This was very sophisticated work for an automated robot at that time. In 1974, Cincinnati Milacron introduced its first computer-controlled industrial robot. Called "The Tomorrow Tool," or T^3, it could lift over 100 lb as well as track moving objects on an assembly line.

During the 1970s a great deal of research work focused on the use of external sensors to facilitate manipulative operations. At Stanford, Bolles and Paul [1973], using both visual and force feedback, demonstrated a computer-controlled Stanford arm connected to a PDP-10 computer for assembling automotive water pumps. At about the same time, Will and Grossman [1975] at IBM developed a computer-controlled manipulator with touch and force sensors to perform mechanical assembly of a 20-part typewriter. Inoue [1974] at the MIT Artificial Intelligence Laboratory worked on the artificial intelligence aspects of force feedback. A landfall navigation search technique was used to perform initial positioning in a precise assembly task. At the Draper Laboratory Nevins et al. [1974] investigated sensing techniques based on compliance. This work developed into the instrumentation of a passive compliance device called *remote center compliance* (RCC) which was attached to the mounting plate of the last joint of the manipulator for close parts-mating assembly. Bejczy [1974], at the Jet Propulsion Laboratory, implemented a computer-based torque-control technique on his extended Stanford arm for space exploration projects. Since then, various control methods have been proposed for servoing mechanical manipulators.

Today, we view robotics as a much broader field of work than we did just a few years ago, dealing with research and development in a number of interdisciplinary areas, including kinematics, dynamics, planning systems, control, sensing, programming languages, and machine intelligence. These topics, introduced briefly in the following sections, constitute the core of the material in this book.

1.3 ROBOT ARM KINEMATICS AND DYNAMICS

Robot arm kinematics deals with the analytical study of the geometry of motion of a robot arm with respect to a fixed reference coordinate system without regard to the forces/moments that cause the motion. Thus, kinematics deals with the analytical description of the spatial displacement of the robot as a function of time, in particular the relations between the joint-variable space and the position and orientation of the end-effector of a robot arm.

There are two fundamental problems in robot arm kinematics. The first problem is usually referred to as the *direct* (or *forward*) *kinematics* problem, while the second problem is the *inverse kinematics* (or *arm solution*) problem. Since the independent variables in a robot arm are the joint variables, and a task is usually stated in terms of the reference coordinate frame, the inverse kinematics problem is used more frequently. Denavit and Hartenberg [1955] proposed a systematic and generalized approach of utilizing matrix algebra to describe and represent the spatial geometry of the links of a robot arm with respect to a fixed reference frame. This method uses a 4×4 homogeneous transformation matrix to describe the spatial relationship between two adjacent rigid mechanical links and reduces the direct kinematics problem to finding an equivalent 4×4 homogeneous transformation matrix that relates the spatial displacement of the hand coordinate frame to the reference coordinate frame. These homogeneous transformation matrices are also useful in deriving the dynamic equations of motion of a robot arm. In general, the inverse kinematics problem can be solved by several techniques. The most commonly used methods are the matrix algebraic, iterative, or geometric approach. Detailed treatments of direct kinematics and inverse kinematics problems are given in Chap. 2.

Robot arm dynamics, on the other hand, deals with the mathematical formulations of the equations of robot arm motion. The dynamic equations of motion of a manipulator are a set of mathematical equations describing the dynamic behavior of the manipulator. Such equations of motion are useful for computer simulation of the robot arm motion, the design of suitable control equations for a robot arm, and the evaluation of the kinematic design and structure of a robot arm. The actual dynamic model of an arm can be obtained from known physical laws such as the laws of newtonian and lagrangian mechanics. This leads to the development of dynamic equations of motion for the various articulated joints of the manipulator in terms of specified geometric and inertial parameters of the links. Conventional approaches like the Lagrange-Euler and the Newton-Euler formulations can then be applied systematically to develop the actual robot arm motion equations. Detailed discussions of robot arm dynamics are presented in Chap. 3.

1.4 MANIPULATOR TRAJECTORY PLANNING AND MOTION CONTROL

With the knowledge of kinematics and dynamics of a serial link manipulator, one would like to servo the manipulator's joint actuators to accomplish a desired task by controlling the manipulator to follow a desired path. Before moving the robot arm, it is of interest to know whether there are any obstacles present in the path that the robot arm has to traverse (obstacle constraint) and whether the manipulator hand needs to traverse along a specified path (path constraint). The control problem of a manipulator can be conveniently divided into two coherent subproblems—the motion (or trajectory) planning subproblem and the motion control subproblem.

The space curve that the manipulator hand moves along from an initial location (position and orientation) to the final location is called the *path*. The trajectory planning (or trajectory planner) interpolates and/or approximates the desired path by a class of polynomial functions and generates a sequence of time-based "control set points" for the control of the manipulator from the initial location to the destination location. Chapter 4 discusses the various trajectory planning schemes for obstacle-free motion, as well as the formalism for describing desired manipulator motion in terms of sequences of points in space through which the manipulator must pass and the space curve that it traverses.

In general, the motion control problem consists of (1) obtaining dynamic models of the manipulator, and (2) using these models to determine control laws or strategies to achieve the desired system response and performance. Since the first part of the control problem is discussed extensively in Chap. 3, Chap. 5 concentrates on the second part of the control problem. From the control analysis point of view, the movement of a robot arm is usually performed in two distinct control phases. The first is the gross motion control in which the arm moves from an initial position/orientation to the vicinity of the desired target position/orientation along a planned trajectory. The second is the fine motion control in which the end-effector of the arm dynamically interacts with the object using sensory feedback information from the sensors to complete the task.

Current industrial approaches to robot arm control treat each joint of the robot arm as a simple joint servomechanism. The servomechanism approach models the varying dynamics of a manipulator inadequately because it neglects the motion and configuration of the whole arm mechanism. These changes in the parameters of the controlled system sometimes are significant enough to render conventional feedback control strategies ineffective. The result is reduced servo response speed and damping, limiting the precision and speed of the end-effector and making it appropriate only for limited-precision tasks. Manipulators controlled in this manner move at slow speeds with unnecessary vibrations. Any significant performance gain in this and other areas of robot arm control require the consideration of more efficient dynamic models, sophisticated control approaches, and the use of dedicated computer architectures and parallel processing techniques. Chapter 5 focuses on deriving gross motion control laws and strategies which utilize the dynamic models discussed in Chap. 3 to efficiently control a manipulator.

1.5 ROBOT SENSING

The use of external sensing mechanisms allows a robot to interact with its environment in a flexible manner. This is in contrast to preprogrammed operations in which a robot is "taught" to perform repetitive tasks via a set of programmed functions. Although the latter is by far the most predominant form of operation of present industrial robots, the use of sensing technology to endow machines with a greater degree of intelligence in dealing with their environment is indeed an active topic of research and development in the robotics field.

The function of robot sensors may be divided into two principal categories: *internal state* and *external state*. Internal state sensors deal with the detection of variables such as arm joint position, which are used for robot control. External state sensors, on the other hand, deal with the detection of variables such as range, proximity, and touch. External sensing, the topic of Chaps. 6 through 8, is used for robot guidance, as well as for object identification and handling. The focus of Chap. 6 is on range, proximity, touch, and force-torque sensing. Vision sensors and techniques are discussed in detail in Chaps. 7 and 8. Although proximity, touch, and force sensing play a significant role in the improvement of robot performance, vision is recognized as the most powerful of robot sensory capabilities. Robot vision may be defined as the process of extracting, characterizing, and interpreting information from images of a three-dimensional world. This process, also commonly referred to as *machine* or *computer vision,* may be subdivided into six principal areas: (1) sensing, (2) preprocessing, (3) segmentation, (4) description, (5) recognition, and (6) interpretation.

It is convenient to group these various areas of vision according to the sophistication involved in their implementation. We consider three levels of processing: low-, medium-, and high-level vision. While there are no clearcut boundaries between these subdivisions, they do provide a useful framework for categorizing the various processes that are inherent components of a machine vision system. In our discussion, we shall treat sensing and preprocessing as low-level vision functions. This will take us from the image formation process itself to compensations such as noise reduction, and finally to the extraction of primitive image features such as intensity discontinuities. We will associate with medium-level vision those processes that extract, characterize, and label components in an image resulting from low-level vision. In terms of our six subdivisions, we will treat segmentation, description, and recognition of individual objects as medium-level vision functions. High-level vision refers to processes that attempt to emulate cognition. The material in Chap. 7 deals with sensing, preprocessing, and with concepts and techniques required to implement low-level vision functions. Topics in higher-level vision are discussed in Chap. 8.

1.6 ROBOT PROGRAMMING LANGUAGES

One major obstacle in using manipulators as general-purpose assembly machines is the lack of suitable and efficient communication between the user and the robotic system so that the user can direct the manipulator to accomplish a given task. There are several ways to communicate with a robot, and the three major approaches to achieve it are discrete word recognition, teach and playback, and high-level programming languages.

Current state-of-the-art speech recognition is quite primitive and generally speaker-dependent. It can recognize a set of discrete words from a limited vocabulary and usually requires the user to pause between words. Although it is now possible to recognize words in real time due to faster computer components and efficient processing algorithms, the usefulness of discrete word recognition to describe a task is limited. Moreover, it requires a large memory space to store speech data, and it usually requires a training period to build up speech templates for recognition.

The method of teach and playback involves teaching the robot by leading it through the motions to be performed. This is usually accomplished in the following steps: (1) leading the robot in slow motion using manual control through the entire assembly task, with the joint angles of the robot at appropriate locations being recorded in order to replay the motion; (2) editing and playing back the taught motion; and (3) if the taught motion is correct, then the robot is run at an appropriate speed in a repetitive motion. This method is also known as guiding and is the most commonly used approach in present-day industrial robots.

A more general approach to solve the human-robot communication problem is the use of high-level programming. Robots are commonly used in areas such as arc welding, spot welding, and paint spraying. These tasks require no interaction between the robot and the environment and can be easily programmed by guiding. However, the use of robots to perform assembly tasks generally requires high-level programming techniques. This effort is warranted because the manipulator is usually controlled by a computer, and the most effective way for humans to communicate with computers is through a high-level programming language. Furthermore, using programs to describe assembly tasks allows a robot to perform different jobs by simply executing the appropriate program. This increases the flexibility and versatility of the robot. Chapter 9 discusses the use of high-level programming techniques for achieving effective communication with a robotic system.

1.7 ROBOT INTELLIGENCE

A basic problem in robotics is *planning* motions to solve some prespecified task, and then *controlling* the robot as it executes the commands necessary to achieve those actions. Here, planning means deciding on a course of action before acting. This action synthesis part of the robot problem can be solved by a problem-solving system that will achieve some stated goal, given some initial situation. A plan is, thus, a representation of a course of action for achieving a stated goal.

Research on robot problem solving has led to many ideas about problem-solving systems in artificial intelligence. In a typical formulation of a robot problem we have a robot that is equipped with sensors and a set of primitive actions that it can perform in some easy-to-understand world. Robot actions change one state, or configuration, of the world into another. In the "blocks world," for example, we imagine a world of several labeled blocks resting on a table or on each other and a robot consisting of a TV camera and a movable arm and hand that is able to pick up and move blocks. In some situations, the robot is a mobile vehicle with a TV camera that performs tasks such as pushing objects from place to place in an environment containing other objects.

In Chap. 10, we introduce several basic methods for problem solving and their applications to robot planning. The discussion emphasizes the problem-solving or planning aspect of a robot. A robot planner attempts to find a path from our initial robot world to a final robot world. The path consists of a sequence of operations that are considered primitive to the system. A solution to a problem could be the basis of a corresponding sequence of physical actions in the physical world. Robot planning, which provides the intelligence and problem-solving capability to a robot system, is still a very active area of research. For real-time robot applications, we still need powerful and efficient planning algorithms that will be executed by high-speed special-purpose computer systems.

1.8 REFERENCES

The general references cited below are representative of publications dealing with topics of interest in robotics and related fields. References given at the end of later chapters are keyed to specific topics discussed in the text. The bibliography at the end of the book is organized in alphabetical order by author, and it contains all the pertinent information for each reference cited in the text.

Some of the major journals and conference proceedings that routinely contain articles on various aspects of robotics include: *IEEE Journal of Robotics and Automation*; *International Journal of Robotics Research*; *Journal of Robotic Systems*; *Robotica*; *IEEE Transactions on Systems, Man and Cybernetics*; *Artificial Intelligence*; *IEEE Transactions on Pattern Analysis and Machine Intelligence*; *Computer Graphics, Vision, and Image Processing*; *Proceedings of the International Symposium on Industrial Robots*; *Proceedings of the International Joint Conference on Artificial Intelligence*; *Proceedings of IEEE International Conference on Robotics and Automation*; *IEEE Transactions on Automatic Control*; *Mechanism and Machine Theory*; *Proceedings of the Society of Photo-Optical and Instrumentation Engineers*; *ASME Journal of Mechanical Design*; *ASME Journal of Applied Mechanics*; *ASME Journal of Dynamic Systems, Measurement and Control*; and *ASME Journal of Mechanisms, Transmissions, and Automation in Design*.

Complementary reading for the material in this book may be found in the books by Dodd and Rossol [1979], Engelberger [1980], Paul [1981], Dorf [1983], Snyder [1985], Lee, Gonzalez, and Fu [1986], Tou [1985], and Craig [1986].

TWO

ROBOT ARM KINEMATICS

And see! she stirs, she starts,
she moves, she seems to feel
the thrill of life!
Henry Wadsworth Longfellow

2.1 INTRODUCTION

A mechanical manipulator can be modeled as an open-loop articulated chain with several rigid bodies (links) connected in series by either revolute or prismatic joints driven by actuators. One end of the chain is attached to a supporting base while the other end is free and attached with a tool (the end-effector) to manipulate objects or perform assembly tasks. The relative motion of the joints results in the motion of the links that positions the hand in a desired orientation. In most robotic applications, one is interested in the spatial description of the end-effector of the manipulator with respect to a fixed reference coordinate system.

Robot arm kinematics deals with the analytical study of the geometry of motion of a robot arm with respect to a fixed reference coordinate system as a function of time without regard to the forces/moments that cause the motion. Thus, it deals with the analytical description of the spatial displacement of the robot as a function of time, in particular the relations between the joint-variable space and the position and orientation of the end-effector of a robot arm. This chapter addresses two fundamental questions of both theoretical and practical interest in robot arm kinematics:

1. For a given manipulator, given the joint angle vector $\mathbf{q}(t) = (q_1(t), q_2(t), \ldots, q_n(t))^T$ and the geometric link parameters, where n is the number of degrees of freedom, what is the position and orientation of the end-effector of the manipulator with respect to a reference coordinate system?
2. Given a desired position and orientation of the end-effector of the manipulator and the geometric link parameters with respect to a reference coordinate system, can the manipulator reach the desired prescribed manipulator hand position and orientation? And if it can, how many different manipulator configurations will satisfy the same condition?

The first question is usually referred to as the *direct (or forward) kinematics* problem, while the second question is the *inverse kinematics (or arm solution)* problem.

12

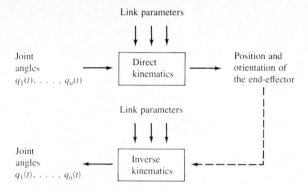

Figure 2.1 The direct and inverse kinematics problems.

Since the independent variables in a robot arm are the joint variables and a task is usually stated in terms of the reference coordinate frame, the inverse kinematics problem is used more frequently. A simple block diagram indicating the relationship between these two problems is shown in Fig. 2.1.

Since the links of a robot arm may rotate and/or translate with respect to a reference coordinate frame, the total spatial displacement of the end-effector is due to the angular rotations and linear translations of the links. Denavit and Hartenberg [1955] proposed a systematic and generalized approach of utilizing matrix algebra to describe and represent the spatial geometry of the links of a robot arm with respect to a fixed reference frame. This method uses a 4 × 4 homogeneous transformation matrix to describe the spatial relationship between two adjacent rigid mechanical links and reduces the direct kinematics problem to finding an equivalent 4 × 4 homogeneous transformation matrix that relates the spatial displacement of the "hand coordinate frame" to the reference coordinate frame. These homogeneous transformation matrices are also useful in deriving the dynamic equations of motion of a robot arm.

In general, the inverse kinematics problem can be solved by several techniques. Most commonly used methods are the matrix algebraic, iterative, or geometric approaches. A geometric approach based on the link coordinate systems and the manipulator configuration will be presented in obtaining a closed form joint solution for simple manipulators with rotary joints. Then a more general approach using 4 × 4 homogeneous matrices will be explored in obtaining a joint solution for simple manipulators.

2.2 THE DIRECT KINEMATICS PROBLEM

Vector and matrix algebra† are utilized to develop a systematic and generalized approach to describe and represent the location of the links of a robot arm with

† Vectors are represented in lowercase bold letters; matrices are in uppercase bold.

respect to a fixed reference frame. Since the links of a robot arm may rotate and/ or translate with respect to a reference coordinate frame, a body-attached coordinate frame will be established along the joint axis for each link. The direct kinematics problem is reduced to finding a transformation matrix that relates the body-attached coordinate frame to the reference coordinate frame. A 3 × 3 rotation matrix is used to describe the rotational operations of the body-attached frame with respect to the reference frame. The homogeneous coordinates are then used to represent position vectors in a three-dimensional space, and the rotation matrices will be expanded to 4 × 4 homogeneous transformation matrices to include the translational operations of the body-attached coordinate frames. This matrix representation of a rigid mechanical link to describe the spatial geometry of a robot arm was first used by Denavit and Hartenberg [1955]. The advantage of using the Denavit-Hartenberg representation of linkages is its algorithmic universality in deriving the kinematic equation of a robot arm.

2.2.1 Rotation Matrices

A 3 × 3 rotation matrix can be defined as a transformation matrix which operates on a position vector in a three-dimensional euclidean space and maps its coordinates expressed in a rotated coordinate system $OUVW$ (body-attached frame) to a reference coordinate system $OXYZ$. In Fig. 2.2, we are given two right-hand rectangular coordinate systems, namely, the $OXYZ$ coordinate system with $OX, OY,$ and OZ as its coordinate axes and the $OUVW$ coordinate system with $OU, OV,$ and OW as its coordinate axes. Both coordinate systems have their origins coincident at point O. The $OXYZ$ coordinate system is fixed in the three-dimensional space and is considered to be the reference frame. The $OUVW$ coordinate frame is rotating with respect to the reference frame $OXYZ$. Physically, one can consider the $OUVW$ coordinate system to be a body-attached coordinate frame. That is, it is

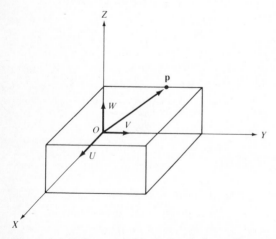

Figure 2.2 Reference and body-attached coordinate systems.

permanently and conveniently attached to the rigid body (e.g., an aircraft or a link of a robot arm) and moves together with it. Let $(\mathbf{i}_x, \mathbf{j}_y, \mathbf{k}_z)$ and $(\mathbf{i}_u, \mathbf{j}_v, \mathbf{k}_w)$ be the unit vectors along the coordinate axes of the *OXYZ* and *OUVW* systems, respectively. A point **p** in the space can be represented by its coordinates with respect to both coordinate systems. For ease of discussion, we shall assume that **p** is at rest and fixed with respect to the *OUVW* coordinate frame. Then the point **p** can be represented by its coordinates with respect to the *OUVW* and *OXYZ* coordinate systems, respectively, as

$$\mathbf{p}_{uvw} = (p_u, p_v, p_w)^T \quad \text{and} \quad \mathbf{p}_{xyz} = (p_x, p_y, p_z)^T \quad (2.2\text{-}1)$$

where $\mathbf{p}_{xyz}$ and $\mathbf{p}_{uvw}$ represent the same point **p** in the space with reference to different coordinate systems, and the superscript *T* on vectors and matrices denotes the transpose operation.

We would like to find a 3×3 transformation matrix **R** that will transform the coordinates of $\mathbf{p}_{uvw}$ to the coordinates expressed with respect to the *OXYZ* coordinate system, after the *OUVW* coordinate system has been rotated. That is,

$$\mathbf{p}_{xyz} = \mathbf{R}\mathbf{p}_{uvw} \quad (2.2\text{-}2)$$

Note that physically the point $\mathbf{p}_{uvw}$ has been rotated together with the *OUVW* coordinate system.

Recalling the definition of the components of a vector, we have

$$\mathbf{p}_{uvw} = p_u\mathbf{i}_u + p_v\mathbf{j}_v + p_w\mathbf{k}_w \quad (2.2\text{-}3)$$

where p_x, p_y, and p_z represent the components of **p** along the *OX, OY,* and *OZ* axes, respectively, or the projections of **p** onto the respective axes. Thus, using the definition of a scalar product and Eq. (2.2-3),

$$p_x = \mathbf{i}_x \cdot \mathbf{p} = \mathbf{i}_x \cdot \mathbf{i}_u p_u + \mathbf{i}_x \cdot \mathbf{j}_v p_v + \mathbf{i}_x \cdot \mathbf{k}_w p_w$$

$$p_y = \mathbf{j}_y \cdot \mathbf{p} = \mathbf{j}_y \cdot \mathbf{i}_u p_u + \mathbf{j}_y \cdot \mathbf{j}_v p_v + \mathbf{j}_y \cdot \mathbf{k}_w p_w \quad (2.2\text{-}4)$$

$$p_z = \mathbf{k}_z \cdot \mathbf{p} = \mathbf{k}_z \cdot \mathbf{i}_u p_u + \mathbf{k}_z \cdot \mathbf{j}_v p_v + \mathbf{k}_z \cdot \mathbf{k}_w p_w$$

or expressed in matrix form,

$$\begin{bmatrix} p_x \\ p_y \\ p_z \end{bmatrix} = \begin{bmatrix} \mathbf{i}_x \cdot \mathbf{i}_u & \mathbf{i}_x \cdot \mathbf{j}_v & \mathbf{i}_x \cdot \mathbf{k}_w \\ \mathbf{j}_y \cdot \mathbf{i}_u & \mathbf{j}_y \cdot \mathbf{j}_v & \mathbf{j}_y \cdot \mathbf{k}_w \\ \mathbf{k}_z \cdot \mathbf{i}_u & \mathbf{k}_z \cdot \mathbf{j}_v & \mathbf{k}_z \cdot \mathbf{k}_w \end{bmatrix} \begin{bmatrix} p_u \\ p_v \\ p_w \end{bmatrix} \quad (2.2\text{-}5)$$

Using this notation, the matrix $\mathbf{R}$ in Eq. (2.2-2) is given by

$$\mathbf{R} = \begin{bmatrix} \mathbf{i}_x \cdot \mathbf{i}_u & \mathbf{i}_x \cdot \mathbf{j}_v & \mathbf{i}_x \cdot \mathbf{k}_w \\ \mathbf{j}_y \cdot \mathbf{i}_u & \mathbf{j}_y \cdot \mathbf{j}_v & \mathbf{j}_y \cdot \mathbf{k}_w \\ \mathbf{k}_z \cdot \mathbf{i}_u & \mathbf{k}_z \cdot \mathbf{j}_v & \mathbf{k}_z \cdot \mathbf{k}_w \end{bmatrix} \tag{2.2-6}$$

Similarly, one can obtain the coordinates of $\mathbf{p}_{uvw}$ from the coordinates of $\mathbf{p}_{xyz}$:

$$\mathbf{p}_{uvw} = \mathbf{Q}\mathbf{p}_{xyz} \tag{2.2-7}$$

or

$$\begin{bmatrix} p_u \\ p_v \\ p_w \end{bmatrix} = \begin{bmatrix} \mathbf{i}_u \cdot \mathbf{i}_x & \mathbf{i}_u \cdot \mathbf{j}_y & \mathbf{i}_u \cdot \mathbf{k}_z \\ \mathbf{j}_v \cdot \mathbf{i}_x & \mathbf{j}_v \cdot \mathbf{j}_y & \mathbf{j}_v \cdot \mathbf{k}_z \\ \mathbf{k}_w \cdot \mathbf{i}_x & \mathbf{k}_w \cdot \mathbf{j}_y & \mathbf{k}_w \cdot \mathbf{k}_z \end{bmatrix} \begin{bmatrix} p_x \\ p_y \\ p_z \end{bmatrix} \tag{2.2-8}$$

Since dot products are commutative, one can see from Eqs. (2.2-6) to (2.2-8) that

$$\mathbf{Q} = \mathbf{R}^{-1} = \mathbf{R}^T \tag{2.2-9}$$

and

$$\mathbf{QR} = \mathbf{R}^T\mathbf{R} = \mathbf{R}^{-1}\mathbf{R} = \mathbf{I}_3 \tag{2.2-10}$$

where $\mathbf{I}_3$ is a 3×3 identity matrix. The transformation given in Eq. (2.2-2) or (2.2-7) is called an *orthogonal* transformation and since the vectors in the dot products are all unit vectors, it is also called an *orthonormal* transformation.

The primary interest in developing the above transformation matrix is to find the rotation matrices that represent rotations of the *OUVW* coordinate system about each of the three principal axes of the reference coordinate system *OXYZ*. If the *OUVW* coordinate system is rotated an α angle about the *OX* axis to arrive at a new location in the space, then the point $\mathbf{p}_{uvw}$ having coordinates $(p_u, p_v, p_w)^T$ with respect to the *OUVW* system will have different coordinates $(p_x, p_y, p_z)^T$ with respect to the reference system *OXYZ*. The necessary transformation matrix $\mathbf{R}_{x,\alpha}$ is called the rotation matrix about the *OX* axis with α angle. $\mathbf{R}_{x,\alpha}$ can be derived from the above transformation matrix concept, that is

$$\mathbf{p}_{xyz} = \mathbf{R}_{x,\alpha}\,\mathbf{p}_{uvw} \tag{2.2-11}$$

with $\mathbf{i}_x \equiv \mathbf{i}_u$, and

$$\mathbf{R}_{x,\alpha} = \begin{bmatrix} \mathbf{i}_x \cdot \mathbf{i}_u & \mathbf{i}_x \cdot \mathbf{j}_v & \mathbf{i}_x \cdot \mathbf{k}_w \\ \mathbf{j}_y \cdot \mathbf{i}_u & \mathbf{j}_y \cdot \mathbf{j}_v & \mathbf{j}_y \cdot \mathbf{k}_w \\ \mathbf{k}_z \cdot \mathbf{i}_u & \mathbf{k}_z \cdot \mathbf{j}_v & \mathbf{k}_z \cdot \mathbf{k}_w \end{bmatrix} = \begin{bmatrix} 1 & 0 & 0 \\ 0 & \cos\alpha & -\sin\alpha \\ 0 & \sin\alpha & \cos\alpha \end{bmatrix} \tag{2.2-12}$$

Similarly, the 3×3 rotation matrices for rotation about the *OY* axis with ϕ angle and about the *OZ* axis with θ angle are, respectively (see Fig. 2.3),

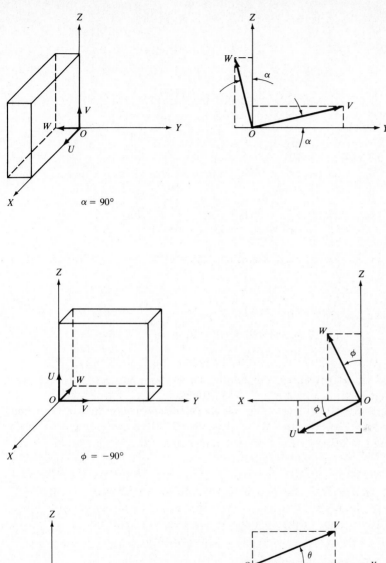

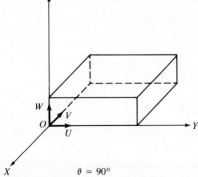

Figure 2.3 Rotating coordinate systems.

$$\mathbf{R}_{y,\phi} = \begin{bmatrix} \cos\phi & 0 & \sin\phi \\ 0 & 1 & 0 \\ -\sin\phi & 0 & \cos\phi \end{bmatrix} \qquad \mathbf{R}_{z,\theta} = \begin{bmatrix} \cos\theta & -\sin\theta & 0 \\ \sin\theta & \cos\theta & 0 \\ 0 & 0 & 1 \end{bmatrix} \qquad (2.2\text{-}13)$$

The matrices $\mathbf{R}_{x,\alpha}$, $\mathbf{R}_{y,\phi}$, and $\mathbf{R}_{z,\theta}$ are called the *basic rotation matrices*. Other finite rotation matrices can be obtained from these matrices.

Example: Given two points $\mathbf{a}_{uvw} = (4, 3, 2)^T$ and $\mathbf{b}_{uvw} = (6, 2, 4)^T$ with respect to the rotated $OUVW$ coordinate system, determine the corresponding points $\mathbf{a}_{xyz}$, $\mathbf{b}_{xyz}$ with respect to the reference coordinate system if it has been rotated 60° about the OZ axis.

SOLUTION: $\quad \mathbf{a}_{xyz} = \mathbf{R}_{z,60°} \mathbf{a}_{uvw} \qquad$ and $\qquad \mathbf{b}_{xyz} = \mathbf{R}_{z,60°} \mathbf{b}_{uvw}$

$$\mathbf{a}_{xyz} = \begin{bmatrix} 0.500 & -0.866 & 0 \\ 0.866 & 0.500 & 0 \\ 0 & 0 & 1 \end{bmatrix} \begin{bmatrix} 4 \\ 3 \\ 2 \end{bmatrix}$$

$$= \begin{bmatrix} 4(0.5)+3(-0.866)+2(0) \\ 4(0.866)+3(0.5)+2(0) \\ 4(0)+3(0)+2(1) \end{bmatrix} = \begin{bmatrix} -0.598 \\ 4.964 \\ 2.0 \end{bmatrix}$$

$$\mathbf{b}_{xyz} = \begin{bmatrix} 0.500 & -0.866 & 0 \\ 0.866 & 0.500 & 0 \\ 0 & 0 & 1 \end{bmatrix} \begin{bmatrix} 6 \\ 2 \\ 4 \end{bmatrix} = \begin{bmatrix} 1.268 \\ 6.196 \\ 4.0 \end{bmatrix}$$

Thus, $\mathbf{a}_{xyz}$ and $\mathbf{b}_{xyz}$ are equal to $(-0.598, 4.964, 2.0)^T$ and $(1.268, 6.196, 4.0)^T$, respectively, when expressed in terms of the reference coordinate system. $\qquad \square$

Example: If $\mathbf{a}_{xyz} = (4, 3, 2)^T$ and $\mathbf{b}_{xyz} = (6, 2, 4)^T$ are the coordinates with respect to the reference coordinate system, determine the corresponding points $\mathbf{a}_{uvw}$, $\mathbf{b}_{uvw}$ with respect to the rotated $OUVW$ coordinate system if it has been rotated 60° about the OZ axis.

SOLUTION: $\quad \mathbf{a}_{uvw} = (\mathbf{R}_{z,60})^T \mathbf{a}_{xyz} \qquad$ and $\qquad \mathbf{b}_{uvw} = (\mathbf{R}_{z,60})^T \mathbf{b}_{xyz}$

$$\mathbf{a}_{uvw} = \begin{bmatrix} 0.500 & 0.866 & 0 \\ -0.866 & 0.500 & 0 \\ 0 & 0 & 1 \end{bmatrix} \begin{bmatrix} 4 \\ 3 \\ 2 \end{bmatrix} = \begin{bmatrix} 4(0.5)+3(0.866)+2(0) \\ 4(-0.866)+3(0.5)+2(0) \\ 4(0)+3(0)+2(1) \end{bmatrix}$$

$$= \begin{bmatrix} 4.598 \\ -1.964 \\ 2.0 \end{bmatrix}$$

$$\mathbf{b}_{uvw} = \begin{bmatrix} 0.500 & 0.866 & 0 \\ -0.866 & 0.500 & 0 \\ 0 & 0 & 1 \end{bmatrix} \begin{bmatrix} 6 \\ 2 \\ 4 \end{bmatrix} = \begin{bmatrix} 4.732 \\ -4.196 \\ 4.0 \end{bmatrix} \qquad \square$$

2.2.2 Composite Rotation Matrix

Basic rotation matrices can be multiplied together to represent a sequence of finite rotations about the principal axes of the *OXYZ* coordinate system. Since matrix multiplications do not commute, the order or sequence of performing rotations is important. For example, to develop a rotation matrix representing a rotation of α angle about the *OX* axis followed by a rotation of θ angle about the *OZ* axis followed by a rotation of ϕ angle about the *OY* axis, the resultant rotation matrix representing these rotations is

$$\mathbf{R} = \mathbf{R}_{y,\phi}\,\mathbf{R}_{z,\theta}\,\mathbf{R}_{x,\alpha} = \begin{bmatrix} C\phi & 0 & S\phi \\ 0 & 1 & 0 \\ -S\phi & 0 & C\phi \end{bmatrix} \begin{bmatrix} C\theta & -S\theta & 0 \\ S\theta & C\theta & 0 \\ 0 & 0 & 1 \end{bmatrix} \begin{bmatrix} 1 & 0 & 0 \\ 0 & C\alpha & -S\alpha \\ 0 & S\alpha & C\alpha \end{bmatrix}$$

$$= \begin{bmatrix} C\phi C\theta & S\phi S\alpha - C\phi S\theta C\alpha & C\phi S\theta S\alpha + S\phi C\alpha \\ S\theta & C\theta C\alpha & -C\theta S\alpha \\ -S\phi C\theta & S\phi S\theta C\alpha + C\phi S\alpha & C\phi C\alpha - S\phi S\theta S\alpha \end{bmatrix} \qquad (2.2\text{-}14)$$

where $C\phi \equiv \cos\phi$; $S\phi \equiv \sin\phi$; $C\theta \equiv \cos\theta$; $S\theta \equiv \sin\theta$; $C\alpha \equiv \cos\alpha$; $S\alpha \equiv \sin\alpha$. That is different from the rotation matrix which represents a rotation of ϕ angle about the *OY* axis followed by a rotation of θ angle about the *OZ* axis followed by a rotation of α angle about the *OX* axis. The resultant rotation matrix is:

$$\mathbf{R} = \mathbf{R}_{x,\alpha}\,\mathbf{R}_{z,\theta}\,\mathbf{R}_{y,\phi} = \begin{bmatrix} 1 & 0 & 0 \\ 0 & C\alpha & -S\alpha \\ 0 & S\alpha & C\alpha \end{bmatrix} \begin{bmatrix} C\theta & -S\theta & 0 \\ S\theta & C\theta & 0 \\ 0 & 0 & 1 \end{bmatrix} \begin{bmatrix} C\phi & 0 & S\phi \\ 0 & 1 & 0 \\ -S\phi & 0 & C\phi \end{bmatrix}$$

$$= \begin{bmatrix} C\theta C\phi & -S\theta & C\theta S\phi \\ C\alpha S\theta C\phi + S\alpha S\phi & C\alpha C\theta & C\alpha S\theta S\phi - S\alpha C\phi \\ S\alpha S\theta C\phi - C\alpha S\phi & S\alpha C\theta & S\alpha S\theta S\phi + C\alpha C\phi \end{bmatrix} \qquad (2.2\text{-}15)$$

In addition to rotating about the principal axes of the reference frame *OXYZ*, the rotating coordinate system *OUVW* can also rotate about its own principal axes. In this case, the resultant or composite rotation matrix may be obtained from the following simple rules:

1. Initially both coordinate systems are coincident, hence the rotation matrix is a 3×3 identity matrix, $\mathbf{I}_3$.
2. If the rotating coordinate system $OUVW$ is rotating about one of the principal axes of the $OXYZ$ frame, then *premultiply* the previous (resultant) rotation matrix with an appropriate basic rotation matrix.
3. If the rotating coordinate system $OUVW$ is rotating about its own principal axes, then *postmultiply* the previous (resultant) rotation matrix with an appropriate basic rotation matrix.

Example: Find the resultant rotation matrix that represents a rotation of ϕ angle about the OY axis followed by a rotation of θ angle about the OW axis followed by a rotation of α angle about the OU axis.

SOLUTION:

$$\mathbf{R} = \mathbf{R}_{y,\phi}\, \mathbf{I}_3\, \mathbf{R}_{w,\theta}\, \mathbf{R}_{u,\alpha} = \mathbf{R}_{y,\phi}\, \mathbf{R}_{w,\theta}\, \mathbf{R}_{u,\alpha}$$

$$= \begin{bmatrix} C\phi & 0 & S\phi \\ 0 & 1 & 0 \\ -S\phi & 0 & C\phi \end{bmatrix} \begin{bmatrix} C\theta & -S\theta & 0 \\ S\theta & C\theta & 0 \\ 0 & 0 & 1 \end{bmatrix} \begin{bmatrix} 1 & 0 & 0 \\ 0 & C\alpha & -S\alpha \\ 0 & S\alpha & C\alpha \end{bmatrix}$$

$$= \begin{bmatrix} C\phi C\theta & S\phi S\alpha - C\phi S\theta C\alpha & C\phi S\theta S\alpha + S\phi C\alpha \\ S\theta & C\theta C\alpha & -C\theta S\alpha \\ -S\phi C\theta & S\phi S\theta C\alpha + C\phi S\alpha & C\phi C\alpha - S\phi S\theta S\alpha \end{bmatrix}$$

Note that this example is chosen so that the resultant matrix is the same as Eq. (2.2-14), but the sequence of rotations is different from the one that generates Eq. (2.2-14). ☐

2.2.3 Rotation Matrix About an Arbitrary Axis

Sometimes the rotating coordinate system $OUVW$ may rotate ϕ angle about an arbitrary axis $\mathbf{r}$ which is a unit vector having components of r_x, r_y, and r_z and passing through the origin O. The advantage is that for certain angular motions the $OUVW$ frame can make one rotation about the axis $\mathbf{r}$ instead of several rotations about the principal axes of the $OUVW$ and/or $OXYZ$ coordinate frames. To derive this rotation matrix $\mathbf{R}_{r,\phi}$, we can first make some rotations about the principal axes of the $OXYZ$ frame to align the axis $\mathbf{r}$ with the OZ axis. Then make the rotation about the $\mathbf{r}$ axis with ϕ angle and rotate about the principal axes of the $OXYZ$ frame again to return the $\mathbf{r}$ axis back to its original location. With reference to Fig. 2.4, aligning the OZ axis with the $\mathbf{r}$ axis can be done by rotating about the OX axis with α angle (the axis $\mathbf{r}$ is in the XZ plane), followed by a rotation of $-\beta$ angle about the OY axis (the axis $\mathbf{r}$ now aligns with the OZ axis). After the rotation of ϕ angle about the OZ or $\mathbf{r}$ axis, reverse the above sequence of rotations with their respective opposite angles. The resultant rotation matrix is

$$\mathbf{R}_{r,\phi} = \mathbf{R}_{x,-\alpha} \mathbf{R}_{y,\beta} \mathbf{R}_{z,\phi} \mathbf{R}_{y,-\beta} \mathbf{R}_{x,\alpha}$$

$$= \begin{bmatrix} 1 & 0 & 0 \\ 0 & C\alpha & S\alpha \\ 0 & -S\alpha & C\alpha \end{bmatrix} \begin{bmatrix} C\beta & 0 & S\beta \\ 0 & 1 & 0 \\ -S\beta & 0 & C\beta \end{bmatrix} \begin{bmatrix} C\phi & -S\phi & 0 \\ S\phi & C\phi & 0 \\ 0 & 0 & 1 \end{bmatrix}$$

$$\times \begin{bmatrix} C\beta & 0 & -S\beta \\ 0 & 1 & 0 \\ S\beta & 0 & C\beta \end{bmatrix} \begin{bmatrix} 1 & 0 & 0 \\ 0 & C\alpha & -S\alpha \\ 0 & S\alpha & C\alpha \end{bmatrix}$$

From Fig. 2.4, we easily find that

$$\sin \alpha = \frac{r_y}{\sqrt{r_y^2 + r_z^2}} \qquad \cos \alpha = \frac{r_z}{\sqrt{r_y^2 + r_z^2}}$$

$$\sin \beta = r_x \qquad \cos \beta = \sqrt{r_y^2 + r_z^2}$$

Substituting into the above equation,

$$\mathbf{R}_{r,\phi} = \begin{bmatrix} r_x^2 V\phi + C\phi & r_x r_y V\phi - r_z S\phi & r_x r_z V\phi + r_y S\phi \\ r_x r_y V\phi + r_z S\phi & r_y^2 V\phi + C\phi & r_y r_z V\phi - r_x S\phi \\ r_x r_z V\phi - r_y S\phi & r_y r_z V\phi + r_x S\phi & r_z^2 V\phi + C\phi \end{bmatrix} \qquad (2.2\text{-}16)$$

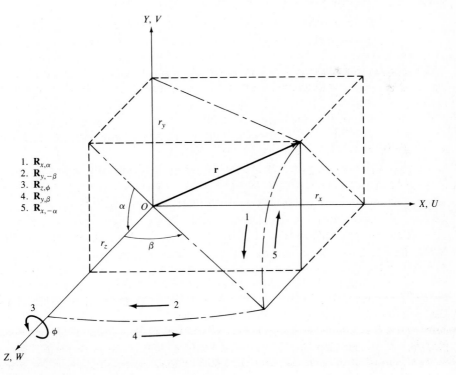

Figure 2.4 Rotation about an arbitrary axis.

where $V\phi = \text{vers } \phi = 1 - \cos\phi$. This is a very useful rotation matrix.

Example: Find the rotation matrix $\mathbf{R}_{r,\phi}$ that represents the rotation of ϕ angle about the vector $\mathbf{r} = (1, 1, 1)^T$.

SOLUTION: Since the vector $\mathbf{r}$ is not a unit vector, we need to normalize it and find its components along the principal axes of the $OXYZ$ frame. Therefore,

$$r_x = \frac{1}{\sqrt{r_x^2 + r_y^2 + r_z^2}} = \frac{1}{\sqrt{3}} \quad r_y = \frac{1}{\sqrt{3}} \quad r_z = \frac{1}{\sqrt{3}}$$

Substituting into Eq. (2.2-16), we obtain the $\mathbf{R}_{r,\phi}$ matrix:

$$\mathbf{R}_{r,\phi} = \begin{bmatrix} \tfrac{1}{3}V\phi + C\phi & \tfrac{1}{3}V\phi - \frac{1}{\sqrt{3}}S\phi & \tfrac{1}{3}V\phi + \frac{1}{\sqrt{3}}S\phi \\[2mm] \tfrac{1}{3}V\phi + \frac{1}{\sqrt{3}}S\phi & \tfrac{1}{3}V\phi + C\phi & \tfrac{1}{3}V\phi - \frac{1}{\sqrt{3}}S\phi \\[2mm] \tfrac{1}{3}V\phi - \frac{1}{\sqrt{3}}S\phi & \tfrac{1}{3}V\phi + \frac{1}{\sqrt{3}}S\phi & \tfrac{1}{3}V\phi + C\phi \end{bmatrix}$$

$\square$

2.2.4 Rotation Matrix with Euler Angles Representation

The matrix representation for rotation of a rigid body simplifies many operations, but it needs nine elements to completely describe the orientation of a rotating rigid body. It does not lead directly to a complete set of generalized coordinates. Such a set of generalized coordinates can describe the orientation of a rotating rigid body with respect to a reference coordinate frame. They can be provided by three angles called Euler angles ϕ, θ, and ψ. Although Euler angles describe the orientation of a rigid body with respect to a fixed reference frame, there are many different types of Euler angle representations. The three most widely used Euler angles representations are tabulated in Table 2.1.

The first Euler angle representation in Table 2.1 is usually associated with gyroscopic motion. This representation is usually called the eulerian angles, and corresponds to the following sequence of rotations (see Fig. 2.5):

Table 2.1 Three types of Euler angle representations

	Eulerian angles system I	Euler angles system II	Roll, pitch, and yaw system III
Sequence of rotations	ϕ about OZ axis θ about OU axis ψ about OW axis	ϕ about OZ axis θ about OV axis ψ about OW axis	ψ about OX axis θ about OY axis ϕ about OZ axis

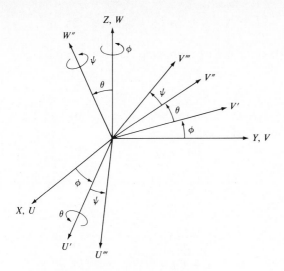

Figure 2.5 Eulerian angles system I.

1. A rotation of ϕ angle about the OZ axis ($\mathbf{R}_{z,\phi}$)
2. A rotation of θ angle about the rotated OU axis ($\mathbf{R}_{u,\theta}$)
3. Finally a rotation of ψ angle about the rotated OW axis ($\mathbf{R}_{w,\psi}$)

The resultant eulerian rotation matrix is

$$\mathbf{R}_{\phi,\theta,\psi} = \mathbf{R}_{z,\phi}\,\mathbf{R}_{u,\theta}\,\mathbf{R}_{w,\psi}$$

$$= \begin{bmatrix} C\phi & -S\phi & 0 \\ S\phi & C\phi & 0 \\ 0 & 0 & 1 \end{bmatrix} \begin{bmatrix} 1 & 0 & 0 \\ 0 & C\theta & -S\theta \\ 0 & S\theta & C\theta \end{bmatrix} \begin{bmatrix} C\psi & -S\psi & 0 \\ S\psi & C\psi & 0 \\ 0 & 0 & 1 \end{bmatrix}$$

$$= \begin{bmatrix} C\phi C\psi - S\phi C\theta S\psi & -C\phi S\psi - S\phi C\theta C\psi & S\phi S\theta \\ S\phi C\psi + C\phi C\theta S\psi & -S\phi S\psi + C\phi C\theta C\psi & -C\phi S\theta \\ S\theta S\psi & S\theta C\psi & C\theta \end{bmatrix} \qquad (2.2\text{-}17)$$

The above eulerian angle rotation matrix $\mathbf{R}_{\phi,\theta,\psi}$ can also be specified in terms of the rotations about the principal axes of the reference coordinate system: a rotation of ψ angle about the OZ axis followed by a rotation of θ angle about the OX axis and finally a rotation of ϕ angle about the OZ axis.

With reference to Fig. 2.6, another set of Euler angles ϕ, θ, and ψ representation corresponds to the following sequence of rotations:

1. A rotation of ϕ angle about the OZ axis ($\mathbf{R}_{z,\phi}$)
2. A rotation of θ angle about the rotated OV axis ($\mathbf{R}_{v,\theta}$)
3. Finally a rotation of ψ angle about the rotated OW axis ($\mathbf{R}_{w,\psi}$)

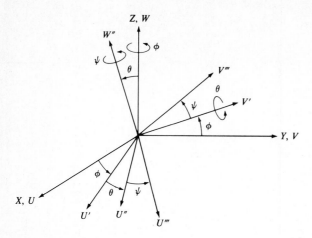

Figure 2.6 Eulerian angles system II.

The resultant rotation matrix is

$$\mathbf{R}_{\phi,\theta,\psi} = \mathbf{R}_{z,\phi}\, \mathbf{R}_{v,\theta}\, \mathbf{R}_{w,\psi}$$

$$= \begin{bmatrix} C\phi & -S\phi & 0 \\ S\phi & C\phi & 0 \\ 0 & 0 & 1 \end{bmatrix} \begin{bmatrix} C\theta & 0 & S\theta \\ 0 & 1 & 0 \\ -S\theta & 0 & C\theta \end{bmatrix} \begin{bmatrix} C\psi & -S\psi & 0 \\ S\psi & C\psi & 0 \\ 0 & 0 & 1 \end{bmatrix}$$

$$= \begin{bmatrix} C\phi C\theta C\psi - S\phi S\psi & -C\phi C\theta S\psi - S\phi C\psi & C\phi S\theta \\ S\phi C\theta C\psi + C\phi S\psi & -S\phi C\theta S\psi + C\phi C\psi & S\phi S\theta \\ -S\theta C\psi & S\theta S\psi & C\theta \end{bmatrix} \quad (2.2\text{-}18)$$

The above Euler angle rotation matrix $\mathbf{R}_{\phi,\theta,\psi}$ can also be specified in terms of the rotations about the principal axes of the reference coordinate system: a rotation of ψ angle about the OZ axis followed by a rotation of θ angle about the OY axis and finally a rotation of ϕ angle about the OZ axis.

Another set of Euler angles representation for rotation is called *roll, pitch,* and *yaw* (RPY). This is mainly used in aeronautical engineering in the analysis of space vehicles. They correspond to the following rotations in sequence:

1. A rotation of ψ about the OX axis ($\mathbf{R}_{x,\psi}$)—yaw
2. A rotation of θ about the OY axis ($\mathbf{R}_{y,\theta}$)—pitch
3. A rotation of ϕ about the OZ axis ($\mathbf{R}_{z,\phi}$)—roll

The resultant rotation matrix is

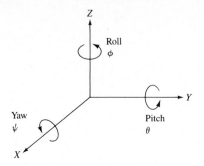

Figure 2.7 Roll, pitch and yaw.

$$\mathbf{R}_{\phi,\theta,\psi} = \mathbf{R}_{z,\phi}\,\mathbf{R}_{y,\theta}\,\mathbf{R}_{x,\psi} = \begin{bmatrix} C\phi & -S\phi & 0 \\ S\phi & C\phi & 0 \\ 0 & 0 & 1 \end{bmatrix} \begin{bmatrix} C\theta & 0 & S\theta \\ 0 & 1 & 0 \\ -S\theta & 0 & C\theta \end{bmatrix} \begin{bmatrix} 1 & 0 & 0 \\ 0 & C\psi & -S\psi \\ 0 & S\psi & C\psi \end{bmatrix}$$

$$= \begin{bmatrix} C\phi C\theta & C\phi S\theta S\psi - S\phi C\psi & C\phi S\theta C\psi + S\phi S\psi \\ S\phi C\theta & S\phi S\theta S\psi + C\phi C\psi & S\phi S\theta C\psi - C\phi S\psi \\ -S\theta & C\theta S\psi & C\theta C\psi \end{bmatrix} \qquad (2.2\text{-}19)$$

The above rotation matrix $\mathbf{R}_{\phi,\theta,\psi}$ for roll, pitch, and yaw can be specified in terms of the rotations about the principal axes of the reference coordinate system and the rotating coordinate system: a rotation of ϕ angle about the OZ axis followed by a rotation of θ angle about the rotated OV axis and finally a rotation of ϕ angle about the rotated OU axis (see Fig. 2.7).

2.2.5 Geometric Interpretation of Rotation Matrices

It is worthwhile to interpret the basic rotation matrices geometrically. Let us choose a point $\mathbf{p}$ fixed in the $OUVW$ coordinate system to be $(1, 0, 0)^T$, that is, $\mathbf{p}_{uvw} \equiv \mathbf{i}_u$. Then the first column of the rotation matrix represents the coordinates of this point with respect to the $OXYZ$ coordinate system. Similarly, choosing $\mathbf{p}$ to be $(0, 1, 0)^T$ and $(0, 0, 1)^T$, one can identify that the second- and third-column elements of a rotation matrix represent the OV and OW axes, respectively, of the $OUVW$ coordinate system with respect to the $OXYZ$ coordinate system. Thus, given a reference frame $OXYZ$ and a rotation matrix, the column vectors of the rotation matrix represent the principal axes of the $OUVW$ coordinate system with respect to the reference frame and one can draw the location of all the principal axes of the $OUVW$ coordinate frame with respect to the reference frame. In other words, a rotation matrix geometrically represents the principal axes of the rotated coordinate system with respect to the reference coordinate system.

Since the inverse of a rotation matrix is equivalent to its transpose, the row vectors of the rotation matrix represent the principal axes of the reference system *OXYZ* with respect to the rotated coordinate system *OUVW*. This geometric interpretation of the rotation matrices is an important concept that provides insight into many robot arm kinematics problems. Several useful properties of rotation matrices are listed as follows:

1. Each column vector of the rotation matrix is a representation of the rotated axis unit vector expressed in terms of the axis unit vectors of the reference frame, and each row vector is a representation of the axis unit vector of the reference frame expressed in terms of the rotated axis unit vectors of the *OUVW* frame.
2. Since each row and column is a unit vector representation, the magnitude of each row and column should be equal to 1. This is a direct property of ortho-normal coordinate systems. Furthermore, the determinant of a rotation matrix is +1 for a right-hand coordinate system and −1 for a left-hand coordinate system.
3. Since each row is a vector representation of orthonormal vectors, the inner product (dot product) of each row with each other row equals zero. Similarly, the inner product of each column with each other column equals zero.
4. The inverse of a rotation matrix is the transpose of the rotation matrix.

$$\mathbf{R}^{-1} = \mathbf{R}^T \qquad \text{and} \qquad \mathbf{R}\,\mathbf{R}^T = \mathbf{I}_3$$

where $\mathbf{I}_3$ is a 3×3 identity matrix.

Properties 3 and 4 are especially useful in checking the results of rotation matrix multiplications, and in determining an erroneous row or column vector.

Example: If the *OU, OV,* and *OW* coordinate axes were rotated with α angle about the *OX* axis, what would the representation of the coordinate axes of the reference frame be in terms of the rotated coordinate system *OUVW*?

SOLUTION: The new coordinate axis unit vectors become $\mathbf{i}_u = (1, 0, 0)^T$, $\mathbf{j}_v = (0, 1, 0)^T$, and $\mathbf{k}_w = (0, 0, 1)^T$ since they are expressed in terms of themselves. The original unit vectors are then

$$\mathbf{i}_x = 1\mathbf{i}_u + 0\mathbf{j}_v + 0\mathbf{k}_w = (1, 0, 0)^T$$

$$\mathbf{j}_y = 0\mathbf{i}_u + \cos\alpha\mathbf{j}_v - \sin\alpha\mathbf{k}_w = (0, \cos\alpha, -\sin\alpha)^T$$

$$\mathbf{k}_z = 0\mathbf{i}_u + \sin\alpha\mathbf{j}_v + \cos\alpha\mathbf{k}_w = (0, \sin\alpha, \cos\alpha)^T$$

Applying property 1 and considering these as rows of the rotation matrix, the $\mathbf{R}_{x,\alpha}$ matrix can be reconstructed as

$$\mathbf{R}_{x,\alpha} = \begin{bmatrix} 1 & 0 & 0 \\ 0 & \cos\alpha & \sin\alpha \\ 0 & -\sin\alpha & \cos\alpha \end{bmatrix}$$

which is the same as the transpose of Eq. (2.2-12). ☐

2.2.6 Homogeneous Coordinates and Transformation Matrix

Since a 3×3 rotation matrix does not give us any provision for translation and scaling, a fourth coordinate or component is introduced to a position vector $\mathbf{p} = (p_x, p_y, p_z)^T$ in a three-dimensional space which makes it $\hat{\mathbf{p}} = (wp_x, wp_y, wp_z, w)^T$. We say that the position vector $\hat{\mathbf{p}}$ is expressed in *homogeneous coordinates*. In this section, we use a "hat" (i.e., $\hat{\mathbf{p}}$) to indicate the representation of a cartesian vector in homogeneous coordinates. Later, if no confusion exists, these "hats" will be lifted. The concept of a homogeneous-coordinate representation of points in a three-dimensional euclidean space is useful in developing matrix transformations that include rotation, translation, scaling, and perspective transformation. In general, the representation of an N-component position vector by an $(N+1)$-component vector is called *homogeneous coordinate representation*. In a homogeneous coordinate representation, the transformation of an N-dimensional vector is performed in the $(N+1)$-dimensional space, and the physical N-dimensional vector is obtained by dividing the homogeneous coordinates by the $(N+1)$th coordinate, w. Thus, in a three-dimensional space, a position vector $\mathbf{p} = (p_x, p_y, p_z)^T$ is represented by an augmented vector $(wp_x, wp_y, wp_z, w)^T$ in the homogeneous coordinate representation. The physical coordinates are related to the homogeneous coordinates as follows:

$$p_x = \frac{wp_x}{w} \qquad p_y = \frac{wp_y}{w} \qquad p_z = \frac{wp_z}{w}$$

There is no unique homogeneous coordinates representation for a position vector in the three-dimensional space. For example, $\hat{\mathbf{p}}_1 = (w_1 p_x, w_1 p_y, w_1 p_z, w_1)^T$ and $\hat{\mathbf{p}}_2 = (w_2 p_x, w_2 p_y, w_2 p_z, w_2)^T$ are all homogeneous coordinates representing the same position vector $\mathbf{p} = (p_x, p_y, p_z)^T$. Thus, one can view the the fourth component of the homogeneous coordinates, w, as a scale factor. If this coordinate is unity ($w = 1$), then the transformed homogeneous coordinates of a position vector are the same as the physical coordinates of the vector. In robotics applications, this scale factor will always be equal to 1, although it is commonly used in computer graphics as a universal scale factor taking on any positive values.

The homogeneous transformation matrix is a 4×4 matrix which maps a position vector expressed in homogeneous coordinates from one coordinate system to another coordinate system. A homogeneous transformation matrix can be considered to consist of four submatrices:

$$
\mathbf{T} =
\begin{bmatrix}
\mathbf{R}_{3\times3} & | & \mathbf{p}_{3\times1} \\
- & | & - \\
\mathbf{f}_{1\times3} & | & 1\times1
\end{bmatrix}
=
\begin{bmatrix}
\text{rotation matrix} & | & \text{position vector} \\
- & | & - \\
\text{perspective transformation} & | & \text{scaling}
\end{bmatrix}
\tag{2.2-20}
$$

The upper left 3×3 submatrix represents the rotation matrix; the upper right 3×1 submatrix represents the position vector of the origin of the rotated coordinate system with respect to the reference system; the lower left 1×3 submatrix represents perspective transformation; and the fourth diagonal element is the global scaling factor. The homogeneous transformation matrix can be used to explain the geometric relationship between the body-attached frame $OUVW$ and the reference coordinate system $OXYZ$.

If a position vector $\mathbf{p}$ in a three-dimensional space is expressed in homogeneous coordinates [i.e., $\hat{\mathbf{p}} = (p_x, \ p_y, \ p_z, \ 1)^T$], then using the transformation matrix concept, a 3×3 rotation matrix can be extended to a 4×4 homogeneous transformation matrix $\mathbf{T}_{rot}$ for pure rotation operations. Thus, Eqs. (2.2-12) and (2.2-13), expressed as homogeneous rotation matrices, become

$$
\mathbf{T}_{x,\alpha} =
\begin{bmatrix}
1 & 0 & 0 & 0 \\
0 & \cos\alpha & -\sin\alpha & 0 \\
0 & \sin\alpha & \cos\alpha & 0 \\
0 & 0 & 0 & 1
\end{bmatrix}
\qquad
\mathbf{T}_{y,\phi} =
\begin{bmatrix}
\cos\phi & 0 & \sin\phi & 0 \\
0 & 1 & 0 & 0 \\
-\sin\phi & 0 & \cos\phi & 0 \\
0 & 0 & 0 & 1
\end{bmatrix}
$$

$$
\mathbf{T}_{z,\theta} =
\begin{bmatrix}
\cos\theta & -\sin\theta & 0 & 0 \\
\sin\theta & \cos\theta & 0 & 0 \\
0 & 0 & 1 & 0 \\
0 & 0 & 0 & 1
\end{bmatrix}
\tag{2.2-21}
$$

These 4×4 rotation matrices are called the *basic homogeneous rotation matrices*.

The upper right 3×1 submatrix of the homogeneous transformation matrix has the effect of translating the $OUVW$ coordinate system which has axes parallel to the reference coordinate system $OXYZ$ but whose origin is at $(dx, \ dy, \ dz)$ of the reference coordinate system:

$$
\mathbf{T}_{tran} =
\begin{bmatrix}
1 & 0 & 0 & dx \\
0 & 1 & 0 & dy \\
0 & 0 & 1 & dz \\
0 & 0 & 0 & 1
\end{bmatrix}
\tag{2.2-22}
$$

This 4×4 transformation matrix is called the *basic homogeneous translation matrix*.

The lower left 1×3 submatrix of the homogeneous transformation matrix represents perspective transformation, which is useful for computer vision and the calibration of camera models, as discussed in Chap. 7. In the present discussion, the elements of this submatrix are set to zero to indicate null perspective transformation.

The principal diagonal elements of a homogeneous transformation matrix produce local and global scaling. The first three diagonal elements produce local stretching or scaling, as in

$$
\begin{bmatrix} a & 0 & 0 & 0 \\ 0 & b & 0 & 0 \\ 0 & 0 & c & 0 \\ 0 & 0 & 0 & 1 \end{bmatrix} \begin{bmatrix} x \\ y \\ z \\ 1 \end{bmatrix} = \begin{bmatrix} ax \\ by \\ cz \\ 1 \end{bmatrix} \tag{2.2-23}
$$

Thus, the coordinate values are stretched by the scalars a, b, and c, respectively. Note that the basic rotation matrices, $\mathbf{T}_{rot}$, do not produce any local scaling effect.

The fourth diagonal element produces global scaling as in

$$
\begin{bmatrix} 1 & 0 & 0 & 0 \\ 0 & 1 & 0 & 0 \\ 0 & 0 & 1 & 0 \\ 0 & 0 & 0 & s \end{bmatrix} \begin{bmatrix} x \\ y \\ z \\ 1 \end{bmatrix} = \begin{bmatrix} x \\ y \\ z \\ s \end{bmatrix} \tag{2.2-24}
$$

where $s > 0$. The physical cartesian coordinates of the vector are

$$
p_x = \frac{x}{s} \qquad p_y = \frac{y}{s} \qquad p_z = \frac{z}{s} \qquad w = \frac{s}{s} = 1 \tag{2.2-25}
$$

Therefore, the fourth diagonal element in the homogeneous transformation matrix has the effect of globally reducing the coordinates if $s > 1$ and of enlarging the coordinates if $0 < s < 1$.

In summary, a 4×4 homogeneous transformation matrix maps a vector expressed in homogeneous coordinates with respect to the $OUVW$ coordinate system to the reference coordinate system $OXYZ$. That is, with $w = 1$,

$$
\hat{\mathbf{p}}_{xyz} = \mathbf{T} \hat{\mathbf{p}}_{uvw} \tag{2.2-26a}
$$

and

$$
\mathbf{T} = \begin{bmatrix} n_x & s_x & a_x & p_x \\ n_y & s_y & a_y & p_y \\ n_z & s_z & a_z & p_z \\ 0 & 0 & 0 & 1 \end{bmatrix} = \begin{bmatrix} \mathbf{n} & \mathbf{s} & \mathbf{a} & \mathbf{p} \\ 0 & 0 & 0 & 1 \end{bmatrix} \tag{2.2-26b}
$$

2.2.7 Geometric Interpretation of Homogeneous Transformation Matrices

In general, a homogeneous transformation matrix for a three-dimensional space can be represented as in Eq. (2.2-26b). Let us choose a point **p** fixed in the $OUVW$ coordinate system and expressed in homogeneous coordinates as $(0, 0, 0, 1)^T$; that is, $\mathbf{p}_{uvw}$ is the origin of the $OUVW$ coordinate system. Then the upper right 3×1 submatrix indicates the position of the origin of the $OUVW$ frame with respect to the $OXYZ$ reference coordinate frame. Next, let us choose the point **p** to be $(1, 0, 0\ 1)^T$; that is $\mathbf{p}_{uvw} \equiv \mathbf{i}_u$. Furthermore, we assume that the origins of both coordinate systems coincide at a point O. This has the effect of making the elements in the upper right 3×1 submatrix a null vector. Then the first column (or **n** vector) of the homogeneous transformation matrix represents the coordinates of the OU axis of $OUVW$ with respect to the $OXYZ$ coordinate system. Similarly, choosing **p** to be $(0, 1, 0, 1)^T$ and $(0, 0, 1, 1)^T$, one can identify that the second-column (or **s** vector) and third-column (or **a** vector) elements of the homogeneous transformation matrix represent the OV and OW axes, respectively, of the $OUVW$ coordinate system with respect to the reference coordinate system. Thus, given a reference frame $OXYZ$ and a homogeneous transformation matrix **T**, the column vectors of the rotation submatrix represent the principal axes of the $OUVW$ coordinate system with respect to the reference coordinate frame, and one can draw the orientation of all the principal axes of the $OUVW$ coordinate frame with respect to the reference coordinate frame. The fourth-column vector of the homogeneous transformation matrix represents the position of the origin of the $OUVW$ coordinate system with respect to the reference system. In other words, a homogeneous transformation matrix geometrically represents the *location* of a rotated coordinate system (position and orientation) with respect to a reference coordinate system.

Since the inverse of a rotation submatrix is equivalent to its transpose, the row vectors of a rotation submatrix represent the principal axes of the reference coordinate system with respect to the rotated coordinate system $OUVW$. However, the inverse of a homogeneous transformation matrix is *not* equivalent to its transpose. The position of the origin of the reference coordinate system with respect to the $OUVW$ coordinate system can only be found after the inverse of the homogeneous transformation matrix is determined. In general, the inverse of a homogeneous transformation matrix can be found to be

$$
\mathbf{T}^{-1} = \begin{bmatrix} n_x & n_y & n_z & -\mathbf{n}^T\mathbf{p} \\ s_x & s_y & s_z & -\mathbf{s}^T\mathbf{p} \\ a_x & a_y & a_z & -\mathbf{a}^T\mathbf{p} \\ 0 & 0 & 0 & 1 \end{bmatrix} = \begin{bmatrix} & & & -\mathbf{n}^T\mathbf{p} \\ & \mathbf{R}^T_{3\times 3} & & -\mathbf{s}^T\mathbf{p} \\ & & & -\mathbf{a}^T\mathbf{p} \\ 0 & 0 & 0 & 1 \end{bmatrix} \quad (2.2\text{-}27)
$$

Thus, from Eq. (2.2-27), the column vectors of the inverse of a homogeneous transformation matrix represent the principal axes of the reference system with respect to the rotated coordinate system $OUVW$, and the upper right 3×1 submatrix represents the position of the origin of the reference frame with respect to the *OUVW* system. This geometric interpretation of the homogeneous transformation matrices is an important concept used frequently throughout this book.

2.2.8 Composite Homogeneous Transformation Matrix

The homogeneous rotation and translation matrices can be multiplied together to obtain a composite homogeneous transformation matrix (we shall call it the **T** matrix). However, since matrix multiplication is not commutative, careful attention must be paid to the order in which these matrices are multiplied. The following rules are useful for finding a composite homogeneous transformation matrix:

1. Initially both coordinate systems are coincident, hence the homogeneous transformation matrix is a 4 × 4 identity matrix, $\mathbf{I}_4$.
2. If the rotating coordinate system *OUVW* is rotating/translating about the principal axes of the *OXYZ* frame, then *premultiply* the previous (resultant) homogeneous transformation matrix with an appropriate basic homogeneous rotation/translation matrix.
3. If the rotating coordinate system *OUVW* is rotating/translating about its own principal axes, then *postmultiply* the previous (resultant) homogeneous transformation matrix with an appropriate basic homogeneous rotation/translation matrix.

Example: Two points $\mathbf{a}_{uvw} = (4, 3, 2)^T$ and $\mathbf{b}_{uvw} = (6, 2, 4)^T$ are to be translated a distance $+5$ units along the *OX* axis and -3 units along the *OZ* axis. Using the appropriate homogeneous transformation matrix, determine the new points $\mathbf{a}_{xyz}$ and $\mathbf{b}_{xyz}$.

SOLUTION:

$$
\hat{\mathbf{a}}_{xyz} =
\begin{bmatrix}
1 & 0 & 0 & 5 \\
0 & 1 & 0 & 0 \\
0 & 0 & 1 & -3 \\
0 & 0 & 0 & 1
\end{bmatrix}
\begin{bmatrix}
4 \\ 3 \\ 2 \\ 1
\end{bmatrix}
=
\begin{bmatrix}
4(1) + 1(5) \\
3(1) + 1(0) \\
2(1) + 1(-3) \\
1(1)
\end{bmatrix}
=
\begin{bmatrix}
9 \\ 3 \\ -1 \\ 1
\end{bmatrix}
$$

$$
\hat{\mathbf{b}}_{xyz} =
\begin{bmatrix}
1 & 0 & 0 & 5 \\
0 & 1 & 0 & 0 \\
0 & 0 & 1 & -3 \\
0 & 0 & 0 & 1
\end{bmatrix}
\begin{bmatrix}
6 \\ 2 \\ 4 \\ 1
\end{bmatrix}
=
\begin{bmatrix}
11 \\ 2 \\ 1 \\ 1
\end{bmatrix}
$$

The translated points are $\mathbf{a}_{xyz} = (9, 3, -1)^T$ and $\mathbf{b}_{xyz} = (11, 2, 1)^T$. □

Example: A **T** matrix is to be determined that represents a rotation of α angle about the *OX* axis, followed by a translation of b units along the rotated *OV* axis.

SOLUTION: This problem can be tricky but illustrates some of the fundamental components of the **T** matrix. Two approaches will be utilized, an unorthodox approach which is illustrative, and the orthodox approach, which is simpler. After the rotation $\mathbf{T}_{x,\,\alpha}$, the rotated OV axis is (in terms of the unit vectors $\mathbf{i}_x$, $\mathbf{j}_y$, $\mathbf{k}_z$ of the reference system) $\mathbf{j}_v = \cos\alpha\,\mathbf{j}_y + \sin\alpha\,\mathbf{k}_z$; i.e., column 2 of Eq. (2.2-21). Thus, a translation along the rotated OV axis of b units is $b\,\mathbf{j}_v = b\cos\alpha\,\mathbf{j}_y + b\sin\alpha\,\mathbf{k}_z$. So the **T** matrix is

$$
\mathbf{T} = \mathbf{T}_{v,\,b}\,\mathbf{T}_{x,\,\alpha} =
\begin{bmatrix}
1 & 0 & 0 & 0 \\
0 & 1 & 0 & b\cos\alpha \\
0 & 0 & 1 & b\sin\alpha \\
0 & 0 & 0 & 1
\end{bmatrix}
\begin{bmatrix}
1 & 0 & 0 & 0 \\
0 & \cos\alpha & -\sin\alpha & 0 \\
0 & \sin\alpha & \cos\alpha & 0 \\
0 & 0 & 0 & 1
\end{bmatrix}
$$

$$
=
\begin{bmatrix}
1 & 0 & 0 & 0 \\
0 & \cos\alpha & -\sin\alpha & b\cos\alpha \\
0 & \sin\alpha & \cos\alpha & b\sin\alpha \\
0 & 0 & 0 & 1
\end{bmatrix}
$$

In the orthodox approach, following the rules as stated earlier, one should realize that since the $\mathbf{T}_{x,\,\alpha}$ matrix will rotate the OY axis to the OV axis, then translation along the OV axis will accomplish the same goal, that is,

$$
\mathbf{T} = \mathbf{T}_{x,\,\alpha}\,\mathbf{T}_{v,\,b} =
\begin{bmatrix}
1 & 0 & 0 & 0 \\
0 & \cos\alpha & -\sin\alpha & 0 \\
0 & \sin\alpha & \cos\alpha & 0 \\
0 & 0 & 0 & 1
\end{bmatrix}
\begin{bmatrix}
1 & 0 & 0 & 0 \\
0 & 1 & 0 & b \\
0 & 0 & 1 & 0 \\
0 & 0 & 0 & 1
\end{bmatrix}
$$

$$
=
\begin{bmatrix}
1 & 0 & 0 & 0 \\
0 & \cos\alpha & -\sin\alpha & b\cos\alpha \\
0 & \sin\alpha & \cos\alpha & b\sin\alpha \\
0 & 0 & 0 & 1
\end{bmatrix}
$$

□

Example: Find a homogeneous transformation matrix **T** that represents a rotation of α angle about the OX axis, followed by a translation of a units along the OX axis, followed by a translation of d units along the OZ axis, followed by a rotation of θ angle about the OZ axis.

SOLUTION:

$$
\mathbf{T} = \mathbf{T}_{z,\,\theta}\,\mathbf{T}_{z,\,d}\,\mathbf{T}_{x,\,a}\,\mathbf{T}_{x,\,\alpha}
$$

$$
= \begin{bmatrix} \cos\theta & -\sin\theta & 0 & 0 \\ \sin\theta & \cos\theta & 0 & 0 \\ 0 & 0 & 1 & 0 \\ 0 & 0 & 0 & 1 \end{bmatrix} \begin{bmatrix} 1 & 0 & 0 & 0 \\ 0 & 1 & 0 & 0 \\ 0 & 0 & 1 & d \\ 0 & 0 & 0 & 1 \end{bmatrix} \begin{bmatrix} 1 & 0 & 0 & a \\ 0 & 1 & 0 & 0 \\ 0 & 0 & 1 & 0 \\ 0 & 0 & 0 & 1 \end{bmatrix} \begin{bmatrix} 1 & 0 & 0 & 0 \\ 0 & \cos\alpha & -\sin\alpha & 0 \\ 0 & \sin\alpha & \cos\alpha & 0 \\ 0 & 0 & 0 & 1 \end{bmatrix}
$$

$$
= \begin{bmatrix} \cos\theta & -\cos\alpha\sin\theta & \sin\alpha\sin\theta & a\cos\theta \\ \sin\theta & \cos\alpha\cos\theta & -\sin\alpha\cos\theta & a\sin\theta \\ 0 & \sin\alpha & \cos\alpha & d \\ 0 & 0 & 0 & 1 \end{bmatrix} \qquad \square
$$

We have identified two coordinate systems, the fixed reference coordinate frame $OXYZ$ and the moving (translating and rotating) coordinate frame $OUVW$. To describe the spatial displacement relationship between these two coordinate systems, a 4×4 homogeneous transformation matrix is used. Homogeneous transformation matrices have the combined effect of rotation, translation, perspective, and global scaling when operating on position vectors expressed in homogeneous coordinates.

If these two coordinate systems are assigned to each link of a robot arm, say link $i-1$ and link i, respectively, then the link $i-1$ coordinate system is the reference coordinate system and the link i coordinate system is the moving coordinate system, when joint i is activated. Using the **T** matrix, we can specify a point $\mathbf{p}_i$ at rest in link i and expressed in the link i (or $OUVW$) coordinate system in terms of the link $i-1$ (or $OXYZ$) coordinate system as

$$\mathbf{p}_{i-1} = \mathbf{T}\,\mathbf{p}_i \qquad (2.2\text{-}28)$$

where

> $\mathbf{T} = 4 \times 4$ homogeneous transformation matrix relating the two coordinate systems
>
> $\mathbf{p}_i = 4 \times 1$ augmented position vector $(x_i, y_i, z_i, 1)^T$ representing a point in the link i coordinate system expressed in homogeneous coordinates
>
> $\mathbf{p}_{i-1} =$ is the 4×1 augmented position vector $(x_{i-1}, y_{i-1}, z_{i-1}, 1)^T$ representing the same point $\mathbf{p}_i$ in terms of the link $i-1$ coordinate system

2.2.9 Links, Joints, and Their Parameters

A mechanical manipulator consists of a sequence of rigid bodies, called links, connected by either revolute or prismatic joints (see Fig. 2.8). Each joint-link pair constitutes 1 degree of freedom. Hence, for an N degree of freedom manipulator, there are N joint-link pairs with link 0 (not considered part of the robot) attached to a supporting base where an inertial coordinate frame is usually established for this dynamic system, and the last link is attached with a tool. The joints and links

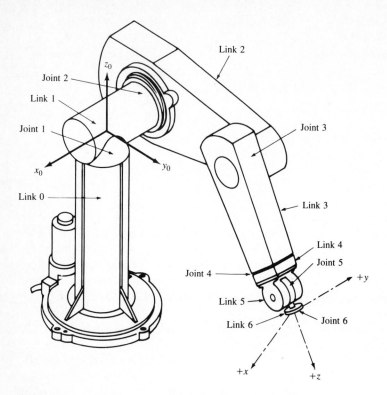

Figure 2.8 A PUMA robot arm illustrating joints and links.

are numbered outwardly from the base; thus, joint 1 is the point of connection between link 1 and the supporting base. Each link is connected to, at most, two others so that no closed loops are formed.

In general, two links are connected by a *lower-pair* joint which has two surfaces sliding over one another while remaining in contact. Only six different lower-pair joints are possible: revolute (rotary), prismatic (sliding), cylindrical, spherical, screw, and planar (see Fig. 2.9). Of these, only rotary and prismatic joints are common in manipulators.

A joint axis (for joint i) is established at the connection of two links (see Fig. 2.10). This joint axis will have two normals connected to it, one for each of the links. The relative position of two such connected links (link $i - 1$ and link i) is given by d_i which is the distance measured along the joint axis between the normals. The joint angle θ_i between the normals is measured in a plane normal to the joint axis. Hence, d_i and θ_i may be called the *distance* and the *angle* between the adjacent links, respectively. They determine the relative position of neighboring links.

A link i ($i = 1, \ldots, 6$) is connected to, at most, two other links (e.g., link $i - 1$ and link $i + 1$); thus, two joint axes are established at both ends of the connection. The significance of links, from a kinematic perspective, is that they maintain a fixed configuration between their joints which can be characterized by two

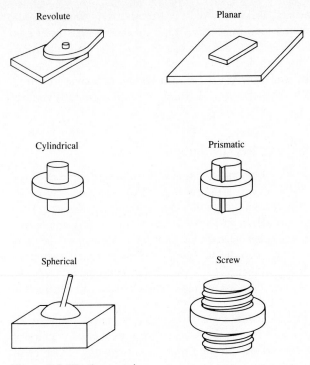

Figure 2.9 The lower pair.

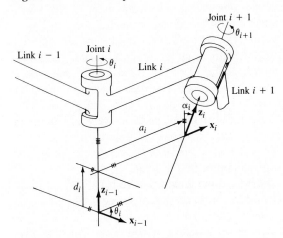

Figure 2.10 Link coordinate system and its parameters.

parameters: a_i and α_i. The parameter a_i is the shortest distance measured along the common normal between the joint axes (i.e., the z_{i-1} and z_i axes for joint i and joint $i + 1$, respectively), and α_i is the angle between the joint axes measured in a plane perpendicular to a_i. Thus, a_i and α_i may be called the *length* and the *twist angle* of the link i, respectively. They determine the structure of link i.

In summary, four parameters, a_i, α_i, d_i, and θ_i, are associated with each link of a manipulator. If a sign convention for each of these parameters has been established, then these parameters constitute a sufficient set to completely determine the kinematic configuration of each link of a robot arm. Note that these four parameters come in pairs: the link parameters (a_i, α_i) which determine the structure of the link and the joint parameters (d_i, θ_i) which determine the relative position of neighboring links.

2.2.10 The Denavit-Hartenberg Representation

To describe the translational and rotational relationships between adjacent links, Denavit and Hartenberg [1955] proposed a matrix method of systematically establishing a coordinate system (body-attached frame) to each link of an articulated chain. The Denavit-Hartenberg (D-H) representation results in a 4×4 homogeneous transformation matrix representing each link's coordinate system at the joint with respect to the previous link's coordinate system. Thus, through sequential transformations, the end-effector expressed in the "hand coordinates" can be transformed and expressed in the "base coordinates" which make up the inertial frame of this dynamic system.

An orthonormal cartesian coordinate system $(\mathbf{x}_i, \mathbf{y}_i, \mathbf{z}_i)$† can be established for each link at its joint axis, where $i = 1, 2, \ldots, n$ (n = number of degrees of freedom) plus the base coordinate frame. Since a rotary joint has only 1 degree of freedom, each $(\mathbf{x}_i, \mathbf{y}_i, \mathbf{z}_i)$ coordinate frame of a robot arm corresponds to joint $i + 1$ and is fixed in link i. When the joint actuator activates joint i, link i will move with respect to link $i - 1$. Since the ith coordinate system is fixed in link i, it moves together with the link i. Thus, the nth coordinate frame moves with the hand (link n). The base coordinates are defined as the 0th coordinate frame $(\mathbf{x}_0, \mathbf{y}_0, \mathbf{z}_0)$ which is also the inertial coordinate frame of the robot arm. Thus, for a six-axis PUMA-like robot arm, we have seven coordinate frames, namely, $(\mathbf{x}_0, \mathbf{y}_0, \mathbf{z}_0)$, $(\mathbf{x}_1, \mathbf{y}_1, \mathbf{z}_1)$, $\ldots$, $(\mathbf{x}_6, \mathbf{y}_6, \mathbf{z}_6)$.

Every coordinate frame is determined and established on the basis of three rules:

1. The $\mathbf{z}_{i-1}$ axis lies along the axis of motion of the ith joint.
2. The $\mathbf{x}_i$ axis is normal to the $\mathbf{z}_{i-1}$ axis, and pointing away from it.
3. The $\mathbf{y}_i$ axis completes the right-handed coordinate system as required.

By these rules, one is free to choose the location of coordinate frame 0 anywhere in the supporting base, as long as the $\mathbf{z}_0$ axis lies along the axis of motion of the first joint. The last coordinate frame (nth frame) can be placed anywhere in the hand, as long as the $\mathbf{x}_n$ axis is normal to the $\mathbf{z}_{n-1}$ axis.

The D-H representation of a rigid link depends on four geometric parameters associated with each link. These four parameters completely describe any revolute

---

† $(\mathbf{x}_i, \mathbf{y}_i, \mathbf{z}_i)$ actually represent the unit vectors along the principal axes of the coordinate frame i, respectively, but are used here to denote the coordinate frame i.

or prismatic joint. Referring to Fig. 2.10, these four parameters are defined as follows:

θ_i is the joint angle from the x_{i-1} axis to the x_i axis about the z_{i-1} axis (using the right-hand rule).

d_i is the distance from the origin of the $(i-1)$th coordinate frame to the intersection of the z_{i-1} axis with the x_i axis along the z_{i-1} axis.

a_i is the offset distance from the intersection of the z_{i-1} axis with the x_i axis to the origin of the ith frame along the x_i axis (or the shortest distance between the z_{i-1} and z_i axes).

α_i is the offset angle from the z_{i-1} axis to the z_i axis about the x_i axis (using the right-hand rule).

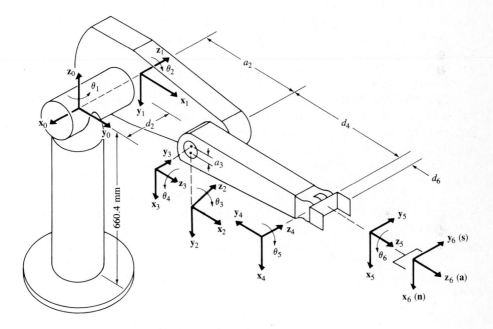

PUMA robot arm link coordinate parameters				
Joint i	θ_i	α_i	a_i	d_i	Joint range
1	90	−90	0	0	−160 to +160
2	0	0	431.8 mm	149.09 mm	−225 to 45
3	90	90	−20.32 mm	0	−45 to 225
4	0	−90	0	433.07 mm	−110 to 170
5	0	90	0	0	−100 to 100
6	0	0	0	56.25 mm	−266 to 266

Figure 2.11 Establishing link coordinate systems for a PUMA robot.

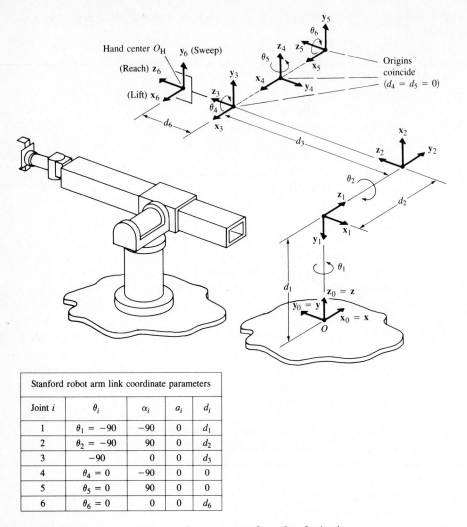

Stanford robot arm link coordinate parameters			
Joint i	θ_i	α_i	a_i	d_i
---	---	---	---	---
1	$\theta_1 = -90$	-90	0	d_1
2	$\theta_2 = -90$	90	0	d_2
3	-90	0	0	d_3
4	$\theta_4 = 0$	-90	0	0
5	$\theta_5 = 0$	90	0	0
6	$\theta_6 = 0$	0	0	d_6

Figure 2.12 Establishing link coordinate systems for a Stanford robot.

For a rotary joint, d_i, a_i, and α_i are the joint parameters and remain constant for a robot, while θ_i is the joint variable that changes when link i moves (or rotates) with respect to link $i - 1$. For a prismatic joint, θ_i, a_i, and α_i are the joint parameters and remain constant for a robot, while d_i is the joint variable. For the remainder of this book, *joint variable* refers to θ_i (or d_i), that is, the varying quantity, and *joint parameters* refer to the remaining three geometric constant values (d_i, a_i, α_i) for a rotary joint, or (θ_i, a_i, α_i) for a prismatic joint.

With the above three basic rules for establishing an orthonormal coordinate system for each link and the geometric interpretation of the joint and link parameters, a procedure for establishing *consistent* orthonormal coordinate systems for a robot is outlined in Algorithm 2.1. Examples of applying this algorithm to a six-axis PUMA-like robot arm and a Stanford arm are given in Figs. 2.11 and 2.12, respectively.

Algorithm 2.1: Link Coordinate System Assignment. Given an n degree of freedom robot arm, this algorithm assigns an orthonormal coordinate system to each link of the robot arm according to arm configurations similar to those of human arm geometry. The labeling of the coordinate systems begins from the supporting base to the end-effector of the robot arm. The relations between adjacent links can be represented by a 4×4 homogeneous transformation matrix. The significance of this assignment is that it will aid the development of a consistent procedure for deriving the joint solution as discussed in the later sections. (Note that the assignment of coordinate systems is not unique.)

D1. *Establish the base coordinate system.* Establish a right-handed orthonormal coordinate system $(\mathbf{x}_0, \mathbf{y}_0, \mathbf{z}_0)$ at the supporting base with the $\mathbf{z}_0$ axis lying along the axis of motion of joint 1 and pointing toward the shoulder of the robot arm. The $\mathbf{x}_0$ and $\mathbf{y}_0$ axes can be conveniently established and are normal to the $\mathbf{z}_0$ axis.

D2. *Initialize and loop.* For each i, $i = 1, \ldots, n - 1$, perform steps D3 to D6.

D3. *Establish joint axis.* Align the $\mathbf{z}_i$ with the axis of motion (rotary or sliding) of joint $i + 1$. For robots having left-right arm configurations, the $\mathbf{z}_1$ and $\mathbf{z}_2$ axes are pointing away from the shoulder and the "trunk" of the robot arm.

D4. *Establish the origin of the ith coordinate system.* Locate the origin of the ith coordinate system at the intersection of the $\mathbf{z}_i$ and $\mathbf{z}_{i-1}$ axes or at the intersection of common normal between the $\mathbf{z}_i$ and $\mathbf{z}_{i-1}$ axes and the $\mathbf{z}_i$ axis.

D5. *Establish $\mathbf{x}_i$ axis.* Establish $\mathbf{x}_i = \pm (\mathbf{z}_{i-1} \times \mathbf{z}_i) / \|\mathbf{z}_{i-1} \times \mathbf{z}_i\|$ or along the common normal between the $\mathbf{z}_{i-1}$ and $\mathbf{z}_i$ axes when they are parallel.

D6. *Establish $\mathbf{y}_i$ axis.* Assign $\mathbf{y}_i = + (\mathbf{z}_i \times \mathbf{x}_i) / \|\mathbf{z}_i \times \mathbf{x}_i\|$ to complete the right-handed coordinate system. (Extend the $\mathbf{z}_i$ and the $\mathbf{x}_i$ axes if necessary for steps D9 to D12).

D7. *Establish the hand coordinate system.* Usually the nth joint is a rotary joint. Establish $\mathbf{z}_n$ along the direction of $\mathbf{z}_{n-1}$ axis and pointing away from the robot. Establish $\mathbf{x}_n$ such that it is normal to both $\mathbf{z}_{n-1}$ and $\mathbf{z}_n$ axes. Assign $\mathbf{y}_n$ to complete the right-handed coordinate system. (See Sec. 2.2.11 for more detail.)

D8. *Find joint and link parameters.* For each i, $i = 1, \ldots, n$, perform steps D9 to D12.

D9. *Find d_i.* d_i is the distance from the origin of the $(i - 1)$th coordinate system to the intersection of the $\mathbf{z}_{i-1}$ axis and the $\mathbf{x}_i$ axis along the $\mathbf{z}_{i-1}$ axis. It is the joint variable if joint i is prismatic.

D10. *Find a_i.* a_i is the distance from the intersection of the $\mathbf{z}_{i-1}$ axis and the $\mathbf{x}_i$ axis to the origin of the ith coordinate system along the $\mathbf{x}_i$ axis.

D11. *Find θ_i.* θ_i is the angle of rotation from the $\mathbf{x}_{i-1}$ axis to the $\mathbf{x}_i$ axis about the $\mathbf{z}_{i-1}$ axis. It is the joint variable if joint i is rotary.

D12. *Find α_i.* α_i is the angle of rotation from the $\mathbf{z}_{i-1}$ axis to the $\mathbf{z}_i$ axis about the $\mathbf{x}_i$ axis.

Once the D-H coordinate system has been established for each link, a homogeneous transformation matrix can easily be developed relating the ith coordinate frame to the $(i - 1)$th coordinate frame. Looking at Fig. 2.10, it is obvious that a point $\mathbf{r}_i$ expressed in the ith coordinate system may be expressed in the $(i - 1)$th coordinate system as $\mathbf{r}_{i-1}$ by performing the following successive transformations:

1. Rotate about the $\mathbf{z}_{i-1}$ axis an angle of θ_i to align the $\mathbf{x}_{i-1}$ axis with the $\mathbf{x}_i$ axis ($\mathbf{x}_{i-1}$ axis is parallel to $\mathbf{x}_i$ and pointing in the same direction).
2. Translate along the $\mathbf{z}_{i-1}$ axis a distance of d_i to bring the $\mathbf{x}_{i-1}$ and $\mathbf{x}_i$ axes into coincidence.
3. Translate along the $\mathbf{x}_i$ axis a distance of a_i to bring the two origins as well as the $\mathbf{x}$ axis into coincidence.
4. Rotate about the $\mathbf{x}_i$ axis an angle of α_i to bring the two coordinate systems into coincidence.

Each of these four operations can be expressed by a basic homogeneous rotation-translation matrix and the product of these four basic homogeneous transformation matrices yields a composite homogeneous transformation matrix, $^{i-1}\mathbf{A}_i$, known as the D-H transformation matrix for adjacent coordinate frames, i and $i - 1$. Thus,

$$^{i-1}\mathbf{A}_i = \mathbf{T}_{z,d}\,\mathbf{T}_{z,\theta}\,\mathbf{T}_{x,a}\,\mathbf{T}_{x,\alpha}$$

$$= \begin{bmatrix} 1 & 0 & 0 & 0 \\ 0 & 1 & 0 & 0 \\ 0 & 0 & 1 & d_i \\ 0 & 0 & 0 & 1 \end{bmatrix} \begin{bmatrix} \cos\theta_i & -\sin\theta_i & 0 & 0 \\ \sin\theta_i & \cos\theta_i & 0 & 0 \\ 0 & 0 & 1 & 0 \\ 0 & 0 & 0 & 1 \end{bmatrix} \begin{bmatrix} 1 & 0 & 0 & a_i \\ 0 & 1 & 0 & 0 \\ 0 & 0 & 1 & 0 \\ 0 & 0 & 0 & 1 \end{bmatrix} \begin{bmatrix} 1 & 0 & 0 & 0 \\ 0 & \cos\alpha_i & -\sin\alpha_i & 0 \\ 0 & \sin\alpha_i & \cos\alpha_i & 0 \\ 0 & 0 & 0 & 1 \end{bmatrix}$$

$$= \begin{bmatrix} \cos\theta_i & -\cos\alpha_i \sin\theta_i & \sin\alpha_i \sin\theta_i & a_i \cos\theta_i \\ \sin\theta_i & \cos\alpha_i \cos\theta_i & -\sin\alpha_i \cos\theta_i & a_i \sin\theta_i \\ 0 & \sin\alpha_i & \cos\alpha_i & d_i \\ 0 & 0 & 0 & 1 \end{bmatrix} \qquad (2.2\text{-}29)$$

Using Eq. (2.2-27), the inverse of this transformation can be found to be

$$[^{i-1}\mathbf{A}_i]^{-1} = {}^i\mathbf{A}_{i-1} = \begin{bmatrix} \cos\theta_i & \sin\theta_i & 0 & -a_i \\ -\cos\alpha_i \sin\theta_i & \cos\alpha_i \cos\theta_i & \sin\alpha_i & -d_i \sin\alpha_i \\ \sin\alpha_i \sin\theta_i & -\sin\alpha_i \cos\theta_i & \cos\alpha_i & -d_i \cos\alpha_i \\ 0 & 0 & 0 & 1 \end{bmatrix}$$

$$(2.2\text{-}30)$$

where α_i, a_i, d_i are constants while θ_i is the joint variable for a revolute joint.

For a prismatic joint, the joint variable is d_i, while α_i, a_i, and θ_i are constants. In this case, $^{i-1}\mathbf{A}_i$ becomes

$$^{i-1}\mathbf{A}_i = \mathbf{T}_{z,\theta}\,\mathbf{T}_{z,d}\,\mathbf{T}_{x,\alpha} = \begin{bmatrix} \cos\theta_i & -\cos\alpha_i\,\sin\theta_i & \sin\alpha_i\,\sin\theta_i & 0 \\ \sin\theta_i & \cos\alpha_i\,\cos\theta_i & -\sin\alpha_i\,\cos\theta_i & 0 \\ 0 & \sin\alpha_i & \cos\alpha_i & d_i \\ 0 & 0 & 0 & 1 \end{bmatrix} \qquad (2.2\text{-}31)$$

and its inverse is

$$[^{i-1}\mathbf{A}_i]^{-1} = {}^{i}\mathbf{A}_{i-1} = \begin{bmatrix} \cos\theta_i & \sin\theta_i & 0 & 0 \\ -\cos\alpha_i\,\sin\theta_i & \cos\alpha_i\,\cos\theta_i & \sin\alpha_i & -d_i\,\sin\alpha_i \\ \cos\alpha_i\,\sin\theta_i & -\sin\alpha_i\,\cos\theta_i & \cos\alpha_i & -d_i\,\cos\alpha_i \\ 0 & 0 & 0 & 1 \end{bmatrix}$$
$$(2.2\text{-}32)$$

Using the $^{i-1}\mathbf{A}_i$ matrix, one can relate a point $\mathbf{p}_i$ at rest in link i, and expressed in homogeneous coordinates with respect to coordinate system i, to the coordinate system $i-1$ established at link $i-1$ by

$$\mathbf{p}_{i-1} = {}^{i-1}\mathbf{A}_i\,\mathbf{p}_i \qquad (2.2\text{-}33)$$

where $\mathbf{p}_{i-1} = (x_{i-1}, y_{i-1}, z_{i-1}, 1)^T$ and $\mathbf{p}_i = (x_i, y_i, z_i, 1)^T$.

The six $^{i-1}\mathbf{A}_i$ transformation matrices for the six-axis PUMA robot arm have been found on the basis of the coordinate systems established in Fig. 2.11. These $^{i-1}\mathbf{A}_i$ matrices are listed in Fig. 2.13.

2.2.11 Kinematic Equations for Manipulators

The homogeneous matrix $^0\mathbf{T}_i$ which specifies the location of the ith coordinate frame with respect to the base coordinate system is the chain product of successive coordinate transformation matrices of $^{i-1}\mathbf{A}_i$, and is expressed as

$$^0\mathbf{T}_i = {}^0\mathbf{A}_1\,{}^1\mathbf{A}_2 \cdots {}^{i-1}\mathbf{A}_i = \prod_{j=1}^{i} {}^{j-1}\mathbf{A}_j \qquad \text{for } i = 1, 2, \ldots, n$$

$$= \begin{bmatrix} \mathbf{x}_i & \mathbf{y}_i & \mathbf{z}_i & \mathbf{p}_i \\ 0 & 0 & 0 & 1 \end{bmatrix} = \begin{bmatrix} {}^0\mathbf{R}_i & {}^0\mathbf{p}_i \\ 0 & 1 \end{bmatrix} \qquad (2.2\text{-}34)$$

where

$[\mathbf{x}_i, \mathbf{y}_i, \mathbf{z}_i] =$ orientation matrix of the ith coordinate system established at link i with respect to the base coordinate system. It is the upper left 3×3 partitioned matrix of $^0\mathbf{T}_i$.

$\mathbf{p}_i =$ position vector which points from the origin of the base coordinate system to the origin of the ith coordinate system. It is the upper right 3×1 partitioned matrix of $^0\mathbf{T}_i$.

$$
{}^{i-1}\mathbf{A}_i = \begin{bmatrix} \cos\theta_i & -\cos\alpha_i \sin\theta_i & \sin\alpha_i \sin\theta_i & a_i \cos\theta_i \\ \sin\theta_i & \cos\alpha_i \cos\theta_i & -\sin\alpha_i \cos\theta_i & a_i \sin\theta_i \\ 0 & \sin\alpha_i & \cos\alpha_i & d_i \\ 0 & 0 & 0 & 1 \end{bmatrix}
$$

$$
{}^{0}\mathbf{A}_1 = \begin{bmatrix} C_1 & 0 & -S_1 & 0 \\ S_1 & 0 & C_1 & 0 \\ 0 & -1 & 0 & 0 \\ 0 & 0 & 0 & 1 \end{bmatrix}
\qquad
{}^{1}\mathbf{A}_2 = \begin{bmatrix} C_2 & -S_2 & 0 & a_2 C_2 \\ S_2 & C_2 & 0 & a_2 S_2 \\ 0 & 0 & 1 & d_2 \\ 0 & 0 & 0 & 1 \end{bmatrix}
$$

$$
{}^{2}\mathbf{A}_3 = \begin{bmatrix} C_3 & 0 & S_3 & a_3 C_3 \\ S_3 & 0 & -C_3 & a_3 S_3 \\ 0 & 1 & 0 & 0 \\ 0 & 0 & 0 & 1 \end{bmatrix}
\qquad
{}^{3}\mathbf{A}_4 = \begin{bmatrix} C_4 & 0 & -S_4 & 0 \\ S_4 & 0 & C_4 & 0 \\ 0 & -1 & 0 & d_4 \\ 0 & 0 & 0 & 1 \end{bmatrix}
$$

$$
{}^{4}\mathbf{A}_5 = \begin{bmatrix} C_5 & 0 & S_5 & 0 \\ S_5 & 0 & -C_5 & 0 \\ 0 & 1 & 0 & 0 \\ 0 & 0 & 0 & 1 \end{bmatrix}
\qquad
{}^{5}\mathbf{A}_6 = \begin{bmatrix} C_6 & -S_6 & 0 & 0 \\ S_6 & C_6 & 0 & 0 \\ 0 & 0 & 1 & d_6 \\ 0 & 0 & 0 & 1 \end{bmatrix}
$$

$$
\mathbf{T}_1 \equiv {}^{0}\mathbf{A}_1\,{}^{1}\mathbf{A}_2\,{}^{2}\mathbf{A}_3 = \begin{bmatrix} C_1 C_{23} & -S_1 & C_1 S_{23} & a_2 C_1 C_2 + a_3 C_1 C_{23} - d_2 S_1 \\ S_1 C_{23} & C_1 & S_1 S_{23} & a_2 S_1 C_2 + a_3 S_1 C_{23} + d_2 C_1 \\ -S_{23} & 0 & C_{23} & -a_2 S_2 - a_3 S_{23} \\ 0 & 0 & 0 & 1 \end{bmatrix}
$$

$$
\mathbf{T}_2 \equiv {}^{3}\mathbf{A}_4\,{}^{4}\mathbf{A}_5\,{}^{5}\mathbf{A}_6 = \begin{bmatrix} C_4 C_5 C_6 - S_4 S_6 & -C_4 C_5 S_6 - S_4 C_6 & C_4 S_5 & d_6 C_4 S_5 \\ S_4 C_5 C_6 + C_4 S_6 & -S_4 C_5 S_6 + C_4 C_6 & S_4 S_5 & d_6 S_4 S_5 \\ -S_5 C_6 & S_5 S_6 & C_5 & d_6 C_5 + d_4 \\ 0 & 0 & 0 & 1 \end{bmatrix}
$$

where $C_i \equiv \cos\theta_i$; $S_i \equiv \sin\theta_i$; $C_{ij} \equiv \cos(\theta_i + \theta_j)$; $S_{ij} \equiv \sin(\theta_i + \theta_j)$.

Figure 2.13 PUMA link coordinate transformation matrices.

Specifically, for $i = 6$, we obtain the $\mathbf{T}$ matrix, $\mathbf{T} = {}^{0}\mathbf{A}_6$, which specifies the position and orientation of the endpoint of the manipulator with respect to the base coordinate system. This $\mathbf{T}$ matrix is used so frequently in robot arm kinematics that it is called the "arm matrix." Consider the $\mathbf{T}$ matrix to be of the form

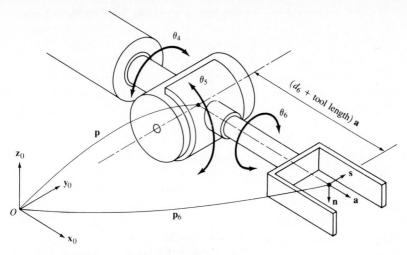

Figure 2.14 Hand coordinate system and [**n**, **s**, **a**].

$$
\mathbf{T} = \begin{bmatrix} \mathbf{x}_6 & \mathbf{y}_6 & \mathbf{z}_6 & \mathbf{p}_6 \\ 0 & 0 & 0 & 1 \end{bmatrix} = \begin{bmatrix} {}^0\mathbf{R}_6 & {}^0\mathbf{p}_6 \\ 0 & 1 \end{bmatrix} = \begin{bmatrix} \mathbf{n} & \mathbf{s} & \mathbf{a} & \mathbf{p} \\ 0 & 0 & 0 & 1 \end{bmatrix}
$$

$$
= \begin{bmatrix} n_x & s_x & a_x & p_x \\ n_y & s_y & a_y & p_y \\ n_z & s_z & a_z & p_z \\ 0 & 0 & 0 & 1 \end{bmatrix} \tag{2.2-35}
$$

where (see Fig. 2.14)

n = normal vector of the hand. Assuming a parallel-jaw hand, it is orthogonal
   to the fingers of the robot arm.

s = sliding vector of the hand. It is pointing in the direction of the finger
   motion as the gripper opens and closes.

a = approach vector of the hand. It is pointing in the direction normal to the
   palm of the hand (i.e., normal to the tool mounting plate of the arm).

p = position vector of the hand. It points from the origin of the base
   coordinate system to the origin of the hand coordinate system, which is
   usually located at the center point of the fully closed fingers.

If the manipulator is related to a reference coordinate frame by a transformation **B** and has a tool attached to its last joint's mounting plate described by **H**, then the endpoint of the tool can be related to the reference coordinate frame by multiplying the matrices **B**, ${}^0\mathbf{T}_6$, and **H** together as

$$
{}^{\text{ref}}\mathbf{T}_{\text{tool}} = \mathbf{B}\,{}^0\mathbf{T}_6\,\mathbf{H} \tag{2.2-36}
$$

Note that $\mathbf{H} \equiv {}^6\mathbf{A}_{\text{tool}}$ and $\mathbf{B} \equiv {}^{\text{ref}}\mathbf{A}_0$.

The direct kinematics solution of a six-link manipulator is, therefore, simply a matter of calculating $\mathbf{T} = {}^{0}\mathbf{A}_6$ by chain multiplying the six ${}^{i-1}\mathbf{A}_i$ matrices and evaluating each element in the $\mathbf{T}$ matrix. Note that the direct kinematics solution yields a unique $\mathbf{T}$ matrix for a given $\mathbf{q} = (q_1, q_2, \ldots, q_6)^T$ and a given set of coordinate systems, where $q_i = \theta_i$ for a rotary joint and $q_i = d_i$ for a prismatic joint. The only constraints are the physical bounds of θ_i for each joint of the robot arm. The table in Fig. 2.11 lists the joint constraints of a PUMA 560 series robot based on the coordinate system assigned in Fig. 2.11.

Having obtained all the coordinate transformation matrices ${}^{i-1}\mathbf{A}_i$ for a robot arm, the next task is to find an efficient method to compute $\mathbf{T}$ on a general purpose digital computer. The most efficient method is by multiplying all six ${}^{i-1}\mathbf{A}_i$ matrices together manually and evaluating the elements of $\mathbf{T}$ matrix out explicitly on a computer program. The disadvantages of this method are (1) it is laborious to multiply all six ${}^{i-1}\mathbf{A}_i$ matrices together, and (2) the arm matrix is applicable only to a particular robot for a specific set of coordinate systems (it is not flexible enough). On the other extreme, one can input all six ${}^{i-1}\mathbf{A}_i$ matrices and let the computer do the multiplication. This method is very flexible but at the expense of computation time as the fourth row of ${}^{i-1}\mathbf{A}_i$ consists mostly of zero elements.

A method that has both fast computation and flexibility is to "hand" multiply the first three ${}^{i-1}\mathbf{A}_i$ matrices together to form $\mathbf{T}_1 = {}^{0}\mathbf{A}_1 {}^{1}\mathbf{A}_2 {}^{2}\mathbf{A}_3$ and also the last three ${}^{i-1}\mathbf{A}_i$ matrices together to form $\mathbf{T}_2 = {}^{3}\mathbf{A}_4 {}^{4}\mathbf{A}_5 {}^{5}\mathbf{A}_6$, which is a fairly straightforward task. Then, we express the elements of $\mathbf{T}_1$ and $\mathbf{T}_2$ out in a computer program explicitly and let the computer multiply them together to form the resultant arm matrix $\mathbf{T} = \mathbf{T}_1 \mathbf{T}_2$.

For a PUMA 560 series robot, $\mathbf{T}_1$ is found from Fig. 2.13 to be

$$
\mathbf{T}_1 = {}^{0}\mathbf{A}_3 = {}^{0}\mathbf{A}_1 {}^{1}\mathbf{A}_2 {}^{2}\mathbf{A}_3
$$

$$
= \begin{bmatrix}
C_1 C_{23} & -S_1 & C_1 S_{23} & a_2 C_1 C_2 + a_3 C_1 C_{23} - d_2 S_1 \\
S_1 C_{23} & C_1 & S_1 S_{23} & a_2 S_1 C_2 + a_3 S_1 C_{23} + d_2 C_1 \\
-S_{23} & 0 & C_{23} & -a_2 S_2 - a_3 S_{23} \\
0 & 0 & 0 & 1
\end{bmatrix}
\tag{2.2-37}
$$

and the $\mathbf{T}_2$ matrix is found to be

$$
\mathbf{T}_2 = {}^{3}\mathbf{A}_6 = {}^{3}\mathbf{A}_4 {}^{4}\mathbf{A}_5 {}^{5}\mathbf{A}_6
$$

$$
= \begin{bmatrix}
C_4 C_5 C_6 - S_4 S_6 & -C_4 C_5 S_6 - S_4 C_6 & C_4 S_5 & d_6 C_4 S_5 \\
S_4 C_5 C_6 + C_4 S_6 & -S_4 C_5 S_6 + C_4 C_6 & S_4 S_5 & d_6 S_4 S_5 \\
-S_5 C_6 & S_5 S_6 & C_5 & d_6 C_5 + d_4 \\
0 & 0 & 0 & 1
\end{bmatrix}
\tag{2.2-38}
$$

where $C_{ij} \equiv \cos(\theta_i + \theta_j)$ and $S_{ij} \equiv \sin(\theta_i + \theta_j)$.

The arm matrix **T** for the PUMA robot arm shown in Fig. 2.11 is found to be

$$\mathbf{T} = \mathbf{T}_1 \mathbf{T}_2 = {}^0\mathbf{A}_1\, {}^1\mathbf{A}_2\, {}^2\mathbf{A}_3\, {}^3\mathbf{A}_4\, {}^4\mathbf{A}_5\, {}^5\mathbf{A}_6 = \begin{bmatrix} n_x & s_x & a_x & p_x \\ n_y & s_y & a_y & p_y \\ n_z & s_z & a_z & p_z \\ 0 & 0 & 0 & 1 \end{bmatrix} \qquad (2.2\text{-}39)$$

where

$$n_x = C_1[C_{23}(C_4 C_5 C_6 - S_4 S_6) - S_{23} S_5 C_6] - S_1(S_4 C_5 C_6 + C_4 S_6)$$

$$n_y = S_1[C_{23}(C_4 C_5 C_6 - S_4 S_6) - S_{23} S_5 C_6] + C_1(S_4 C_5 C_6 + C_4 S_6) \qquad (2.2\text{-}40)$$

$$n_z = -S_{23}[C_4 C_5 C_6 - S_4 S_6] - C_{23} S_5 C_6$$

$$s_x = C_1[-C_{23}(C_4 C_5 S_6 + S_4 C_6) + S_{23} S_5 S_6] - S_1(-S_4 C_5 S_6 + C_4 C_6)$$

$$s_y = S_1[-C_{23}(C_4 C_5 S_6 + S_4 C_6) + S_{23} S_5 S_6] + C_1(-S_4 C_5 S_6 + C_4 C_6)$$

$$s_z = S_{23}(C_4 C_5 S_6 + S_4 C_6) + C_{23} S_5 S_6 \qquad (2.2\text{-}41)$$

$$a_x = C_1(C_{23} C_4 S_5 + S_{23} C_5) - S_1 S_4 S_5$$

$$a_y = S_1(C_{23} C_4 S_5 + S_{23} C_5) + C_1 S_4 S_5 \qquad (2.2\text{-}42)$$

$$a_z = -S_{23} C_4 S_5 + C_{23} C_5$$

$$p_x = C_1[d_6(C_{23} C_4 S_5 + S_{23} C_5) + S_{23} d_4 + a_3 C_{23} + a_2 C_2] - S_1(d_6 S_4 S_5 + d_2)$$

$$p_y = S_1[d_6(C_{23} C_4 S_5 + S_{23} C_5) + S_{23} d_4 + a_3 C_{23} + a_2 C_2] + C_1(d_6 S_4 S_5 + d_2)$$

$$p_z = d_6(C_{23} C_5 - S_{23} C_4 S_5) + C_{23} d_4 - a_3 S_{23} - a_2 S_2 \qquad (2.2\text{-}43)$$

As a check, if $\theta_1 = 90°, \theta_2 = 0°, \theta_3 = 90°, \theta_4 = 0°, \theta_5 = 0°, \theta_6 = 0°$, then the **T** matrix is

$$\mathbf{T} = \begin{bmatrix} 0 & -1 & 0 & -149.09 \\ 0 & 0 & 1 & 921.12 \\ -1 & 0 & 0 & 20.32 \\ 0 & 0 & 0 & 1 \end{bmatrix}$$

which agrees with the coordinate systems established in Fig. 2.11.

From Eqs. (2.2-40) through (2.2-43), the arm matrix **T** requires 12 transcendental function calls, 40 multiplications and 20 additions if we only compute the upper right 3×3 submatrix of **T** and the normal vector **n** is found from the cross-product of $(\mathbf{n} = \mathbf{s} \times \mathbf{a})$. Furthermore, if we combine d_6 with the tool length of the terminal device, then $d_6 = 0$ and the new tool length will be increased by d_6 unit. This reduces the computation to 12 transcendental function calls, 35 multiplications, and 16 additions.

Example: A robot work station has been set up with a TV camera (see the figure). The camera can see the origin of the base coordinate system where a six-joint robot is attached. It can also see the center of an object (assumed to be a cube) to be manipulated by the robot. If a local coordinate system has been established at the center of the cube, this object as seen by the camera can be represented by a homogeneous transformation matrix $\mathbf{T}_1$. If the origin of the base coordinate system as seen by the camera can also be expressed by a homogeneous transformation matrix $\mathbf{T}_2$ and

$$
\mathbf{T}_1 = \begin{bmatrix} 0 & 1 & 0 & 1 \\ 1 & 0 & 0 & 10 \\ 0 & 0 & -1 & 9 \\ 0 & 0 & 0 & 1 \end{bmatrix}
\qquad
\mathbf{T}_2 = \begin{bmatrix} 1 & 0 & 0 & -10 \\ 0 & -1 & 0 & 20 \\ 0 & 0 & -1 & 10 \\ 0 & 0 & 0 & 1 \end{bmatrix}
$$

(a) What is the position of the center of the cube with respect to the base coordinate system?

(b) Assume that the cube is within the arm's reach. What is the orientation matrix $[\mathbf{n}, \mathbf{s}, \mathbf{a}]$ if you want the gripper (or finger) of the hand to be aligned with the **y** axis of the object and at the same time pick up the object from the top?

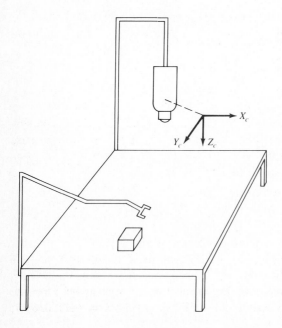

SOLUTION:

$$
{}^{camera}\mathbf{T}_{cube} \equiv \mathbf{T}_1 = \begin{bmatrix} 0 & 1 & 0 & 1 \\ 1 & 0 & 0 & 10 \\ 0 & 0 & -1 & 9 \\ 0 & 0 & 0 & 1 \end{bmatrix}
$$

and

$$
{}^{camera}\mathbf{T}_{base} \equiv \mathbf{T}_2 = \begin{bmatrix} 1 & 0 & 0 & -10 \\ 0 & -1 & 0 & 20 \\ 0 & 0 & -1 & 10 \\ 0 & 0 & 0 & 1 \end{bmatrix}
$$

To find ${}^{base}\mathbf{T}_{cube}$, we use the "chain product" rule:

$$
{}^{base}\mathbf{T}_{cube} = {}^{base}\mathbf{T}_{camera}\,{}^{camera}\mathbf{T}_{cube} = (\mathbf{T}_2)^{-1}\,\mathbf{T}_1
$$

Using Eq. (2.2-27) to invert the $\mathbf{T}_2$ matrix, we obtain the resultant transformation matrix:

$$
{}^{base}\mathbf{T}_{cube} = \begin{bmatrix} 1 & 0 & 0 & 10 \\ 0 & -1 & 0 & 20 \\ 0 & 0 & -1 & 10 \\ 0 & 0 & 0 & 1 \end{bmatrix} \begin{bmatrix} 0 & 1 & 0 & 1 \\ 1 & 0 & 0 & 10 \\ 0 & 0 & -1 & 9 \\ 0 & 0 & 0 & 1 \end{bmatrix}
$$

$$
= \begin{bmatrix} 0 & 1 & 0 & 11 \\ -1 & 0 & 0 & 10 \\ 0 & 0 & 1 & 1 \\ 0 & 0 & 0 & 1 \end{bmatrix}
$$

Therefore, the cube is at location $(11, 10, 1)^T$ from the base coordinate system. Its $\mathbf{x}, \mathbf{y}$, and $\mathbf{z}$ axes are parallel to the $-\mathbf{y}, \mathbf{x}$, and $\mathbf{z}$ axes of the base coordinate system, respectively.

To find $[\mathbf{n}, \mathbf{s}, \mathbf{a}]$, we make use of

$$
{}^{0}\mathbf{T}_6 = \begin{bmatrix} \mathbf{n} & \mathbf{s} & \mathbf{a} & \mathbf{p} \\ 0 & 0 & 0 & 1 \end{bmatrix}
$$

where $\mathbf{p} = (11, 10, 1)^T$ from the above solution. From the above figure, we want the approach vector $\mathbf{a}$ to align with the negative direction of the OZ axis of the base coordinate system [i.e., $\mathbf{a} = (0, 0, -1)^T$]; the $\mathbf{s}$ vector can be aligned in either direction of the $\mathbf{y}$ axis of ${}^{base}\mathbf{T}_{cube}$ [i.e., $\mathbf{s} = (\pm 1, 0, 0)^T$];

and the **n** vector can be obtained from the cross product of **s** and **a**:

$$
\mathbf{n} = \begin{vmatrix} \mathbf{i} & \mathbf{j} & \mathbf{k} \\ s_x & s_y & s_z \\ a_x & a_y & a_z \end{vmatrix} = \begin{vmatrix} \mathbf{i} & \mathbf{j} & \mathbf{k} \\ \pm 1 & 0 & 0 \\ 0 & 0 & -1 \end{vmatrix} = \begin{bmatrix} 0 \\ \pm 1 \\ 0 \end{bmatrix}
$$

Therefore, the orientation matrix $[\mathbf{n}, \mathbf{s}, \mathbf{a}]$ is found to be

$$
[\mathbf{n}, \mathbf{s}, \mathbf{a}] = \begin{bmatrix} 0 & 1 & 0 \\ +1 & 0 & 0 \\ 0 & 0 & -1 \end{bmatrix} \quad \text{or} \quad \begin{bmatrix} 0 & -1 & 0 \\ -1 & 0 & 0 \\ 0 & 0 & -1 \end{bmatrix} \qquad \square
$$

2.2.12 Other Specifications of the Location of the End-Effector

In previous sections, we analyzed the translations and rotations of rigid bodies (or links) and introduced the homogeneous transformation matrix for describing the position and orientation of a link coordinate frame. Of particular interest is the arm matrix ${}^{0}\mathbf{T}_6$ which describes the position and orientation of the hand with respect to the base coordinate frame. The upper left 3×3 submatrix of ${}^{0}\mathbf{T}_6$ describes the orientation of the hand. This rotation submatrix is equivalent to ${}^{0}\mathbf{R}_6$. There are other specifications which can be used to describe the location of the end-effector.

Euler Angle Representation for Orientation. As indicated in Sec. 2.2.4, this matrix representation for rotation of a rigid body simplifies many operations, but it does not lead directly to a complete set of generalized coordinates. Such a set of generalized coordinates can be provided by three Euler angles (ϕ, θ, and ψ).

Using the rotation matrix with eulerian angle representation as in Eq. (2.2-17), the arm matrix ${}^{0}\mathbf{T}_6$ can be expressed as:

$$
{}^{0}\mathbf{T}_6 = \begin{bmatrix} C\phi C\psi - S\phi C\theta S\psi & -C\phi S\psi - S\phi C\theta C\psi & S\phi S\theta & p_x \\ S\phi C\psi + C\phi C\theta S\psi & -S\phi S\psi + C\phi C\theta C\psi & -C\phi S\theta & p_y \\ S\theta S\psi & S\theta C\psi & C\theta & p_z \\ 0 & 0 & 0 & 1 \end{bmatrix} \qquad (2.2\text{-}44)
$$

Another advantage of using Euler angle representation for the orientation is that the storage for the position and orientation of an object is reduced to a six-element vector $XYZ\phi\theta\psi$. From this vector, one can construct the arm matrix ${}^{0}\mathbf{T}_6$ by Eq. (2.2-44).

Roll, Pitch, and Yaw Representation for Orientation. Another set of Euler angle representation for rotation is roll, pitch, and yaw (RPY). Again, using Eq.

(2.2-19), the rotation matrix representing roll, pitch, and yaw can be used to obtain the arm matrix $^0T_6$ as:

$$
^0T_6 = \begin{bmatrix}
C\phi C\theta & C\phi S\theta S\psi - S\phi C\psi & C\phi S\theta C\psi + S\phi S\psi & p_x \\
S\phi C\theta & S\phi S\theta S\psi + C\phi C\psi & S\phi S\theta C\psi - C\phi S\psi & p_y \\
-S\theta & C\theta S\psi & C\theta C\psi & p_z \\
0 & 0 & 0 & 1
\end{bmatrix} \qquad (2.2\text{-}45)
$$

As discussed in Chap. 1, there are different types of robot arms according to their joint motion (*XYZ*, cylindrical, spherical, and articulated arm). Thus, one can specify the position of the hand $(p_x, p_y, p_z)^T$ in other coordinates such as cylindrical or spherical. The resultant arm transformation matrix can be obtained by

$$
^0T_6 = \begin{bmatrix}
1 & 0 & 0 & p_x \\
0 & 1 & 0 & p_y \\
0 & 0 & 1 & p_z \\
0 & 0 & 0 & 1
\end{bmatrix}
\begin{bmatrix}
 & & & 0 \\
 & ^0R_6 & & 0 \\
 & & & 0 \\
0 & 0 & 0 & 1
\end{bmatrix} \qquad (2.2\text{-}46)
$$

where $^0R_6$ = rotation matrix expressed in either Euler angles or $[\mathbf{n}, \mathbf{s}, \mathbf{a}]$ or roll, pitch, and yaw.

Cylindrical Coordinates for Positioning Subassembly. In a cylindrical coordinate representation, the position of the end-effector can be specified by the following translations/rotations (see Fig. 2.15):

1. A translation of r unit along the *OX* axis $(\mathbf{T}_{x,r})$
2. A rotation of α angle about the *OZ* axis $(\mathbf{T}_{z,\alpha})$
3. A translation of d unit along the *OZ* axis $(\mathbf{T}_{z,d})$

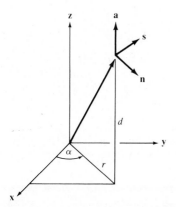

Figure 2.15 Cylindrical coordinate system representation.

The homogeneous transformation matrix that represents the above operations can be expressed as:

$$\mathbf{T}_{\text{cylindrical}} = \mathbf{T}_{z,d}\, \mathbf{T}_{z,\alpha}\, \mathbf{T}_{x,r} = \begin{bmatrix} 1 & 0 & 0 & 0 \\ 0 & 1 & 0 & 0 \\ 0 & 0 & 1 & d \\ 0 & 0 & 0 & 1 \end{bmatrix} \begin{bmatrix} C\alpha & -S\alpha & 0 & 0 \\ S\alpha & C\alpha & 0 & 0 \\ 0 & 0 & 1 & 0 \\ 0 & 0 & 0 & 1 \end{bmatrix}$$

$$\times \begin{bmatrix} 1 & 0 & 0 & r \\ 0 & 1 & 0 & 0 \\ 0 & 0 & 1 & 0 \\ 0 & 0 & 0 & 1 \end{bmatrix} = \begin{bmatrix} C\alpha & -S\alpha & 0 & rC\alpha \\ S\alpha & C\alpha & 0 & rS\alpha \\ 0 & 0 & 1 & d \\ 0 & 0 & 0 & 1 \end{bmatrix} \qquad (2.2\text{-}47)$$

Since we are only interested in the position vectors (i.e., the fourth column of $\mathbf{T}_{\text{cylindrical}}$), the arm matrix ${}^{0}\mathbf{T}_6$ can be obtained utilizing Eq. (2.2-46).

$$ {}^{0}\mathbf{T}_6 = \begin{bmatrix} 1 & 0 & 0 & rC\alpha \\ 0 & 1 & 0 & rS\alpha \\ 0 & 0 & 1 & d \\ 0 & 0 & 0 & 1 \end{bmatrix} \begin{bmatrix} & & & 0 \\ & {}^{0}\mathbf{R}_6 & & 0 \\ & & & 0 \\ 0 & 0 & 0 & 1 \end{bmatrix} \qquad (2.2\text{-}48)$$

and $p_x \equiv rC\alpha$, $p_y \equiv rS\alpha$, $p_z \equiv d$.

Spherical Coordinates for Positioning Subassembly. We can also utilize the spherical coordinate system for specifying the position of the end-effector. This involves the following translations/rotations (see Fig. 2.16):

1. A translation of r unit along the OZ axis ($\mathbf{T}_{z,r}$)
2. A rotation of β angle about the OY axis ($\mathbf{T}_{y,\beta}$)
3. A rotation of α angle about the OZ axis ($\mathbf{T}_{z,\alpha}$)

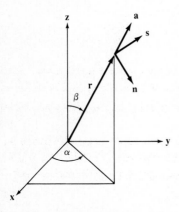

Figure 2.16 Spherical coordinate system representation.

The transformation matrix for the above operations is

$$
\mathbf{T}_{sph} = \mathbf{T}_{z,\alpha}\,\mathbf{R}_{y,\beta}\,\mathbf{T}_{z,r} =
\begin{bmatrix}
C\alpha & -S\alpha & 0 & 0 \\
S\alpha & C\alpha & 0 & 0 \\
0 & 0 & 1 & 0 \\
0 & 0 & 0 & 1
\end{bmatrix}
\begin{bmatrix}
C\beta & 0 & S\beta & 0 \\
0 & 1 & 0 & 0 \\
-S\beta & 0 & C\beta & 0 \\
0 & 0 & 0 & 1
\end{bmatrix}
$$

$$
\times
\begin{bmatrix}
1 & 0 & 0 & 0 \\
0 & 1 & 0 & 0 \\
0 & 0 & 1 & r \\
0 & 0 & 0 & 1
\end{bmatrix}
=
\begin{bmatrix}
C\alpha C\beta & -S\alpha & C\alpha S\beta & rC\alpha S\beta \\
S\alpha C\beta & C\alpha & S\alpha S\beta & rS\alpha S\beta \\
-S\beta & 0 & C\beta & rC\beta \\
0 & 0 & 0 & 1
\end{bmatrix}
\quad (2.2\text{-}49)
$$

Again, our interest is the position vector with respect to the base coordinate system, therefore, the arm matrix $^0\mathbf{T}_6$ whose position vector is expressed in spherical coordinates and the orientation matrix is expressed in $[\mathbf{n}, \mathbf{s}, \mathbf{a}]$ or Euler angles or roll, pitch, and yaw can be obtained by:

$$
^0\mathbf{T}_6 =
\begin{bmatrix}
1 & 0 & 0 & rC\alpha S\beta \\
0 & 1 & 0 & rS\alpha S\beta \\
0 & 0 & 1 & rC\beta \\
0 & 0 & 0 & 1
\end{bmatrix}
\begin{bmatrix}
 & & & 0 \\
 & ^0\mathbf{R}_6 & & 0 \\
 & & & 0 \\
0 & 0 & 0 & 1
\end{bmatrix}
\quad (2.2\text{-}50)
$$

where $p_x \equiv rC\alpha S\beta$, $p_y \equiv rS\alpha S\beta$, $p_z \equiv rC\beta$.

In summary, there are several methods (or coordinate systems) that one can choose to describe the position and orientation of the end-effector. For positioning, the position vector can be expressed in cartesian $(p_x, p_y, p_z)^T$, cylindrical $(rC\alpha, rS\alpha, d)^T$, or spherical $(rC\alpha S\beta, rS\alpha S\beta, rC\beta)^T$ terms. For describing the orientation of the end-effector with respect to the base coordinate system, we have cartesian $[\mathbf{n}, \mathbf{s}, \mathbf{a}]$, Euler angles (ϕ, θ, ψ), and (roll, pitch, and yaw). The result of the above discussion is tabulated in Table 2.2.

Table 2.2 Various positioning/orientation representations

Positioning	Orientation
Cartesian $(p_x, p_y, p_z)^T$	Cartesian $[\mathbf{n}, \mathbf{s}, \mathbf{a}]$
Cylindrical $(rC\alpha, rS\alpha, d)^T$	Euler angles (ϕ, θ, ψ)
Spherical $(rC\alpha S\beta, rS\alpha S\beta, rC\beta)^T$	Roll, pitch, and yaw

$$
\mathbf{T}_{position} =
\begin{bmatrix}
1 & 0 & 0 & p_x \\
0 & 1 & 0 & p_y \\
0 & 0 & 1 & p_z \\
0 & 0 & 0 & 1
\end{bmatrix}
\qquad
\mathbf{T}_{rot} =
\begin{bmatrix}
 & & & 0 \\
 [\mathbf{n}, \mathbf{s}, \mathbf{a}] & \text{ or } & \mathbf{R}_{\phi,\theta,\psi} & 0 \\
 & & & 0 \\
0 & 0 & 0 & 1
\end{bmatrix}
$$

$$
^0\mathbf{T}_6 = \mathbf{T}_{position}\,\mathbf{T}_{rot}
$$

2.2.13 Classification of Manipulators

A manipulator consists of a group of rigid bodies or links, with the first link connected to a supporting base and the last link containing the terminal device (or tool). In addition, each link is connected to, at most, two others so that closed loops are not formed. We made the assumption that the connection between links (the joints) have only 1 degree of freedom. With this restriction, two types of joints are of interest: revolute (or rotary) and prismatic. A revolute joint only permits rotation about an axis, while the prismatic joint allows sliding along an axis with no rotation (sliding with rotation is called a *screw joint*). These links are connected and powered in such a way that they are forced to move relative to one another in order to position the end-effector (a hand or tool) in a particular *position* and *orientation*.

Hence, a manipulator, considered to be a combination of links and joints, with the first link connected to ground and the last link containing the "hand," may be classified by the *type of joints* and their *order* (from the base to the hand). With this convention, the PUMA robot arm may be classified as 6R and the Stanford arm as 2R-P-3R, where R is a revolute joint and P is a prismatic joint.

2.3 THE INVERSE KINEMATICS PROBLEM

This section addresses the second problem of robot arm kinematics: the inverse kinematics or arm solution for a six-joint manipulator. Computer-based robots are usually servoed in the joint-variable space, whereas objects to be manipulated are usually expressed in the world coordinate system. In order to control the position and orientation of the end-effector of a robot to reach its object, the inverse kinematics solution is more important. In other words, given the position and orientation of the end-effector of a six-axis robot arm as $^0T_6$ and its joint and link parameters, we would like to find the corresponding joint angles $q = (q_1, q_2, q_3, q_4, q_5, q_6)^T$ of the robot so that the end-effector can be positioned as desired.

In general, the inverse kinematics problem can be solved by various methods, such as inverse transform (Paul et al. [1981]), screw algebra (Kohli and Soni [1975]), dual matrices (Denavit [1956]), dual quaternian (Yang and Freudenstein [1964]), iterative (Uicker et al. [1964]), and geometric approaches (Lee and Ziegler [1984]). Pieper [1968] presented the kinematics solution for any 6 degree of freedom manipulator which has revolute or prismatic pairs for the first three joints and the joint axes of the last three joints intersect at a point. The solution can be expressed as a fourth-degree polynomial in one unknown, and closed form solution for the remaining unknowns. Paul et al. [1981] presented an inverse transform technique using the 4 × 4 homogeneous transformation matrices in solving the kinematics solution for the same class of simple manipulators as discussed by Pieper. Although the resulting solution is correct, it suffers from the fact that the solution does not give a clear indication on how to select an appropriate solution from the several possible solutions for a particular arm configuration. The user often needs to rely on his or her intuition to pick the right answer. We shall discuss Pieper's approach in solving the inverse solution for Euler angles. Uicker et al. [1964] and Milenkovic and Huang [1983] presented iterative solutions for most industrial robots. The iterative solution often requires more computation and it does not guarantee convergence to the correct solution, especially in the singular and degenerate cases. Furthermore, as with the inverse transform technique, there is no indication on how to choose the correct solution for a particular arm configuration.

It is desirable to find a closed-form arm solution for manipulators. Fortunately, most of the commercial robots have either one of the following sufficient conditions which make the closed-form arm solution possible:

1. Three adjacent joint axes intersecting
2. Three adjacent joint axes parallel to one another

Both PUMA and Stanford robot arms satisfy the first condition while ASEA and MINIMOVER robot arms satisfy the second condition for finding the closed-form solution.

From Eq. (2.2-39), we have the arm transformation matrix given as

$$
\mathbf{T}_6 = \begin{bmatrix} n_x & s_x & a_x & p_x \\ n_y & s_y & a_y & p_y \\ n_z & s_z & a_z & p_z \\ 0 & 0 & 0 & 1 \end{bmatrix} = {}^0\mathbf{A}_1\,{}^1\mathbf{A}_2\,{}^2\mathbf{A}_3\,{}^3\mathbf{A}_4\,{}^4\mathbf{A}_5\,{}^5\mathbf{A}_6 \qquad (2.3\text{-}1)
$$

The above equation indicates that the arm matrix $\mathbf{T}$ is a function of sine and cosine of $\theta_1, \theta_2, \ldots, \theta_6$. For example, for a PUMA robot arm, equating the elements of the matrix equations as in Eqs. (2.2-40) to (2.2-43), we have twelve equations with six unknowns (joint angles) and these equations involve complex trigonometric functions. Since we have more equations than unknowns, one can immediately conclude that multiple solutions exist for a PUMA-like robot arm. We shall explore two methods for finding the inverse solution: inverse transform technique for finding Euler angles solution, which can also be used to find the joint solution of a PUMA-like robot arm, and a geometric approach which provides more insight into solving simple manipulators with rotary joints.

2.3.1 Inverse Transform Technique for Euler Angles Solution

In this section, we shall show the basic concept of the inverse transform technique by applying it to solve for the Euler angles. Since the 3×3 rotation matrix can be expressed in terms of the Euler angles (ϕ, θ, ψ) as in Eq. (2.2-17), and given

$$
\begin{bmatrix} n_x & s_x & a_x \\ n_y & s_y & a_y \\ n_z & s_z & a_z \end{bmatrix} = \mathbf{R}_{z,\,\phi}\,\mathbf{R}_{u,\,\theta}\,\mathbf{R}_{w,\,\psi}
$$

$$
= \begin{bmatrix} C\phi C\psi - S\phi C\theta S\psi & -C\phi S\psi - S\phi C\theta C\psi & S\phi S\theta \\ S\phi C\psi + C\phi C\theta S\psi & -S\phi S\psi + C\phi C\theta C\psi & -C\phi S\theta \\ S\theta S\psi & S\theta C\psi & C\theta \end{bmatrix}
$$

(2.3-2)

we would like to find the corresponding value of ϕ, θ, ψ. Equating the elements of the above matrix equation, we have:

$$n_x = C\phi C\psi - S\phi C\theta S\psi \qquad (2.3\text{-}3a)$$

$$n_y = S\phi C\psi + C\phi C\theta S\psi \qquad (2.3\text{-}3b)$$

$$n_z = S\theta S\psi \qquad (2.3\text{-}3c)$$

$$s_x = -C\phi S\psi - S\phi C\theta C\psi \qquad (2.3\text{-}3d)$$

$$s_y = -S\phi S\psi + C\phi C\theta C\psi \qquad (2.3\text{-}3e)$$

$$s_z = S\theta C\psi \qquad (2.3\text{-}3f)$$

$$a_x = S\phi S\theta \qquad (2.3\text{-}3g)$$

$$a_y = -C\phi S\theta \qquad (2.3\text{-}3h)$$

$$a_z = C\theta \qquad (2.3\text{-}3i)$$

Using Eqs. (2.3-3i), (2.3-3f), and (2.3-3h), a solution to the above nine equations is:

$$\theta = \cos^{-1}(a_z) \qquad (2.3\text{-}4)$$

$$\psi = \cos^{-1}\left[\frac{s_z}{S\theta}\right] \qquad (2.3\text{-}5)$$

$$\phi = \cos^{-1}\left[\frac{-a_y}{S\theta}\right] \qquad (2.3\text{-}6)$$

The above solution is inconsistent and ill-conditioned because:

1. The arc cosine function does not behave well as its accuracy in determining the angle is dependent on the angle. That is, $\cos(\theta) = \cos(-\theta)$.
2. When $\sin(\theta)$ approaches zero, that is, $\theta \approx 0°$ or $\theta \approx \pm 180°$, Eqs. (2.3-5) and (2.3-6) give inaccurate solutions or are undefined.

We must, therefore, find a more consistent approach to determining the Euler angle solution and a more consistent arc trigonometric function in evaluating the angle solution. In order to evaluate θ for $-\pi \leqslant \theta \leqslant \pi$, an arc tangent function, atan2 (y, x), which returns $\tan^{-1}(y/x)$ adjusted to the proper quadrant will be used. It is defined as:

$$\theta = \text{atan2}\,(y, x) = \begin{cases} 0° \leqslant \theta \leqslant 90° & \text{for } +x \text{ and } +y \\ 90° \leqslant \theta \leqslant 180° & \text{for } -x \text{ and } +y \\ -180° \leqslant \theta \leqslant -90° & \text{for } -x \text{ and } -y \\ -90° \leqslant \theta \leqslant 0° & \text{for } +x \text{ and } -y \end{cases} \qquad (2.3\text{-}7)$$

Using the arc tangent function (atan2) with two arguments, we shall take a look at a general solution proposed by Paul et al. [1981].

From the matrix equation in Eq. (2.3-2), the elements of the matrix on the left hand side (LHS) of the matrix equation are given, while the elements of the three matrices on the right-hand side (RHS) are unknown and they are dependent on ϕ, θ, ψ. Paul et al. [1981] suggest *premultiplying* the above matrix equation by its unknown inverse transforms successively and from the elements of the resultant matrix equation determine the unknown angle. That is, we move one unknown (by its inverse transform) from the RHS of the matrix equation to the LHS and solve for the unknown, then move the next unknown to the LHS, and repeat the process until all the unknowns are solved.

Premultiplying the above matrix equation by $\mathbf{R}_{z,\phi}^{-1}$, we have one unknown (ϕ) on the LHS and two unknowns (θ, ψ) on the RHS of the matrix equation, thus we have

$$\begin{bmatrix} C\phi & S\phi & 0 \\ -S\phi & C\phi & 0 \\ 0 & 0 & 1 \end{bmatrix} \begin{bmatrix} n_x & s_x & a_x \\ n_y & s_y & a_y \\ n_z & s_z & a_z \end{bmatrix} = \begin{bmatrix} 1 & 0 & 0 \\ 0 & C\theta & -S\theta \\ 0 & S\theta & C\theta \end{bmatrix} \begin{bmatrix} C\psi & -S\psi & 0 \\ S\psi & C\psi & 0 \\ 0 & 0 & 1 \end{bmatrix}$$

or

$$\begin{bmatrix} C\phi n_x + S\phi n_y & C\phi s_x + S\phi s_y & C\phi a_x + S\phi a_y \\ -S\phi n_x + C\phi n_y & -S\phi s_x + C\phi s_y & -S\phi a_x + C\phi a_y \\ n_z & s_z & a_z \end{bmatrix} = \begin{bmatrix} C\psi & -S\psi & 0 \\ C\theta S\psi & C\theta C\psi & -S\theta \\ S\theta S\psi & S\theta C\psi & C\theta \end{bmatrix}$$

$$(2.3\text{-}8)$$

Equating the (1, 3) elements of both matrices in Eq. (2.3-8), we have:

$$C\phi a_x + S\phi a_y = 0 \qquad (2.3\text{-}9)$$

which gives

$$\phi = \tan^{-1}\left[\frac{a_x}{-a_y}\right] = \text{atan2}\,(a_x, -a_y) \qquad (2.3\text{-}10)$$

Equating the (1, 1) and (1, 2) elements of the both matrices, we have:

$$C\psi = C\phi n_x + S\phi n_y \tag{2.3-11a}$$

$$S\psi = -C\phi s_x - S\phi s_y \tag{2.3-11b}$$

which lead to the solution for ψ,

$$\psi = \tan^{-1}\left[\frac{S\psi}{C\psi}\right] = \tan^{-1}\left[\frac{-C\phi s_x - S\phi s_y}{C\phi n_x + S\phi n_y}\right]$$

$$= \operatorname{atan2}(-C\phi s_x - S\phi s_y, C\phi n_x + S\phi n_y) \tag{2.3-12}$$

Equating the (2, 3) and (3, 3) elements of the both matrices, we have:

$$S\theta = S\phi a_x - C\phi a_y$$

$$C\theta = a_z \tag{2.3-13}$$

which gives us the solution for θ,

$$\theta = \tan^{-1}\left[\frac{S\theta}{C\theta}\right] = \tan^{-1}\left[\frac{S\phi a_x - C\phi a_y}{a_z}\right] = \operatorname{atan2}(S\phi a_x - C\phi a_y, a_z) \tag{2.3-14}$$

Since the concept of inverse transform technique is to move one unknown to the LHS of the matrix equation at a time and solve for the unknown, we can try to solve the above matrix equation for ϕ, θ, ψ by *postmultiplying* the above matrix equation by its inverse transform $\mathbf{R}_{w,\psi}^{-1}$

$$\begin{bmatrix} n_x & s_x & a_x \\ n_y & s_y & a_y \\ n_z & s_z & a_z \end{bmatrix} \begin{bmatrix} C\psi & S\psi & 0 \\ -S\psi & C\psi & 0 \\ 0 & 0 & 1 \end{bmatrix} = \begin{bmatrix} C\phi & -S\phi & 0 \\ S\phi & C\phi & 0 \\ 0 & 0 & 1 \end{bmatrix} \begin{bmatrix} 1 & 0 & 0 \\ 0 & C\theta & -S\theta \\ 0 & S\theta & C\theta \end{bmatrix}$$

Multiplying the matrices out, we have,

$$\begin{bmatrix} n_x C\psi - s_x S\psi & n_x S\psi + s_x C\psi & a_x \\ n_y C\psi - s_y S\psi & n_y S\psi + s_y C\psi & a_y \\ n_z C\psi - s_z S\psi & n_z S\psi + s_z C\psi & a_z \end{bmatrix} = \begin{bmatrix} C\phi & -S\phi C\theta & S\phi S\theta \\ S\phi & C\phi C\theta & -C\phi S\theta \\ 0 & S\theta & C\theta \end{bmatrix} \tag{2.3-15}$$

Again equating the (3, 1) elements of both matrices in the above matrix equation, we have

$$n_z C\psi - s_z S\psi = 0 \tag{2.3-16}$$

which gives

$$\psi = \tan^{-1}\left[\frac{n_z}{s_z}\right] = \text{atan2}\,(n_z,\,s_z) \tag{2.3-17}$$

Equating the (3, 2) and (3, 3) elements of both matrices, we have:

$$S\theta = n_z S\psi + s_z C\psi \tag{2.3-18a}$$

$$C\theta = a_z \tag{2.3-18b}$$

which leads us to the solution for θ,

$$\theta = \tan^{-1}\left[\frac{n_z S\psi + s_z C\psi}{a_z}\right] = \text{atan2}\,(n_z S\psi + s_z C\psi,\,a_z) \tag{2.3-19}$$

Equating the (1, 1) and (2, 1) elements of both matrices, we have

$$C\phi = n_x C\psi - s_x S\psi \tag{2.3-20a}$$

$$S\phi = n_y C\psi - s_y S\psi \tag{2.3-20b}$$

which gives

$$\phi = \tan^{-1}\left[\frac{n_y C\psi - s_y S\psi}{n_x C\psi - s_x S\psi}\right]$$

$$= \text{atan2}\,(n_y C\psi - s_y S\psi,\, n_x C\psi - s_x S\psi) \tag{2.3-21}$$

Whether one should premultiply or postmultiply a given matrix equation is up to the user's discretion and it depends on the intuition of the user.

Let us apply this inverse transform technique to solve the Euler angles for a PUMA robot arm (OAT solution of a PUMA robot). PUMA robots use the symbols O, A, T to indicate the Euler angles and their definitions are given as follows (with reference to Fig. 2.17):

O (*orientation*) is the angle formed from the y_0 axis to the projection of the tool **a** axis on the XY plane about the z_0 axis.

A (*altitude*) is the angle formed from the XY plane to the tool **a** axis about the **s** axis of the tool.

T (*tool*) is the angle formed from the XY plane to the tool **s** axis about the **a** axis of the tool.

Initially the tool coordinate system (or the hand coordinate system) is aligned with the base coordinate system of the robot as shown in Fig. 2.18. That is, when $O = A = T = 0°$, the hand points in the negative y_0 axis with the fingers in a horizontal plane, and the **s** axis is pointing to the positive x_0 axis. The necessary

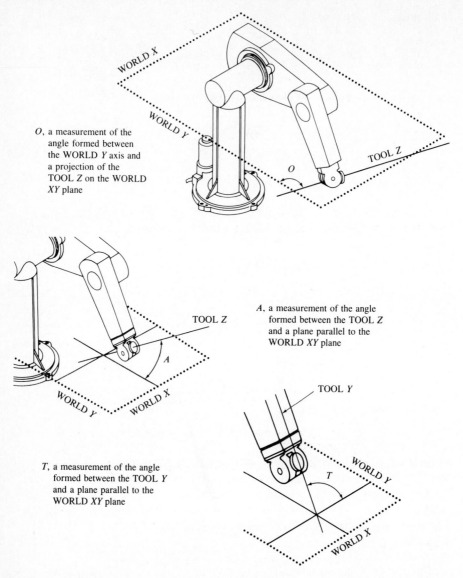

O, a measurement of the angle formed between the WORLD *Y* axis and a projection of the TOOL *Z* on the WORLD *XY* plane

A, a measurement of the angle formed between the TOOL *Z* and a plane parallel to the WORLD *XY* plane

T, a measurement of the angle formed between the TOOL *Y* and a plane parallel to the WORLD *XY* plane

Figure 2.17 Definition of Euler angles *O*, *A*, and *T*. (Taken from PUMA robot manual 398H.)

transform that describes the orientation of the hand coordinate system ($\mathbf{n}$, $\mathbf{s}$, $\mathbf{a}$) with respect to the base coordinate system ($\mathbf{x}_0$, $\mathbf{y}_0$, $\mathbf{z}_0$) is given by

$$
\begin{bmatrix}
0 & 1 & 0 \\
0 & 0 & -1 \\
-1 & 0 & 0
\end{bmatrix}
\qquad (2.3\text{-}22)
$$

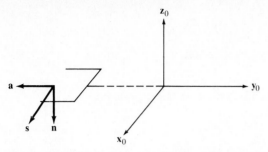

Figure 2.18 Initial alignment of tool coordinate system.

From the definition of the *OAT* angles and the initial alignment matrix [Eq. (2.3-22)], the relationship between the hand transform and the *OAT* angle is given by

$$
\begin{bmatrix} n_x & s_x & a_x \\ n_y & s_y & a_y \\ n_z & s_z & a_z \end{bmatrix} = \mathbf{R}_{z,O} \begin{bmatrix} 0 & 1 & 0 \\ 0 & 0 & -1 \\ -1 & 0 & 0 \end{bmatrix} \mathbf{R}_{s,A} \mathbf{R}_{a,T}
$$

$$
= \begin{bmatrix} CO & -SO & 0 \\ SO & CO & 0 \\ 0 & 0 & 1 \end{bmatrix} \begin{bmatrix} 0 & 1 & 0 \\ 0 & 0 & -1 \\ -1 & 0 & 0 \end{bmatrix} \begin{bmatrix} CA & 0 & SA \\ 0 & 1 & 0 \\ -SA & 0 & CA \end{bmatrix} \begin{bmatrix} CT & -ST & 0 \\ ST & CT & 0 \\ 0 & 0 & 1 \end{bmatrix}
$$

Postmultiplying the above matrix equation by the inverse transform of $\mathbf{R}_{a,T}$,

$$
\begin{bmatrix} n_x & s_x & a_x \\ n_y & s_y & a_y \\ n_z & s_z & a_z \end{bmatrix} \begin{bmatrix} CT & ST & 0 \\ -ST & CT & 0 \\ 0 & 0 & 1 \end{bmatrix} = \begin{bmatrix} CO & -SO & 0 \\ SO & CO & 0 \\ 0 & 0 & 1 \end{bmatrix} \begin{bmatrix} 0 & 1 & 0 \\ 0 & 0 & -1 \\ -1 & 0 & 0 \end{bmatrix}
$$

$$
\times \begin{bmatrix} CA & 0 & SA \\ 0 & 1 & 0 \\ -SA & 0 & CA \end{bmatrix}
$$

and multiplying the matrices out, we have:

$$
\begin{bmatrix} n_x CT - s_x ST & n_x ST + s_x CT & a_x \\ n_y CT - s_y ST & n_y ST + s_y CT & a_y \\ n_z CT - s_z ST & n_z ST + s_z CT & a_z \end{bmatrix} = \begin{bmatrix} -SOSA & CO & SOCA \\ COSA & SO & -COCA \\ -CA & 0 & -SA \end{bmatrix}
$$

$$
\tag{2.3-23}
$$

Equating the (3, 2) elements of the above matrix equation, we have:

$$
n_z ST + s_z CT = 0 \tag{2.3-24}
$$

which gives the solution of T,

$$T = \tan^{-1}\left(\frac{s_z}{-n_z}\right) = \text{atan2}\,(s_z, -n_z) \tag{2.3-25}$$

Equating the (3, 1) and (3, 3) elements of the both matrices, we have:

$$SA = -a_z \tag{2.3-26a}$$

and

$$CA = -n_z CT + s_z ST \tag{2.3-26b}$$

then the above equations give

$$A = \tan^{-1}\left(\frac{-a_z}{-n_z CT + s_z ST}\right) = \text{atan2}\,(-a_z, -n_z CT + s_z ST) \tag{2.3-27}$$

Equating the (1, 2) and (2, 2) elements of the both matrices, we have:

$$CO = n_x ST + s_x CT \tag{2.3-28a}$$

$$SO = n_y ST + s_y CT \tag{2.3-28b}$$

which give the solution of O,

$$O = \tan^{-1}\left(\frac{n_y ST + s_y CT}{n_x ST + s_x CT}\right)$$

$$= \text{atan2}\,(n_y ST + s_y CT, n_x ST + s_x CT) \tag{2.3-29}$$

The above premultiplying or postmultiplying of the unknown inverse transforms can also be applied to find the joint solution of a PUMA robot. Details about the PUMA robot arm joint solution can be found in Paul et al. [1981].

Although the inverse transform technique provides a general approach in determining the joint solution of a manipulator, it does not give a clear indication on how to select an appropriate solution from the several possible solutions for a particular arm configuration. This has to rely on the user's geometric intuition. Thus, a geometric approach is more useful in deriving a consistent joint-angle solution given the arm matrix as in Eq. (2.2-39), and it provides a means for the user to select a unique solution for a particular arm configuration. This approach is presented in Sec. 2.3.2.

2.3.2 A Geometric Approach

This section presents a geometric approach to solving the inverse kinematics problem of six-link manipulators with rotary joints. The discussion focuses on a PUMA-like manipulator. Based on the link coordinate systems and human arm geometry, various arm configurations of a PUMA-like robot (Fig. 2.11) can be identified with the assistance of three configuration indicators (ARM, ELBOW, and WRIST)—two associated with the solution of the first three joints and the other with the last three joints. For a six-axis PUMA-like robot arm, there are four possible solutions to the first three joints and for each of these four solutions there are two possible solutions to the last three joints. The first two configuration indicators allow one to determine one solution from the possible four solutions for the first three joints. Similarly, the third indicator selects a solution from the possible two solutions for the last three joints. The arm configuration indicators are prespecified by a user for finding the inverse solution. The solution is calculated in two stages. First, a position vector pointing from the shoulder to the wrist is derived. This is used to derive the solution of each joint i ($i = 1, 2, 3$) for the first three joints by looking at the projection of the position vector onto the $x_{i-1} y_{i-1}$ plane. The last three joints are solved using the calculated joint solution from the first three joints, the orientation submatrices of ${}^{0}T_i$ and ${}^{i-1}A_i$ ($i = 4, 5, 6$), and the projection of the link coordinate frames onto the $x_{i-1} y_{i-1}$ plane. From the geometry, one can easily find the arm solution consistently. As a verification of the joint solution, the arm configuration indicators can be determined from the corresponding decision equations which are functions of the joint angles. With appropriate modification and adjustment, this approach can be generalized to solve the inverse kinematics problem of most present day industrial robots with rotary joints.

If we are given ${}^{ref}T_{tool}$, then we can find ${}^{0}T_6$ by premultiplying and post-multiplying ${}^{ref}T_{tool}$ by B^{-1} and H^{-1}, respectively, and the joint-angle solution can be applied to ${}^{0}T_6$ as desired.

$$
{}^{0}T_6 \equiv T = B^{-1}\,{}^{ref}T_{tool}\,H^{-1} = \begin{bmatrix} n_x & s_x & a_x & p_x \\ n_y & s_y & a_y & p_y \\ n_z & s_z & a_z & p_z \\ 0 & 0 & 0 & 1 \end{bmatrix} \tag{2.3-30}
$$

Definition of Various Arm Configurations. For the PUMA robot arm shown in Fig. 2.11 (and other rotary robot arms), various arm configurations are defined according to human arm geometry and the link coordinate systems which are established using Algorithm 2.1 as (Fig. 2.19)

RIGHT (shoulder) ARM: Positive θ_2 moves the wrist in the *positive* z_0 direction while joint 3 is not activated.
LEFT (shoulder) ARM: Positive θ_2 moves the wrist in the *negative* z_0 direction while joint 3 is not activated.
ABOVE ARM (elbow above wrist): Position of the wrist of the

$\left. \begin{matrix} \text{RIGHT} \\ \text{LEFT} \end{matrix} \right\}$ arm with respect to the shoulder coordinate system has

$$\left\{ \begin{array}{l} \text{negative} \\ \text{positive} \end{array} \right\} \text{coordinate value along the } y_2 \text{ axis.}$$

BELOW ARM (elbow below wrist): Position of the wrist of the

$$\left\{ \begin{array}{l} \text{RIGHT} \\ \text{LEFT} \end{array} \right\} \text{arm with respect to the shoulder coordinate system has}$$

$$\left\{ \begin{array}{l} \text{positive} \\ \text{negative} \end{array} \right\} \text{coordinate value along the } y_2 \text{ axis.}$$

WRIST DOWN: The s unit vector of the hand coordinate system and the y_5 unit vector of the (x_5, y_5, z_5) coordinate system have a positive dot product.

WRIST UP: The s unit vector of the hand coordinate system and the y_5 unit vector of the (x_5, y_5, z_5) coordinate system have a negative dot product.

(Note that the definition of the arm configurations with respect to the link coordinate systems may have to be slightly modified if one uses different link coordinate systems.)

With respect to the above definition of various arm configurations, two arm configuration *indicators* (ARM and ELBOW) are defined for each arm configuration. These two indicators are combined to give one solution out of the possible four joint solutions for the first three joints. For each of the four arm configurations (Fig. 2.19) defined by these two indicators, the third indicator (WRIST) gives one of the two possible joint solutions for the last three joints. These three indicators can be defined as:

$$\text{ARM} = \left\{ \begin{array}{ll} +1 & \text{RIGHT arm} \\ -1 & \text{LEFT arm} \end{array} \right. \tag{2.3-31}$$

$$\text{ELBOW} = \left\{ \begin{array}{ll} +1 & \text{ABOVE arm} \\ -1 & \text{BELOW arm} \end{array} \right. \tag{2.3-32}$$

$$\text{WRIST} = \left\{ \begin{array}{ll} +1 & \text{WRIST DOWN} \\ -1 & \text{WRIST UP} \end{array} \right. \tag{2.3-33}$$

In addition to these indicators, the user can define a "FLIP" toggle as:

$$\text{FLIP} = \left\{ \begin{array}{ll} +1 & \text{Flip the wrist orientation} \\ -1 & \text{Do not flip the wrist orientation} \end{array} \right. \tag{2.3-34}$$

The signed values of these indicators and the toggle are prespecified by a user for finding the inverse kinematics solution. These indicators can also be set from the knowledge of the joint angles of the robot arm using the corresponding decision equations. We shall later give the decision equations that determine these indicator

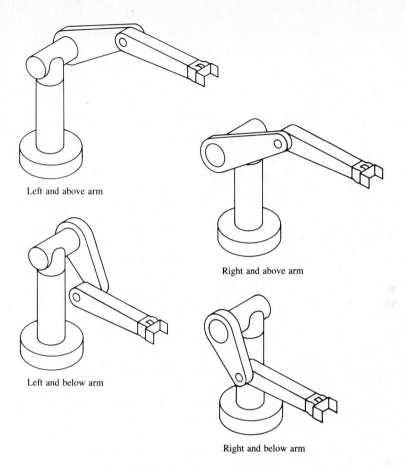

Left and above arm

Right and above arm

Left and below arm

Right and below arm

Figure 2.19 Definition of various arm configurations.

values. The decision equations can be used as a verification of the inverse kinematics solution.

Arm Solution for the First Three Joints. From the kinematics diagram of the PUMA robot arm in Fig. 2.11, we define a position vector $\mathbf{p}$ which points from the origin of the shoulder coordinate system ($\mathbf{x}_0$, $\mathbf{y}_0$, $\mathbf{z}_0$) to the point where the last three joint axes intersect as (see Fig. 2.14):

$$\mathbf{p} = \mathbf{p}_6 - d_6\mathbf{a} = (p_x, \ p_y, \ p_z)^T \tag{2.3-35}$$

which corresponds to the position vector of $^0\mathbf{T}_4$:

$$\begin{bmatrix} p_x \\ p_y \\ p_z \end{bmatrix} = \begin{bmatrix} C_1(a_2C_2 + a_3C_{23} + d_4S_{23}) - d_2S_1 \\ S_1(a_2C_2 + a_3C_{23} + d_4S_{23}) + d_2C_1 \\ d_4C_{23} - a_3S_{23} - a_2S_2 \end{bmatrix} \tag{2.3-36}$$

Joint 1 solution. If we project the position vector **p** onto the $x_0 y_0$ plane as in Fig. 2.20, we obtain the following equations for solving θ_1:

$$\theta_1^L = \phi - \alpha \qquad \theta_1^R = \pi + \phi + \alpha \qquad (2.3\text{-}37)$$

$$r = \sqrt{p_x^2 + p_y^2 - d_2^2} \qquad R = \sqrt{p_x^2 + p_y^2} \qquad (2.3\text{-}38)$$

$$\sin \phi = \frac{p_y}{R} \qquad \cos \phi = \frac{p_x}{R} \qquad (2.3\text{-}39)$$

$$\sin \alpha = \frac{d_2}{R} \qquad \cos \alpha = \frac{r}{R} \qquad (2.3\text{-}40)$$

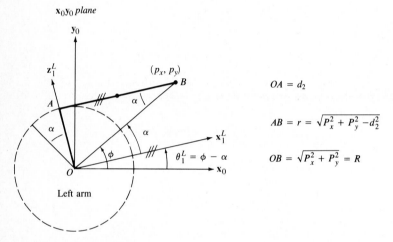

$$OA = d_2$$

$$AB = r = \sqrt{P_x^2 + P_y^2 - d_2^2}$$

$$OB = \sqrt{P_x^2 + P_y^2} = R$$

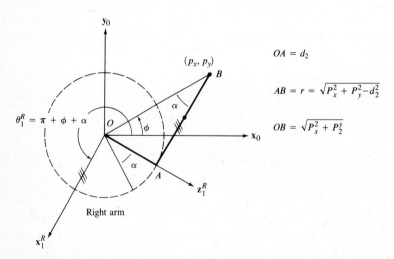

$$OA = d_2$$

$$AB = r = \sqrt{P_x^2 + P_y^2 - d_2^2}$$

$$OB = \sqrt{P_x^2 + P_2^y}$$

Figure 2.20 Solution for joint 1.

where the superscripts L and R on joint angles indicate the LEFT/RIGHT arm configurations. From Eqs. (2.3-37) to (2.3-40), we obtain the sine and cosine functions of θ_1 for LEFT/RIGHT arm configurations:

$$\sin \theta_1^L = \sin (\phi - \alpha) = \sin \phi \cos \alpha - \cos \phi \sin \alpha = \frac{p_y r - p_x d_2}{R^2} \qquad (2.3\text{-}41)$$

$$\cos \theta_1^L = \cos (\phi - \alpha) = \cos \phi \cos \alpha + \sin \phi \sin \alpha = \frac{p_x r + p_y d_2}{R^2} \qquad (2.3\text{-}42)$$

$$\sin \theta_1^R = \sin (\pi + \phi + \alpha) = \frac{-p_y r - p_x d_2}{R^2} \qquad (2.3\text{-}43)$$

$$\cos \theta_1^R = \cos (\pi + \phi + \alpha) = \frac{-p_x r + p_y d_2}{R^2} \qquad (2.3\text{-}44)$$

Combining Eqs. (2.3-41) to (2.3-44) and using the ARM indicator to indicate the LEFT/RIGHT arm configuration, we obtain the sine and cosine functions of θ_1, respectively:

$$\sin \theta_1 = \frac{-\,\mathrm{ARM}\; p_y \sqrt{p_x^2 + p_y^2 - d_2^2} - p_x d_2}{p_x^2 + p_y^2} \qquad (2.3\text{-}45)$$

$$\cos \theta_1 = \frac{-\,\mathrm{ARM}\; p_x \sqrt{p_x^2 + p_y^2 - d_2^2} + p_y d_2}{p_x^2 + p_y^2} \qquad (2.3\text{-}46)$$

where the positive square root is taken in these equations and ARM is defined as in Eq. (2.3-31). In order to evaluate θ_1 for $-\pi \leqslant \theta_1 \leqslant \pi$, an arc tangent function as defined in Eq. (2.3-7) will be used. From Eqs. (2.3-45) and (2.3-46), and using Eq. (2.3-7), θ_1 is found to be:

$$\theta_1 = \tan^{-1} \left[\frac{\sin \theta_1}{\cos \theta_1} \right]$$

$$= \tan^{-1} \left[\frac{-\,\mathrm{ARM}\; p_y \sqrt{p_x^2 + p_y^2 - d_2^2} - p_x d_2}{-\,\mathrm{ARM}\; p_x \sqrt{p_x^2 + p_y^2 - d_2^2} + p_y d_2} \right] \qquad -\pi \leqslant \theta_1 \leqslant \pi$$

$$(2.3\text{-}47)$$

Joint 2 solution. To find joint 2, we project the position vector **p** onto the $x_1 y_1$ plane as shown in Fig. 2.21. From Fig. 2.21, we have four different arm configurations. Each arm configuration corresponds to different values of joint 2 as shown in Table 2.3, where $0° \leqslant \alpha \leqslant 360°$ and $0° \leqslant \beta \leqslant 90°$.

Table 2.3 Various arm configurations for joint 2

Arm configurations	θ_2	ARM	ELBOW	ARM $\cdot$ ELBOW
LEFT and ABOVE arm	$\alpha - \beta$	-1	$+1$	-1
LEFT and BELOW arm	$\alpha + \beta$	-1	-1	$+1$
RIGHT and ABOVE arm	$\alpha + \beta$	$+1$	$+1$	$+1$
RIGHT and BELOW arm	$\alpha - \beta$	$+1$	-1	-1

From the above table, θ_2 can be expressed in one equation for different arm and elbow configurations using the ARM and ELBOW indicators as:

$$\theta_2 = \alpha + (\text{ARM} \cdot \text{ELBOW})\beta = \alpha + K \cdot \beta \tag{2.3-48}$$

where the combined arm configuration indicator $K = \text{ARM} \cdot \text{ELBOW}$ will give an appropriate signed value and the "dot" represents a multiplication operation on the indicators. From the arm geometry in Fig. 2.21, we obtain:

$$R = \sqrt{p_x^2 + p_y^2 + p_z^2 - d_2^2} \qquad r = \sqrt{p_x^2 + p_y^2 - d_2^2} \tag{2.3-49}$$

$$\sin \alpha = -\frac{p_z}{R} = -\frac{p_z}{\sqrt{p_x^2 + p_y^2 + p_z^2 - d_2^2}} \tag{2.3-50}$$

$$\cos \alpha = -\frac{\text{ARM} \cdot r}{R} = -\frac{\text{ARM} \cdot \sqrt{p_x^2 + p_y^2 - d_2^2}}{\sqrt{p_x^2 + p_y^2 + p_z^2 - d_2^2}} \tag{2.3-51}$$

$$\cos \beta = \frac{a_2^2 + R^2 - (d_4^2 + a_3^2)}{2a_2 R} \tag{2.3-52}$$

$$= \frac{p_x^2 + p_y^2 + p_z^2 + a_2^2 - d_2^2 - (d_4^2 + a_3^2)}{2a_2\sqrt{p_x^2 + p_y^2 + p_z^2 - d_2^2}}$$

$$\sin \beta = \sqrt{1 - \cos^2 \beta} \tag{2.3-53}$$

From Eqs. (2.3-48) to (2.3-53), we can find the sine and cosine functions of θ_2:

$$\sin \theta_2 = \sin (\alpha + K \cdot \beta) = \sin \alpha \cos (K \cdot \beta) + \cos \alpha \sin (K \cdot \beta)$$

$$= \sin \alpha \cos \beta + (\text{ARM} \cdot \text{ELBOW}) \cos \alpha \sin \beta \tag{2.3-54}$$

$$\cos \theta_2 = \cos (\alpha + K \cdot \beta)$$

$$= \cos \alpha \cos \beta - (\text{ARM} \cdot \text{ELBOW}) \sin \alpha \sin \beta \tag{2.3-55}$$

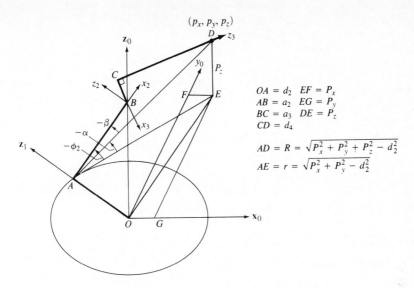

Figure 2.21 Solution for joint 2.

From Eqs. (2.3-54) and (2.3-55), we obtain the solution for θ_2:

$$\theta_2 = \tan^{-1}\left[\frac{\sin\theta_2}{\cos\theta_2}\right] \qquad -\pi \leqslant \theta_2 \leqslant \pi \qquad (2.3\text{-}56)$$

Joint 3 solution. For joint 3, we project the position vector **p** onto the $\mathbf{x}_2\mathbf{y}_2$ plane as shown in Fig. 2.22. From Fig. 2.22, we again have four different arm configurations. Each arm configuration corresponds to different values of joint 3 as shown in Table 2.4, where $(^2\mathbf{p}_4)_y$ is the y component of the position vector from the origin of $(\mathbf{x}_2, \mathbf{y}_2, \mathbf{z}_2)$ to the point where the last three joint axes intersect.

From the arm geometry in Fig. 2.22, we obtain the following equations for finding the solution for θ_3:

$$R = \sqrt{p_x^2 + p_y^2 + p_z^2 - d_2^2} \qquad (2.3\text{-}57)$$

$$\cos\phi = \frac{a_2^2 + (d_4^2 + a_3^2) - R^2}{2a_2\sqrt{d_4^2 + a_3^2}} \qquad (2.3\text{-}58)$$

$$\sin\phi = \text{ARM} \cdot \text{ELBOW} \sqrt{1 - \cos^2\phi}$$

$$\sin\beta = \frac{d_4}{\sqrt{d_4^2 + a_3^2}} \qquad \cos\beta = \frac{|a_3|}{\sqrt{d_4^2 + a_3^2}} \qquad (2.3\text{-}59)$$

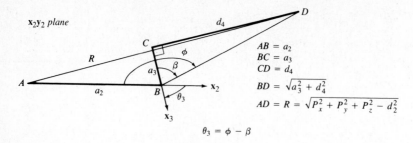

Left and below arm

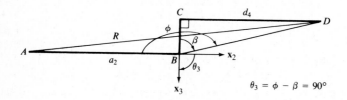

Left and below arm

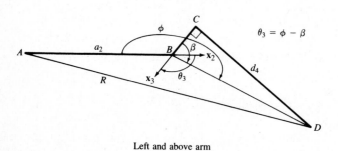

Left and above arm

Figure 2.22 Solution for joint 3.

From Table 2.4, we can express θ_3 in one equation for different arm configurations:

$$\theta_3 = \phi - \beta \tag{2.3-60}$$

From Eq. (2.3-60), the sine and cosine functions of θ_3 are, respectively,

$$\sin \theta_3 = \sin (\phi - \beta) = \sin \phi \cos \beta - \cos \phi \sin \beta \tag{2.3-61}$$

$$\cos \theta_3 = \cos (\phi - \beta) = \cos \phi \cos \beta + \sin \phi \sin \beta \tag{2.3-62}$$

From Eqs. (2.3-61) and (2.3-62), and using Eqs. (2.3-57) to (2.3-59), we find the solution for θ_3:

Table 2.4 Various arm configurations for joint 3

Arm configurations	$(^2\mathbf{p}_4)_y$	θ_3	ARM	ELBOW	ARM · ELBOW
LEFT and ABOVE arm	$\geqslant 0$	$\phi - \beta$	-1	$+1$	-1
LEFT and BELOW arm	$\leqslant 0$	$\phi - \beta$	-1	-1	$+1$
RIGHT and ABOVE arm	$\leqslant 0$	$\phi - \beta$	$+1$	$+1$	$+1$
RIGHT and BELOW arm	$\geqslant 0$	$\phi - \beta$	$+1$	-1	-1

$$\theta_3 = \tan^{-1}\left[\frac{\sin\theta_3}{\cos\theta_3}\right] \qquad -\pi \leqslant \theta_3 \leqslant \pi \qquad (2.3\text{-}63)$$

Arm Solution for the Last Three Joints. Knowing the first three joint angles, we can evaluate the $^0\mathbf{T}_3$ matrix which is used extensively to find the solution of the last three joints. The solution of the last three joints of a PUMA robot arm can be found by setting these joints to meet the following criteria:

1. Set joint 4 such that a rotation about joint 5 will align the axis of motion of joint 6 with the given approach vector (**a** of **T**).
2. Set joint 5 to align the axis of motion of joint 6 with the approach vector.
3. Set joint 6 to align the given orientation vector (or sliding vector or $\mathbf{y}_6$) and normal vector.

Mathematically the above criteria respectively mean:

$$\mathbf{z}_4 = \frac{\pm(\mathbf{z}_3 \times \mathbf{a})}{\|\mathbf{z}_3 \times \mathbf{a}\|} \qquad \text{given } \mathbf{a} = (a_x, a_y, a_z)^T \qquad (2.3\text{-}64)$$

$$\mathbf{a} = \mathbf{z}_5 \qquad \text{given } \mathbf{a} = (a_x, a_y, a_z)^T \qquad (2.3\text{-}65)$$

$$\mathbf{s} = \mathbf{y}_6 \qquad \text{given } \mathbf{s} = (s_x, s_y, s_z)^T \text{ and } \mathbf{n} = (n_x, n_y, n_z)^T \qquad (2.3\text{-}66)$$

In Eq. (2.3-64), the vector cross product may be taken to be positive or negative. As a result, there are two possible solutions for θ_4. If the vector cross product is zero (i.e., $\mathbf{z}_3$ is parallel to $\mathbf{a}$), it indicates the degenerate case. This happens when the axes of rotation for joint 4 and joint 6 are parallel. It indicates that at this particular arm configuration, a five-axis robot arm rather than a six-axis one would suffice.

Joint 4 solution. Both orientations of the wrist (UP and DOWN) are defined by looking at the orientation of the hand coordinate frame ($\mathbf{n}$, $\mathbf{s}$, $\mathbf{a}$) with respect to the ($\mathbf{x}_5$, $\mathbf{y}_5$, $\mathbf{z}_5$) coordinate frame. The sign of the vector cross product in Eq. (2.3-64) cannot be determined without referring to the orientation of either the $\mathbf{n}$ or $\mathbf{s}$ unit vector with respect to the $\mathbf{x}_5$ or $\mathbf{y}_5$ unit vector, respectively, which have a fixed relation with respect to the z_4 unit vector from the assignment of the link coordinate frames. (From Fig. 2.11, we have the z_4 unit vector pointing at the same direction as the y_5 unit vector.)

We shall start with the assumption that the vector cross product in Eq. (2.3-64) has a positive sign. This can be indicated by an orientation indicator Ω which is defined as:

$$\Omega = \begin{cases} 0 & \text{if in the degenerate case} \\ s \cdot y_5 & \text{if } s \cdot y_5 \neq 0 \\ n \cdot y_5 & \text{if } s \cdot y_5 = 0 \end{cases} \tag{2.3-67}$$

From Fig. 2.11, $y_5 = z_4$, and using Eq. (2.3-64), the orientation indicator Ω can be rewritten as:

$$\Omega = \begin{cases} 0 & \text{if in the degenerate case} \\ s \cdot \dfrac{(z_3 \times a)}{\|z_3 \times a\|} & \text{if } s \cdot (z_3 \times a) \neq 0 \\ n \cdot \dfrac{(z_3 \times a)}{\|z_3 \times a\|} & \text{if } s \cdot (z_3 \times a) = 0 \end{cases} \tag{2.3-68}$$

If our assumption of the sign of the vector cross product in Eq. (2.3-64) is not correct, it will be corrected later using the combination of the WRIST indicator and the orientation indicator Ω. The Ω is used to indicate the initial orientation of the z_4 unit vector (positive direction) from the link coordinate systems assignment, while the WRIST indicator specifies the user's preference of the orientation of the wrist subsystem according to the definition given in Eq. (2.3-33). If both the orientation Ω and the WRIST indicators have the same sign, then the assumption of the sign of the vector cross product in Eq. (2.3-64) is correct. Various wrist orientations resulting from the combination of the various values of the WRIST and orientation indicators are tabulated in Table 2.5.

Table 2.5 Various orientations for the wrist

Wrist orientation	$\Omega = s \cdot y_5$ or $n \cdot y_5$	WRIST	$M = $ WRIST sign (Ω)
DOWN	≥ 0	$+1$	$+1$
DOWN	< 0	$+1$	-1
UP	≥ 0	-1	-1
UP	< 0	-1	$+1$

Again looking at the projection of the coordinate frame $(\mathbf{x}_4, \mathbf{y}_4, \mathbf{z}_4)$ on the $\mathbf{x}_3 \mathbf{y}_3$ plane and from Table 2.5 and Fig. 2.23, it can be shown that the following are true (see Fig. 2.23):

$$\sin \theta_4 = -M(\mathbf{z}_4 \cdot \mathbf{x}_3) \qquad \cos \theta_4 = M(\mathbf{z}_4 \cdot \mathbf{y}_3) \qquad (2.3\text{-}69)$$

where $\mathbf{x}_3$ and $\mathbf{y}_3$ are the x and y column vectors of $^0\mathbf{T}_3$, respectively, $M =$ WRIST sign (Ω), and the sign function is defined as:

$$\text{sign}(x) = \begin{cases} +1 & \text{if } x \geqslant 0 \\ -1 & \text{if } x < 0 \end{cases} \qquad (2.3\text{-}70)$$

Thus, the solution for θ_4 with the orientation and WRIST indicators is:

$$\theta_4 = \tan^{-1} \left[\frac{\sin \theta_4}{\cos \theta_4} \right]$$

$$= \tan^{-1} \left[\frac{M(C_1 a_y - S_1 a_x)}{M(C_1 C_{23} a_x + S_1 C_{23} a_y - S_{23} a_z)} \right] \qquad -\pi \leqslant \theta_4 \leqslant \pi \qquad (2.3\text{-}71)$$

If the degenerate case occurs, any convenient value may be chosen for θ_4 as long as the orientation of the wrist (UP/DOWN) is satisfied. This can always be ensured by setting θ_4 equals to the current value of θ_4. In addition to this, the user can turn on the FLIP toggle to obtain the other solution of θ_4, that is, $\theta_4 = \theta_4 + 180°$.

Joint 5 solution. To find θ_5, we use the criterion that aligns the axis of rotation of joint 6 with the approach vector (or $\mathbf{a} = \mathbf{z}_5$). Looking at the projection of the

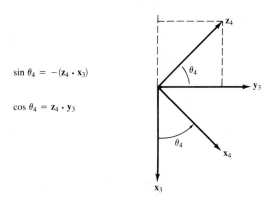

$$\sin \theta_4 = -(\mathbf{z}_4 \cdot \mathbf{x}_3)$$

$$\cos \theta_4 = \mathbf{z}_4 \cdot \mathbf{y}_3$$

Figure 2.23 Solution for joint 4.

coordinate frame $(\mathbf{x}_5, \mathbf{y}_5, \mathbf{z}_5)$ on the $\mathbf{x}_4\mathbf{y}_4$ plane, it can be shown that the following are true (see Fig. 2.24):

$$\sin\theta_5 = \mathbf{a} \cdot \mathbf{x}_4 \qquad \cos\theta_5 = -(\mathbf{a} \cdot \mathbf{y}_4) \qquad (2.3\text{-}72)$$

where $\mathbf{x}_4$ and $\mathbf{y}_4$ are the x and y column vectors of $^0T_4$, respectively, and $\mathbf{a}$ is the approach vector. Thus, the solution for θ_5 is:

$$\theta_5 = \tan^{-1}\left[\frac{\sin\theta_5}{\cos\theta_5}\right] \qquad -\pi \leqslant \theta_5 \leqslant \pi$$

$$= \tan^{-1}\left[\frac{(C_1 C_{23} C_4 - S_1 S_4)a_x + (S_1 C_{23} C_4 + C_1 S_4)a_y - C_4 S_{23} a_z}{C_1 S_{23} a_x + S_1 S_{23} a_y + C_{23} a_z}\right] \qquad (2.3\text{-}73)$$

If $\theta_5 \approx 0$, then the degenerate case occurs.

Joint 6 solution. Up to now, we have aligned the axis of joint 6 with the approach vector. Next, we need to align the orientation of the gripper to ease picking up the object. The criterion for doing this is to set $\mathbf{s} = \mathbf{y}_6$. Looking at the projection of the hand coordinate frame $(\mathbf{n}, \mathbf{s}, \mathbf{a})$ on the $\mathbf{x}_5\mathbf{y}_5$ plane, it can be shown that the following are true (see Fig. 2.25):

$$\sin\theta_6 = \mathbf{n} \cdot \mathbf{y}_5 \qquad \cos\theta_6 = \mathbf{s} \cdot \mathbf{y}_5 \qquad (2.3\text{-}74)$$

where $\mathbf{y}_5$ is the y column vector of $^0T_5$ and $\mathbf{n}$ and $\mathbf{s}$ are the normal and sliding vectors of $^0T_6$, respectively. Thus, the solution for θ_6 is:

$$\theta_6 = \tan^{-1}\left[\frac{\sin\theta_6}{\cos\theta_6}\right] \qquad -\pi \leqslant \theta_6 \leqslant \pi$$

$$= \tan^{-1}\left[\frac{(-S_1 C_4 - C_1 C_{23} S_4)n_x + (C_1 C_4 - S_1 C_{23} S_4)n_y + (S_4 S_{23})n_z}{(-S_1 C_4 - C_1 C_{23} S_4)s_x + (C_1 C_4 - S_1 C_{23} S_4)s_y + (S_4 S_{23})s_z}\right]$$
$$(2.3\text{-}75)$$

$$\sin\theta_5 = \mathbf{a} \cdot \mathbf{x}_4$$

$$\cos\theta_5 = -(\mathbf{a} \cdot \mathbf{y}_4)$$

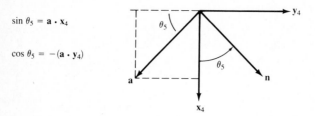

Figure 2.24 Solution for joint 5.

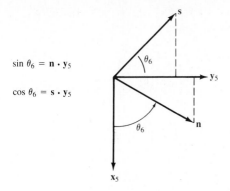

$\sin \theta_6 = \mathbf{n} \cdot \mathbf{y}_5$

$\cos \theta_6 = \mathbf{s} \cdot \mathbf{y}_5$

Figure 2.25 Solution for joint 6.

The above derivation of the inverse kinematics solution of a PUMA robot arm is based on the geometric interpretation of the position of the endpoint of link 3 and the hand (or tool) orientation requirement. There is one pitfall in the above derivation for θ_4, θ_5, and θ_6. The criterion for setting the axis of motion of joint 5 equal to the cross product of $\mathbf{z}_3$ and $\mathbf{a}$ may not be valid when $\sin \theta_5 \approx 0$, which means that $\theta_5 \approx 0$. In this case, the manipulator becomes *degenerate* with both the axes of motion of joints 4 and 6 aligned. In this state, only the sum of θ_4 and θ_6 is significant. If the degenerate case occurs, then we are free to choose any value for θ_4, and usually its current value is used and then we would like to have $\theta_4 + \theta_6$ equal to the total angle required to align the sliding vector $\mathbf{s}$ and the normal vector $\mathbf{n}$. If the FLIP toggle is on (i.e., FLIP = 1), then $\theta_4 = \theta_4 + \pi$, $\theta_5 = -\theta_5$, and $\theta_6 = \theta_6 + \pi$.

In summary, there are eight solutions to the inverse kinematics problem of a six-joint PUMA-like robot arm. The first three-joint solution $(\theta_1, \theta_2, \theta_3)$ positions the arm while the last three-joint solution, $(\theta_4, \theta_5, \theta_6)$, provides appropriate orientation for the hand. There are four solutions for the first three-joint solutions—two for the right shoulder arm configuration and two for the left shoulder arm configuration. For each arm configuration, Eqs. (2.3-47), (2.3-56), (2.3-63), (2.3-71), (2.3-73), and (2.3-75) give one set of solutions $(\theta_1, \theta_2, \theta_3, \theta_4, \theta_5, \theta_6)$ and $(\theta_1, \theta_2, \theta_3, \theta_4 + \pi, -\theta_5, \theta_6 + \pi)$ (with the FLIP toggle on) gives another set of solutions.

Decision Equations for the Arm Configuration Indicators. The solution for the PUMA-like robot arm derived in the previous section is not unique and depends on the arm configuration indicators specified by the user. These arm configuration indicators (ARM, ELBOW, and WRIST) can also be determined from the joint angles. In this section, we derive the respective decision equation for each arm configuration indicator. The signed value of the decision equation (positive, zero, or negative) provides an indication of the arm configuration as defined in Eqs. (2.3-31) to (2.3-33).

For the ARM indicator, following the definition of the RIGHT/LEFT arm, a decision equation for the ARM indicator can be found to be:

$$g(\theta, \mathbf{p}) = \mathbf{z}_0 \cdot \frac{\mathbf{z}_1 \times \mathbf{p}'}{\|\mathbf{z}_1 \times \mathbf{p}'\|} = \mathbf{z}_0 \cdot \begin{vmatrix} \mathbf{i} & \mathbf{j} & \mathbf{k} \\ -\sin\theta_1 & \cos\theta_1 & 0 \\ p_x & p_y & 0 \end{vmatrix} \frac{1}{\|\mathbf{z}_1 \times \mathbf{p}'\|}$$

$$= \frac{-p_y \sin\theta_1 - p_x \cos\theta_1}{\|\mathbf{z}_1 \times \mathbf{p}'\|} \tag{2.3-76}$$

where $\mathbf{p}' = (p_x, p_y, 0)^T$ is the projection of the position vector $\mathbf{p}$ [Eq. (2.3-36)] onto the $\mathbf{x}_0 \mathbf{y}_0$ plane, $\mathbf{z}_1 = (-\sin\theta_1, \cos\theta_1, 0)^T$ from the third column vector of $^0\mathbf{T}_1$, and $\mathbf{z}_0 = (0, 0, 1)^T$. We have the following possibilities:

1. If $g(\theta, \mathbf{p}) > 0$, then the arm is in the RIGHT arm configuration.
2. If $g(\theta, \mathbf{p}) < 0$, then the arm is in the LEFT arm configuration.
3. If $g(\theta, \mathbf{p}) = 0$, then the criterion for finding the LEFT/RIGHT arm configuration cannot be uniquely determined. The arm is within the inner cylinder of radius d_2 in the workspace (see Fig. 2.19). In this case, it is default to the RIGHT arm (ARM = +1).

Since the denominator of the above decision equation is always positive, the determination of the LEFT/RIGHT arm configuration is reduced to checking the sign of the numerator of $g(\theta, \mathbf{p})$:

$$\text{ARM} = \text{sign}\,[g(\theta, \mathbf{p})] = \text{sign}\,(-p_x \cos\theta_1 - p_y \sin\theta_1) \tag{2.3-77}$$

where the sign function is defined in Eq. (2.3-70). Substituting the x and y components of $\mathbf{p}$ from Eq. (2.3-36), Eq. (2.3-77) becomes:

$$\text{ARM} = \text{sign}\,[g(\theta, \mathbf{p})] = \text{sign}\,[g(\theta)] = \text{sign}\,(-d_4 S_{23} - a_3 C_{23} - a_2 C_2) \tag{2.3-78}$$

Hence, from the decision equation in Eq. (2.3-78), one can relate its signed value to the ARM indicator for the RIGHT/LEFT arm configuration as:

$$\text{ARM} = \text{sign}\,(-d_4 S_{23} - a_3 C_{23} - a_2 C_2) = \begin{cases} +1 & \Rightarrow \text{RIGHT arm} \\ -1 & \Rightarrow \text{LEFT arm} \end{cases} \tag{2.3-79}$$

For the ELBOW arm indicator, we follow the definition of ABOVE/BELOW arm to formulate the corresponding decision equation. Using $(^2\mathbf{p}_4)_y$ and the ARM indicator in Table 2.4, the decision equation for the ELBOW indicator is based on the sign of the y component of the position vector of $^2\mathbf{A}_3\,^3\mathbf{A}_4$ and the ARM indicator:

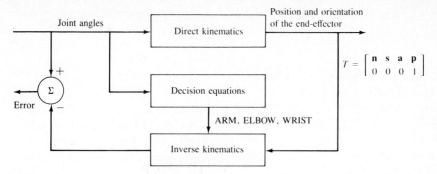

Figure 2.26 Computer simulation of joint solution.

$$\text{ELBOW} = \text{ARM} \cdot \text{sign}\,(d_4 C_3 - a_3 S_3) = \begin{cases} +1 & \Rightarrow \text{ELBOW above wrist} \\ -1 & \Rightarrow \text{ELBOW below wrist} \end{cases} \quad (2.3\text{-}80)$$

For the WRIST indicator, we follow the definition of DOWN/UP wrist to obtain a positive dot product of the $\mathbf{s}$ and $\mathbf{y}_5$ (or $\mathbf{z}_4$) unit vectors:

$$\text{WRIST} = \begin{cases} +1 & \text{if } \mathbf{s} \cdot \mathbf{z}_4 > 0 \\ -1 & \text{if } \mathbf{s} \cdot \mathbf{z}_4 < 0 \end{cases} = \text{sign}\,(\mathbf{s} \cdot \mathbf{z}_4) \quad (2.3\text{-}81)$$

If $\mathbf{s} \cdot \mathbf{z}_4 = 0$, then the WRIST indicator can be found from:

$$\text{WRIST} = \begin{cases} +1 & \text{if } \mathbf{n} \cdot \mathbf{z}_4 > 0 \\ -1 & \text{if } \mathbf{n} \cdot \mathbf{z}_4 < 0 \end{cases} = \text{sign}\,(\mathbf{n} \cdot \mathbf{z}_4) \quad (2.3\text{-}82)$$

Combining Eqs. (2.3-81) and (2.3-82), we have

$$\text{WRIST} = \begin{cases} \text{sign}\,(\mathbf{s} \cdot \mathbf{z}_4) & \text{if } \mathbf{s} \cdot \mathbf{z}_4 \neq 0 \\ \text{sign}\,(\mathbf{n} \cdot \mathbf{z}_4) & \text{if } \mathbf{s} \cdot \mathbf{z}_4 = 0 \end{cases} = \begin{cases} +1 & \Rightarrow \text{WRIST DOWN} \\ -1 & \Rightarrow \text{WRIST UP} \end{cases}$$
$$(2.3\text{-}83)$$

These decision equations provide a verification of the arm solution. We use them to preset the arm configuration in the direct kinematics and then use the arm configuration indicators to find the inverse kinematics solution (see Fig. 2.26).

Computer Simulation. A computer program can be written to verify the validity of the inverse solution of the PUMA robot arm shown in Fig. 2.11. The software initially generates all the locations in the workspace of the robot within the joint angles limits. They are inputed into the direct kinematics routine to obtain the arm matrix **T**. These joint angles are also used to compute the decision equations to obtain the three arm configuration indicators. These indicators together with the arm matrix **T** are fed into the inverse solution routine to obtain the joint angle solution which should agree to the joint angles fed into the direct kinematics routine previously. A computer simulation block diagram is shown in Fig. 2.26.

2.4 CONCLUDING REMARKS

We have discussed both direct and inverse kinematics in this chapter. The parameters of robot arm links and joints are defined and a 4×4 homogeneous transformation matrix is introduced to describe the location of a link with respect to a fixed coordinate frame. The forward kinematic equations for a six-axis PUMA-like robot arm are derived.

The inverse kinematics problem is introduced and the inverse transform technique is used to determine the Euler angle solution. This technique can also be used to find the inverse solution of simple robots. However, it does not provide geometric insight to the problem. Thus, a geometric approach is introduced to find the inverse solution of a six-joint robot arm with rotary joints. The inverse solution is determined with the assistance of three arm configuration indicators (ARM, ELBOW, and WRIST). There are eight solutions to a six-joint PUMA-like robot arm—four solutions for the first three joints and for each arm configuration, two more solutions for the last three joints. The validity of the forward and inverse kinematics solution can be verified by computer simulation. The geometric approach, with appropriate modification and adjustment, can be generalized to other simple industrial robots with rotary joints. The kinematics concepts covered in this chapter will be used extensively in Chap. 3 for deriving the equations of motion that describe the dynamic behavior of a robot arm.

REFERENCES

Further reading on matrices can be found in Bellman [1970], Frazer et al. [1960], and Gantmacher [1959]. Utilization of matrices to describe the location of a rigid mechanical link can be found in the paper by Denavit and Hartenberg [1955] and in their book (Hartenberg and Denavit [1964]). Further reading about homogeneous coordinates can be found in Duda and Hart [1973] and Newman and Sproull [1979]. The discussion on kinematics is an extension of a paper by Lee [1982]. More discussion in kinematics can be found in Hartenberg and Denavit [1964] and Suh and Radcliffe [1978]. Although matrix representation of linkages presents a systematic approach to solving the forward kinematics problem, the vector approach to the kinematics problem presents a more concise representation of linkages. This is discussed in a paper by Chase [1963]. Other robotics books that discuss the kinematics problem are Paul [1981], Lee, Gonzalez, and Fu [1986], and Snyder [1985].

Pieper [1968] in his doctoral dissertation utilized an algebraic approach to solve the inverse kinematics problem. The discussion of the inverse transform technique in finding the arm solution was based on the paper by Paul et al. [1981]. The geometric approach to solving the inverse kinematics for a six-link manipulator with rotary joints was based on the paper by Lee and Ziegler [1984]. The arm solution of a Stanford robot arm can be found in a report by Lewis [1974]. Other techniques in solving the inverse kinematics can be found in articles by Denavit [1956], Kohli and Soni [1975], Yang and Freudenstein [1964], Yang [1969], Yuan and Freudenstein [1971], Duffy and Rooney [1975], Uicker et al. [1964]. Finally, the tutorial book edited by Lee, Gonzalez, and Fu [1986] contains numerous recent papers on robotics.

PROBLEMS

2.1 What is the rotation matrix for a rotation of 30° about the OZ axis, followed by a rotation of 60° about the OX axis, followed by a rotation of 90° about the OY axis?

2.2 What is the rotation matrix for a rotation of ϕ angle about the OX axis, followed by a rotation of ψ angle about the OW axis, followed by a rotation of θ angle about the OY axis?

2.3 Find another sequence of rotations that is different from Prob. 2.2, but which results in the same rotation matrix.

2.4 Derive the formula for $\sin(\phi + \theta)$ and $\cos(\phi + \theta)$ by expanding symbolically two rotations of ϕ and θ using the rotation matrix concepts discussed in this chapter.

2.5 Determine a **T** matrix that represents a rotation of α angle about the OX axis, followed by a translation of b unit of distance along the OZ axis, followed by a rotation of ϕ angle about the OV axis.

2.6 For the figure shown below, find the 4 × 4 homogeneous transformation matrices ${}^{i-1}\mathbf{A}_i$ and ${}^{0}\mathbf{A}_i$ for $i = 1, 2, 3, 4, 5$.

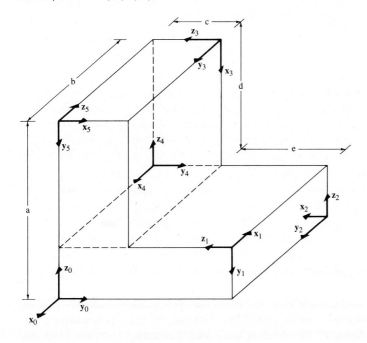

2.7 For the figure shown below, find the 4×4 homogeneous transformation matrices $^{i-1}\mathbf{A}_i$ and $^0\mathbf{A}_i$ for $i = 1, 2, 3, 4$.

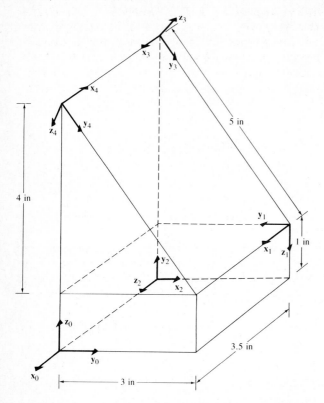

2.8 A robot workstation has been set up with a TV camera, as shown in the example in Sec. 2.2.11. The camera can see the origin of the base coordinate system where a six-link robot arm is attached, and also the center of a cube to be manipulated by the robot. If a local coordinate system has been established at the center of the cube, then this object, as seen by the camera, can be represented by a homogeneous transformation matrix $\mathbf{T}_1$. Also, the origin of the base coordinate system as seen by the camera can be expressed by a homogeneous transformation matrix $\mathbf{T}_2$, where

$$
\mathbf{T}_1 = \begin{bmatrix} 0 & 1 & 0 & 1 \\ 1 & 0 & 0 & 10 \\ 0 & 0 & -1 & 9 \\ 0 & 0 & 0 & 1 \end{bmatrix} \quad \text{and} \quad \mathbf{T}_2 = \begin{bmatrix} 1 & 0 & 0 & -10 \\ 0 & -1 & 0 & 20 \\ 0 & 0 & -1 & 10 \\ 0 & 0 & 0 & 1 \end{bmatrix}
$$

(a) Unfortunately, after the equipment has been set up and these coordinate systems have been taken, someone rotates the camera 90° about the **z** axis of the camera. What is the position/orientation of the camera with respect to the robot's base coordinate system? (b) After you have calculated the answer for question (a), the same person rotated the object 90° about the **x** axis of the object and translated it 4 units of distance along the rotated **y** axis. What is the position/orientation of the object with respect to the robot's base coordinate system? To the rotated camera coordinate system?

2.9 We have discussed a geometric approach for finding the inverse kinematic solution of a PUMA robot arm. Find the computational requirements of the joint solution in terms of multiplication and addition operations and the number of transcendental calls (if the same term appears twice, the computation should be counted once only).

2.10 Establish orthonormal link coordinate systems (x_i, y_i, z_i) for $i = 1, 2, \ldots, 6$ for the PUMA 260 robot arm shown in the figure below and complete the table.

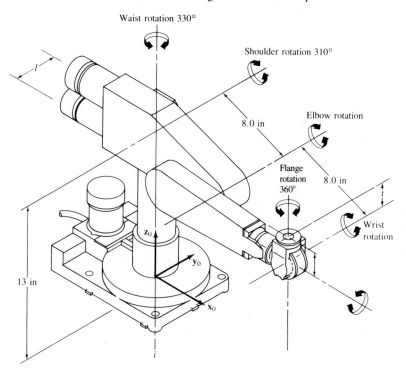

PUMA robot arm link coordinate parameters

Joint i	θ_i	α_i	a_i	d_i
1				
2				
3				
4				
5				
6				

2.11 Establish orthonormal link coordinate systems $(\mathbf{x}_i, \mathbf{y}_i, \mathbf{z}_i)$ for $i = 1, 2, \ldots, 5$ for the MINIMOVER robot arm shown in the figure below and complete the table.

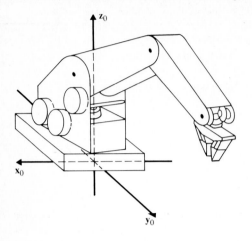

MINIMOVER robot arm link coordinate parameters

Joint i	θ_i	α_i	a_i	d_i
1				
2				
3				
4				
5				

2.12 A Stanford robot arm has moved to the position shown in the figure below. The joint variables at this position are: $\mathbf{q} = (90°, -120°, 22 \text{ cm}, 0°, 70°, 90°)^T$. Establish the orthonormal link coordinate systems $(\mathbf{x}_i, \mathbf{y}_i, \mathbf{z}_i)$ for $i = 1, 2, \ldots, 6$, for this arm and complete the table.

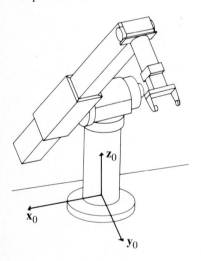

Stanford arm link coordinate parameters

Joint i	θ_i	α_i	a_i	d_i
1				
2				
3				
4				
5				
6				

2.13 Using the six $^{i-1}\mathbf{A}_i$ matrices ($i = 1, 2, \ldots, 6$) of the PUMA robot arm in Fig. 2.13, find its *position* error at the end of link 3 due to the measurement error of the first three joint angles ($\Delta\theta_1, \Delta\theta_2, \Delta\theta_3$). A first-order approximation solution is adequate.

2.14 Repeat Prob. 2.13 for the Stanford arm shown in Fig. 2.12.

2.15 A two degree-of-freedom manipulator is shown in the figure below. Given that the length of each link is 1 m, establish its link coordinate frames and find $^{0}A_{1}$ and $^{1}A_{2}$. Find the inverse kinematics solution for this manipulator.

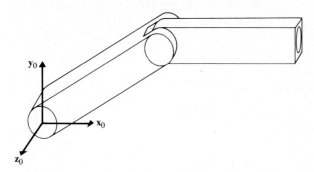

2.16 For the PUMA robot arm shown in Fig. 2.11, assume that we have found the first three joint solution $(\theta_{1}, \theta_{2}, \theta_{3})$ correctly and that we are given $^{i-1}A_{i}$, $i = 1, 2, \ldots, 6$ and $^{0}T_{6}$. Use the inverse transformation technique to find the solution for the last three joint angles $(\theta_{4}, \theta_{5}, \theta_{6})$. Compare your solution with the one given in Eqs. (2.3-71), (2.3-73), and (2.3-75).

2.17 For the Stanford robot arm shown in Fig. 2.12, derive the solution of the first three joint angles. You may use any method that you feel comfortable with.

2.18 Repeat Prob. 2.16 for the Stanford arm shown in Fig. 2.12.

THREE

ROBOT ARM DYNAMICS

The inevitable comes to pass by effort.
Oliver Wendell Holmes

3.1 INTRODUCTION

Robot arm dynamics deals with the mathematical formulations of the equations of robot arm motion. The dynamic equations of motion of a manipulator are a set of mathematical equations describing the dynamic behavior of the manipulator. Such equations of motion are useful for computer simulation of the robot arm motion, the design of suitable control equations for a robot arm, and the evaluation of the kinematic design and structure of a robot arm. In this chapter, we shall concentrate on the formulation, characteristics, and properties of the dynamic equations of motion that are suitable for control purposes. The purpose of manipulator control is to maintain the dynamic response of a computer-based manipulator in accordance with some prespecified system performance and desired goals. In general, the dynamic performance of a manipulator directly depends on the efficiency of the control algorithms and the dynamic model of the manipulator. The control problem consists of obtaining dynamic models of the physical robot arm system and then specifying corresponding control laws or strategies to achieve the desired system response and performance. This chapter deals mainly with the former part of the manipulator control problem; that is, modeling and evaluating the dynamical properties and behavior of computer-controlled robots.

The actual dynamic model of a robot arm can be obtained from known physical laws such as the laws of newtonian mechanics and lagrangian mechanics. This leads to the development of the dynamic equations of motion for the various articulated joints of the manipulator in terms of specified geometric and inertial parameters of the links. Conventional approaches like the Lagrange-Euler (L-E) and Newton-Euler (N-E) formulations could then be applied systematically to develop the actual robot arm motion equations. Various forms of robot arm motion equations describing the rigid-body robot arm dynamics are obtained from these two formulations, such as Uicker's Lagrange-Euler equations (Uicker [1965], Bejczy [1974]), Hollerbach's Recursive-Lagrange (R-L) equations (Hollerbach [1980]), Luh's Newton-Euler equations (Luh et al. [1980a]), and Lee's generalized d'Alembert (G-D) equations (Lee et al. [1983]). These motion equations are "equivalent" to each other in the sense that they describe the dynamic behavior of the same physical robot manipulator. However, the structure of these equations

82

may differ as they are obtained for various reasons and purposes. Some are obtained to achieve fast computation time in evaluating the nominal joint torques in servoing a manipulator, others are obtained to facilitate control analysis and synthesis, and still others are obtained to improve computer simulation of robot motion.

The derivation of the dynamic model of a manipulator based on the L-E formulation is simple and systematic. Assuming rigid body motion, the resulting equations of motion, excluding the dynamics of electronic control devices, backlash, and gear friction, are a set of second-order coupled nonlinear differential equations. Bejczy [1974], using the 4×4 homogeneous transformation matrix representation of the kinematic chain and the lagrangian formulation, has shown that the dynamic motion equations for a six-joint Stanford robot arm are highly nonlinear and consist of inertia loading, coupling reaction forces between joints (Coriolis and centrifugal), and gravity loading effects. Furthermore, these torques/forces depend on the manipulator's physical parameters, instantaneous joint configuration, joint velocity and acceleration, and the load it is carrying. The L-E equations of motion provide explicit state equations for robot dynamics and can be utilized to analyze and design advanced joint-variable space control strategies. To a lesser extent, they are being used to solve for the *forward dynamics* problem, that is, given the desired torques/forces, the dynamic equations are used to solve for the joint accelerations which are then integrated to solve for the generalized coordinates and their velocities; or for the *inverse dynamics* problem, that is, given the desired generalized coordinates and their first two time derivatives, the generalized forces/torques are computed. In both cases, it may be required to compute the dynamic coefficients D_{ik}, h_{ikm}, and c_i defined in Eqs. (3.2-31), (3.2-33), and (3.2-34), respectively. Unfortunately, the computation of these coefficients requires a fair amount of arithmetic operations. Thus, the L-E equations are very difficult to utilize for real-time control purposes unless they are simplified.

As an alternative to deriving more efficient equations of motion, attention was turned to develop efficient algorithms for computing the generalized forces/torques based on the N-E equations of motion (Armstrong [1979], Orin et al. [1979], Luh et al. [1980a]). The derivation is simple, but messy, and involves vector cross-product terms. The resulting dynamic equations, excluding the dynamics of the control device, backlash, and gear friction, are a set of forward and backward recursive equations. This set of recursive equations can be applied to the robot links sequentially. The forward recursion propagates kinematics information—such as linear velocities, angular velocities, angular accelerations, and linear accelerations at the center of mass of each link—from the inertial coordinate frame to the hand coordinate frame. The backward recursion propagates the forces and moments exerted on each link from the end-effector of the manipulator to the base reference frame. The most significant result of this formulation is that the computation time of the generalized forces/torques is found linearly proportional to the number of joints of the robot arm and independent of the robot arm configuration. With this algorithm, one can implement simple real-time control of a robot arm in the joint-variable space.

The inefficiency of the L-E equations of motion arises partly from the 4×4 homogeneous matrices describing the kinematic chain, while the efficiency of the N-E formulation is based on the vector formulation and its recursive nature. To further improve the computation time of the lagrangian formulation, Hollerbach [1980] has exploited the recursive nature of the lagrangian formulation. However, the recursive equations destroy the "structure" of the dynamic model which is quite useful in providing insight for designing the controller in state space. For state-space control analysis, one would like to obtain an explicit set of closed-form differential equations (state equations) that describe the dynamic behavior of a manipulator. In addition, the interaction and coupling reaction forces in the equations should be easily identified so that an appropriate controller can be designed to compensate for their effects (Huston and Kelly [1982]). Another approach for obtaining an efficient set of explicit equations of motion is based on the generalized d'Alembert principle to derive the equations of motion which are expressed explicitly in vector-matrix form suitable for control analysis. In addition to allowing faster computation of the dynamic coefficients than the L-E equations of motion, the G-D equations of motion explicitly identify the contributions of the *translational* and *rotational* effects of the links. Such information is useful for designing a controller in state space. The computational efficiency is achieved from a compact formulation using Euler transformation matrices (or rotation matrices) and relative position vectors between joints.

In this chapter, the L-E, N-E, and G-D equations of robot arm motion are derived and discussed, and the motion equations of a two-link manipulator are worked out to illustrate the use of these equations. Since the computation of the dynamic coefficients of the equations of motion is important both in control analysis and computer simulation, the mathematical operations and their computational issues for these motion equations are tabulated. The computation of the applied forces/torques from the generalized d'Alembert equations of motion is of order $O(n^3)$, while the L-E equations are of order $O(n^4)$ [or of order $O(n^3)$ if optimized] and the N-E equations are of order $O(n)$, where n is the number of degrees of freedom of the robot arm.

3.2 LAGRANGE-EULER FORMULATION

The general motion equations of a manipulator can conveniently be expressed through the direct application of the Lagrange-Euler formulation to nonconservative systems. Many investigators utilize the Denavit-Hartenberg matrix representation to describe the spatial displacement between the neighboring link coordinate frames to obtain the link kinematic information, and they employ the lagrangian dynamics technique to derive the dynamic equations of a manipulator. The direct application of the lagrangian dynamics formulation, together with the Denavit-Hartenberg link coordinate representation, results in a convenient and compact algorithmic description of the manipulator equations of motion. The algorithm is expressed by matrix operations and facilitates both analysis and computer implementation. The evaluation of the dynamic and control equations in functionally explicit terms will be based on the compact matrix algorithm derived in this section.

The derivation of the dynamic equations of an n degrees of freedom manipulator is based on the understanding of:

1. The 4×4 homogeneous coordinate transformation matrix, $^{i-1}\mathbf{A}_i$, which describes the spatial relationship between the ith and the $(i-1)$th link coordinate frames. It relates a point fixed in link i expressed in homogeneous coordinates with respect to the ith coordinate system to the $(i-1)$th coordinate system.
2. The Lagrange-Euler equation

$$
\frac{d}{dt} \left(\frac{\partial L}{\partial \dot{q}_i} \right) - \frac{\partial L}{\partial q_i} = \tau_i \qquad i = 1, 2, \ldots, n \qquad (3.2-1)
$$

where

L = lagrangian function = kinetic energy K $-$ potential energy P
K = total kinetic energy of the robot arm
P = total potential energy of the robot arm
q_i = generalized coordinates of the robot arm
$\dot{q}_i$ = first time derivative of the generalized coordinate, q_i
τ_i = generalized force (or torque) applied to the system at joint i to drive link i

From the above Lagrange-Euler equation, one is required to properly choose a set of *generalized coordinates* to describe the system. Generalized coordinates are used as a convenient set of coordinates which completely describe the location (position and orientation) of a system with respect to a reference coordinate frame. For a simple manipulator with rotary-prismatic joints, various sets of generalized coordinates are available to describe the manipulator. However, since the angular positions of the joints are readily available because they can be measured by potentiometers or encoders or other sensing devices, they provide a natural correspondence with the generalized coordinates. This, in effect, corresponds to the generalized coordinates with the joint variable defined in each of the 4×4 link coordinate transformation matrices. Thus, in the case of a rotary joint, $q_i \equiv \theta_i$, the joint angle span of the joint; whereas for a prismatic joint, $q_i \equiv d_i$, the distance traveled by the joint.

The following derivation of the equations of motion of an n degrees of freedom manipulator is based on the homogeneous coordinate transformation matrices developed in Chap. 2.

3.2.1 Joint Velocities of a Robot Manipulator

The Lagrange-Euler formulation requires knowledge of the kinetic energy of the physical system, which in turn requires knowledge of the velocity of each joint. In this section, the velocity of a point fixed in link i will be derived and the effects of the motion of other joints on all the points in this link will be explored.

With reference to Fig. 3.1, let $^i\mathbf{r}_i$ be a point fixed and at rest in a link i and expressed in homogeneous coordinates with respect to the ith link coordinate frame,

$$^i\mathbf{r}_i = \begin{bmatrix} x_i \\ y_i \\ z_i \\ 1 \end{bmatrix} = (x_i,\ y_i,\ z_i,\ 1)^T \tag{3.2-2}$$

Let $^0\mathbf{r}_i$ be the same point $^i\mathbf{r}_i$ with respect to the base coordinate frame, $^{i-1}\mathbf{A}_i$ the homogeneous coordinate transformation matrix which relates the spatial displacement of the ith link coordinate frame to the $(i-1)$th link coordinate frame, and $^0\mathbf{A}_i$ the coordinate transformation matrix which relates the ith coordinate frame to the base coordinate frame; then $^0\mathbf{r}_i$ is related to the point $^i\mathbf{r}_i$ by

$$^0\mathbf{r}_i = {}^0\mathbf{A}_i\ {}^i\mathbf{r}_i \tag{3.2-3}$$

where

$$^0\mathbf{A}_i = {}^0\mathbf{A}_1\ {}^1\mathbf{A}_2 \cdots {}^{i-1}\mathbf{A}_i \tag{3.2-4}$$

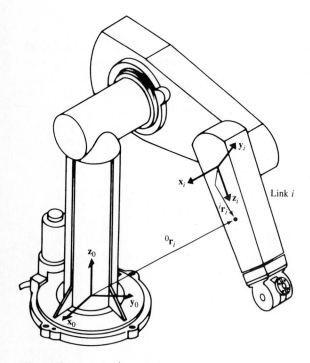

Figure 3.1 A point $^i\mathbf{r}_i$ in link i.

If joint i is revolute, it follows from Eq. (2.2-29) that the general form of $^{i-1}\mathbf{A}_i$ is given by

$$^{i-1}\mathbf{A}_i = \begin{bmatrix} \cos\theta_i & -\cos\alpha_i \sin\theta_i & \sin\alpha_i \sin\theta_i & a_i \cos\theta_i \\ \sin\theta_i & \cos\alpha_i \cos\theta_i & -\sin\alpha_i \cos\theta_i & a_i \sin\theta_i \\ 0 & \sin\alpha_i & \cos\alpha_i & d_i \\ 0 & 0 & 0 & 1 \end{bmatrix} \quad (3.2\text{-}5)$$

or, if joint i is prismatic, from Eq. (2.2-31), the general form of $^{i-1}\mathbf{A}_i$ is

$$^{i-1}\mathbf{A}_i = \begin{bmatrix} \cos\theta_i & -\cos\alpha_i \sin\theta_i & \sin\alpha_i \sin\theta_i & 0 \\ \sin\theta_i & \cos\alpha_i \cos\theta_i & -\sin\alpha_i \cos\theta_i & 0 \\ 0 & \sin\alpha_i & \cos\alpha_i & d_i \\ 0 & 0 & 0 & 1 \end{bmatrix} \quad (3.2\text{-}6)$$

In general, all the nonzero elements in the matrix $^0\mathbf{A}_i$ are a function of $(\theta_1, \theta_2, \ldots, \theta_i)$, and α_i, a_i, d_i are known parameters from the kinematic structure of the arm and θ_i or d_i is the joint variable of joint i. In order to derive the equations of motion that are applicable to both revolute and prismatic joints, we shall use the variable q_i to represent the generalized coordinate of joint i which is either θ_i (for a rotary joint) or d_i (for a prismatic joint).

Since the point $^i\mathbf{r}_i$ is at rest in link i, and assuming rigid body motion, other points as well as the point $^i\mathbf{r}_i$ fixed in the link i and expressed with respect to the ith coordinate frame will have zero velocity with respect to the ith coordinate frame (which is not an inertial frame). The velocity of $^i\mathbf{r}_i$ expressed in the base coordinate frame (which is an inertial frame) can be expressed as

$$^0\mathbf{v}_i \equiv \mathbf{v}_i = \frac{d}{dt}(^0\mathbf{r}_i) = \frac{d}{dt}(^0\mathbf{A}_i \, ^i\mathbf{r}_i)$$

$$= {}^0\dot{\mathbf{A}}_1 \, ^1\mathbf{A}_2 \cdots {}^{i-1}\mathbf{A}_i \, ^i\mathbf{r}_i + {}^0\mathbf{A}_1 \, ^1\dot{\mathbf{A}}_2 \cdots {}^{i-1}\mathbf{A}_i \, ^i\mathbf{r}_i + \cdots$$

$$+ {}^0\mathbf{A}_1 \cdots {}^{i-1}\dot{\mathbf{A}}_i \, ^i\mathbf{r}_i + {}^0\mathbf{A}_i \, ^i\dot{\mathbf{r}}_i = \left[\sum_{j=1}^{i} \frac{\partial^0\mathbf{A}_i}{\partial q_j} \dot{q}_j \right] ^i\mathbf{r}_i \quad (3.2\text{-}7)$$

The above compact form is obtained because $^i\dot{\mathbf{r}}_i = 0$. The partial derivative of $^0\mathbf{A}_i$ with respect to q_j can be easily calculated with the help of a matrix $\mathbf{Q}_i$ which, for a revolute joint, is defined as

$$\mathbf{Q}_i = \begin{bmatrix} 0 & -1 & 0 & 0 \\ 1 & 0 & 0 & 0 \\ 0 & 0 & 0 & 0 \\ 0 & 0 & 0 & 0 \end{bmatrix} \quad (3.2\text{-}8a)$$

and, for a prismatic joint, as

$$
\mathbf{Q}_i = \begin{bmatrix} 0 & 0 & 0 & 0 \\ 0 & 0 & 0 & 0 \\ 0 & 0 & 0 & 1 \\ 0 & 0 & 0 & 0 \end{bmatrix} \tag{3.2-8b}
$$

It then follows that

$$
\frac{\partial\, ^{i-1}\mathbf{A}_i}{\partial q_i} = \mathbf{Q}_i\, ^{i-1}\mathbf{A}_i \tag{3.2-9}
$$

For example, for a robot arm with all rotary joints, $q_i = \theta_i$, and using Eq. (3.2-5),

$$
\frac{\partial\, ^{i-1}\mathbf{A}_i}{\partial \theta_i} = \begin{bmatrix} -\sin\theta_i & -\cos\alpha_i\cos\theta_i & \sin\alpha_i\cos\theta_i & -a_i\sin\theta_i \\ \cos\theta_i & -\cos\alpha_i\sin\theta_i & \sin\alpha_i\sin\theta_i & a_i\cos\theta_i \\ 0 & 0 & 0 & 0 \\ 0 & 0 & 0 & 0 \end{bmatrix}
$$

$$
= \begin{bmatrix} 0 & -1 & 0 & 0 \\ 1 & 0 & 0 & 0 \\ 0 & 0 & 0 & 0 \\ 0 & 0 & 0 & 0 \end{bmatrix} \begin{bmatrix} \cos\theta_i & -\cos\alpha_i\sin\theta_i & \sin\alpha_i\sin\theta_i & a_i\cos\theta_i \\ \sin\theta_i & \cos\alpha_i\cos\theta_i & -\sin\alpha_i\cos\theta_i & a_i\sin\theta_i \\ 0 & \sin\alpha_i & \cos\alpha_i & d_i \\ 0 & 0 & 0 & 1 \end{bmatrix}
$$

$$
\equiv \mathbf{Q}_i\, ^{i-1}\mathbf{A}_i
$$

Hence, for $i = 1, 2, \ldots, n$,

$$
\frac{\partial\, ^0\mathbf{A}_i}{\partial q_j} = \begin{cases} ^0\mathbf{A}_1\, ^1\mathbf{A}_2 \cdots\, ^{j-2}\mathbf{A}_{j-1}\mathbf{Q}_j\, ^{j-1}\mathbf{A}_j \cdots\, ^{i-1}\mathbf{A}_i & \text{for } j \leqslant i \\ 0 & \text{for } j > i \end{cases} \tag{3.2-10}
$$

Eq. (3.2-10) can be interpreted as the effect of the motion of joint j on all the points on link i. In order to simplify notations, let us define $\mathbf{U}_{ij} \triangleq \partial\, ^0\mathbf{A}_i/\partial q_j$, then Eq. (3.2-10) can be written as follows for $i = 1, 2, \ldots, n$,

$$
\mathbf{U}_{ij} = \begin{cases} ^0\mathbf{A}_{j-1}\mathbf{Q}_j\, ^{j-1}\mathbf{A}_i & \text{for } j \leqslant i \\ 0 & \text{for } j > i \end{cases} \tag{3.2-11}
$$

Using this notation, $\mathbf{v}_i$ can be expressed as

$$
\mathbf{v}_i = \left(\sum_{j=1}^{i} \mathbf{U}_{ij}\dot{q}_j \right)\, ^i\mathbf{r}_i \tag{3.2-12}
$$

It is worth pointing out that the partial derivative of $^{i-1}\mathbf{A}_i$ with respect to q_i results in a matrix that does not retain the structure of a homogeneous coordinate transformation matrix. For a rotary joint, the effect of premultiplying $^{i-1}\mathbf{A}_i$ by $\mathbf{Q}_i$ is equivalent to interchanging the elements of the first two rows of $^{i-1}\mathbf{A}_i$, negating all the elements of the first row, and zeroing out all the elements of the third and fourth rows. For a prismatic joint, the effect is to replace the elements of the third row with the fourth row of $^{i-1}\mathbf{A}_i$ and zeroing out the elements in the other rows. The advantage of using the $\mathbf{Q}_i$ matrices is that we can still use the $^{i-1}\mathbf{A}_i$ matrices and apply the above operations to $^{i-1}\mathbf{A}_i$ when premultiplying it with the $\mathbf{Q}_i$.

Next, we need to find the interaction effects between joints as

$$
\frac{\partial \mathbf{U}_{ij}}{\partial q_k} \triangleq \mathbf{U}_{ijk} =
\begin{cases}
^0\mathbf{A}_{j-1}\mathbf{Q}_j{}^{j-1}\mathbf{A}_{k-1}\mathbf{Q}_k{}^{k-1}\mathbf{A}_i & i \geqslant k \geqslant j \\
^0\mathbf{A}_{k-1}\mathbf{Q}_k{}^{k-1}\mathbf{A}_{j-1}\mathbf{Q}_j{}^{j-1}\mathbf{A}_i & i \geqslant j \geqslant k \\
0 & i < j \text{ or } i < k
\end{cases}
\tag{3.2-13}
$$

For example, for a robot arm with all rotary joints, $i = j = k = 1$ and $q_1 = \theta_1$, so that

$$
\frac{\partial \mathbf{U}_{11}}{\partial \theta_1} = \frac{\partial}{\partial \theta_1}(\mathbf{Q}_1{}^0\mathbf{A}_1) = \mathbf{Q}_1\mathbf{Q}_1{}^0\mathbf{A}_1
$$

Eq. (3.2-13) can be interpreted as the interaction effects of the motion of joint j and joint k on all the points on link i.

3.2.2 Kinetic Energy of a Robot Manipulator

After obtaining the joint velocity of each link, we need to find the kinetic energy of link i. Let K_i be the kinetic energy of link i, $i = 1, 2, \ldots, n$, as expressed in the base coordinate system, and let dK_i be the kinetic energy of a particle with differential mass dm in link i; then

$$
dK_i = \tfrac{1}{2}(\dot{x}_i^2 + \dot{y}_i^2 + \dot{z}_i^2)\, dm
$$

$$
= \tfrac{1}{2} \text{ trace } (\mathbf{v}_i\mathbf{v}_i^T)\, dm = \tfrac{1}{2} \text{ Tr } (\mathbf{v}_i\mathbf{v}_i^T)\, dm
\tag{3.2-14}
$$

where a trace operator[†] instead of a vector dot product is used in the above equation to form the tensor from which the link inertia matrix (or pseudo-inertia matrix) $\mathbf{J}_i$ can be obtained. Substituting $\mathbf{v}_i$ from Eq. (3.2-12), the kinetic energy of the differential mass is

[†] $\text{Tr } \mathbf{A} \triangleq \sum\limits_{i=1}^{n} a_{ii}$.

$$
dK_i = \tfrac{1}{2} \operatorname{Tr} \left[\sum_{p=1}^{i} \mathbf{U}_{ip} \dot{q}_p \, {}^i\mathbf{r}_i \left(\sum_{r=1}^{i} \mathbf{U}_{ir} \dot{q}_r \, {}^i\mathbf{r}_i \right)^T \right] dm
$$

$$
= \tfrac{1}{2} \operatorname{Tr} \left[\sum_{p=1}^{i} \sum_{r=1}^{i} \mathbf{U}_{ip} \, {}^i\mathbf{r}_i \, {}^i\mathbf{r}_i^T \mathbf{U}_{ir}^T \dot{q}_p \dot{q}_r \right] dm
$$

$$
= \tfrac{1}{2} \operatorname{Tr} \left[\sum_{p=1}^{i} \sum_{r=1}^{i} \mathbf{U}_{ip} ({}^i\mathbf{r}_i \, dm \, {}^i\mathbf{r}_i^T) \, \mathbf{U}_{ir}^T \dot{q}_p \dot{q}_r \right] \tag{3.2-15}
$$

The matrix $\mathbf{U}_{ij}$ is the rate of change of the points ($ {}^i\mathbf{r}_i $) on link i relative to the base coordinate frame as q_j changes. It is constant for all points on link i and independent of the mass distribution of the link i. Also $\dot{q}_i$ are independent of the mass distribution of link i, so summing all the kinetic energies of all links and putting the integral inside the bracket,

$$
K_i = \int dK_i = \tfrac{1}{2} \operatorname{Tr} \left[\sum_{p=1}^{i} \sum_{r=1}^{i} \mathbf{U}_{ip} \left(\int {}^i\mathbf{r}_i \, {}^i\mathbf{r}_i^T \, dm \right) \mathbf{U}_{ir}^T \dot{q}_p \dot{q}_r \right] \tag{3.2-16}
$$

The integral term inside the bracket is the inertia of all the points on link i, hence,

$$
\mathbf{J}_i = \int {}^i\mathbf{r}_i \, {}^i\mathbf{r}_i^T \, dm =
\begin{bmatrix}
\int x_i^2 \, dm & \int x_i y_i \, dm & \int x_i z_i \, dm & \int x_i \, dm \\
\int x_i y_i \, dm & \int y_i^2 \, dm & \int y_i z_i \, dm & \int y_i \, dm \\
\int x_i z_i \, dm & \int y_i z_i \, dm & \int z_i^2 \, dm & \int z_i \, dm \\
\int x_i \, dm & \int y_i \, dm & \int z_i \, dm & \int dm
\end{bmatrix} \tag{3.2-17}
$$

where $ {}^i\mathbf{r}_i = (x_i, y_i, z_i, 1)^T $ as defined before. If we use inertia tensor I_{ij} which is defined as

$$
I_{ij} = \int \left[\delta_{ij} \left(\sum_{k} x_k^2 \right) - x_i x_j \right] dm
$$

where the indices i, j, k indicate principal axes of the ith coordinate frame and δ_{ij} is the so-called Kronecker delta, then $\mathbf{J}_i$ can be expressed in inertia tensor as

$$
\mathbf{J}_i =
\begin{bmatrix}
\dfrac{-I_{xx} + I_{yy} + I_{zz}}{2} & I_{xy} & I_{xz} & m_i \bar{x}_i \\[2ex]
I_{xy} & \dfrac{I_{xx} - I_{yy} + I_{zz}}{2} & I_{yz} & m_i \bar{y}_i \\[2ex]
I_{xz} & I_{yz} & \dfrac{I_{xx} + I_{yy} - I_{zz}}{2} & m_i \bar{z}_i \\[2ex]
m_i \bar{x}_i & m_i \bar{y}_i & m_i \bar{z}_i & m_i
\end{bmatrix} \tag{3.2-18}
$$

or using the radius of gyration of the rigid body m_i in the $(\mathbf{x}_i, \mathbf{y}_i, \mathbf{z}_i)$ coordinate system, $\mathbf{J}_i$ can be expressed as

$$
\mathbf{J}_i = m_i
\begin{bmatrix}
\dfrac{-k_{i11}^2 + k_{i22}^2 + k_{i33}^2}{2} & k_{i12}^2 & k_{i13}^2 & \bar{x}_i \\[2ex]
k_{i12}^2 & \dfrac{k_{i11}^2 - k_{i22}^2 + k_{i33}^2}{2} & k_{i23}^2 & \bar{y}_i \\[2ex]
k_{i13}^2 & k_{i23}^2 & \dfrac{k_{i11}^2 + k_{i22}^2 - k_{i33}^2}{2} & \bar{z}_i \\[2ex]
\bar{x}_i & \bar{y}_i & \bar{z}_i & 1
\end{bmatrix}
$$

(3.2-19)

where k_{i23} is the radius of gyration of link i about the **yz** axes and ${}^i\bar{\mathbf{r}}_i = (\bar{x}_i, \bar{y}_i, \bar{z}_i, 1)^T$ is the center of mass vector of link i from the ith link coordinate frame and expressed in the ith link coordinate frame. Hence, the total kinetic energy K of a robot arm is

$$
K = \sum_{i=1}^{n} K_i = \tfrac{1}{2} \sum_{i=1}^{n} \mathrm{Tr} \left[\sum_{p=1}^{i} \sum_{r=1}^{i} \mathbf{U}_{ip} \mathbf{J}_i \mathbf{U}_{ir}^T \dot{q}_p \, \dot{q}_r \right]
$$

$$
= \tfrac{1}{2} \sum_{i=1}^{n} \sum_{p=1}^{i} \sum_{r=1}^{i} [\mathrm{Tr}\,(\mathbf{U}_{ip} \mathbf{J}_i \mathbf{U}_{ir}^T)\dot{q}_p \, \dot{q}_r]
\tag{3.2-20}
$$

which is a *scalar* quantity. Note that the $\mathbf{J}_i$ are dependent on the mass distribution of link i and not their position or rate of motion and are expressed with respect to the ith coordinate frame. Hence, the $\mathbf{J}_i$ need be computed only *once* for evaluating the kinetic energy of a robot arm.

3.2.3 Potential Energy of a Robot Manipulator

Let the total potential energy of a robot arm be P and let each of its link's potential energy be P_i:

$$
P_i = - m_i \mathbf{g}\, {}^0\bar{\mathbf{r}}_i = -m_i \mathbf{g}({}^0\mathbf{A}_i\, {}^i\bar{\mathbf{r}}_i) \qquad i = 1, 2, \ldots, n
\tag{3.2-21}
$$

and the total potential energy of the robot arm can be obtained by summing all the potential energies in each link,

$$
P = \sum_{i=1}^{n} P_i = \sum_{i=1}^{n} - m_i \mathbf{g} \,({}^0\mathbf{A}_i\, {}^i\bar{\mathbf{r}}_i)
\tag{3.2-22}
$$

where $\mathbf{g} = (g_x, g_y, g_z, 0)$ is a gravity row vector expressed in the base coordinate system. For a level system, $\mathbf{g} = (0, 0, -|g|, 0)$ and g is the gravitational constant ($g = 9.8062$ m/sec^2).

3.2.4 Motion Equations of a Manipulator

From Eqs. (3.2-20) and (3.2-22), the lagrangian function $L = K - P$ is given by

$$L = \frac{1}{2} \sum_{i=1}^{n} \sum_{j=1}^{i} \sum_{k=1}^{i} [\text{Tr}\,(\mathbf{U}_{ij}\,\mathbf{J}_i\,\mathbf{U}_{ik}^T)\dot{q}_j\,\dot{q}_k] + \sum_{i=1}^{n} m_i \mathbf{g}(^0\mathbf{A}_i\,{}^i\bar{\mathbf{r}}_i) \qquad (3.2\text{-}23)$$

Applying the Lagrange-Euler formulation to the lagrangian function of the robot arm [Eq. (3.2-23)] yields the necessary generalized torque τ_i for joint i actuator to drive the ith link of the manipulator,

$$\tau_i = \frac{d}{dt}\left[\frac{\partial L}{\partial \dot{q}_i}\right] - \frac{\partial L}{\partial q_i}$$

$$= \sum_{j=i}^{n} \sum_{k=1}^{j} \text{Tr}\,(\mathbf{U}_{jk}\,\mathbf{J}_j\,\mathbf{U}_{ji}^T)\,\ddot{q}_k + \sum_{j=i}^{n} \sum_{k=1}^{j} \sum_{m=1}^{j} \text{Tr}\,(\mathbf{U}_{jkm}\,\mathbf{J}_j\,\mathbf{U}_{ji}^T)\,\dot{q}_k\dot{q}_m - \sum_{j=i}^{n} m_j \mathbf{g}\mathbf{U}_{ji}\,{}^j\bar{\mathbf{r}}_j$$

$$(3.2\text{-}24)$$

for $i = 1, 2, \ldots, n$. The above equation can be expressed in a much simpler matrix notation form as

$$\tau_i = \sum_{k=1}^{n} D_{ik}\ddot{q}_k + \sum_{k=1}^{n} \sum_{m=1}^{n} h_{ikm}\dot{q}_k\dot{q}_m + c_i \qquad i = 1, 2, \ldots, n \quad (3.2\text{-}25)$$

or in a matrix form as

$$\tau(t) = \mathbf{D}(\mathbf{q}(t))\,\ddot{\mathbf{q}}(t) + \mathbf{h}(\mathbf{q}(t), \dot{\mathbf{q}}(t)) + \mathbf{c}(\mathbf{q}(t)) \qquad (3.2\text{-}26)$$

where

$\tau(t) = n \times 1$ generalized torque vector applied at joints $i = 1, 2, \ldots, n$; that is,

$$\tau(t) = (\tau_1(t), \tau_2(t), \ldots, \tau_n(t))^T \qquad (3.2\text{-}27)$$

$\mathbf{q}(t) =$ an $n \times 1$ vector of the joint variables of the robot arm and can be expressed as

$$\mathbf{q}(t) = (q_1(t), q_2(t), \ldots, q_n(t))^T \qquad (3.2\text{-}28)$$

$\dot{\mathbf{q}}(t) =$ an $n \times 1$ vector of the joint velocity of the robot arm and can be expressed as

$$\dot{\mathbf{q}}(t) = (\dot{q}_1(t), \dot{q}_2(t), \ldots, \dot{q}_n(t))^T \qquad (3.2\text{-}29)$$

$\ddot{\mathbf{q}}(t) =$ an $n \times 1$ vector of the acceleration of the joint variables $\mathbf{q}(t)$ and can be expressed as

$$\ddot{\mathbf{q}}(t) = (\ddot{q}_1(t), \ddot{q}_2(t), \ldots, \ddot{q}_n(t))^T \qquad (3.2\text{-}30)$$

$\mathbf{D}(\mathbf{q})$ = an $n \times n$ inertial acceleration-related symmetric matrix whose elements are

$$D_{ik} = \sum_{j=\max(i,k)}^{n} \text{Tr } (\mathbf{U}_{jk} \mathbf{J}_j \mathbf{U}_{ji}^T) \qquad i, k = 1, 2, \ldots, n \qquad (3.2\text{-}31)$$

$\mathbf{h}(\mathbf{q}, \dot{\mathbf{q}})$ = an $n \times 1$ nonlinear Coriolis and centrifugal force vector whose elements are

$$\mathbf{h}(\mathbf{q}, \dot{\mathbf{q}}) = (h_1, h_2, \ldots, h_n)^T$$

where
$$h_i = \sum_{k=1}^{n} \sum_{m=1}^{n} h_{ikm} \dot{q}_k \dot{q}_m \qquad i = 1, 2, \ldots, n \qquad (3.2\text{-}32)$$

and
$$h_{ikm} = \sum_{j=\max(i,k,m)}^{n} \text{Tr } (\mathbf{U}_{jkm} \mathbf{J}_j \mathbf{U}_{ji}^T) \qquad i, k, m = 1, 2, \ldots, n \qquad (3.2\text{-}33)$$

$\mathbf{c}(\mathbf{q})$ = an $n \times 1$ gravity loading force vector whose elements are

$$\mathbf{c}(\mathbf{q}) = (c_1, c_2, \ldots, c_n)^T$$

where
$$c_i = \sum_{j=i}^{n} (-m_j \mathbf{g} \mathbf{U}_{ji} \,^j\bar{\mathbf{r}}_j) \qquad i = 1, 2, \ldots, n \qquad (3.2\text{-}34)$$

3.2.5 Motion Equations of a Robot Arm with Rotary Joints

If the equations given by Eqs. (3.2-26) to (3.2-34) are expanded for a six-axis robot arm with rotary joints, then the following terms that form the dynamic motion equations are obtained:

The Acceleration-Related Symmetric Matrix, $\mathbf{D}(\theta)$. From Eq. (3.2-31), we have

$$\mathbf{D}(\theta) = \begin{bmatrix} D_{11} & D_{12} & D_{13} & D_{14} & D_{15} & D_{16} \\ D_{12} & D_{22} & D_{23} & D_{24} & D_{25} & D_{26} \\ D_{13} & D_{23} & D_{33} & D_{34} & D_{35} & D_{36} \\ D_{14} & D_{24} & D_{34} & D_{44} & D_{45} & D_{46} \\ D_{15} & D_{25} & D_{35} & D_{45} & D_{55} & D_{56} \\ D_{16} & D_{26} & D_{36} & D_{46} & D_{56} & D_{66} \end{bmatrix} \qquad (3.2\text{-}35)$$

where

$$D_{11} = \text{Tr } (\mathbf{U}_{11}\mathbf{J}_1\mathbf{U}_{11}^T) + \text{Tr } (\mathbf{U}_{21}\mathbf{J}_2\mathbf{U}_{21}^T) + \text{Tr } (\mathbf{U}_{31}\mathbf{J}_3\mathbf{U}_{31}^T) + \text{Tr } (\mathbf{U}_{41}\mathbf{J}_4\mathbf{U}_{41}^T)$$

$$+ \text{Tr } (\mathbf{U}_{51}\mathbf{J}_5\mathbf{U}_{51}^T) + \text{Tr } (\mathbf{U}_{61}\mathbf{J}_6\mathbf{U}_{61}^T)$$

$$D_{12} = D_{21} = \text{Tr } (\mathbf{U}_{22}\mathbf{J}_2\mathbf{U}_{21}^T) + \text{Tr } (\mathbf{U}_{32}\mathbf{J}_3\mathbf{U}_{31}^T) + \text{Tr } (\mathbf{U}_{42}\mathbf{J}_4\mathbf{U}_{41}^T)$$

$$+ \text{Tr } (\mathbf{U}_{52}\mathbf{J}_5\mathbf{U}_{51}^T) + \text{Tr } (\mathbf{U}_{62}\mathbf{J}_6\mathbf{U}_{61}^T)$$

$$D_{13} = D_{31} = \text{Tr } (\mathbf{U}_{33}\mathbf{J}_3\mathbf{U}_{31}^T) + \text{Tr } (\mathbf{U}_{43}\mathbf{J}_4\mathbf{U}_{41}^T) + \text{Tr } (\mathbf{U}_{53}\mathbf{J}_5\mathbf{U}_{51}^T) + \text{Tr } (\mathbf{U}_{63}\mathbf{J}_6\mathbf{U}_{61}^T)$$

$$D_{14} = D_{41} = \text{Tr } (\mathbf{U}_{44}\mathbf{J}_4\mathbf{U}_{41}^T) + \text{Tr } (\mathbf{U}_{54}\mathbf{J}_5\mathbf{U}_{51}^T) + \text{Tr } (\mathbf{U}_{64}\mathbf{J}_6\mathbf{U}_{61}^T)$$

$$D_{15} = D_{51} = \text{Tr } (\mathbf{U}_{55}\mathbf{J}_5\mathbf{U}_{51}^T) + \text{Tr } (\mathbf{U}_{65}\mathbf{J}_6\mathbf{U}_{61}^T)$$

$$D_{16} = D_{61} = \text{Tr } (\mathbf{U}_{66}\mathbf{J}_6\mathbf{U}_{61}^T)$$

$$D_{22} = \text{Tr } (\mathbf{U}_{22}\mathbf{J}_2\mathbf{U}_{22}^T) + \text{Tr } (\mathbf{U}_{32}\mathbf{J}_3\mathbf{U}_{32}^T) + \text{Tr } (\mathbf{U}_{42}\mathbf{J}_4\mathbf{U}_{42}^T)$$

$$+ \text{Tr } (\mathbf{U}_{52}\mathbf{J}_5\mathbf{U}_{52}^T) + \text{Tr } (\mathbf{U}_{62}\mathbf{J}_6\mathbf{U}_{62}^T)$$

$$D_{23} = D_{32} = \text{Tr } (\mathbf{U}_{33}\mathbf{J}_3\mathbf{U}_{32}^T) + \text{Tr } (\mathbf{U}_{43}\mathbf{J}_4\mathbf{U}_{42}^T) + \text{Tr } (\mathbf{U}_{53}\mathbf{J}_5\mathbf{U}_{52}^T) + \text{Tr } (\mathbf{U}_{63}\mathbf{J}_6\mathbf{U}_{62}^T)$$

$$D_{24} = D_{42} = \text{Tr } (\mathbf{U}_{44}\mathbf{J}_4\mathbf{U}_{42}^T) + \text{Tr } (\mathbf{U}_{54}\mathbf{J}_5\mathbf{U}_{52}^T) + \text{Tr } (\mathbf{U}_{64}\mathbf{J}_6\mathbf{U}_{62}^T)$$

$$D_{25} = D_{52} = \text{Tr } (\mathbf{U}_{55}\mathbf{J}_5\mathbf{U}_{52}^T) + \text{Tr } (\mathbf{U}_{65}\mathbf{J}_6\mathbf{U}_{62}^T)$$

$$D_{26} = D_{62} = \text{Tr } (\mathbf{U}_{66}\mathbf{J}_6\mathbf{U}_{62}^T)$$

$$D_{33} = \text{Tr } (\mathbf{U}_{33}\mathbf{J}_3\mathbf{U}_{33}^T) + \text{Tr } (\mathbf{U}_{43}\mathbf{J}_4\mathbf{U}_{43}^T) + \text{Tr } (\mathbf{U}_{53}\mathbf{J}_5\mathbf{U}_{53}^T) + \text{Tr } (\mathbf{U}_{63}\mathbf{J}_6\mathbf{U}_{63}^T)$$

$$D_{34} = D_{43} = \text{Tr } (\mathbf{U}_{44}\mathbf{J}_4\mathbf{U}_{43}^T) + \text{Tr } (\mathbf{U}_{54}\mathbf{J}_5\mathbf{U}_{53}^T) + \text{Tr } (\mathbf{U}_{64}\mathbf{J}_6\mathbf{U}_{63}^T)$$

$$D_{35} = D_{53} = \text{Tr } (\mathbf{U}_{55}\mathbf{J}_5\mathbf{U}_{53}^T) + \text{Tr } (\mathbf{U}_{65}\mathbf{J}_6\mathbf{U}_{63}^T)$$

$$D_{36} = D_{63} = \text{Tr } (\mathbf{U}_{66}\mathbf{J}_6\mathbf{U}_{63}^T)$$

$$D_{44} = \text{Tr } (\mathbf{U}_{44}\mathbf{J}_4\mathbf{U}_{44}^T) + \text{Tr } (\mathbf{U}_{54}\mathbf{J}_5\mathbf{U}_{54}^T) + \text{Tr } (\mathbf{U}_{64}\mathbf{J}_6\mathbf{U}_{64}^T)$$

$$D_{45} = D_{54} = \text{Tr } (\mathbf{U}_{55}\mathbf{J}_5\mathbf{U}_{54}^T) + \text{Tr } (\mathbf{U}_{65}\mathbf{J}_6\mathbf{U}_{64}^T)$$

$$D_{46} = D_{64} = \text{Tr } (\mathbf{U}_{66}\mathbf{J}_6\mathbf{U}_{64}^T)$$

$$D_{55} = \text{Tr } (\mathbf{U}_{55}\mathbf{J}_5\mathbf{U}_{55}^T) + \text{Tr } (\mathbf{U}_{65}\mathbf{J}_6\mathbf{U}_{65}^T)$$

$$D_{56} = D_{65} = \text{Tr } (\mathbf{U}_{66}\mathbf{J}_6\mathbf{U}_{65}^T)$$

$$D_{66} = \text{Tr } (\mathbf{U}_{66}\mathbf{J}_6\mathbf{U}_{66}^T)$$

The Coriolis and Centrifugal Terms, $\mathbf{h}(\theta, \dot{\theta})$. The velocity-related coefficients in the Coriolis and centrifugal terms in Eqs. (3.2-32) and (3.2-33) can be expressed separately by a 6×6 symmetric matrix denoted by $\mathbf{H}_{i,v}$ and defined in the following way:

$$\mathbf{H}_{i,v} = \begin{bmatrix} h_{i11} & h_{i12} & h_{i13} & h_{i14} & h_{i15} & h_{i16} \\ h_{i12} & h_{i22} & h_{i23} & h_{i24} & h_{i25} & h_{i26} \\ h_{i13} & h_{i23} & h_{i33} & h_{i34} & h_{i35} & h_{i36} \\ h_{i14} & h_{i24} & h_{i34} & h_{i44} & h_{i45} & h_{i46} \\ h_{i15} & h_{i25} & h_{i35} & h_{i45} & h_{i55} & h_{i56} \\ h_{i16} & h_{i26} & h_{i36} & h_{i46} & h_{i56} & h_{i66} \end{bmatrix} \qquad i = 1, 2, \dots, 6 \tag{3.2-36}$$

Let the velocity of the six joint variables be expressed by a six-dimensional column vector denoted by $\dot{\theta}$:

$$\dot{\theta}(t) = [\dot{\theta}_1(t), \dot{\theta}_2(t), \dot{\theta}_3(t), \dot{\theta}_4(t), \dot{\theta}_5(t), \dot{\theta}_6(t)]^T \tag{3.2-37}$$

Then, Eq. (3.2-32) can be expressed in the following compact matrix-vector product form:

$$h_i = \dot{\theta}^T \mathbf{H}_{i,v} \dot{\theta} \tag{3.2-38}$$

where the subscript i refers to the joint ($i = 1, \dots, 6$) at which the velocity-induced torques or forces are "felt."

The expression given by Eq. (3.2-38) is a component in a six-dimensional column vector denoted by $\mathbf{h}(\theta, \dot{\theta})$:

$$\mathbf{h}(\theta, \dot{\theta}) = \begin{bmatrix} h_1 \\ h_2 \\ h_3 \\ h_4 \\ h_5 \\ h_6 \end{bmatrix} = \begin{bmatrix} \dot{\theta}^T \mathbf{H}_{1,v} \dot{\theta} \\ \dot{\theta}^T \mathbf{H}_{2,v} \dot{\theta} \\ \dot{\theta}^T \mathbf{H}_{3,v} \dot{\theta} \\ \dot{\theta}^T \mathbf{H}_{4,v} \dot{\theta} \\ \dot{\theta}^T \mathbf{H}_{5,v} \dot{\theta} \\ \dot{\theta}^T \mathbf{H}_{6,v} \dot{\theta} \end{bmatrix} \tag{3.2-39}$$

The Gravity Terms, $\mathbf{c}(\theta)$. From Eq. (3.2-34) we have

$$\mathbf{c}(\theta) = (c_1, c_2, c_3, c_4, c_5, c_6)^T \tag{3.2-40}$$

where

$$c_1 = -(m_1 \mathbf{g} \mathbf{U}_{11}{}^1\bar{\mathbf{r}}_1 + m_2 \mathbf{g} \mathbf{U}_{21}{}^2\bar{\mathbf{r}}_2 + m_3 \mathbf{g} \mathbf{U}_{31}{}^3\bar{\mathbf{r}}_3 + m_4 \mathbf{g} \mathbf{U}_{41}{}^4\bar{\mathbf{r}}_4$$

$$+ m_5 \mathbf{g} \mathbf{U}_{51}{}^5\bar{\mathbf{r}}_5 + m_6 \mathbf{g} \mathbf{U}_{61}{}^6\bar{\mathbf{r}}_6)$$

$$c_2 = -(m_2 \mathbf{g} \mathbf{U}_{22}{}^2\bar{\mathbf{r}}_2 + m_3 \mathbf{g} \mathbf{U}_{32}{}^3\bar{\mathbf{r}}_3 + m_4 \mathbf{g} \mathbf{U}_{42}{}^4\bar{\mathbf{r}}_4 + m_5 \mathbf{g} \mathbf{U}_{52}{}^5\bar{\mathbf{r}}_5 + m_6 \mathbf{g} \mathbf{U}_{62}{}^6\bar{\mathbf{r}}_6)$$

$$c_3 = -(m_3 \mathbf{g} \mathbf{U}_{33} \, {}^3\bar{\mathbf{r}}_3 + m_4 \mathbf{g} \mathbf{U}_{43} \, {}^4\bar{\mathbf{r}}_4 + m_5 \mathbf{g} \mathbf{U}_{53} \, {}^5\bar{\mathbf{r}}_5 + m_6 \mathbf{g} \mathbf{U}_{63} \, {}^6\bar{\mathbf{r}}_6)$$

$$c_4 = -(m_4 \mathbf{g} \mathbf{U}_{44} \, {}^4\bar{\mathbf{r}}_4 + m_5 \mathbf{g} \mathbf{U}_{54} \, {}^5\bar{\mathbf{r}}_5 + m_6 \mathbf{g} \mathbf{U}_{64} \, {}^6\bar{\mathbf{r}}_6)$$

$$c_5 = -(m_5 \mathbf{g} \mathbf{U}_{55} \, {}^5\bar{\mathbf{r}}_5 + m_6 \mathbf{g} \mathbf{U}_{65} \, {}^6\bar{\mathbf{r}}_6)$$

$$c_6 = - m_6 \mathbf{g} \mathbf{U}_{66} \, {}^6\bar{\mathbf{r}}_6$$

The coefficients c_i, D_{ik}, and h_{ikm} in Eqs. (3.2-31) to (3.2-34) are functions of both the joint variables and inertial parameters of the manipulator, and sometimes are called the *dynamic coefficients* of the manipulator. The physical meaning of these dynamic coefficients can easily be seen from the Lagrange-Euler equations of motion given by Eqs. (3.2-26) to (3.2-34):

1. The coefficient c_i represents the gravity loading terms due to the links and is defined by Eq. (3.2-34).
2. The coefficient D_{ik} is related to the acceleration of the joint variables and is defined by Eq. (3.2-31). In particular, for $i = k$, D_{ii} is related to the acceleration of joint i where the driving torque τ_i acts, while for $i \neq k$, D_{ik} is related to the reaction torque (or force) induced by the acceleration of joint k and acting at joint i, or vice versa. Since the inertia matrix is symmetric and $\text{Tr}(\mathbf{A}) = \text{Tr}(\mathbf{A}^T)$, it can be shown that $D_{ik} = D_{ki}$.
3. The coefficient h_{ikm} is related to the velocity of the joint variables and is defined by Eqs. (3.2-32) and (3.2-33). The last two indices, km, are related to the velocities of joints k and m, whose dynamic interplay induces a reaction torque (or force) at joint i. Thus, the first index i is always related to the joint where the velocity-induced reaction torques (or forces) are "felt." In particular, for $k = m$, h_{ikk} is related to the centrifugal force generated by the angular velocity of joint k and "felt" at joint i, while for $k \neq m$, h_{ikm} is related to the Coriolis force generated by the velocities of joints k and m and "felt" at joint i. It is noted that, for a given i, we have $h_{ikm} = h_{imk}$ which is apparent by physical reasoning.

In evaluating these coefficients, it is worth noting that some of the coefficients may be zero for the following reasons:

1. The particular kinematic design of a manipulator can eliminate some dynamic coupling (D_{ij} and h_{ikm} coefficients) between joint motions.
2. Some of the velocity-related dynamic coefficients have only a dummy existence in Eqs. (3.2-32) and (3.2-33); that is, they are physically nonexistent. (For instance, the centrifugal force will not interact with the motion of that joint which generates it, that is, $h_{iii} = 0$ always; however, it can interact with motions at the other joints in the chain, that is, we can have $h_{jii} \neq 0$.)
3. Due to particular variations in the link configuration during motion, some dynamic coefficients may become zero at particular instants of time.

Table 3.1 Computational complexity of Lagrange-Euler equations of motion†

Lagrange-Euler formulation	Multiplications†	Additions
$^{j}\mathbf{A}_i$	$32n(n-1)$	$24n(n-1)$
$-m_j\mathbf{g}\mathbf{U}_{ji}{}^{j}\bar{\mathbf{r}}_j$	$4n(9n-7)$	$n\dfrac{51n-45}{2}$
$\displaystyle\sum_{j=i}^{n} m_j\mathbf{g}\mathbf{U}_{ji}{}^{j}\bar{\mathbf{r}}_j$	0	$\tfrac{1}{2}n(n-1)$
$\mathrm{Tr}\,[\mathbf{U}_{kj}\mathbf{J}_k(\mathbf{U}_{ki})^T]$	$(128/3)\,n(n+1)(n+2)$	$(65/2)\,n(n+1)(n+2)$
$\displaystyle\sum_{k=\max(i,j)}^{n}\mathrm{Tr}\,[\mathbf{U}_{kj}\mathbf{J}_k(\mathbf{U}_{ki})^T]$	0	$(1/6)\,n(n-1)(n+1)$
$\mathrm{Tr}\,[\mathbf{U}_{mjk}\mathbf{J}_m(\mathbf{U}_{mi})^T]$	$(128/3)\,n^2(n+1)(n+2)$	$(65/2)\,n^2(n+1)(n+2)$
$\displaystyle\sum_{m=\max(i,j,k)}^{n}\mathrm{Tr}\,[\mathbf{U}_{mjk}\mathbf{J}_m(\mathbf{U}_{mi})^T]$	0	$(1/6)\,n^2(n-1)(n+1)$
$\tau = \mathbf{D}(\mathbf{q})\ddot{\mathbf{q}} + \mathbf{h}(\mathbf{q},\dot{\mathbf{q}}) + \mathbf{c}(\mathbf{q})$	$(128/3)\,n^4 + (512/3)\,n^3$ $+ (844/3)\,n^2 + (76/3)\,n$	$(98/3)\,n^4 + (781/6)\,n^3$ $+ (637/3)\,n^2 + (107/6)\,n$

† n = number of degrees of freedom of the robot arm.

The motion equations of a manipulator as given by Eqs. (3.2-26) to (3.2-34) are coupled, nonlinear, second-order ordinary differential equations. These equations are in symbolic differential equation form and they include all inertial, centrifugal and Coriolis, and gravitational effects of the links. For a given set of applied torques $\tau_i(i = 1, 2, \ldots, n)$ as a function of time, Eq. (3.2-26) should be integrated simultaneously to obtain the actual motion of the manipulator in terms of the time history of the joint variables $\mathbf{q}(t)$. Then, the time history of the joint variables can be transformed to obtain the time history of the hand motion (hand trajectory) by using the appropriate homogeneous transformation matrices. Or, if the time history of the joint variables, the joint velocities, and the joint accelerations is known ahead of time from a trajectory planning program, then Eqs. (3.2-26) to (3.2-34) can be utilized to compute the applied torques $\tau(t)$ as a function of time which is required to produce the particular planned manipulator motion. This is known as *open-loop control*. However, closed-loop control is more desirable for an autonomous robotic system. This subject will be discussed in Chap. 5.

Because of its matrix structure, the L-E equations of motion are appealing from the closed-loop control viewpoint in that they give a set of state equations as in Eq. (3.2-26). This form allows design of a control law that easily compensates for all the nonlinear effects. Quite often in designing a feedback controller for a manipulator, the dynamic coefficients are used to minimize the nonlinear effects of the reaction forces (Markiewicz [1973]).

It is of interest to evaluate the computational complexities inherent in obtaining the coefficients in Eqs. (3.2-31) to (3.2-34). Table 3.1 summarizes the computational complexities of the L-E equations of motion in terms of required mathematical operations (multiplications and additions) that are required to compute Eq. (3.2-26) for every set point in the trajectory. Computationally, these equations of motion are extremely inefficient as compared with other formulations. In the next section, we shall develop the motion equations of a robot arm which will prove to be more efficient in computing the nominal torques.

3.2.6 A Two-Link Manipulator Example

To show how to use the L-E equations of motion in Eqs. (3.2-26) to (3.2-34), an example is worked out in this section for a two-link manipulator with revolute joints, as shown in Fig. 3.2. All the rotation axes at the joints are along the **z** axis normal to the paper surface. The physical dimensions such as location of center of mass, mass of each link, and coordinate systems are shown below. We would like to derive the motion equations for the above two-link robot arm using Eqs. (3.2-26) to (3.2-34).

We assume the following: joint variables $= \theta_1$, θ_2; mass of the links $= m_1$, m_2; link parameters $= \alpha_1 = \alpha_2 = 0$; $d_1 = d_2 = 0$; and $a_1 = a_2 = l$. Then, from Fig. 3.2, and the discussion in the previous section, the homogeneous coordinate transformation matrices $^{i-1}\mathbf{A}_i$ $(i = 1, 2)$ are obtained as

$$
^0\mathbf{A}_1 = \begin{bmatrix} C_1 & -S_1 & 0 & lC_1 \\ S_1 & C_1 & 0 & lS_1 \\ 0 & 0 & 1 & 0 \\ 0 & 0 & 0 & 1 \end{bmatrix} \qquad
^1\mathbf{A}_2 = \begin{bmatrix} C_2 & -S_2 & 0 & lC_2 \\ S_2 & C_2 & 0 & lS_2 \\ 0 & 0 & 1 & 0 \\ 0 & 0 & 0 & 1 \end{bmatrix}
$$

$$
^0\mathbf{A}_2 = {^0}\mathbf{A}_1 {^1}\mathbf{A}_2 = \begin{bmatrix} C_{12} & -S_{12} & 0 & l(C_{12} + C_1) \\ S_{12} & C_{12} & 0 & l(S_{12} + S_1) \\ 0 & 0 & 1 & 0 \\ 0 & 0 & 0 & 1 \end{bmatrix}
$$

where $C_i = \cos \theta_i$; $S_i = \sin \theta_i$; $C_{ij} = \cos (\theta_i + \theta_j)$; $S_{ij} = \sin (\theta_i + \theta_j)$. From the definition of the $\mathbf{Q}_i$ matrix, for a rotary joint i, we have:

$$
\mathbf{Q}_i = \begin{bmatrix} 0 & -1 & 0 & 0 \\ 1 & 0 & 0 & 0 \\ 0 & 0 & 0 & 0 \\ 0 & 0 & 0 & 0 \end{bmatrix}
$$

Then, using Eq. (3.2-11), we have

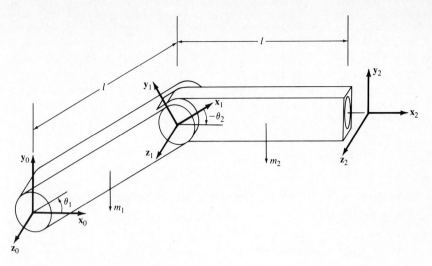

Figure 3.2 A two-link manipulator.

$$
\mathbf{U}_{11} = \frac{\partial\,^0\mathbf{A}_1}{\partial\theta_1} = \mathbf{Q}_1\,^0\mathbf{A}_1 =
\begin{bmatrix}
0 & -1 & 0 & 0 \\
1 & 0 & 0 & 0 \\
0 & 0 & 0 & 0 \\
0 & 0 & 0 & 0
\end{bmatrix}
\begin{bmatrix}
C_1 & -S_1 & 0 & lC_1 \\
S_1 & C_1 & 0 & lS_1 \\
0 & 0 & 1 & 0 \\
0 & 0 & 0 & 1
\end{bmatrix}
$$

$$
=
\begin{bmatrix}
-S_1 & -C_1 & 0 & -lS_1 \\
C_1 & -S_1 & 0 & lC_1 \\
0 & 0 & 0 & 0 \\
0 & 0 & 0 & 0
\end{bmatrix}
$$

Similarly, for $\mathbf{U}_{21}$ and $\mathbf{U}_{22}$ we have

$$
\mathbf{U}_{21} = \frac{\partial\,^0\mathbf{A}_2}{\partial\theta_1} = \mathbf{Q}_1\,^0\mathbf{A}_2 =
\begin{bmatrix}
0 & -1 & 0 & 0 \\
1 & 0 & 0 & 0 \\
0 & 0 & 0 & 0 \\
0 & 0 & 0 & 0
\end{bmatrix}
\begin{bmatrix}
C_{12} & -S_{12} & 0 & l(C_{12} + C_1) \\
S_{12} & C_{12} & 0 & l(S_{12} + S_1) \\
0 & 0 & 1 & 0 \\
0 & 0 & 0 & 1
\end{bmatrix}
$$

$$
=
\begin{bmatrix}
-S_{12} & -C_{12} & 0 & -l(S_{12} + S_1) \\
C_{12} & -S_{12} & 0 & l(C_{12} + C_1) \\
0 & 0 & 0 & 0 \\
0 & 0 & 0 & 0
\end{bmatrix}
$$

$$
\mathbf{U}_{22} = \frac{\partial\,^0\mathbf{A}_2}{\partial\theta_2}
$$

$$
= {}^0\mathbf{A}_1\mathbf{Q}_2\,^1\mathbf{A}_2
$$

$$
= \begin{bmatrix} C_1 & -S_1 & 0 & lC_1 \\ S_1 & C_1 & 0 & lS_1 \\ 0 & 0 & 1 & 0 \\ 0 & 0 & 0 & 1 \end{bmatrix} \begin{bmatrix} 0 & -1 & 0 & 0 \\ 1 & 0 & 0 & 0 \\ 0 & 0 & 0 & 0 \\ 0 & 0 & 0 & 0 \end{bmatrix} \begin{bmatrix} C_2 & -S_2 & 0 & lC_2 \\ S_2 & C_2 & 0 & lS_2 \\ 0 & 0 & 1 & 0 \\ 0 & 0 & 0 & 1 \end{bmatrix}
$$

$$
= \begin{bmatrix} -S_{12} & -C_{12} & 0 & -lS_{12} \\ C_{12} & -S_{12} & 0 & lC_{12} \\ 0 & 0 & 0 & 0 \\ 0 & 0 & 0 & 0 \end{bmatrix}
$$

From Eq. (3.2-18), assuming all the products of inertia are zero, we can derive the pseudo-inertia matrix J_i:

$$
J_1 = \begin{bmatrix} \tfrac{1}{3}m_1 l^2 & 0 & 0 & -\tfrac{1}{2}m_1 l \\ 0 & 0 & 0 & 0 \\ 0 & 0 & 0 & 0 \\ -\tfrac{1}{2}m_1 l & 0 & 0 & m_1 \end{bmatrix} \qquad J_2 = \begin{bmatrix} \tfrac{1}{3}m_2 l^2 & 0 & 0 & -\tfrac{1}{2}m_2 l \\ 0 & 0 & 0 & 0 \\ 0 & 0 & 0 & 0 \\ -\tfrac{1}{2}m_2 l & 0 & 0 & m_2 \end{bmatrix}
$$

Then, using Eq. (3.2-31), we have

$$
D_{11} = \mathrm{Tr}\,(U_{11}J_1 U_{11}^T) + \mathrm{Tr}\,(U_{21}J_2 U_{21}^T)
$$

$$
= \mathrm{Tr}\left\{ \begin{bmatrix} -S_1 & -C_1 & 0 & -lS_1 \\ C_1 & -S_1 & 0 & lC_1 \\ 0 & 0 & 0 & 0 \\ 0 & 0 & 0 & 0 \end{bmatrix} \begin{bmatrix} \tfrac{1}{3}m_1 l^2 & 0 & 0 & -\tfrac{1}{2}m_1 l \\ 0 & 0 & 0 & 0 \\ 0 & 0 & 0 & 0 \\ -\tfrac{1}{2}m_1 l & 0 & 0 & m_1 \end{bmatrix} U_{11}^T \right\}
$$

$$
+ \mathrm{Tr}\left\{ \begin{bmatrix} -S_{12} & -C_{12} & 0 & -l(S_{12} + S_1) \\ C_{12} & -S_{12} & 0 & l(C_{12} + C_1) \\ 0 & 0 & 0 & 0 \\ 0 & 0 & 0 & 0 \end{bmatrix} \begin{bmatrix} \tfrac{1}{3}m_2 l^2 & 0 & 0 & -\tfrac{1}{2}m_2 l \\ 0 & 0 & 0 & 0 \\ 0 & 0 & 0 & 0 \\ -\tfrac{1}{2}m_2 l & 0 & 0 & m_2 \end{bmatrix} U_{21}^T \right\}
$$

$$
= \tfrac{1}{3}m_1 l^2 + \tfrac{4}{3}m_2 l^2 + m_2 C_2 l^2
$$

For D_{12} we have

$$
D_{12} = D_{21} = \mathrm{Tr}(U_{22}J_2 U_{21}^T)
$$

$$
= \mathrm{Tr}\left\{ \begin{bmatrix} -S_{12} & -C_{12} & 0 & -lS_{12} \\ C_{12} & -S_{12} & 0 & lC_{12} \\ 0 & 0 & 0 & 0 \\ 0 & 0 & 0 & 0 \end{bmatrix} \begin{bmatrix} \tfrac{1}{3}m_2 l^2 & 0 & 0 & -\tfrac{1}{2}m_2 l \\ 0 & 0 & 0 & 0 \\ 0 & 0 & 0 & 0 \\ -\tfrac{1}{2}m_2 l & 0 & 0 & m_2 \end{bmatrix} U_{21}^T \right\}
$$

$$
= m_2 l^2(-\tfrac{1}{6} + \tfrac{1}{2} + \tfrac{1}{2}C_2) = \tfrac{1}{3}m_2 l^2 + \tfrac{1}{2}m_2 l^2 C_2
$$

For D_{22} we have

$$D_{22} = \text{Tr}\,(\mathbf{U}_{22}\mathbf{J}_2\mathbf{U}_{22}^T)$$

$$= \text{Tr}\left\{\begin{bmatrix} -S_{12} & -C_{12} & 0 & -lS_{12} \\ C_{12} & -S_{12} & 0 & lC_{12} \\ 0 & 0 & 0 & 0 \\ 0 & 0 & 0 & 0 \end{bmatrix}\begin{bmatrix} \tfrac{1}{3}m_2 l^2 & 0 & 0 & -\tfrac{1}{2}m_2 l \\ 0 & 0 & 0 & 0 \\ 0 & 0 & 0 & 0 \\ -\tfrac{1}{2}m_2 l & 0 & 0 & m_2 \end{bmatrix}\mathbf{U}_{22}^T\right\}$$

$$= \tfrac{1}{3}m_2 l^2 S_{12}^2 + \tfrac{1}{3}m_2 l^2 C_{12}^2 = \tfrac{1}{3}m_2 l^2$$

To derive the Coriolis and centrifugal terms, we use Eq. (3.2-32). For $i = 1$, using Eq. (3.2-32), we have

$$h_1 = \sum_{k=1}^{2}\sum_{m=1}^{2} h_{1km}\dot{\theta}_k\dot{\theta}_m = h_{111}\dot{\theta}_1^2 + h_{112}\dot{\theta}_1\dot{\theta}_2 + h_{121}\dot{\theta}_1\dot{\theta}_2 + h_{122}\dot{\theta}_2^2$$

Using Eq. (3.2-33), we can obtain the value of h_{ikm}. Therefore, the above value which corresponds to joint 1 is

$$h_1 = -\tfrac{1}{2}m_2 S_2 l^2 \dot{\theta}_2^2 - m_2 S_2 l^2 \dot{\theta}_1\dot{\theta}_2$$

Similarly, for $i = 2$ we have

$$h_2 = \sum_{k=1}^{2}\sum_{m=1}^{2} h_{2km}\dot{\theta}_k\dot{\theta}_m = h_{211}\dot{\theta}_1^2 + h_{212}\dot{\theta}_1\dot{\theta}_2 + h_{221}\dot{\theta}_2\dot{\theta}_1 + h_{222}\dot{\theta}_2^2$$

$$= \tfrac{1}{2}m_2 S_2 l^2 \dot{\theta}_1^2$$

Therefore,

$$\mathbf{h}(\theta, \dot{\theta}) = \begin{bmatrix} -\tfrac{1}{2}\,m_2 S_2 l^2 \dot{\theta}_2^2 - m_2 S_2 l^2 \dot{\theta}_1\dot{\theta}_2 \\ \tfrac{1}{2}\,m_2 S_2 l^2 \dot{\theta}_1^2 \end{bmatrix}$$

Next, we need to derive the gravity-related terms, $\mathbf{c} = (c_1, c_2)^T$. Using Eq. (3.2-34), we have:

$$c_1 = -(m_1\mathbf{g}\mathbf{U}_{11}\,{}^1\bar{\mathbf{r}}_1 + m_2\mathbf{g}\mathbf{U}_{21}\,{}^2\bar{\mathbf{r}}_2)$$

$$= -m_1(0, -g, 0, 0)\begin{bmatrix} -S_1 & -C_1 & 0 & -lS_1 \\ C_1 & -S_1 & 0 & lC_1 \\ 0 & 0 & 0 & 0 \\ 0 & 0 & 0 & 0 \end{bmatrix}\begin{bmatrix} -\dfrac{l}{2} \\ 0 \\ 0 \\ 1 \end{bmatrix}$$

$$- m_2 (0, -g, 0, 0) \begin{bmatrix} -S_{12} & -C_{12} & 0 & -l(S_{12} + S_1) \\ C_{12} & -S_{12} & 0 & l(C_{12} + C_1) \\ 0 & 0 & 0 & 0 \\ 0 & 0 & 0 & 0 \end{bmatrix} \begin{bmatrix} -\dfrac{l}{2} \\ 0 \\ 0 \\ 1 \end{bmatrix}$$

$$= \tfrac{1}{2} m_1 glC_1 + \tfrac{1}{2} m_2 glC_{12} + m_2 glC_1$$

$$c_2 = -m_2 \mathbf{g} \mathbf{U}_{22}{}^2 \bar{\mathbf{r}}_2$$

$$= -m_2 (0, -g, 0, 0) \begin{bmatrix} -S_{12} & -C_{12} & 0 & -lS_{12} \\ C_{12} & -S_{12} & 0 & lC_{12} \\ 0 & 0 & 0 & 0 \\ 0 & 0 & 0 & 0 \end{bmatrix} \begin{bmatrix} -\dfrac{l}{2} \\ 0 \\ 0 \\ 1 \end{bmatrix}$$

$$= -m_2 (\tfrac{1}{2} glC_{12} - glC_{12})$$

Hence, we obtain the gravity matrix terms:

$$\mathbf{c}(\boldsymbol{\theta}) = \begin{bmatrix} c_1 \\ c_2 \end{bmatrix} = \begin{bmatrix} \tfrac{1}{2} m_1 glC_1 + \tfrac{1}{2} m_2 glC_{12} + m_2 glC_1 \\ \tfrac{1}{2} m_2 glC_{12} \end{bmatrix}$$

Finally, the Lagrange-Euler equations of motion for the two-link manipulator are found to be

$$\boldsymbol{\tau}(\mathbf{t}) = \mathbf{D}(\boldsymbol{\theta}) \ddot{\boldsymbol{\theta}}(\mathbf{t}) + \mathbf{h}(\boldsymbol{\theta}, \dot{\boldsymbol{\theta}}) + \mathbf{c}(\boldsymbol{\theta})$$

$$\begin{bmatrix} \tau_1 \\ \tau_2 \end{bmatrix} = \begin{bmatrix} \tfrac{1}{3} m_1 l^2 + \tfrac{4}{3} m_2 l^2 + m_2 C_2 l^2 & \tfrac{1}{3} m_2 l^2 + \tfrac{1}{2} m_2 l^2 C_2 \\ \tfrac{1}{3} m_2 l^2 + \tfrac{1}{2} m_2 l^2 C_2 & \tfrac{1}{3} m_2 l^2 \end{bmatrix} \begin{bmatrix} \ddot{\theta}_1 \\ \ddot{\theta}_2 \end{bmatrix}$$

$$+ \begin{bmatrix} -\tfrac{1}{2} m_2 S_2 l^2 \dot{\theta}_2^2 - m_2 S_2 l^2 \dot{\theta}_1 \dot{\theta}_2 \\ \tfrac{1}{2} m_2 S_2 l^2 \dot{\theta}_1^2 \end{bmatrix}$$

$$+ \begin{bmatrix} \tfrac{1}{2} m_1 glC_1 + \tfrac{1}{2} m_2 glC_{12} + m_2 glC_1 \\ \tfrac{1}{2} m_2 glC_{12} \end{bmatrix}$$

3.3 NEWTON-EULER FORMULATION

In the previous sections, we have derived a set of nonlinear second-order differential equations from the Lagrange-Euler formulation that describe the dynamic behavior of a robot arm. The use of these equations to compute the nominal joint torques from the given joint positions, velocities, and accelerations for each trajectory set point in real time has been a computational bottleneck in open-loop control. The problem is due mainly to the inefficiency of the Lagrange-Euler equations of motion, which use the 4×4 homogeneous transformation matrices. In order to perform real-time control, a simplified robot arm dynamic model has been proposed which ignores the Coriolis and centrifugal forces. This reduces the computation time for the joint torques to an affordable limit (e.g., less than 10 ms for each trajectory point using a PDP 11/45 computer). However, the Coriolis and centrifugal forces are significant in the joint torques when the arm is moving at fast speeds. Thus, the simplified robot arm dynamics restricts a robot arm motion to slow speeds which are not desirable in the typical manufacturing environment. Furthermore, the errors in the joint torques resulting from ignoring the Coriolis and centrifugal forces cannot be corrected with feedback control when the arm is moving at fast speeds because of excessive requirements on the corrective torques.

As an alternative to deriving more efficient equations of motion, several investigators turned to Newton's second law and developed various forms of Newton-Euler equations of motion for an open kinematic chain (Armstrong [1979], Orin et al. [1979], Luh et al. [1980a], Walker and Orin [1982]). This formulation when applied to a robot arm results in a set of forward and backward recursive equations with "messy" vector cross-product terms. The most significant aspect of this formulation is that the computation time of the applied torques can be reduced significantly to allow real-time control. The derivation is based on the d'Alembert principle and a set of mathematical equations that describe the kinematic relation of the moving links of a robot arm with respect to the base coordinate system. In order to understand the Newton-Euler formulation, we need to review some concepts in moving and rotating coordinate systems.

3.3.1 Rotating Coordinate Systems

In this section, we shall develop the necessary mathematical relation between a rotating coordinate system and a fixed inertial coordinate frame, and then extend the concept to include a discussion of the relationship between a moving coordinate system (rotating and translating) and an inertial frame. From Fig. 3.3, two right-handed coordinate systems, an unstarred coordinate system $OXYZ$ (inertial frame) and a starred coordinate system $OX^*Y^*Z^*$ (rotating frame), whose origins are coincident at a point O, and the axes OX^*, OY^*, OZ^* are rotating relative to the axes OX, OY, OZ, respectively. Let $(\mathbf{i}, \mathbf{j}, \mathbf{k})$ and $(\mathbf{i}^*, \mathbf{j}^*, \mathbf{k}^*)$ be their respective unit vectors along the principal axes. A point $\mathbf{r}$ fixed and at rest in the starred coordinate system can be expressed in terms of its components on either set of axes:

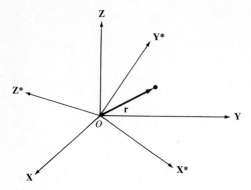

Figure 3.3 The rotating coordinate system.

$$\mathbf{r} = x\mathbf{i} + y\mathbf{j} + z\mathbf{k} \tag{3.3-1}$$

or
$$\mathbf{r} = x^*\mathbf{i}^* + y^*\mathbf{j}^* + z^*\mathbf{k}^* \tag{3.3-2}$$

We would like to evaluate the time derivative of the point $\mathbf{r}$, and because the coordinate systems are rotating with respect to each other, the time derivative of $\mathbf{r}(t)$ can be taken with respect to two different coordinate systems. Let us distinguish these two time derivatives by noting the following notation:

$\dfrac{d(\ )}{dt} \triangleq$ time derivative with respect to the fixed reference coordinate

system which is fixed $=$ time derivative of $\mathbf{r}(t)$ (3.3-3)

$\dfrac{d^*(\ )}{dt} \triangleq$ time derivative with respect to the starred coordinate

system which is rotating $=$ starred derivative of $\mathbf{r}(t)$ (3.3-4)

Then, using Eq. (3.3-1), the time derivative of $\mathbf{r}(t)$ can be expressed as

$$\frac{d\mathbf{r}}{dt} = \dot{x}\mathbf{i} + \dot{y}\mathbf{j} + \dot{z}\mathbf{k} + x\frac{d\mathbf{i}}{dt} + y\frac{d\mathbf{j}}{dt} + z\frac{d\mathbf{k}}{dt}$$

$$= \dot{x}\mathbf{i} + \dot{y}\mathbf{j} + \dot{z}\mathbf{k} \tag{3.3-5}$$

and, using Eq. (3.3-2), the starred derivative of $\mathbf{r}(t)$ is

$$\frac{d^*\mathbf{r}}{dt} = \dot{x}^*\mathbf{i}^* + \dot{y}^*\mathbf{j}^* + \dot{z}^*\mathbf{k}^* + x^*\frac{d^*\mathbf{i}^*}{dt} + y^*\frac{d^*\mathbf{j}^*}{dt} + z^*\frac{d^*\mathbf{k}^*}{dt}$$

$$= \dot{x}^*\mathbf{i}^* + \dot{y}^*\mathbf{j}^* + \dot{z}^*\mathbf{k}^* \tag{3.3-6}$$

Using Eqs. (3.3-2) and (3.3-6), the time derivative of $\mathbf{r}(t)$ can be expressed as

$$\frac{d\mathbf{r}}{dt} = \dot{x}^*\mathbf{i}^* + \dot{y}^*\mathbf{j}^* + \dot{z}^*\mathbf{k}^* + x^*\frac{d\mathbf{i}^*}{dt} + y^*\frac{d\mathbf{j}^*}{dt} + z^*\frac{d\mathbf{k}^*}{dt}$$

$$= \frac{d^*\mathbf{r}}{dt} + x^*\frac{d\mathbf{i}^*}{dt} + y^*\frac{d\mathbf{j}^*}{dt} + z^*\frac{d\mathbf{k}^*}{dt} \tag{3.3-7}$$

In evaluating this derivative, we encounter the difficulty of finding $d\mathbf{i}^*/dt$, $d\mathbf{j}^*/dt$, and $d\mathbf{k}^*/dt$ because the unit vectors, $\mathbf{i}^*$, $\mathbf{j}^*$, and $\mathbf{k}^*$, are rotating with respect to the unit vectors $\mathbf{i}$, $\mathbf{j}$, and $\mathbf{k}$.

In order to find a relationship between the starred and unstarred derivatives, let us suppose that the starred coordinate system is rotating about some axis OQ passing through the origin O, with angular velocity $\boldsymbol{\omega}$ (see Fig. 3.4), then the angular velocity $\boldsymbol{\omega}$ is defined as a vector of magnitude ω directed along the axis OQ in the direction of a right-hand rotation with the starred coordinate system. Consider a vector $\mathbf{s}$ at rest in the starred coordinate system. Its starred derivative is zero, and we would like to show that its unstarred derivative is

$$\frac{d\mathbf{s}}{dt} = \boldsymbol{\omega} \times \mathbf{s} \tag{3.3-8}$$

Since the time derivative of a vector can be expressed as

$$\frac{d\mathbf{s}}{dt} = \lim_{\Delta t \to 0} \frac{\mathbf{s}(t + \Delta t) - \mathbf{s}(t)}{\Delta t} \tag{3.3-9}$$

we can verify the correctness of Eq. (3.3-8) by showing that

$$\frac{d\mathbf{s}}{dt} = \boldsymbol{\omega} \times \mathbf{s} = \lim_{\Delta t \to 0} \frac{\mathbf{s}(t + \Delta t) - \mathbf{s}(t)}{\Delta t} \tag{3.3-10}$$

With reference to Fig. 3.4, and recalling that a vector has magnitude and direction, we need to verify the correctness of Eq. (3.3-10) both in direction and magnitude. The magnitude of $d\mathbf{s}/dt$ is

$$\left| \frac{d\mathbf{s}}{dt} \right| = |\boldsymbol{\omega} \times \mathbf{s}| = \omega s \sin\theta \tag{3.3-11}$$

The above equation is correct because if Δt is small, then

$$|\Delta\mathbf{s}| = (s \sin\theta)(\omega\,\Delta t) \tag{3.3-12}$$

which is obvious in Fig. 3.4. The direction of $\boldsymbol{\omega} \times \mathbf{s}$ can be found from the definition of the vector cross product to be perpendicular to $\mathbf{s}$ and in the plane of the circle as shown in Fig. 3.4.

If Eq. (3.3-8) is applied to the unit vectors $(\mathbf{i}^*, \mathbf{j}^*, \mathbf{k}^*)$, then Eq. (3.3-7) becomes

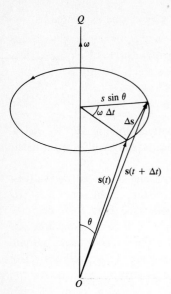

Figure 3.4 Time derivative of a rotating coordinate system.

$$\frac{d\mathbf{r}}{dt} = \frac{d^*\mathbf{r}}{dt} + x^*(\omega \times \mathbf{i}^*) + y^*(\omega \times \mathbf{j}^*) + z^*(\omega \times \mathbf{k}^*)$$

$$= \frac{d^*\mathbf{r}}{dt} + \omega \times \mathbf{r} \qquad (3.3\text{-}13)$$

This is the fundamental equation establishing the relationship between time derivatives for rotating coordinate systems. Taking the derivative of right- and left-hand sides of Eq. (3.3-13) and applying Eq. (3.3-8) again to $\mathbf{r}$ and $d^*\mathbf{r}/dt$, we obtain the second time derivative of the vector $\mathbf{r}(t)$:

$$\frac{d^2\mathbf{r}}{dt^2} = \frac{d}{dt}\left[\frac{d^*\mathbf{r}}{dt}\right] + \omega \times \frac{d\mathbf{r}}{dt} + \frac{d\omega}{dt} \times \mathbf{r}$$

$$= \frac{d^{*2}\mathbf{r}}{dt^2} + \omega \times \frac{d^*\mathbf{r}}{dt} + \omega \times \left[\frac{d^*\mathbf{r}}{dt} + \omega \times \mathbf{r}\right] + \frac{d\omega}{dt} \times \mathbf{r}$$

$$= \frac{d^{*2}\mathbf{r}}{dt^2} + 2\omega \times \frac{d^*\mathbf{r}}{dt} + \omega \times (\omega \times \mathbf{r}) + \frac{d\omega}{dt} \times \mathbf{r} \qquad (3.3\text{-}14)$$

Equation (3.3-14) is called the *Coriolis theorem*. The first term on the right-hand side of the equation is the acceleration relative to the starred coordinate system. The second term is called the *Coriolis acceleration*. The third term is called the *centripetal* (toward the center) *acceleration* of a point in rotation about an axis. One can verify that $\omega \times (\omega \times \mathbf{r})$ points directly toward and perpendicular to the axis of rotation. The last term vanishes for a constant angular velocity of rotation about a fixed axis.

3.3.2 Moving Coordinate Systems

Let us extend the above rotating coordinate systems concept further to include the translation motion of the starred coordinate system with respect to the unstarred coordinate system. From Fig. 3.5, the starred coordinate system $O^*X^*Y^*Z^*$ is rotating and translating with respect to the unstarred coordinate system $OXYZ$ which is an inertial frame. A particle $\mathbf{p}$ with mass m is located by vectors $\mathbf{r}^*$ and $\mathbf{r}$ with respect to the origins of the coordinate frames $O^*X^*Y^*Z^*$ and $OXYZ$, respectively. Origin O^* is located by a vector $\mathbf{h}$ with respect to the origin O. The relation between the position vectors $\mathbf{r}$ and $\mathbf{r}^*$ is given by (Fig. 3.5)

$$\mathbf{r} = \mathbf{r}^* + \mathbf{h} \tag{3.3-15}$$

If the starred coordinate system $O^*X^*Y^*Z^*$ is moving (rotating and translating) with respect to the unstarred coordinate system $OXYZ$, then

$$\mathbf{v}(t) \triangleq \frac{d\mathbf{r}}{dt} = \frac{d\mathbf{r}^*}{dt} + \frac{d\mathbf{h}}{dt} = \mathbf{v}^* + \mathbf{v}_h \tag{3.3-16}$$

where $\mathbf{v}^*$ and $\mathbf{v}$ are the velocities of the moving particle $\mathbf{p}$ relative to the coordinate frames $O^*X^*Y^*Z^*$ and $OXYZ$, respectively, and $\mathbf{v}_h$ is the velocity of the starred coordinate system $O^*X^*Y^*Z^*$ relative to the unstarred coordinate system $OXYZ$. Using Eq. (3.3-13), Eq. (3.3-16) can be expressed as

$$\mathbf{v}(t) = \frac{d\mathbf{r}^*}{dt} + \frac{d\mathbf{h}}{dt} = \frac{d^*\mathbf{r}^*}{dt} + \omega \times \mathbf{r}^* + \frac{d\mathbf{h}}{dt} \tag{3.3-17}$$

Similarly, the acceleration of the particle $\mathbf{p}$ with respect to the unstarred coordinate system is

$$\mathbf{a}(t) = \frac{d\mathbf{v}(t)}{dt} = \frac{d^2\mathbf{r}^*}{dt^2} + \frac{d^2\mathbf{h}}{dt^2} = \mathbf{a}^* + \mathbf{a}_h \tag{3.3-18}$$

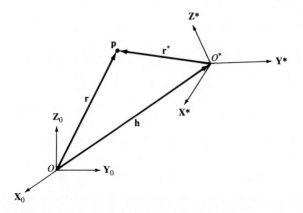

Figure 3.5 Moving coordinate system.

where $\mathbf{a}^*$ and $\mathbf{a}$ are the accelerations of the moving particle $\mathbf{p}$ relative to the coordinate frames $O^*X^*Y^*Z^*$ and $OXYZ$, respectively, and $\mathbf{a}_h$ is the acceleration of the starred coordinate system $O^*X^*Y^*Z^*$ relative to the unstarred coordinate system $OXYZ$. Using Eq. (3.3-14), Eq. (3.3-17) can be expressed as

$$\mathbf{a}(t) = \frac{d^{*2}\mathbf{r}^*}{dt^2} + 2\boldsymbol{\omega} \times \frac{d^*\mathbf{r}^*}{dt} + \boldsymbol{\omega} \times (\boldsymbol{\omega} \times \mathbf{r}^*) + \frac{d\boldsymbol{\omega}}{dt} \times \mathbf{r}^* + \frac{d^2\mathbf{h}}{dt^2} \qquad (3.3\text{-}19)$$

With this introduction to moving coordinate systems, we would like to apply this concept to the link coordinate systems that we established for a robot arm to obtain the kinematics information of the links, and then apply the d'Alembert principle to these translating and/or rotating coordinate systems to derive the motion equations of the robot arm.

3.3.3 Kinematics of the Links

The objective of this section is to derive a set of mathematical equations that, based on the moving coordinate systems described in Sec. 3.3.2, describe the kinematic relationship of the moving-rotating links of a robot arm with respect to the base coordinate system.

With reference to Fig. 3.6, recall that an orthonormal coordinate system $(\mathbf{x}_{i-1}, \mathbf{y}_{i-1}, \mathbf{z}_{i-1})$ is established at joint i. Coordinate system $(\mathbf{x}_0, \mathbf{y}_0, \mathbf{z}_0)$ is then the base coordinate system while the coordinate systems $(\mathbf{x}_{i-1}, \mathbf{y}_{i-1}, \mathbf{z}_{i-1})$ and $(\mathbf{x}_i, \mathbf{y}_i, \mathbf{z}_i)$ are attached to link $i-1$ with origin O^* and link i with origin O', respectively. Origin O' is located by a position vector $\mathbf{p}_i$ with respect to the origin O and by a position vector $\mathbf{p}_i^*$ from the origin O^* with respect to the base coordinate system. Origin O^* is located by a position vector $\mathbf{p}_{i-1}$ from the origin O with respect to the base coordinate system.

Let $\mathbf{v}_{i-1}$ and $\boldsymbol{\omega}_{i-1}$ be the linear and angular velocities of the coordinate system $(\mathbf{x}_{i-1}, \mathbf{y}_{i-1}, \mathbf{z}_{i-1})$ with respect to the base coordinate system $(\mathbf{x}_0, \mathbf{y}_0, \mathbf{z}_0)$, respectively. Let $\boldsymbol{\omega}_i$, and $\boldsymbol{\omega}_i^*$ be the angular velocity of O' with respect to $(\mathbf{x}_0, \mathbf{y}_0, \mathbf{z}_0)$ and $(\mathbf{x}_{i-1}, \mathbf{y}_{i-1}, \mathbf{z}_{i-1})$, respectively. Then, the linear velocity $\mathbf{v}_i$ and the angular velocity $\boldsymbol{\omega}_i$ of the coordinate system $(\mathbf{x}_i, \mathbf{y}_i, \mathbf{z}_i)$ with respect to the base coordinate system are [from Eq. (3.3-17)], respectively,

$$\mathbf{v}_i = \frac{d^*\mathbf{p}_i^*}{dt} + \boldsymbol{\omega}_{i-1} \times \mathbf{p}_i^* + \mathbf{v}_{i-1} \qquad (3.3\text{-}20)$$

and $\qquad \boldsymbol{\omega}_i = \boldsymbol{\omega}_{i-1} + \boldsymbol{\omega}_i^* \qquad\qquad\qquad\qquad\qquad\qquad (3.3\text{-}21)$

where $d^*(\ )/dt$ denotes the time derivative with respect to the moving coordinate system $(\mathbf{x}_{i-1}, \mathbf{y}_{i-1}, \mathbf{z}_{i-1})$. The linear acceleration $\dot{\mathbf{v}}_i$ and the angular acceleration $\dot{\boldsymbol{\omega}}_i$ of the coordinate system $(\mathbf{x}_i, \mathbf{y}_i, \mathbf{z}_i)$ with respect to the base coordinate system are [from Eq. (3.3-19)], respectively,

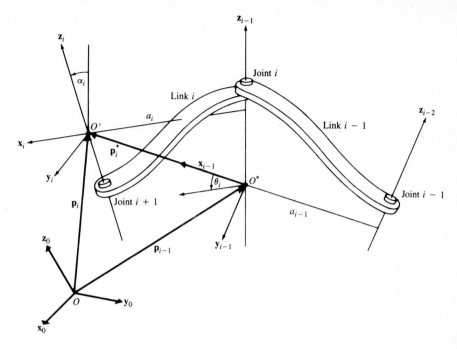

Figure 3.6 Relationship between O, O^* and O' frames.

$$\dot{\mathbf{v}}_i = \frac{d^{*2}\mathbf{p}_i^*}{dt^2} + \dot{\boldsymbol{\omega}}_{i-1} \times \mathbf{p}_i^* + 2\boldsymbol{\omega}_{i-1} \times \frac{d^*\mathbf{p}_i^*}{dt} \qquad (3.3\text{-}22)$$

$$+ \boldsymbol{\omega}_{i-1} \times (\boldsymbol{\omega}_{i-1} \times \mathbf{p}_i^*) + \dot{\mathbf{v}}_{i-1}$$

and

$$\dot{\boldsymbol{\omega}}_i = \dot{\boldsymbol{\omega}}_{i-1} + \dot{\boldsymbol{\omega}}_i^* \qquad (3.3\text{-}23)$$

then, from Eq. (3.3-13), the angular acceleration of the coordinate system $(\mathbf{x}_i, \mathbf{y}_i, \mathbf{z}_i)$ with respect to $(\mathbf{x}_{i-1}, \mathbf{y}_{i-1}, \mathbf{z}_{i-1})$ is

$$\dot{\boldsymbol{\omega}}_i^* = \frac{d^*\boldsymbol{\omega}_i^*}{dt} + \boldsymbol{\omega}_{i-1} \times \boldsymbol{\omega}_i^* \qquad (3.3\text{-}24)$$

therefore, Eq. (3.3-23) can be expressed as

$$\dot{\boldsymbol{\omega}}_i = \dot{\boldsymbol{\omega}}_{i-1} + \frac{d^*\boldsymbol{\omega}_i^*}{dt} + \boldsymbol{\omega}_{i-1} \times \boldsymbol{\omega}_i^* \qquad (3.3\text{-}25)$$

Recalling from the definition of link-joint parameters and the procedure for establishing link coordinate systems for a robot arm, the coordinate systems $(\mathbf{x}_{i-1}, \mathbf{y}_{i-1}, \mathbf{z}_{i-1})$ and $(\mathbf{x}_i, \mathbf{y}_i, \mathbf{z}_i)$ are attached to links $i - 1$ and i, respectively. If link i is translational in the coordinate system $(\mathbf{x}_{i-1}, \mathbf{y}_{i-1}, \mathbf{z}_{i-1})$, it travels in the direction of z_{i-1} with a joint velocity $\dot{q}_i$ relative to link $i-1$. If it is rotational in coordinate system $(x_{i-1}, y_{i-1}, z_{i-1})$, it has an angular velocity of ω_i^* and the angular motion of link i is about the z_{i-1} axis. Therefore,

$$\omega_i^* = \begin{cases} z_{i-1}\dot{q}_i & \text{if link } i \text{ is rotational} \\ 0 & \text{if link } i \text{ is translational} \end{cases} \tag{3.3-26}$$

where $\dot{q}_i$ is the magnitude of angular velocity of link i with respect to the coordinate system $(x_{i-1}, y_{i-1}, z_{i-1})$. Similarly,

$$\frac{d^*\omega_i^*}{dt} = \begin{cases} z_{i-1}\ddot{q}_i & \text{if link } i \text{ is rotational} \\ 0 & \text{if link } i \text{ is translational} \end{cases} \tag{3.3-27}$$

Using Eqs. (3.3-26) and (3.3-27), Eqs. (3.3-21) and (3.3-25) can be expressed, respectively, as

$$\omega_i = \begin{cases} \omega_{i-1} + z_{i-1}\dot{q}_i & \text{if link } i \text{ is rotational} \\ \omega_{i-1} & \text{if link } i \text{ is translational} \end{cases} \tag{3.3-28}$$

$$\dot{\omega}_i = \begin{cases} \dot{\omega}_{i-1} + z_{i-1}\ddot{q}_i + \omega_{i-1} \times (z_{i-1}\dot{q}_i) & \text{if link } i \text{ is rotational} \\ \dot{\omega}_{i-1} & \text{if link } i \text{ is translational} \end{cases} \tag{3.3-29}$$

Using Eq. (3.3-8), the linear velocity and acceleration of link i with respect to link $i-1$ can be obtained, respectively, as

$$\frac{d^*p_i^*}{dt} = \begin{cases} \omega_i^* \times p_i^* & \text{if link } i \text{ is rotational} \\ z_{i-1}\dot{q}_i & \text{if link } i \text{ is translational} \end{cases} \tag{3.3-30}$$

$$\frac{d^{*2}p_i^*}{dt^2} = \begin{cases} \dfrac{d^*\omega_i^*}{dt} \times p_i^* + \omega_i^* \times (\omega_i^* \times p_i^*) & \text{if link } i \text{ is rotational} \\ z_{i-1}\ddot{q}_i & \text{if link } i \text{ is translational} \end{cases} \tag{3.3-31}$$

Therefore, using Eqs. (3.3-30) and (3.3-21), the linear velocity of link i with respect to the reference frame is [from Eq. (3.3-20)]

$$v_i = \begin{cases} \omega_i \times p_i^* + v_{i-1} & \text{if link } i \text{ is rotational} \\ z_{i-1}\dot{q}_i + \omega_i \times p_i^* + v_{i-1} & \text{if link } i \text{ is translational} \end{cases} \tag{3.3-32}$$

Using the following vector cross-product identities,

$$(a \times b) \times c = b(a \cdot c) - a(b \cdot c) \tag{3.3-33}$$

$$\mathbf{a} \times (\mathbf{b} \times \mathbf{c}) = \mathbf{b}(\mathbf{a} \cdot \mathbf{c}) - \mathbf{c}(\mathbf{a} \cdot \mathbf{b}) \tag{3.3-34}$$

and Eqs. (3.3-26) to (3.3-31), the acceleration of link i with respect to the reference system is [from Eq. (3.3-22)]

$$\dot{\mathbf{v}}_i = \begin{cases} \dot{\boldsymbol{\omega}}_i \times \mathbf{p}_i^* + \boldsymbol{\omega}_i \times (\boldsymbol{\omega}_i \times \mathbf{p}_i^*) + \dot{\mathbf{v}}_{i-1} & \text{if link } i \text{ is rotational} \\ \\ \mathbf{z}_{i-1}\ddot{q}_i + \dot{\boldsymbol{\omega}}_i \times \mathbf{p}_i^* + 2\boldsymbol{\omega}_i \times (\mathbf{z}_{i-1}\dot{q}_i) & \text{if link } i \text{ is} \\ + \boldsymbol{\omega}_i \times (\boldsymbol{\omega}_i \times \mathbf{p}_i^*) + \dot{\mathbf{v}}_{i-1} & \text{translational} \end{cases} \tag{3.3-35}$$

Note that $\boldsymbol{\omega}_i = \boldsymbol{\omega}_{i-1}$ if link i is translational in Eq. (3.3-32). Equations (3.3-28), (3.3-29), (3.3-32), and (3.3-35) describe the kinematics information of link i that are useful in deriving the motion equations of a robot arm.

3.3.4 Recursive Equations of Motion for Manipulators

From the above kinematic information of each link, we would like to describe the motion of the robot arm links by applying d'Alembert's principle to each link. d'Alembert's principle applies the conditions of static equilibrium to problems in dynamics by considering both the externally applied driving forces and the reaction forces of mechanical elements which resist motion. d'Alembert's principle applies for all instants of time. It is actually a slightly modified form of Newton's second law of motion, and can be stated as:

> *For any body, the algebraic sum of externally applied forces and the forces resisting motion in any given direction is zero.*

Consider a link i as shown in Fig. 3.7, and let the origin O' be situated at its center of mass. Then, by corresponding the variables defined in Fig. 3.6 with variables defined in Fig. 3.7, the remaining undefined variables, expressed with respect to the base reference system ($\mathbf{x}_0$, $\mathbf{y}_0$, $\mathbf{z}_0$), are:

m_i = total mass of link i

$\bar{\mathbf{r}}_i$ = position of the center of mass of link i from the origin of the base reference frame

$\bar{\mathbf{s}}_i$ = position of the center of mass of link i from the origin of the coordinate system ($\mathbf{x}_i$, $\mathbf{y}_i$, $\mathbf{z}_i$)

$\mathbf{p}_i^*$ = the origin of the ith coordinate frame with respect to the $(i-1)$th coordinate system

$\bar{\mathbf{v}}_i = \dfrac{d\bar{\mathbf{r}}_i}{dt}$, linear velocity of the center of mass of link i

$\bar{\mathbf{a}}_i = \dfrac{d\bar{\mathbf{v}}_i}{dt}$, linear acceleration of the center of mass of link i

$\mathbf{F}_i$ = total external force exerted on link i at the center of mass

$\mathbf{N}_i$ = total external moment exerted on link i at the center of mass
$\mathbf{I}_i$ = inertia matrix of link i about its center of mass with reference to the coordinate system $(\mathbf{x}_0, \mathbf{y}_0, \mathbf{z}_0)$
$\mathbf{f}_i$ = force exerted on link i by link $i - 1$ at the coordinate frame $(\mathbf{x}_{i-1}, \mathbf{y}_{i-1}, \mathbf{z}_{i-1})$ to support link i and the links above it
$\mathbf{n}_i$ = moment exerted on link i by link $i - 1$ at the coordinate frame $(\mathbf{x}_{i-1}, \mathbf{y}_{i-1}, \mathbf{z}_{i-1})$

Then, omitting the viscous damping effects of all the joints, and applying the d'Alembert principle to each link, we have

$$\mathbf{F}_i = \frac{d(m_i \, \bar{\mathbf{v}}_i)}{dt} = m_i \, \bar{\mathbf{a}}_i \tag{3.3-36}$$

and

$$\mathbf{N}_i = \frac{d(\mathbf{I}_i \, \boldsymbol{\omega}_i)}{dt} = \mathbf{I}_i \dot{\boldsymbol{\omega}}_i + \boldsymbol{\omega}_i \times (\mathbf{I}_i \boldsymbol{\omega}_i) \tag{3.3-37}$$

where, using Eqs. (3.3-32) and (3.3-35), the linear velocity and acceleration of the center of mass of link i are, respectively,†

$$\bar{\mathbf{v}}_i = \boldsymbol{\omega}_i \times \bar{\mathbf{s}}_i + \mathbf{v}_i \tag{3.3-38}$$

and

$$\bar{\mathbf{a}}_i = \dot{\boldsymbol{\omega}}_i \times \bar{\mathbf{s}}_i + \boldsymbol{\omega}_i \times (\boldsymbol{\omega}_i \times \bar{\mathbf{s}}_i) + \dot{\mathbf{v}}_i \tag{3.3-39}$$

Then, from Fig. 3.7, and looking at all the forces and moments acting on link i, the total external force $\mathbf{F}_i$ and moment $\mathbf{N}_i$ are those exerted on link i by gravity and neighboring links, link $i - 1$ and link $i + 1$. That is,

$$\mathbf{F}_i = \mathbf{f}_i - \mathbf{f}_{i+1} \tag{3.3-40}$$

and

$$\mathbf{N}_i = \mathbf{n}_i - \mathbf{n}_{i+1} + (\mathbf{p}_{i-1} - \bar{\mathbf{r}}_i) \times \mathbf{f}_i - (\mathbf{p}_i - \bar{\mathbf{r}}_i) \times \mathbf{f}_{i+1} \tag{3.3-41}$$

$$= \mathbf{n}_i - \mathbf{n}_{i+1} + (\mathbf{p}_{i-1} - \bar{\mathbf{r}}_i) \times \mathbf{F}_i - \mathbf{p}_i^* \times \mathbf{f}_{i+1} \tag{3.3-42}$$

Then, the above equations can be rewritten into recursive equations using the fact that $\bar{\mathbf{r}}_i - \mathbf{p}_{i-1} = \mathbf{p}_i^* + \bar{\mathbf{s}}_i$

$$\mathbf{f}_i = \mathbf{F}_i + \mathbf{f}_{i+1} = m_i \bar{\mathbf{a}}_i + \mathbf{f}_{i+1} \tag{3.3-43}$$

and

$$\mathbf{n}_i = \mathbf{n}_{i+1} + \mathbf{p}_i^* \times \mathbf{f}_{i+1} + (\mathbf{p}_i^* + \bar{\mathbf{s}}_i) \times \mathbf{F}_i + \mathbf{N}_i \tag{3.3-44}$$

The above equations are recursive and can be used to derive the forces and moments $(\mathbf{f}_i, \mathbf{n}_i)$ at the links for $i = 1, 2, \ldots, n$ for an n-link manipulator, noting that $\mathbf{f}_{n+1}$ and $\mathbf{n}_{n+1}$ are, respectively, the forces and moments exerted by the manipulator hand upon an external object.

From Chap. 2, the kinematic relationship between the neighboring links and the establishment of coordinate systems show that if joint i is rotational, then it

---

† Here $(\mathbf{x}_i, \mathbf{y}_i, \mathbf{z}_i)$ is the moving-rotating coordinate frame.

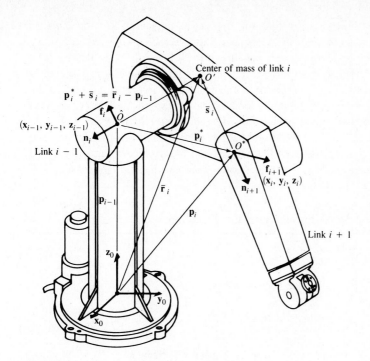

Figure 3.7 Forces and moments on link i.

actually rotates q_i radians in the coordinate system ($\mathbf{x}_{i-1}$, $\mathbf{y}_{i-1}$, $\mathbf{z}_{i-1}$) about the $\mathbf{z}_{i-1}$ axis. Thus, the input torque at joint i is the sum of the projection of $\mathbf{n}_i$ onto the $\mathbf{z}_{i-1}$ axis and the viscous damping moment in that coordinate system. However, if joint i is translational, then it translates q_i unit relative to the coordinate system ($\mathbf{x}_{i-1}$, $\mathbf{y}_{i-1}$, $\mathbf{z}_{i-1}$) along the $\mathbf{z}_{i-1}$ axis. Then, the input force τ_i at that joint is the sum of the projection of $\mathbf{f}_i$ onto the $\mathbf{z}_{i-1}$ axis and the viscous damping force in that coordinate system. Hence, the input torque/force for joint i is

$$
\tau_i =
\begin{cases}
\mathbf{n}_i^T \mathbf{z}_{i-1} + b_i \dot{q}_i & \text{if link } i \text{ is rotational} \\[2mm]
\mathbf{f}_i^T \mathbf{z}_{i-1} + b_i \dot{q}_i & \text{if link } i \text{ is translational}
\end{cases}
\tag{3.3-45}
$$

where b_i is the viscous damping coefficient for joint i in the above equations.

If the supporting base is bolted on the platform and link 0 is stationary, then $\boldsymbol{\omega}_0 = \dot{\boldsymbol{\omega}}_0 = 0$ and $\mathbf{v}_0 = 0$ and $\dot{\mathbf{v}}_0$ (to include gravity) is

$$
\dot{\mathbf{v}}_0 = \mathbf{g} =
\begin{bmatrix}
g_x \\
g_y \\
g_z
\end{bmatrix}
\qquad \text{where} \, |\mathbf{g}| = 9.8062 \text{ m/s}^2
\tag{3.3-46}
$$

In summary, the Newton-Euler equations of motion consist of a set of forward and backward recursive equations. They are Eqs. (3.3-28), (3.3-29), (3.3-35), (3.3-39), and (3.3-43) to (3.3-45) and are listed in Table 3.2. For the forward recursive equations, linear velocity and acceleration, angular velocity and acceleration of each individual link, are propagated from the base reference system to the end-effector. For the backward recursive equations, the torques and forces exerted on each link are computed recursively from the end-effector to the base reference system. Hence, the forward equations propagate kinematics information of each link from the base reference frame to the hand, while the backward equations compute the necessary torques/forces for each joint from the hand to the base reference system.

Table 3.2 Recursive Newton-Euler equations of motion

Forward equations: $i = 1, 2, \ldots, n$

$$
\boldsymbol{\omega}_i = \begin{cases} \boldsymbol{\omega}_{i-1} + \mathbf{z}_{i-1}\dot{q}_i & \text{if link } i \text{ is rotational} \\[2ex] \boldsymbol{\omega}_{i-1} & \text{if link } i \text{ is translational} \end{cases}
$$

$$
\dot{\boldsymbol{\omega}}_i = \begin{cases} \dot{\boldsymbol{\omega}}_{i-1} + \mathbf{z}_{i-1}\ddot{q}_i + \boldsymbol{\omega}_{i-1} \times (\mathbf{z}_{i-1}\dot{q}_i) & \text{if link } i \text{ is rotational} \\[2ex] \dot{\boldsymbol{\omega}}_{i-1} & \text{if link } i \text{ is translational} \end{cases}
$$

$$
\dot{\mathbf{v}}_i = \begin{cases} \dot{\boldsymbol{\omega}}_i \times \mathbf{p}_i^* + \boldsymbol{\omega}_i \times (\boldsymbol{\omega}_i \times \mathbf{p}_i^*) + \dot{\mathbf{v}}_{i-1} & \text{if link } i \text{ is rotational} \\[2ex] \mathbf{z}_{i-1}\ddot{q}_i + \dot{\boldsymbol{\omega}}_i \times \mathbf{p}_i^* + 2\boldsymbol{\omega}_i \times (\mathbf{z}_{i-1}\dot{q}_i) & \\ \quad + \boldsymbol{\omega}_i \times (\boldsymbol{\omega}_i \times \mathbf{p}_i^*) + \dot{\mathbf{v}}_{i-1} & \text{if link } i \text{ is translational} \end{cases}
$$

$$
\bar{\mathbf{a}}_i = \dot{\boldsymbol{\omega}}_i \times \bar{\mathbf{s}}_i + \boldsymbol{\omega}_i \times (\boldsymbol{\omega}_i \times \bar{\mathbf{s}}_i) + \dot{\mathbf{v}}_i
$$

Backward equations: $i = n, n-1, \ldots, 1$

$$
\mathbf{F}_i = m_i \bar{\mathbf{a}}_i
$$

$$
\mathbf{N}_i = \mathbf{I}_i \dot{\boldsymbol{\omega}}_i + \boldsymbol{\omega}_i \times (\mathbf{I}_i \boldsymbol{\omega}_i)
$$

$$
\mathbf{f}_i = \mathbf{F}_i + \mathbf{f}_{i+1}
$$

$$
\mathbf{n}_i = \mathbf{n}_{i+1} + \mathbf{p}_i^* \times \mathbf{f}_{i+1} + (\mathbf{p}_i^* + \bar{\mathbf{s}}_i) \times \mathbf{F}_i + \mathbf{N}_i
$$

$$
\tau_i = \begin{cases} \mathbf{n}_i^T \mathbf{z}_{i-1} + b_i \dot{q}_i & \text{if link } i \text{ is rotational} \\[2ex] \mathbf{f}_i^T \mathbf{z}_{i-1} + b_i \dot{q}_i & \text{if link } i \text{ is translational} \end{cases}
$$

where b_i is the viscous damping coefficient for joint i.

The "usual" initial conditions are $\boldsymbol{\omega}_0 = \dot{\boldsymbol{\omega}}_0 = \mathbf{v}_0 = \mathbf{0}$ and $\dot{\mathbf{v}}_0 = (g_x, g_y, g_z)^T$ (to include gravity), where $|\mathbf{g}| = 9.8062$ m/s^2.

3.3.5 Recursive Equations of Motion of a Link About Its Own Coordinate Frame

The above equations of motion of a robot arm indicate that the resulting N-E dynamic equations, excluding gear friction, are a set of compact forward and backward recursive equations. This set of recursive equations can be applied to the robot links sequentially. The forward recursion propagates kinematics information such as angular velocities, angular accelerations, and linear accelerations from the base reference frame (inertial frame) to the end-effector. The backward recursion propagates the forces exerted on each link from the end-effector of the manipulator to the base reference frame, and the applied joint torques are computed from these forces.

One obvious drawback of the above recursive equations of motion is that all the inertial matrices $\mathbf{I}_i$ and the physical geometric parameters $(\bar{\mathbf{r}}_i, \bar{\mathbf{s}}_i, \mathbf{p}_{i-1}, \mathbf{p}_i^*)$ are referenced to the base coordinate system. As a result, they change as the robot arm is moving. Luh et al. [1980a] improved the above N-E equations of motion by referencing all velocities, accelerations, inertial matrices, location of center of mass of each link, and forces/moments to their own link coordinate systems. Because of the nature of the formulation and the method of systematically computing the joint torques, computations are much simpler. The most important consequence of this modification is that the computation time of the applied torques is found linearly proportional to the number of joints of the robot arm and independent of the robot arm configuration. This enables the implementation of a simple real-time control algorithm for a robot arm in the joint-variable space.

Let $^{i-1}\mathbf{R}_i$ be a 3×3 rotation matrix which transforms any vector with reference to coordinate frame $(\mathbf{x}_i, \mathbf{y}_i, \mathbf{z}_i)$ to the coordinate system $(\mathbf{x}_{i-1}, \mathbf{y}_{i-1}, \mathbf{z}_{i-1})$. This is the upper left 3×3 submatrix of $^{i-1}\mathbf{A}_i$.

It has been shown before that

$$(^{i-1}\mathbf{R}_i)^{-1} = {}^i\mathbf{R}_{i-1} = (^{i-1}\mathbf{R}_i)^T \tag{3.3-47}$$

where

$$^{i-1}\mathbf{R}_i = \begin{bmatrix} \cos\theta_i & -\cos\alpha_i\sin\theta_i & \sin\alpha_i\sin\theta_i \\ \sin\theta_i & \cos\alpha_i\cos\theta_i & -\sin\alpha_i\cos\theta_i \\ 0 & \sin\alpha_i & \cos\alpha_i \end{bmatrix} \tag{3.3-48}$$

and

$$[^{i-1}\mathbf{R}_i]^{-1} = \begin{bmatrix} \cos\theta_i & \sin\theta_i & 0 \\ -\cos\alpha_i\sin\theta_i & \cos\alpha_i\cos\theta_i & \sin\alpha_i \\ \sin\alpha_i\sin\theta_i & -\sin\alpha_i\cos\theta_i & \cos\alpha_i \end{bmatrix} \tag{3.3-49}$$

Instead of computing ω_i, $\dot{\omega}_i$, $\dot{\mathbf{v}}_i$, $\bar{\mathbf{a}}_i$, $\mathbf{p}_i^*$, $\bar{\mathbf{s}}_i$, $\mathbf{F}_i$, $\mathbf{N}_i$, $\mathbf{f}_i$, $\mathbf{n}_i$, and τ_i which are referenced to the base coordinate system, we compute $^i\mathbf{R}_0\omega_i$, $^i\mathbf{R}_0\dot{\omega}_i$, $^i\mathbf{R}_0\dot{\mathbf{v}}_i$, $^i\mathbf{R}_0\bar{\mathbf{a}}_i$, $^i\mathbf{R}_0\mathbf{F}_i$, $^i\mathbf{R}_0\mathbf{N}_i$, $^i\mathbf{R}_0\mathbf{f}_i$, $^i\mathbf{R}_0\mathbf{n}_i$, and $^i\mathbf{R}_0\tau_i$ which are referenced to its own link coordinate system $(\mathbf{x}_i, \mathbf{y}_i, \mathbf{z}_i)$. Hence, Eqs. (3.3-28), (3.3-29), (3.3-35), (3.3-39), (3.3-36), (3.3-37), (3.3-43), (3.3-44), and (3.3-45), respectively, become:

$$
{}^{i}\mathbf{R}_0\omega_i = \begin{cases} {}^{i}\mathbf{R}_{i-1}({}^{i-1}\mathbf{R}_0\omega_{i-1} + \mathbf{z}_0\dot{q}_i) & \text{if link } i \text{ is rotational} \\ \\ {}^{i}\mathbf{R}_{i-1}({}^{i-1}\mathbf{R}_0\omega_{i-1}) & \text{if link } i \text{ is translational} \end{cases} \tag{3.3-50}
$$

$$
{}^{i}\mathbf{R}_0\dot{\omega}_i = \begin{cases} {}^{i}\mathbf{R}_{i-1}[{}^{i-1}\mathbf{R}_0\dot{\omega}_{i-1} + \mathbf{z}_0\ddot{q}_i + ({}^{i-1}\mathbf{R}_0\omega_{i-1}) \times \mathbf{z}_0\dot{q}_i] & \begin{array}{l}\text{if link } i \text{ is} \\ \text{rotational}\end{array} \\ \\ {}^{i}\mathbf{R}_{i-1}({}^{i-1}\mathbf{R}_0\dot{\omega}_{i-1}) & \begin{array}{l}\text{if link } i \text{ is} \\ \text{translational}\end{array} \end{cases} \tag{3.3-51}
$$

$$
{}^{i}\mathbf{R}_0\dot{\mathbf{v}}_i = \begin{cases} ({}^{i}\mathbf{R}_0\dot{\omega}_i) \times ({}^{i}\mathbf{R}_0\mathbf{p}_i^*) + ({}^{i}\mathbf{R}_0\omega_i) \\ \quad \times [({}^{i}\mathbf{R}_0\omega_i) \times ({}^{i}\mathbf{R}_0\mathbf{p}_i^*)] + {}^{i}\mathbf{R}_{i-1}({}^{i-1}\mathbf{R}_0\dot{\mathbf{v}}_{i-1}) & \text{if link } i \text{ is rotational} \\ \\ {}^{i}\mathbf{R}_{i-1}(\mathbf{z}_0\ddot{q}_i + {}^{i-1}\mathbf{R}_0\dot{\mathbf{v}}_{i-1}) + ({}^{i}\mathbf{R}_0\dot{\omega}_i) \times ({}^{i}\mathbf{R}_0\mathbf{p}_i^*) \\ \quad + 2({}^{i}\mathbf{R}_0\omega_i) \times ({}^{i}\mathbf{R}_{i-1}\mathbf{z}_0\dot{q}_i) \\ \quad + ({}^{i}\mathbf{R}_0\omega_i) \times [({}^{i}\mathbf{R}_0\omega_i) \times ({}^{i}\mathbf{R}_0\mathbf{p}_i^*)] & \text{if link } i \text{ is translational} \end{cases} \tag{3.3-52}
$$

$$
{}^{i}\mathbf{R}_0\bar{\mathbf{a}}_i = ({}^{i}\mathbf{R}_0\dot{\omega}_i) \times ({}^{i}\mathbf{R}_0\bar{\mathbf{s}}_i) + ({}^{i}\mathbf{R}_0\omega_i) \times [({}^{i}\mathbf{R}_0\omega_i) \times ({}^{i}\mathbf{R}_0\bar{\mathbf{s}}_i)] + {}^{i}\mathbf{R}_0\dot{\mathbf{v}}_i \tag{3.3-53}
$$

$$
{}^{i}\mathbf{R}_0\mathbf{F}_i = m_i{}^{i}\mathbf{R}_0\bar{\mathbf{a}}_i \tag{3.3-54}
$$

$$
{}^{i}\mathbf{R}_0\mathbf{N}_i = ({}^{i}\mathbf{R}_0\mathbf{I}_i{}^{0}\mathbf{R}_i)({}^{i}\mathbf{R}_0\dot{\omega}_i) + ({}^{i}\mathbf{R}_0\omega_i) \times [({}^{i}\mathbf{R}_0\mathbf{I}_i{}^{0}\mathbf{R}_i)({}^{i}\mathbf{R}_0\omega_i)] \tag{3.3-55}
$$

$$
{}^{i}\mathbf{R}_0\mathbf{f}_i = {}^{i}\mathbf{R}_{i+1}({}^{i+1}\mathbf{R}_0\mathbf{f}_{i+1}) + {}^{i}\mathbf{R}_0\mathbf{F}_i \tag{3.3-56}
$$

$$
{}^{i}\mathbf{R}_0\mathbf{n}_i = {}^{i}\mathbf{R}_{i+1}[{}^{i+1}\mathbf{R}_0\mathbf{n}_{i+1} + ({}^{i+1}\mathbf{R}_0\mathbf{p}_i^*) \times ({}^{i+1}\mathbf{R}_0\mathbf{f}_{i+1})]
$$
$$
\quad + ({}^{i}\mathbf{R}_0\mathbf{p}_i^* + {}^{i}\mathbf{R}_0\bar{\mathbf{s}}_i) \times ({}^{i}\mathbf{R}_0\mathbf{F}_i) + {}^{i}\mathbf{R}_0\mathbf{N}_i \tag{3.3-57}
$$

and
$$
\tau_i = \begin{cases} ({}^{i}\mathbf{R}_0\mathbf{n}_i)^T({}^{i}\mathbf{R}_{i-1}\mathbf{z}_0) + b_i\dot{q}_i & \text{if link } i \text{ is rotational} \\ \\ ({}^{i}\mathbf{R}_0\mathbf{f}_i)^T({}^{i}\mathbf{R}_{i-1}\mathbf{z}_0) + b_i\dot{q}_i & \text{if link } i \text{ is translational} \end{cases} \tag{3.3-58}
$$

where $\mathbf{z}_0 = (0, 0, 1)^T$, ${}^{i}\mathbf{R}_0\bar{\mathbf{s}}_i$ is the center of mass of link i referred to its own link coordinate system $(\mathbf{x}_i, \mathbf{y}_i, \mathbf{z}_i)$, and ${}^{i}\mathbf{R}_0\mathbf{p}_i^*$ is the location of $(\mathbf{x}_i, \mathbf{y}_i, \mathbf{z}_i)$ from the origin of $(\mathbf{x}_{i-1}, \mathbf{y}_{i-1}, \mathbf{z}_{i-1})$ with respect to the ith coordinate frame and is found to be

$$
{}^i\mathbf{R}_0\, \mathbf{p}_i^* \;=\;
\begin{bmatrix}
a_i \\[4pt]
d_i \sin \alpha_i \\[4pt]
d_i \cos \alpha_i
\end{bmatrix}
\tag{3.3-59}
$$

and $({}^i\mathbf{R}_0 \mathbf{I}_i{}^0\mathbf{R}_i)$ is the inertia matrix of link i about its center of mass referred to its own link coordinate system ($\mathbf{x}_i$, $\mathbf{y}_i$, $\mathbf{z}_i$).

Hence, in summary, efficient Newton-Euler equations of motion are a set of forward and backward recursive equations with the dynamics and kinematics of each link referenced to its own coordinate system. A list of the recursive equations are found in Table 3.3.

3.3.6 Computational Algorithm

The Newton-Euler equations of motion represent the most efficient set of computational equations running on a uniprocessor computer at the present time. The computational complexity of the Newton-Euler equations of motion has been tabulated in Table 3.4. The total mathematical operations (multiplications and additions) are proportional to n, the number of degrees of freedom of the robot arm.

Since the equations of motion obtained are recursive in nature, it is advisable to state an algorithmic approach for computing the input joint torque/force for each joint actuator. Such an algorithm is given below.

Algorithm 3.1: Newton-Euler approach. Given an n-link manipulator, this computational procedure generates the nominal joint torque/force for all the joint actuators. Computations are based on the equations in Table 3.3.

Initial conditions:

$n \;=\;$ number of links (n joints)

$\boldsymbol{\omega}_0 = \dot{\boldsymbol{\omega}}_0 = \mathbf{v}_0 = \mathbf{0} \qquad \dot{\mathbf{v}}_0 = \mathbf{g} = (g_x,\, g_y,\, g_z)^T$ where $|\mathbf{g}| = 9.8062$ m/s^2

Joint variables are q_i, $\dot{q}_i$, $\ddot{q}_i$ for $i = 1, 2, \ldots, n$

Link variables are i, $\mathbf{F}_i$, $\mathbf{f}_i$, $\mathbf{n}_i$, τ_i

Forward iterations:

N1. [*Set counter for iteration*] Set $i \leftarrow 1$.

N2. [*Forward iteration for kinematics information*] Compute ${}^i\mathbf{R}_0\,\boldsymbol{\omega}_i$, ${}^i\mathbf{R}_0\dot{\boldsymbol{\omega}}_i$, ${}^i\mathbf{R}_0\dot{\mathbf{v}}_i$, and ${}^i\mathbf{R}_0\bar{\mathbf{a}}_i$ using equations in Table 3.3.

N3. [*Check $i = n$?*] If $i = n$, go to step N4; otherwise set $i \leftarrow i + 1$ and return to step N2.

Table 3.3 Efficient recursive Newton-Euler equations of motion

Forward equations: $i = 1, 2, \ldots, n$

$$
^i\mathbf{R}_0\boldsymbol{\omega}_i =
\begin{cases}
^i\mathbf{R}_{i-1}(^{i-1}\mathbf{R}_0\boldsymbol{\omega}_{i-1} + \mathbf{z}_0\dot{q}_i) & \text{if link } i \text{ is rotational} \\[2ex]
^i\mathbf{R}_{i-1}(^{i-1}\mathbf{R}_0\boldsymbol{\omega}_{i-1}) & \text{if link } i \text{ is translational}
\end{cases}
$$

$$
^i\mathbf{R}_0\dot{\boldsymbol{\omega}}_i =
\begin{cases}
^i\mathbf{R}_{i-1}[^{i-1}\mathbf{R}_0\dot{\boldsymbol{\omega}}_{i-1} + \mathbf{z}_0\ddot{q}_i + (^{i-1}\mathbf{R}_0\boldsymbol{\omega}_{i-1}) \times \mathbf{z}_0\dot{q}_i] & \text{if link } i \text{ is rotational} \\[2ex]
^i\mathbf{R}_{i-1}(^{i-1}\mathbf{R}_0\dot{\boldsymbol{\omega}}_{i-1}) & \text{if link } i \text{ is translational}
\end{cases}
$$

$$
^i\mathbf{R}_0\dot{\mathbf{v}}_i =
\begin{cases}
(^i\mathbf{R}_0\dot{\boldsymbol{\omega}}_i) \times (^i\mathbf{R}_0\mathbf{p}_i^*) + (^i\mathbf{R}_0\boldsymbol{\omega}_i) \times [\,(^i\mathbf{R}_0\boldsymbol{\omega}_i) \times (^i\mathbf{R}_0\mathbf{p}_i^*)\,] \\
\quad + \,^i\mathbf{R}_{i-1}(^{i-1}\mathbf{R}_0\dot{\mathbf{v}}_{i-1}) \quad \text{if link } i \text{ is rotational} \\[2ex]
^i\mathbf{R}_{i-1}(\mathbf{z}_0\ddot{q}_i + \,^{i-1}\mathbf{R}_0\dot{\mathbf{v}}_{i-1}) + (^i\mathbf{R}_0\dot{\boldsymbol{\omega}}_i) \times (^i\mathbf{R}_0\mathbf{p}_i^*) \\
\quad + \, 2(^i\mathbf{R}_0\boldsymbol{\omega}_i) \times (^i\mathbf{R}_{i-1}\mathbf{z}_0\dot{q}_i) \\
\quad + (^i\mathbf{R}_0\boldsymbol{\omega}_i) \times [\,(^i\mathbf{R}_0\boldsymbol{\omega}_i) \times (^i\mathbf{R}_0\mathbf{p}_i^*)\,] \quad \text{if link } i \text{ is translational}
\end{cases}
$$

$$
^i\mathbf{R}_0\bar{\mathbf{a}}_i = (^i\mathbf{R}_0\dot{\boldsymbol{\omega}}_i) \times (^i\mathbf{R}_0\bar{\mathbf{s}}_i) + (^i\mathbf{R}_0\boldsymbol{\omega}_i) \times [(^i\mathbf{R}_0\boldsymbol{\omega}_i) \times (^i\mathbf{R}_0\bar{\mathbf{s}}_i)] + \,^i\mathbf{R}_0\dot{\mathbf{v}}_i
$$

Backward equations: $i = n, n-1, \ldots, 1$

$$
^i\mathbf{R}_0\mathbf{f}_i = \,^i\mathbf{R}_{i+1}(^{i+1}\mathbf{R}_0\mathbf{f}_{i+1}) + m_i{}^i\mathbf{R}_0\bar{\mathbf{a}}_i
$$

$$
^i\mathbf{R}_0\mathbf{n}_i = \,^i\mathbf{R}_{i+1}[^{i+1}\mathbf{R}_0\mathbf{n}_{i+1} + (^{i+1}\mathbf{R}_0\mathbf{p}_i^*) \times (^{i+1}\mathbf{R}_0\mathbf{f}_{i+1})] + (^i\mathbf{R}_0\mathbf{p}_i^* + \,^i\mathbf{R}_0\bar{\mathbf{s}}_i) \times (^i\mathbf{R}_0\mathbf{F}_i)
$$
$$
\quad + (^i\mathbf{R}_0\mathbf{I}_i{}^0\mathbf{R}_i)(^i\mathbf{R}_0\dot{\boldsymbol{\omega}}_i) + (^i\mathbf{R}_0\boldsymbol{\omega}_i) \times [(^i\mathbf{R}_0\mathbf{I}_i{}^0\mathbf{R}_i)(^i\mathbf{R}_0\boldsymbol{\omega}_i)]
$$

$$
\tau_i =
\begin{cases}
(^i\mathbf{R}_0\mathbf{n}_i)^T(^i\mathbf{R}_{i-1}\mathbf{z}_0) + b_i\dot{q}_i & \text{if link } i \text{ is rotational} \\[2ex]
(^i\mathbf{R}_0\mathbf{f}_i)^T(^i\mathbf{R}_{i-1}\mathbf{z}_0) + b_i\dot{q}_i & \text{if link } i \text{ is translational}
\end{cases}
$$

where $\mathbf{z}_0 = (0, 0, 1)^T$ and b_i is the viscous damping coefficient for joint i. The usual initial conditions are $\boldsymbol{\omega}_0 = \dot{\boldsymbol{\omega}}_0 = \mathbf{v}_0 = \mathbf{0}$ and $\dot{\mathbf{v}}_0 = (g_x, g_y, g_z)^T$ (to include gravity), where $|\mathbf{g}| = 9.8062$ m/s^2.

Backward iterations:

N4. [*Set* $\mathbf{f}_{n+1}$ *and* $\mathbf{n}_{n+1}$] Set $\mathbf{f}_{n+1}$ and $\mathbf{n}_{n+1}$ to the required force and moment, respectively, to carry the load. If no load, they are set to zero.

Table 3.4 Breakdown of mathematical operations of the Newton-Euler equations of motion for a PUMA robot arm

Newton-Euler equations of motion	Multiplications	Additions
${}^i\mathbf{R}_0\boldsymbol{\omega}_i$	$9n$†	$7n$
${}^i\mathbf{R}_0\dot{\boldsymbol{\omega}}_i$	$9n$	$9n$
${}^i\mathbf{R}_0\dot{\mathbf{v}}_i$	$27n$	$22n$
${}^i\mathbf{R}_0\bar{\mathbf{a}}_i$	$15n$	$14n$
${}^i\mathbf{R}_0\mathbf{F}_i$	$3n$	0
${}^i\mathbf{R}_0\mathbf{f}_i$	$9(n-1)$	$9n-6$
${}^i\mathbf{R}_0\mathbf{N}_i$	$24n$	$18n$
${}^i\mathbf{R}_0\mathbf{n}_i$	$21n-15$	$24n-15$
Total mathematical operations	$117n-24$	$103n-21$

† n = number of degrees of freedom of the robot arm.

N5. [*Compute joint force/torque*] Compute ${}^i\mathbf{R}_0\mathbf{F}_i$, ${}^i\mathbf{R}_0\mathbf{N}_i$, ${}^i\mathbf{R}_0\mathbf{f}_i$, ${}^i\mathbf{R}_0\mathbf{n}_i$, and τ_i with $\mathbf{f}_{n+1}$ and $\mathbf{n}_{n+1}$ given.

N6. [*Backward iteration*] If $i = 1$, then stop; otherwise set $i \leftarrow i - 1$ and go to step N5.

3.3.7 A Two-Link Manipulator Example

In order to illustrate the use of the N-E equations of motion, the same two-link manipulator with revolute joints as shown in Fig. 3.2 is worked out in this section. All the rotation axes at the joints are along the **z** axis perpendicular to the paper surface. The physical dimensions, center of mass, and mass of each link and coordinate systems are given in Sec. 3.2.6.

First, we obtain the rotation matrices from Fig. 3.2 using Eqs. (3.3-48) and (3.3-49):

$$
{}^0\mathbf{R}_1 = \begin{bmatrix} C_1 & -S_1 & 0 \\ S_1 & C_1 & 0 \\ 0 & 0 & 1 \end{bmatrix} \quad
{}^1\mathbf{R}_2 = \begin{bmatrix} C_2 & -S_2 & 0 \\ S_2 & C_2 & 0 \\ 0 & 0 & 1 \end{bmatrix} \quad
{}^0\mathbf{R}_2 = \begin{bmatrix} C_{12} & -S_{12} & 0 \\ S_{12} & C_{12} & 0 \\ 0 & 0 & 1 \end{bmatrix}
$$

$$
{}^1\mathbf{R}_0 = \begin{bmatrix} C_1 & S_1 & 0 \\ -S_1 & C_1 & 0 \\ 0 & 0 & 1 \end{bmatrix} \quad
{}^2\mathbf{R}_1 = \begin{bmatrix} C_2 & S_2 & 0 \\ -S_2 & C_2 & 0 \\ 0 & 0 & 1 \end{bmatrix} \quad
{}^2\mathbf{R}_0 = \begin{bmatrix} C_{12} & S_{12} & 0 \\ -S_{12} & C_{12} & 0 \\ 0 & 0 & 1 \end{bmatrix}
$$

From the equations in Table 3.3 we assume the following initial conditions:

$$\omega_0 = \dot{\omega}_0 = v_0 = 0 \quad \text{and} \quad \dot{v}_0 = (0, g, 0)^T \quad \text{with } g = 9.8062 \text{ m/s}^2$$

Forward Equations for $i = 1, 2$. Using Eq. (3.3-50), compute the angular velocity for revolute joint for $i = 1, 2$. So for $i = 1$, with $\omega_0 = 0$, we have:

$$^1\mathbf{R}_0\omega_1 = {}^1\mathbf{R}_0(\omega_0 + z_0\dot{\theta}_1)$$

$$= \begin{bmatrix} C_1 & S_1 & 0 \\ -S_1 & C_1 & 0 \\ 0 & 0 & 1 \end{bmatrix} \begin{bmatrix} 0 \\ 0 \\ 1 \end{bmatrix} \dot{\theta}_1 = \begin{bmatrix} 0 \\ 0 \\ 1 \end{bmatrix} \dot{\theta}_1$$

For $i = 2$, we have:

$$^2\mathbf{R}_0\omega_2 = {}^2\mathbf{R}_1({}^1\mathbf{R}_0\omega_1 + z_0\dot{\theta}_2)$$

$$= \begin{bmatrix} C_2 & S_2 & 0 \\ -S_2 & C_2 & 0 \\ 0 & 0 & 1 \end{bmatrix} \left[\begin{bmatrix} 0 \\ 0 \\ 1 \end{bmatrix} \dot{\theta}_1 + \begin{bmatrix} 0 \\ 0 \\ 1 \end{bmatrix} \dot{\theta}_2 \right] = \begin{bmatrix} 0 \\ 0 \\ 1 \end{bmatrix} (\dot{\theta}_1 + \dot{\theta}_2)$$

Using Eq. (3.3-51), compute the angular acceleration for revolute joints for $i = 1, 2$:

For $i = 1$, with $\dot{\omega}_0 = \omega_0 = 0$, we have:

$$^1\mathbf{R}_0\dot{\omega}_1 = {}^1\mathbf{R}_0(\dot{\omega}_0 + z_0\ddot{\theta}_1 + \omega_0 \times z_0\dot{\theta}_1) = (0, 0, 1)^T \ddot{\theta}_1$$

For $i = 2$, we have:

$$^2\mathbf{R}_0\dot{\omega}_2 = {}^2\mathbf{R}_1[{}^1\mathbf{R}_0\dot{\omega}_1 + z_0\ddot{\theta}_2 + ({}^1\mathbf{R}_0\omega_1) \times z_0\dot{\theta}_2] = (0, 0, 1)^T (\ddot{\theta}_1 + \ddot{\theta}_2)$$

Using Eq. (3.3-52), compute the linear acceleration for revolute joints for $i = 1, 2$:

For $i = 1$, with $\dot{v}_0 = (0, g, 0)^T$, we have:

$$^1\mathbf{R}_0\dot{v}_1 = ({}^1\mathbf{R}_0\dot{\omega}_1) \times ({}^1\mathbf{R}_0\mathbf{p}_1^*) + ({}^1\mathbf{R}_0\omega_1) \times [({}^1\mathbf{R}_0\omega_1) \times ({}^1\mathbf{R}_0\mathbf{p}_1^*)] + {}^1\mathbf{R}_0\dot{v}_0$$

$$= \begin{bmatrix} 0 \\ 0 \\ 1 \end{bmatrix} \ddot{\theta}_1 \times \begin{bmatrix} l \\ 0 \\ 0 \end{bmatrix} + \begin{bmatrix} 0 \\ 0 \\ 1 \end{bmatrix} \dot{\theta}_1 \times \left\{ \begin{bmatrix} 0 \\ 0 \\ 1 \end{bmatrix} \dot{\theta}_1 \times \begin{bmatrix} l \\ 0 \\ 0 \end{bmatrix} \right\} + \begin{bmatrix} gS_1 \\ gC_1 \\ 0 \end{bmatrix}$$

$$= \begin{bmatrix} -l\dot{\theta}_1^2 + gS_1 \\ l\ddot{\theta}_1 + gC_1 \\ 0 \end{bmatrix}$$

For $i = 2$, we have:

$${}^2\mathbf{R}_0\dot{\mathbf{v}}_2 = ({}^2\mathbf{R}_0\dot{\boldsymbol{\omega}}_2) \times ({}^2\mathbf{R}_0\mathbf{p}_2^*) + ({}^2\mathbf{R}_0\boldsymbol{\omega}_2) \times [({}^2\mathbf{R}_0\boldsymbol{\omega}_2) \times ({}^2\mathbf{R}_0\mathbf{p}_2^*)] + {}^2\mathbf{R}_1({}^1\mathbf{R}_0\dot{\mathbf{v}}_1)$$

$$= \begin{bmatrix} 0 \\ 0 \\ \ddot{\theta}_1 + \ddot{\theta}_2 \end{bmatrix} \times \begin{bmatrix} l \\ 0 \\ 0 \end{bmatrix} + \begin{bmatrix} 0 \\ 0 \\ \dot{\theta}_1 + \dot{\theta}_2 \end{bmatrix} \times \left\{ \begin{bmatrix} 0 \\ 0 \\ \dot{\theta}_1 + \dot{\theta}_2 \end{bmatrix} \times \begin{bmatrix} l \\ 0 \\ 0 \end{bmatrix} \right\}$$

$$+ \begin{bmatrix} C_2 & S_2 & 0 \\ -S_2 & C_2 & 0 \\ 0 & 0 & 1 \end{bmatrix} \begin{bmatrix} -l\dot{\theta}_1^2 + gS_1 \\ l\ddot{\theta}_1 + gC_1 \\ 0 \end{bmatrix}$$

$$= \begin{bmatrix} l(S_2\ddot{\theta}_1 - C_2\dot{\theta}_1^2 - \dot{\theta}_1^2 - \dot{\theta}_2^2 - 2\dot{\theta}_1\dot{\theta}_2) + gS_{12} \\ l(\ddot{\theta}_1 + \ddot{\theta}_2 + C_2\ddot{\theta}_1 + S_2\dot{\theta}_1^2) + gC_{12} \\ 0 \end{bmatrix}$$

Using Eq. (3.3-53), compute the linear acceleration at the center of mass for links 1 and 2:

For $i = 1$, we have:

$${}^1\mathbf{R}_0\bar{\mathbf{a}}_1 = ({}^1\mathbf{R}_0\dot{\boldsymbol{\omega}}_1) \times ({}^1\mathbf{R}_0\bar{\mathbf{s}}_1) + ({}^1\mathbf{R}_0\boldsymbol{\omega}_1) \times [({}^1\mathbf{R}_0\boldsymbol{\omega}_1) \times ({}^1\mathbf{R}_0\bar{\mathbf{s}}_1)] + {}^1\mathbf{R}_0\dot{\mathbf{v}}_1$$

where

$$\bar{\mathbf{s}}_1 = \begin{bmatrix} -\dfrac{l}{2}C_1 \\ -\dfrac{l}{2}S_1 \\ 0 \end{bmatrix} \qquad {}^1\mathbf{R}_0\bar{\mathbf{s}}_1 = \begin{bmatrix} C_1 & S_1 & 0 \\ -S_1 & C_1 & 0 \\ 0 & 0 & 1 \end{bmatrix} \begin{bmatrix} -\dfrac{l}{2}C_1 \\ -\dfrac{l}{2}S_1 \\ 0 \end{bmatrix} = \begin{bmatrix} -\dfrac{l}{2} \\ 0 \\ 0 \end{bmatrix}$$

Thus,

$${}^1\mathbf{R}_0\bar{\mathbf{a}}_1 = \begin{bmatrix} 0 \\ 0 \\ 1 \end{bmatrix} \ddot{\theta}_1 \times \begin{bmatrix} -\dfrac{l}{2} \\ 0 \\ 0 \end{bmatrix} + \begin{bmatrix} 0 \\ 0 \\ \dot{\theta}_1 \end{bmatrix} \times \left\{ \begin{bmatrix} 0 \\ 0 \\ \dot{\theta}_1 \end{bmatrix} \times \begin{bmatrix} -\dfrac{l}{2} \\ 0 \\ 0 \end{bmatrix} \right\}$$

$$+ \begin{bmatrix} -l\dot{\theta}_1^2 + gS_1 \\ l\ddot{\theta}_1 + gC_1 \\ 0 \end{bmatrix} = \begin{bmatrix} -\dfrac{l}{2}\dot{\theta}_1^2 + gS_1 \\ \dfrac{l}{2}\ddot{\theta}_1 + gC_1 \\ 0 \end{bmatrix}$$

For $i = 2$, we have:

$$^2R_0\bar{a}_2 = (^2R_0\dot{\omega}_2) \times (^2R_0\bar{s}_2) + (^2R_0\omega_2) \times [(^2R_0\omega_2) \times (^2R_0\bar{s}_2)] + {}^2R_0\dot{v}_2$$

where

$$\bar{s}_2 = \begin{bmatrix} -\dfrac{l}{2}C_{12} \\[2mm] -\dfrac{l}{2}S_{12} \\[2mm] 0 \end{bmatrix} \qquad {}^2R_0\bar{s}_2 = \begin{bmatrix} C_{12} & S_{12} & 0 \\ -S_{12} & C_{12} & 0 \\ 0 & 0 & 1 \end{bmatrix} \begin{bmatrix} -\dfrac{l}{2}C_{12} \\[2mm] -\dfrac{l}{2}S_{12} \\[2mm] 0 \end{bmatrix} = \begin{bmatrix} -\dfrac{l}{2} \\[2mm] 0 \\[2mm] 0 \end{bmatrix}$$

Thus,

$$^2R_0\bar{a}_2 = \begin{bmatrix} 0 \\ 0 \\ \ddot{\theta}_1 + \ddot{\theta}_2 \end{bmatrix} \times \begin{bmatrix} -\dfrac{l}{2} \\[2mm] 0 \\[2mm] 0 \end{bmatrix} + \begin{bmatrix} 0 \\ 0 \\ \dot{\theta}_1 + \dot{\theta}_2 \end{bmatrix} \times \begin{bmatrix} 0 \\ 0 \\ \dot{\theta}_1 + \dot{\theta}_2 \end{bmatrix} \times \begin{bmatrix} -\dfrac{l}{2} \\[2mm] 0 \\[2mm] 0 \end{bmatrix}$$

$$+ \begin{bmatrix} l(S_2\ddot{\theta}_1 - C_2\dot{\theta}_1^2 - \dot{\theta}_1^2 - \dot{\theta}_2^2 - 2\dot{\theta}_1\dot{\theta}_2) + gS_{12} \\ l(\ddot{\theta}_1 + \ddot{\theta}_2 + C_2\ddot{\theta}_1 + S_2\dot{\theta}_1^2) + gC_{12} \\ 0 \end{bmatrix}$$

$$= \begin{bmatrix} l(S_2\ddot{\theta}_1 - C_2\dot{\theta}_1^2 - \tfrac{1}{2}\dot{\theta}_1^2 - \tfrac{1}{2}\dot{\theta}_2^2 - \dot{\theta}_1\dot{\theta}_2) + gS_{12} \\ l(C_2\ddot{\theta}_1 + S_2\dot{\theta}_1^2 + \tfrac{1}{2}\ddot{\theta}_1 + \tfrac{1}{2}\ddot{\theta}_2) + gC_{12} \\ 0 \end{bmatrix}$$

Backward Equations for $i = 2, 1$. Assuming no-load conditions, $f_3 = n_3 = 0$. We use Eq. (3.3-56) to compute the force exerted on link i for $i = 2, 1$:

For $i = 2$, with $f_3 = 0$, we have:

$$^2R_0f_2 = {}^2R_3(^3R_0f_3) + {}^2R_0F_2 = {}^2R_0F_2 = m_2{}^2R_0\bar{a}_2$$

$$= \begin{bmatrix} m_2l(S_2\ddot{\theta}_1 - C_2\dot{\theta}_1^2 - \tfrac{1}{2}\dot{\theta}_1^2 - \tfrac{1}{2}\dot{\theta}_2^2 - \dot{\theta}_1\dot{\theta}_2) + gm_2S_{12} \\ m_2l(C_2\ddot{\theta}_1 + S_2\dot{\theta}_1^2 + \tfrac{1}{2}\ddot{\theta}_1 + \tfrac{1}{2}\ddot{\theta}_2) + gm_2C_{12} \\ 0 \end{bmatrix}$$

For $i = 1$, we have:

$$^1R_0f_1 = {}^1R_2(^2R_0f_2) + {}^1R_0F_1$$

$$= \begin{bmatrix} C_2 & -S_2 & 0 \\ S_2 & C_2 & 0 \\ 0 & 0 & 1 \end{bmatrix} \begin{bmatrix} m_2l(S_2\ddot{\theta}_1 - C_2\dot{\theta}_2 - \tfrac{1}{2}\dot{\theta}_1^2 - \tfrac{1}{2}\dot{\theta}_2^2 - \dot{\theta}_1\dot{\theta}_2) + gm_2S_{12} \\ m_2l(C_2\ddot{\theta}_1 + S_2\dot{\theta}_1^2 + \tfrac{1}{2}\ddot{\theta}_1 + \tfrac{1}{2}\ddot{\theta}_2) + gm_2C_{12} \\ 0 \end{bmatrix} + m_1 {}^1R_0\bar{a}_1$$

$$
= \begin{bmatrix}
m_2 l[\, -\dot{\theta}_1^2 - \tfrac{1}{2}C_2(\dot{\theta}_1^2 + \dot{\theta}_2^2) - C_2\dot{\theta}_1\dot{\theta}_2 - \tfrac{1}{2}S_2(\ddot{\theta}_1 + \ddot{\theta}_2)] - m_2 g(C_{12}S_2 - C_2 S_{12}) - \tfrac{1}{2}m_1 l\dot{\theta}_1^2 + m_1 g S_1 \\[2mm]
m_2 l[\, \ddot{\theta}_1 - \tfrac{1}{2}S_2(\dot{\theta}_1^2 + \dot{\theta}_2^2) - S_2\dot{\theta}_1\dot{\theta}_2 + \tfrac{1}{2}C_2(\ddot{\theta}_1 + \ddot{\theta}_2)] + m_2 g C_1 + \tfrac{1}{2}m_1 l\ddot{\theta}_1 + g m_1 C_1 \\[2mm]
0
\end{bmatrix}
$$

Using Eq. (3.3-57), compute the moment exerted on link i for $i = 2, 1$: For $i = 2$, with $\mathbf{n}_3 = \mathbf{0}$, we have:

$$
{}^{2}\mathbf{R}_0\mathbf{n}_2 = ({}^{2}\mathbf{R}_0\mathbf{p}_2^* + {}^{2}\mathbf{R}_0\bar{\mathbf{s}}_2) \times ({}^{2}\mathbf{R}_0\mathbf{F}_2) + {}^{2}\mathbf{R}_0\mathbf{N}_2
$$

where

$$
\mathbf{p}_2^* = \begin{bmatrix} lC_{12} \\ lS_{12} \\ 0 \end{bmatrix}
\qquad
{}^{2}\mathbf{R}_0\,\mathbf{p}_2^* = \begin{bmatrix} C_{12} & S_{12} & 0 \\ -S_{12} & C_{12} & 0 \\ 0 & 0 & 1 \end{bmatrix} \begin{bmatrix} lC_{12} \\ lS_{12} \\ 0 \end{bmatrix} = \begin{bmatrix} l \\ 0 \\ 0 \end{bmatrix}
$$

Thus,

$$
{}^{2}\mathbf{R}_0\mathbf{n}_2 = \begin{bmatrix} \dfrac{l}{2} \\ 0 \\ 0 \end{bmatrix} \times \begin{bmatrix} m_2 l(S_2\ddot{\theta}_1 - C_2\dot{\theta}_1^2 - \tfrac{1}{2}\dot{\theta}_1^2 - \tfrac{1}{2}\dot{\theta}_2^2 - \dot{\theta}_1\dot{\theta}_2) + g m_2 S_{12} \\ m_2 l(C_2\ddot{\theta}_1 + S_2\dot{\theta}_1^2 + \tfrac{1}{2}\ddot{\theta}_1 + \tfrac{1}{2}\ddot{\theta}_2) + g m_2 C_{12} \\ 0 \end{bmatrix}
$$

$$
+ \begin{bmatrix} 0 & 0 & 0 \\ 0 & \tfrac{1}{12}m_2 l^2 & 0 \\ 0 & 0 & \tfrac{1}{12}m_2 l^2 \end{bmatrix} \begin{bmatrix} 0 \\ 0 \\ \ddot{\theta}_1 + \ddot{\theta}_2 \end{bmatrix}
$$

$$
= \begin{bmatrix} 0 \\ 0 \\ \tfrac{1}{3}m_2 l^2\ddot{\theta}_1 + \tfrac{1}{3}m_2 l^2\ddot{\theta}_2 + \tfrac{1}{2}m_2 l^2(C_2\ddot{\theta}_1 + S_2\dot{\theta}_1^2) + \tfrac{1}{2}m_2 g lC_{12} \end{bmatrix}
$$

For $i = 1$, we have:

$$
{}^{1}\mathbf{R}_0\mathbf{n}_1 = {}^{1}\mathbf{R}_2[{}^{2}\mathbf{R}_0\mathbf{n}_2 + ({}^{2}\mathbf{R}_0\mathbf{p}_1^*) \times ({}^{2}\mathbf{R}_0\mathbf{f}_2)] + ({}^{1}\mathbf{R}_0\mathbf{p}_1^* + {}^{1}\mathbf{R}_0\bar{\mathbf{s}}_1) \times ({}^{1}\mathbf{R}_0\mathbf{F}_1) + {}^{1}\mathbf{R}_0\mathbf{N}_1
$$

where

$$
\mathbf{p}_1^* = \begin{bmatrix} lC_1 \\ lS_1 \\ 0 \end{bmatrix}
\qquad
{}^{2}\mathbf{R}_0\,\mathbf{p}_1^* = \begin{bmatrix} lC_2 \\ -lS_2 \\ 0 \end{bmatrix}
\qquad
{}^{1}\mathbf{R}_0\,\mathbf{p}_1^* = \begin{bmatrix} l \\ 0 \\ 0 \end{bmatrix}
$$

Thus,

$$
{}^1\mathbf{R}_0\mathbf{n}_1 = {}^1\mathbf{R}_2({}^2\mathbf{R}_0\mathbf{n}_2) + {}^1\mathbf{R}_2[({}^2\mathbf{R}_0\mathbf{p}_i^*) \times ({}^2\mathbf{R}_0\mathbf{f}_2)] + \left[\frac{l}{2}, 0, 0\right]^T \times {}^1\mathbf{R}_0\mathbf{F}_1
$$

$$
+ {}^1\mathbf{R}_0\mathbf{N}_1
$$

Finally, we obtain the joint torques applied to each of the joint actuators for both links, using Eq. (3.3-58):

For $i = 2$, with $b_2 = 0$, we have:

$$
\tau_2 = ({}^2\mathbf{R}_0\mathbf{n}_2)^T({}^2\mathbf{R}_1\mathbf{z}_0)
$$

$$
= \tfrac{1}{3}m_2 l^2\ddot{\theta}_1 + \tfrac{1}{3}m_2 l^2\ddot{\theta}_2 + \tfrac{1}{2}m_2 l^2 C_2\ddot{\theta}_1 + \tfrac{1}{2}m_2 glC_{12} + \tfrac{1}{2}m_2 l^2 S_2\dot{\theta}_1^2
$$

For $i = 1$, with $b_1 = 0$, we have:

$$
\tau_1 = ({}^1\mathbf{R}_0\mathbf{n}_1)^T({}^1\mathbf{R}_0\mathbf{z}_0)
$$

$$
= \tfrac{1}{3}m_1 l^2\ddot{\theta}_1 + \tfrac{4}{3}m_2 l^2\ddot{\theta}_1 + \tfrac{1}{3}m_2 l^2\ddot{\theta}_2 + m_2 C_2 l^2\ddot{\theta}_1
$$

$$
+ \tfrac{1}{2}m_2 l^2 C_2\ddot{\theta}_2 - m_2 S_2 l^2\dot{\theta}_1\dot{\theta}_2 - \tfrac{1}{2}m_2 S_2 l^2\dot{\theta}_2^2 + \tfrac{1}{2}m_1 glC_1
$$

$$
+ \tfrac{1}{2}m_2 glC_{12} + m_2 glC_1
$$

The above equations of motion agree with those obtained from the Lagrange-Euler formulation in Sec. 3.2.6.

3.4 GENERALIZED d'ALEMBERT EQUATIONS OF MOTION

Computationally, the L-E equations of motion are inefficient due to the 4×4 homogeneous matrix manipulations, while the efficiency of the N-E formulation can be seen from the vector formulation and its recursive nature. In order to obtain an efficient set of closed-form equations of motion, one can utilize the relative position vector and rotation matrix representation to describe the kinematic information of each link, obtain the kinetic and potential energies of the robot arm to form the lagrangian function, and apply the Lagrange-Euler formulation to obtain the equations of motion. In this section, we derive a Lagrange form of d'Alembert equations of motion or generalized d'Alembert equations of motion (G-D). We shall only focus on robot arms with rotary joints.

Assuming that the links of the robot arm are rigid bodies, the angular velocity ω_s of link s with respect to the base coordinate frame can be expressed as a sum of the relative angular velocities from the lower joints (see Fig. 3.8),

$$
\omega_s = \sum_{j=1}^{s} \dot{\theta}_j \mathbf{z}_{j-1} \tag{3.4-1}
$$

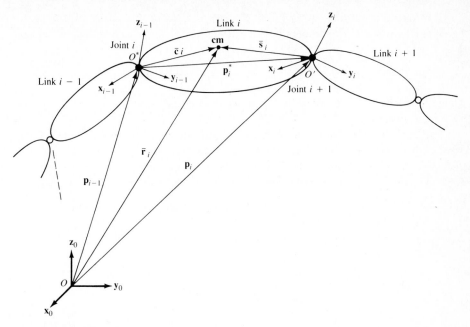

Base coordinate system

Figure 3.8 Vector definition in the generalized d'Alembert equations.

where $\mathbf{z}_{j-1}$ is the axis of rotation of joint j with reference to the base coordinate frame. Premultiplying the above angular velocity by the rotation matrix $^s\mathbf{R}_0$ changes its reference to the coordinate frame of link s; that is,

$$^s\mathbf{R}_0\boldsymbol{\omega}_s = \sum_{j=1}^{s} \dot{\theta}_j \; {}^s\mathbf{R}_0\mathbf{z}_{j-1} \tag{3.4-2}$$

In Fig. 3.8, let $\bar{\mathbf{r}}_s$ be the position vector to the center of mass of link s from the base coordinate frame. This position vector can be expressed as

$$\bar{\mathbf{r}}_s = \sum_{j=1}^{s-1} \mathbf{p}_j^* + \bar{\mathbf{c}}_s \tag{3.4-3}$$

where $\bar{\mathbf{c}}_s$ is the position vector of the center of mass of link s from the $(s-1)$th coordinate frame with reference to the base coordinate frame.

Using Eqs. (3.4-1) to (3.4-3), the linear velocity of link s, $\mathbf{v}_s$, with respect to the base coordinate frame can be computed as a sum of the linear velocities from the lower links,

$$\mathbf{v}_s = \sum_{k=1}^{s-1} \left[\left(\sum_{j=1}^{k} \dot{\theta}_j \mathbf{z}_{j-1} \right) \times \mathbf{p}_k^* \right] + \left(\sum_{j=1}^{s} \dot{\theta}_j \mathbf{z}_{j-1} \right) \times \bar{\mathbf{c}}_s \tag{3.4-4}$$

The kinetic energy of link s ($1 \leqslant s \leqslant n$) with mass m_s can be expressed as the summation of the kinetic energies due to the translational and rotational effects at its center of mass:

$$K_s = (K_s)_{\text{tran}} + (K_s)_{\text{rot}} = \tfrac{1}{2}m_s(\mathbf{v}_s \cdot \mathbf{v}_s) + \tfrac{1}{2}(^s\mathbf{R}_0\boldsymbol{\omega}_s)^T \mathbf{I}_s(^s\mathbf{R}_0\boldsymbol{\omega}_s) \quad (3.4\text{-}5)$$

where $\mathbf{I}_s$ is the inertia tensor of link s about its center of mass expressed in the sth coordinate system.

For ease of discussion and derivation, the equations of motion due to the translational, rotational, and gravitational effects of the links will be considered and treated separately. Applying the Lagrange-Euler formulation to the above translational kinetic energy of link s with respect to the generalized coordinate θ_i ($s \geqslant i$), we have

$$\frac{d}{dt}\left[\frac{\partial(K_s)_{\text{tran}}}{\partial\dot{\theta}_i}\right] - \frac{\partial(K_s)_{\text{tran}}}{\partial\theta_i}$$

$$= \frac{d}{dt}\left[m_s\,\mathbf{v}_s \cdot \frac{\partial\mathbf{v}_s}{\partial\dot{\theta}_i}\right] - m_s\,\mathbf{v}_s \cdot \frac{\partial\mathbf{v}_s}{\partial\theta_i}$$

$$= m_s\,\dot{\mathbf{v}}_s \cdot \frac{\partial\mathbf{v}_s}{\partial\dot{\theta}_i} + m_s\mathbf{v}_s \cdot \frac{d}{dt}\left(\frac{\partial\mathbf{v}_s}{\partial\dot{\theta}_i}\right) - m_s\,\mathbf{v}_s \cdot \frac{\partial\mathbf{v}_s}{\partial\theta_i} \quad (3.4\text{-}6)$$

where

$$\frac{\partial\mathbf{v}_s}{\partial\dot{\theta}_i} = \mathbf{z}_{i-1} \times (\mathbf{p}_i^* + \mathbf{p}_{i+1}^* + \cdots + \mathbf{p}_{s-1}^* + \bar{\mathbf{c}}_s) \quad (3.4\text{-}7)$$

$$= \mathbf{z}_{i-1} \times (\bar{\mathbf{r}}_s - \mathbf{p}_{i-1}) \qquad s \geqslant i$$

Using the identities

$$\frac{d}{dt}\left(\frac{\partial\mathbf{v}_s}{\partial\dot{\theta}_i}\right) = \frac{\partial\dot{\mathbf{v}}_s}{\partial\dot{\theta}_i} \qquad \text{and} \qquad \frac{\partial\dot{\mathbf{v}}_s}{\partial\dot{\theta}_i} = \frac{\partial\mathbf{v}_s}{\partial\theta_i} \quad (3.4\text{-}8)$$

Eq. (3.4-6) becomes

$$\frac{d}{dt}\left[\frac{\partial(K_s)_{\text{tran}}}{\partial\dot{\theta}_i}\right] - \frac{\partial(K_s)_{\text{tran}}}{\partial\theta_i} = m_s\dot{\mathbf{v}}_s \cdot [\mathbf{z}_{i-1} \times (\bar{\mathbf{r}}_s - \mathbf{p}_{i-1})] \quad (3.4\text{-}9)$$

Summing all the links from i to n gives the reaction torques due to the translational effect of all the links,

$$\frac{d}{dt}\left[\frac{\partial(\text{K.E.})_{\text{tran}}}{\partial\dot{\theta}_i}\right] - \frac{\partial(\text{K.E.})_{\text{tran}}}{\partial\theta_i} = \sum_{s=i}^{n}\left\{\frac{d}{dt}\left[\frac{\partial(K_s)_{\text{tran}}}{\partial\dot{\theta}_i}\right] - \frac{\partial(K_s)_{\text{tran}}}{\partial\theta_i}\right\}$$

$$= \sum_{s=i}^{n} m_s\,\dot{\mathbf{v}}_s \cdot [\mathbf{z}_{i-1} \times (\bar{\mathbf{r}}_s - \mathbf{p}_{i-1})] \quad (3.4\text{-}10)$$

where, using Eqs. (3.4-4) and (3.3-8), the acceleration of link s is given by

$$
\dot{\mathbf{v}}_s = \frac{d}{dt} \left\{ \sum_{k=1}^{s-1} \left[\left(\sum_{j=1}^{k} \dot{\theta}_j \mathbf{z}_{j-1} \right) \times \mathbf{p}_k^* \right] + \left(\sum_{j=1}^{s} \dot{\theta}_j \mathbf{z}_{j-1} \right) \times \bar{\mathbf{c}}_s \right\}
$$

$$
= \sum_{k=1}^{s-1} \left[\left(\sum_{j=1}^{k} \ddot{\theta}_j \mathbf{z}_{j-1} \right) \times \mathbf{p}_k^* + \left\{ \left(\sum_{j=1}^{k} \dot{\theta}_j \mathbf{z}_{j-1} \right) \times \left[\left(\sum_{j=1}^{k} \dot{\theta}_j \mathbf{z}_{j-1} \right) \times \mathbf{p}_k^* \right] \right\} \right]
$$

$$
+ \left[\left(\sum_{j=1}^{s} \ddot{\theta}_j \mathbf{z}_{j-1} \right) \times \bar{\mathbf{c}}_s \right] + \left\{ \left(\sum_{j=1}^{s} \dot{\theta}_j \mathbf{z}_{j-1} \right) \times \left[\left(\sum_{j=1}^{s} \dot{\theta}_j \mathbf{z}_{j-1} \right) \times \bar{\mathbf{c}}_s \right] \right\}
$$

$$
+ \sum_{k=2}^{s-1} \left\{ \sum_{p=2}^{k} \left[\left(\sum_{q=1}^{p-1} \dot{\theta}_q \mathbf{z}_{q-1} \right) \times \dot{\theta}_p \mathbf{z}_{p-1} \right] \times \mathbf{p}_k^* \right\}
$$

$$
+ \left\{ \sum_{p=2}^{s} \left[\left(\sum_{q=1}^{p-1} \dot{\theta}_q \mathbf{z}_{q-1} \right) \times \dot{\theta}_p \mathbf{z}_{p-1} \right] \times \bar{\mathbf{c}}_s \right\} \tag{3.4-11}
$$

Next, the kinetic energy due to the rotational effect of link s is:

$$
(K_s)_{\text{rot}} = \tfrac{1}{2} ({}^s\mathbf{R}_0 \boldsymbol{\omega}_s)^T \mathbf{I}_s ({}^s\mathbf{R}_0 \boldsymbol{\omega}_s)
$$

$$
= \frac{1}{2} \left[\sum_{j=1}^{s} \dot{\theta}_j \, {}^s\mathbf{R}_0 \mathbf{z}_{j-1} \right]^T \mathbf{I}_s \left[\sum_{j=1}^{s} \dot{\theta}_j \, {}^s\mathbf{R}_0 \mathbf{z}_{j-1} \right] \tag{3.4-12}
$$

Since

$$
\frac{\partial (K_s)_{\text{rot}}}{\partial \dot{\theta}_i} = ({}^s\mathbf{R}_0 \mathbf{z}_{i-1})^T \mathbf{I}_s \left[\sum_{j=1}^{s} \dot{\theta}_j \, {}^s\mathbf{R}_0 \mathbf{z}_{j-1} \right] \qquad s \geqslant i \tag{3.4-13}
$$

$$
\frac{\partial}{\partial \theta_i} ({}^s\mathbf{R}_0 \mathbf{z}_{j-1}) = {}^s\mathbf{R}_0 \mathbf{z}_{j-1} \times {}^s\mathbf{R}_0 \mathbf{z}_{i-1} \qquad i \geqslant j \tag{3.4-14}
$$

and

$$
\frac{d}{dt} ({}^s\mathbf{R}_0 \mathbf{z}_{i-1}) = \sum_{j=i}^{s} \left[\frac{\partial}{\partial \theta_j} \, {}^s\mathbf{R}_0 \mathbf{z}_{i-1} \right] \frac{d\theta_j}{dt}
$$

$$
= {}^s\mathbf{R}_0 \mathbf{z}_{i-1} \times \left[\sum_{j=i}^{s} \dot{\theta}_j \, {}^s\mathbf{R}_0 \mathbf{z}_{j-1} \right] \tag{3.4-15}
$$

then the time derivative of Eq. (3.4-13) is

$$
\frac{d}{dt}\left[\frac{\partial (K_s)_{\text{rot}}}{\partial \dot{\theta}_i}\right] = \left[\frac{d}{dt}\,{}^s\mathbf{R}_0\mathbf{z}_{i-1}\right]^T \mathbf{I}_s \left[\sum_{j=1}^{s}\dot{\theta}_j\,{}^s\mathbf{R}_0\mathbf{z}_{j-1}\right]
$$

$$
+\,({}^s\mathbf{R}_0\mathbf{z}_{i-1})^T \mathbf{I}_s \left[\sum_{j=1}^{s}\ddot{\theta}_j\,{}^s\mathbf{R}_0\mathbf{z}_{j-1}\right] + ({}^s\mathbf{R}_0\mathbf{z}_{i-1})^T \mathbf{I}_s \left[\sum_{j=1}^{s}\dot{\theta}_j\left[\frac{d}{dt}\,{}^s\mathbf{R}_0\mathbf{z}_{j-1}\right]\right]
$$

$$
= \left[{}^s\mathbf{R}_0\mathbf{z}_{i-1}\times\sum_{j=i}^{s}\dot{\theta}_j\,{}^s\mathbf{R}_0\mathbf{z}_{j-1}\right]^T \mathbf{I}_s \left[\sum_{j=1}^{s}\dot{\theta}_j\,{}^s\mathbf{R}_0\mathbf{z}_{j-1}\right]
$$

$$
+\,({}^s\mathbf{R}_0\mathbf{z}_{i-1})^T \mathbf{I}_s \left[\sum_{j=1}^{s}\ddot{\theta}_j\,{}^s\mathbf{R}_0\mathbf{z}_{j-1}\right]
$$

$$
+\,({}^s\mathbf{R}_0\mathbf{z}_{i-1})^T \mathbf{I}_s \left[\sum_{j=1}^{s}\left[\dot{\theta}_j\,{}^s\mathbf{R}_0\mathbf{z}_{j-1}\times\sum_{k=j+1}^{s}\dot{\theta}_k\,{}^s\mathbf{R}_0\mathbf{z}_{k-1}\right]\right] \tag{3.4-16}
$$

Next, using Eq. (3.4-14), we can find the partial derivative of $(K_s)_{\text{rot}}$ with respect to the generalized coordinate θ_i $(s \geq i)$; that is,

$$
\frac{\partial (K_s)_{\text{rot}}}{\partial \theta_i} = \left[\left[\sum_{j=1}^{i}\dot{\theta}_j\,{}^s\mathbf{R}_0\mathbf{z}_{j-1}\right]\times{}^s\mathbf{R}_0\mathbf{z}_{i-1}\right]^T \mathbf{I}_s \left[\sum_{j=1}^{s}\dot{\theta}_j\,{}^s\mathbf{R}_0\mathbf{z}_{j-1}\right] \tag{3.4-17}
$$

Subtracting Eq. (3.4-17) from Eq. (3.4-16) and summing all the links from i to n gives us the reaction torques due to the rotational effects of all the links,

$$
\frac{d}{dt}\left[\frac{\partial (\text{K.E.})_{\text{rot}}}{\partial \dot{\theta}_i}\right] - \frac{\partial (\text{K.E.})_{\text{rot}}}{\partial \theta_i} = \sum_{s=i}^{n}\left\{\frac{d}{dt}\left[\frac{\partial (K_s)_{\text{rot}}}{\partial \dot{\theta}_i}\right] - \frac{\partial (K_s)_{\text{rot}}}{\partial \theta_i}\right\}
$$

$$
= \sum_{s=i}^{n}\left[({}^s\mathbf{R}_0\mathbf{z}_{i-1})^T \mathbf{I}_s \left[\sum_{j=1}^{s}\ddot{\theta}_j\,{}^s\mathbf{R}_0\mathbf{z}_{j-1}\right]\right.
$$

$$
+\,({}^s\mathbf{R}_0\mathbf{z}_{i-1})^T \mathbf{I}_s \left\{\sum_{j=1}^{s}\left[\dot{\theta}_j\,{}^s\mathbf{R}_0\mathbf{z}_{j-1}\times\left[\sum_{k=j+1}^{s}\dot{\theta}_k\,{}^s\mathbf{R}_0\mathbf{z}_{k-1}\right]\right]\right\}
$$

$$
+\left.\left[{}^s\mathbf{R}_0\mathbf{z}_{i-1}\times\left[\sum_{k=1}^{s}\dot{\theta}_k\,{}^s\mathbf{R}_0\mathbf{z}_{k-1}\right]\right]^T \mathbf{I}_s \left[\sum_{j=1}^{s}\dot{\theta}_j\,{}^s\mathbf{R}_0\mathbf{z}_{j-1}\right]\right]
$$

$$
i = 1, 2, \ldots, n \tag{3.4-18}
$$

The potential energy of the robot arm equals to the sum of the potential energies of each link,

$$\text{P.E.} = \sum_{s=1}^{n} P_s \tag{3.4-19}$$

where P_s is the potential energy of link s given by

$$P_s = -\mathbf{g} \cdot m_s \bar{\mathbf{r}}_s = -\mathbf{g} \cdot m_s(\mathbf{p}_{i-1} + \mathbf{p}_i^* + \cdots + \bar{\mathbf{c}}_s) \tag{3.4-20}$$

where $\mathbf{g} = (g_x, g_y, g_z)^T$ and $|\mathbf{g}| = 9.8062$ m/s^2. Applying the Lagrange-Euler formulation to the potential energy of link s with respect to the generalized coordinate θ_i ($s \geqslant i$), we have

$$\frac{d}{dt}\left[\frac{\partial(P_s)}{\partial\dot{\theta}_i}\right] - \frac{\partial(P_s)}{\partial\theta_i} = -\frac{\partial(P_s)}{\partial\theta_i} = \mathbf{g} \cdot m_s \frac{\partial(\mathbf{p}_{i-1} + \mathbf{p}_i^* + \cdots + \bar{\mathbf{c}}_s)}{\partial\theta_i}$$

$$= \mathbf{g} \cdot m_s \frac{\partial(\bar{\mathbf{r}}_s - \mathbf{p}_{i-1})}{\partial\theta_i} = \mathbf{g} \cdot m_s [\mathbf{z}_{i-1} \times (\bar{\mathbf{r}}_s - \mathbf{p}_{i-1})] \tag{3.4-21}$$

where $\mathbf{p}_{i-1}$ is not a function of θ_i. Summing all the links from i to n gives the reaction torques due to the gravity effects of all the links,

$$\frac{d}{dt}\left[\frac{\partial(\text{P.E.})}{\partial\dot{\theta}_i}\right] - \frac{\partial(\text{P.E.})}{\partial\theta_i} = -\sum_{s=i}^{n}\frac{\partial(P_s)}{\partial\theta_i}$$

$$= \sum_{s=i}^{n} \mathbf{g} \cdot m_s [\mathbf{z}_{i-1} \times (\bar{\mathbf{r}}_s - \mathbf{p}_{i-1})] \tag{3.4-22}$$

The summation of Eqs. (3.4-10), (3.4-18), and (3.4-22) is equal to the generalized applied torque exerted at joint i to drive link i,

$$\tau_i = \left\{\frac{d}{dt}\left[\frac{\partial(\text{K.E.})_{\text{tran}}}{\partial\dot{\theta}_i}\right] - \frac{\partial(\text{K.E.})_{\text{tran}}}{\partial\theta_i}\right\} + \left[\frac{d}{dt}\left(\frac{\partial(\text{K.E.})_{\text{rot}}}{\partial\dot{\theta}_i}\right) - \frac{\partial(\text{K.E.})_{\text{rot}}}{\partial\theta_i}\right] + \frac{\partial(\text{P.E.})}{\partial\theta_i}$$

$$= \sum_{s=i}^{n}\left[m_s\left\{\left[\sum_{k=1}^{s-1}\sum_{j=1}^{k}\ddot{\theta}_j\mathbf{z}_{j-1}\right]\times\mathbf{p}_k^* + \left[\sum_{j=1}^{s}\ddot{\theta}_j\mathbf{z}_{j-1}\right]\times\bar{\mathbf{c}}_s\right\}\cdot[\mathbf{z}_{i-1}\times(\bar{\mathbf{r}}_s - \mathbf{p}_{i-1})]\right]$$

$$+ \sum_{s=i}^{n}\left[({}^s\mathbf{R}_0\mathbf{z}_{i-1})^T\mathbf{I}_s\left[\sum_{j=1}^{s}\ddot{\theta}_j\,{}^s\mathbf{R}_0\mathbf{z}_{j-1}\right]\right]$$

$$+ \sum_{s=i}^{n}\left[m_s\left[\sum_{k=1}^{s-1}\left[\left\{\left[\sum_{p=1}^{k}\dot{\theta}_p\mathbf{z}_{p-1}\right]\times\left[\sum_{q=1}^{k}\dot{\theta}_q\mathbf{z}_{q-1}\right]\times\mathbf{p}_k^*\right]\right.\right.\right.$$

$$+ \left.\left.\left.\left\{\sum_{p=2}^{k}\left[\left[\sum_{q=1}^{p-1}\dot{\theta}_q\mathbf{z}_{q-1}\right]\times\dot{\theta}_p\mathbf{z}_{p-1}\right]\times\mathbf{p}_k^*\right\}\right]\right]\cdot[\mathbf{z}_{i-1}\times(\bar{\mathbf{r}}_s - \mathbf{p}_{i-1})]\right]$$

$$+ \sum_{s=i}^{n} \left[m_s \left(\left\{ \left[\sum_{p=1}^{s} \dot{\theta}_p \mathbf{z}_{p-1} \right] \times \left[\left[\sum_{q=1}^{s} \dot{\theta}_q \mathbf{z}_{q-1} \right] \times \bar{\mathbf{c}}_s \right] \right\} \right. \right.$$

$$+ \left. \left\{ \sum_{p=2}^{s} \left[\left[\sum_{q=1}^{p-1} \dot{\theta}_q \mathbf{z}_{q-1} \right] \times \dot{\theta}_p \mathbf{z}_{p-1} \right] \times \bar{\mathbf{c}}_s \right\} \cdot [\mathbf{z}_{i-1} \times (\bar{\mathbf{r}}_s - \mathbf{p}_{i-1})] \right]$$

$$+ \sum_{s=i}^{n} \left[({}^s\mathbf{R}_0 \mathbf{z}_{i-1})^T \mathbf{I}_s \left\{ \sum_{j=1}^{s} \left[\dot{\theta}_j {}^s\mathbf{R}_0 \mathbf{z}_{j-1} \times \left[\sum_{k=j+1}^{s} \dot{\theta}_k {}^s\mathbf{R}_0 \mathbf{z}_{k-1} \right] \right] \right\} \right.$$

$$+ \left[{}^s\mathbf{R}_0 \mathbf{z}_{i-1} \times \left[\sum_{p=1}^{s} \dot{\theta}_p {}^s\mathbf{R}_0 \mathbf{z}_{p-1} \right] \right]^T \mathbf{I}_s \left(\sum_{q=1}^{s} \dot{\theta}_q {}^s\mathbf{R}_0 \mathbf{z}_{q-1} \right) \right]$$

$$- \mathbf{g} \cdot \left\{ \mathbf{z}_{i-1} \times \left[\sum_{j=i}^{n} m_j (\bar{\mathbf{r}}_j - \mathbf{p}_{i-1}) \right] \right\} \tag{3.4-23}$$

for $i = 1, 2, \ldots, n$.

The above equation can be rewritten in a more "structured" form as (for $i = 1, 2, \ldots, n$):

$$\sum_{j=1}^{n} D_{ij} \ddot{\theta}_j(t) + h_i^{\text{tran}}(\theta, \dot{\theta}) + h_i^{\text{rot}}(\theta, \dot{\theta}) + c_i = \tau_i(t) \tag{3.4-24}$$

where, for $i = 1, 2, \ldots, n$,

$$D_{ij} = D_{ij}^{\text{rot}} + D_{ij}^{\text{tran}} = \sum_{s=j}^{n} [({}^s\mathbf{R}_0 \mathbf{z}_{i-1})^T \mathbf{I}_s ({}^s\mathbf{R}_0 \mathbf{z}_{j-1})]$$

$$+ \sum_{s=j}^{n} \left\{ m_s \left[\mathbf{z}_{j-1} \times \left[\sum_{k=j}^{s-1} \mathbf{p}_k^* + \bar{\mathbf{c}}_s \right] \right] \cdot [\mathbf{z}_{i-1} \times (\bar{\mathbf{r}}_s - \mathbf{p}_{i-1})] \right\} \quad i \leqslant j$$

$$= \sum_{s=j}^{n} [({}^s\mathbf{R}_0 \mathbf{z}_{i-1})^T \mathbf{I}_s ({}^s\mathbf{R}_0 \mathbf{z}_{j-1})]$$

$$+ \sum_{s=j}^{n} \left\{ m_s [\mathbf{z}_{j-1} \times (\bar{\mathbf{r}}_s - \mathbf{p}_{j-1})] \cdot [\mathbf{z}_{i-1} \times (\bar{\mathbf{r}}_s - \mathbf{p}_{i-1})] \right\} \quad i \leqslant j$$

$$\tag{3.4-25}$$

also,

$$
\begin{aligned}
h_i^{\text{tran}}(\boldsymbol{\theta}, \dot{\boldsymbol{\theta}}) = \sum_{s=i}^{n} \Bigg\{ & m_s \left[\sum_{k=1}^{s-1} \left(\left\{ \left(\sum_{p=1}^{k} \dot{\theta}_p \mathbf{z}_{p-1} \right) \times \left[\left(\sum_{q=1}^{k} \dot{\theta}_q \mathbf{z}_{q-1} \right) \times \mathbf{p}_k^* \right] \right\} \right. \right. \\
& + \left. \left. \left\{ \sum_{p=2}^{k} \left[\left(\sum_{q=1}^{p-1} \dot{\theta}_q \mathbf{z}_{q-1} \right) \times \dot{\theta}_p \mathbf{z}_{p-1} \right] \times \mathbf{p}_k^* \right\} \right] \right) \cdot [\mathbf{z}_{i-1} \times (\bar{\mathbf{r}}_s - \mathbf{p}_{i-1})] \Bigg\} \\
& + \sum_{s=i}^{n} \left[m_s \left(\left\{ \left(\sum_{p=1}^{s} \dot{\theta}_p \mathbf{z}_{p-1} \right) \times \left[\left(\sum_{q=1}^{s} \dot{\theta}_q \mathbf{z}_{q-1} \right) \times \bar{\mathbf{c}}_s \right] \right\} \right. \right. \\
& + \left. \left. \left\{ \sum_{p=2}^{s} \left[\left(\sum_{q=1}^{p-1} \dot{\theta}_q \mathbf{z}_{q-1} \right) \times \dot{\theta}_p \mathbf{z}_{p-1} \right] \times \bar{\mathbf{c}}_s \right\} \right) \cdot [\mathbf{z}_{i-1} \times (\bar{\mathbf{r}}_s - \mathbf{p}_{i-1})] \right]
\end{aligned}
$$

(3.4-26)

and

$$
\begin{aligned}
h_i^{\text{rot}}(\boldsymbol{\theta}, \dot{\boldsymbol{\theta}}) = \sum_{s=i}^{n} \Bigg\{ & ({}^s\mathbf{R}_0 \mathbf{z}_{i-1})^T \mathbf{I}_s \left\{ \sum_{j=1}^{s} \left[\dot{\theta}_j \, {}^s\mathbf{R}_0 \mathbf{z}_{j-1} \times \left(\sum_{k=j+1}^{s} \dot{\theta}_k \, {}^s\mathbf{R}_0 \mathbf{z}_{k-1} \right) \right] \right\} \\
& + \left[{}^s\mathbf{R}_0 \mathbf{z}_{i-1} \times \left(\sum_{p=1}^{s} \dot{\theta}_p \, {}^s\mathbf{R}_0 \mathbf{z}_{p-1} \right) \right]^T \mathbf{I}_s \left(\sum_{q=1}^{s} \dot{\theta}_q \, {}^s\mathbf{R}_0 \mathbf{z}_{q-1} \right) \Bigg\}
\end{aligned}
$$

(3.4-27)

Finally, we have

$$
c_i = -\mathbf{g} \cdot \left[\mathbf{z}_{i-1} \times \sum_{j=i}^{n} m_j (\bar{\mathbf{r}}_j - \mathbf{p}_{i-1}) \right]
$$

(3.4-28)

The dynamic coefficients D_{ij} and c_i are functions of both the joint variables and inertial parameters of the manipulator, while the h_i^{tran} and h_i^{rot} are functions of the joint variables, the joint velocities and inertial parameters of the manipulator. These coefficients have the following physical interpretation:

1. The elements of the D_{ij} matrix are related to the inertia of the links in the manipulator. Equation (3.4-25) reveals the acceleration effects of joint j acting on joint i where the driving torque τ_i acts. The first term of Eq. (3.4-25) indicates the inertial effects of moving link j on joint i due to the *rotational* motion of link j, and vice versa. If $i = j$, it is the effective inertias felt at joint i due to the rotational motion of link i; while if $i \neq j$, it is the pseudoproducts of inertia of link j felt at joint i due to the rotational motion of link j. The second term has the same physical meaning except that it is due to the *translational* motion of link j acting on joint i.

2. The $h_i^{\text{tran}}(\theta, \dot{\theta})$ term is related to the velocities of the joint variables. Equation (3.4-26) represents the combined centrifugal and Coriolis reaction torques felt at joint i due to the velocities of joints p and q resulting from the *translational* motion of links p and q. The first and third terms of Eq. (3.4-26) constitute, respectively, the centrifugal and Coriolis reaction forces from all the links below link s and link s itself, due to the translational motion of the links. If $p = q$, then it represents the centrifugal reaction forces felt at joint i. If $p \neq q$, then it indicates the Coriolis forces acting on joint i. The second and fourth terms of Eq. (3.4-26) indicate, respectively, the Coriolis reaction forces contributed from the links below link s and link s itself, due to the translational motion of the links.

3. The $h_i^{\text{rot}}(\theta, \dot{\theta})$ term is also related to the velocities of the joint variables. Similar to $h_i^{\text{tran}}(\theta, \dot{\theta})$, Eq. (3.4-27) reveals the combined centrifugal and Coriolis reaction torques felt at joint i due to the velocities of joints p and q resulting from the *rotational* motion of links p and q. The first term of Eq. (3.4-27) indicates purely the Coriolis reaction forces of joints p and q acting on joint i due to the rotational motion of the links. The second term is the combined centrifugal and Coriolis reaction forces acting on joint i. If $p = q$, then it indicates the centrifugal reaction forces felt at joint i, but if $p \neq q$, then it represents the Coriolis forces acting on joint i due to the rotational motion of the links.

4. The coefficient c_i represents the gravity effects acting on joint i from the links above joint i.

At first sight, Eqs. (3.4-25) to (3.4-28) would seem to require a large amount of computation. However, most of the cross-product terms can be computed very fast. As an indication of their computational complexities, a block diagram explicitly showing the procedure in calculating these coefficients for every set point in the trajectory in terms of multiplication and addition operations is shown in Fig. 3.9. Table 3.5 summarizes the computational complexities of the L-E, N-E, and G-D equations of motion in terms of required mathematical operations per trajectory set point.

Table 3.5 Comparison of robot arm dynamics computational complexities†

Approach	Lagrange-Euler	Newton-Euler	Generalized d'Alembert
Multiplications	$128/_3 n^4 + 512/_3 n^3$ $+ 739/_3 n^2 + 160/_3 n$	$132n$	$13/_6 n^3 + 105/_2 n^2$ $+ 268/_3 n + 69$
Additions	$98/_3 n^4 + 781/_6 n^3$ $+ 559/_3 n^2 + 245/_6 n$	$111n - 4$	$4/_3 n^3 + 44n^2$ $+ 146/_3 n + 45$
Kinematics representation	4×4 Homogeneous matrices	Rotation matrices and position vectors	Rotation matrices and position vectors
Equations of motion	Closed-form differential equations	Recursive equations	Closed-form differential equations

† n = number of degrees of freedom of the robot arm. No effort is spent here to optimize the computation.

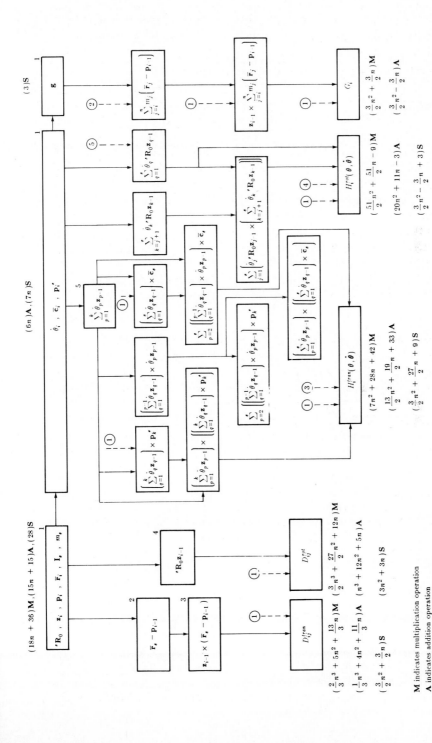

Figure 3.9 Computational procedure for D_{ij}, h_i^{tran}, h_i^{rot}, and c_i.

M indicates multiplication operation

A indicates addition operation

S indicates memory storage requirement

Ⓝ indicates output from block n

3.4.1 An Empirical Method for Obtaining a Simplified Dynamic Model

One of the objectives of developing the G-D equations of motion [Eqs. (3.4-24) to (3.4-28)] is to facilitate the design of a suitable controller for a manipulator in the

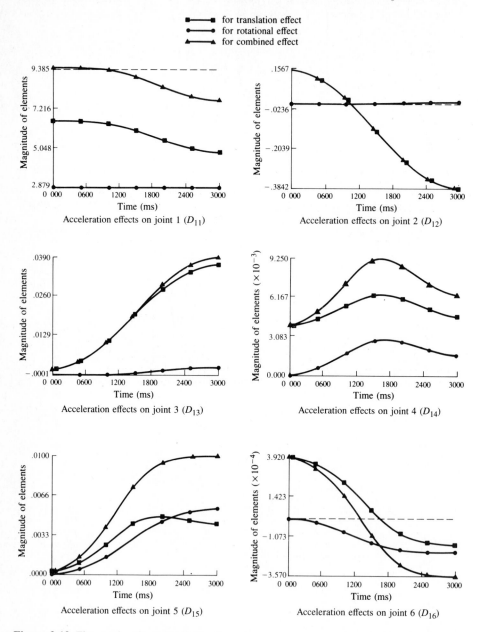

Figure 3.10 The acceleration-related D_{ij} elements.

state space or to obtain an approximate dynamic model for a manipulator. Similar to the L-E equations of motion [Eqs. (3.3-24) and (3.3-25)], the G-D equations of motion are explicitly expressed in vector-matrix form and all the interaction and coupling reaction forces can be easily identified. Furthermore, D_{ij}^{tran}, D_{ij}^{rot}, h_i^{tran}, h_i^{rot}, and c_i can be clearly identified as coming from the translational and the rotational effects of the motion. Comparing the magnitude of the translational and rotational effects for each term of the dynamic equations of motion, the extent of

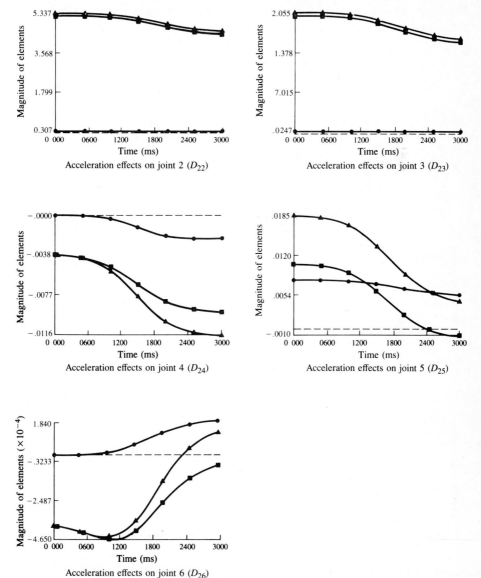

Figure 3.10 (Continued.)

dominance from the translational and rotational effects can be computed for each set point along the trajectory. The less dominant terms or elements can be neglected in calculating the dynamic equations of motion of the manipulator. This greatly aids the construction of a simplified dynamic model for control purpose.

As an example of obtaining a simplified dynamic model for a specific trajectory, we consider a PUMA 560 robot and its dynamic equations of motion along a preplanned trajectory. The D_{ij}^{tran}, D_{ij}^{rot}, h_i^{tran}, and h_i^{rot} elements along the trajectory are computed and plotted in Figs. 3.10 and 3.11. The total number of trajectory set points used is 31. Figure 3.10 shows the acceleration-related elements D_{ij}^{tran} and D_{ij}^{rot}. Figure 3.11 shows the Coriolis and centrifugal elements h_i^{tran} and h_i^{rot}. These figures show the separate and combined effects from the translational and rotational terms.

From Fig. 3.10, we can approximate the elements of the **D** matrix along the trajectory as follows: (1) The translational effect is dominant for the D_{12}, D_{22}, D_{23}, D_{33}, D_{45}, and D_{56} elements. (2) The rotational effect is dominant for the D_{44}, D_{46}, D_{55}, and D_{66} elements. (3) Both translational and rotational effects are dominant for the remaining elements of the **D** matrix. In Fig. 3.11, the elements D_{56}^{tran} and D_{45}^{tran} show a staircase shape which is due primarily to the round-off

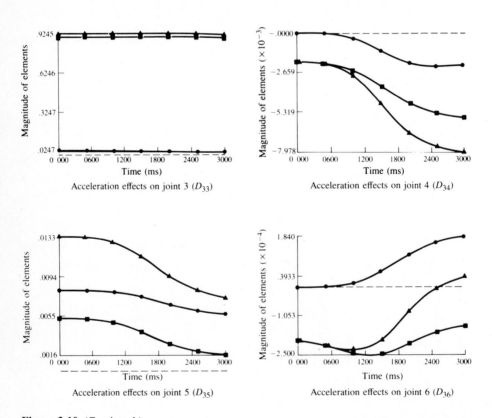

Acceleration effects on joint 3 (D_{33})

Acceleration effects on joint 4 (D_{34})

Acceleration effects on joint 5 (D_{35})

Acceleration effects on joint 6 (D_{36})

Figure 3.10 (Continued.)

error generated by the VAX-11/780 computer used in the simulation. These elements are very small in magnitude when compared with the rotational elements. Similarly, we can approximate the elements of the **h** vector as follows: (1) The translational effect is dominant for the h_1, h_2, and h_3 elements. (2) The rotational effect is dominant for the h_4 element. (3) Both translational and rotational effects are dominant for the h_5 and h_6 elements.

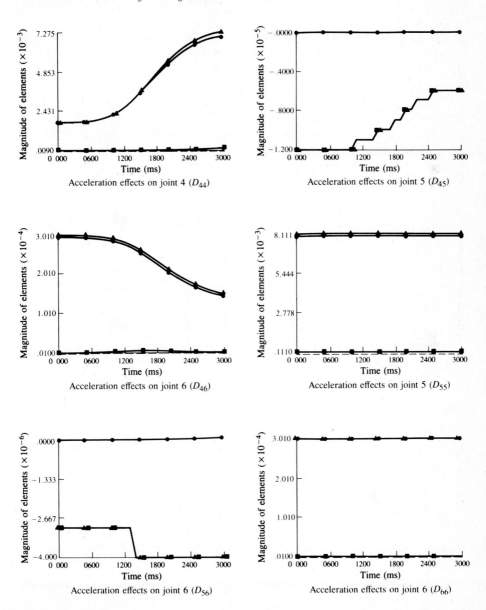

Figure 3.10 (Continued.)

The above simplification depends on the specific trajectory being considered. The resulting simplified model retains most of the major interaction and coupling reaction forces/torques at a reduced computation time, which greatly aids the design of an appropriate law for controlling the robot arm.

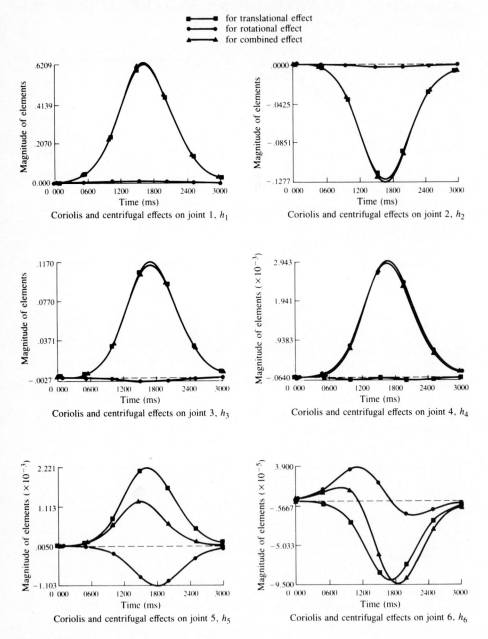

Figure 3.11 The Coriolis and centrifugal terms h_i.

3.4.2 A Two-Link Manipulator Example

Consider the two-link manipulator shown in Fig. 3.2. We would like to derive the generalized d'Alembert equations of motion for it. Letting m_1 and m_2 represent the link masses, and assuming that each link is l units long, yields the following expressions for the link inertia tensors:

$$\mathbf{I}_1 = \begin{bmatrix} 0 & 0 & 0 \\ 0 & \frac{1}{12}m_1 l^2 & 0 \\ 0 & 0 & \frac{1}{12}m_1 l^2 \end{bmatrix} \qquad \mathbf{I}_2 = \begin{bmatrix} 0 & 0 & 0 \\ 0 & \frac{1}{12}m_2 l^2 & 0 \\ 0 & 0 & \frac{1}{12}m_2 l^2 \end{bmatrix}$$

The rotation matrices are, respectively,

$$^0\mathbf{R}_1 = \begin{bmatrix} C_1 & -S_1 & 0 \\ S_1 & C_1 & 0 \\ 0 & 0 & 1 \end{bmatrix} \qquad ^1\mathbf{R}_2 = \begin{bmatrix} C_2 & -S_2 & 0 \\ S_2 & C_2 & 0 \\ 0 & 0 & 1 \end{bmatrix}$$

$$^0\mathbf{R}_2 = {^0\mathbf{R}_1}{^1\mathbf{R}_2} = \begin{bmatrix} C_{12} & -S_{12} & 0 \\ S_{12} & C_{12} & 0 \\ 0 & 0 & 1 \end{bmatrix}$$

and

$$^1\mathbf{R}_0 = (^0\mathbf{R}_1)^T \qquad ^2\mathbf{R}_0 = (^0\mathbf{R}_2)^T$$

where $C_i = \cos\theta_i$, $S_i = \sin\theta_i$, $C_{ij} = \cos(\theta_i + \theta_j)$, and $S_{ij} = \sin(\theta_i + \theta_j)$. The physical parameters of the manipulator such as $\mathbf{p}_i^*$, $\bar{\mathbf{c}}_i$, $\bar{\mathbf{r}}_i$, and $\mathbf{p}_i$ are:

$$\mathbf{p}_1^* = \mathbf{p}_1 = \begin{bmatrix} lC_1 \\ lS_1 \\ 0 \end{bmatrix} \qquad \mathbf{p}_2^* = \begin{bmatrix} lC_{12} \\ lS_{12} \\ 0 \end{bmatrix} \qquad \mathbf{p}_2 = \begin{bmatrix} l(C_1 + C_{12}) \\ l(S_1 + S_{12}) \\ 0 \end{bmatrix}$$

$$\bar{\mathbf{c}}_1 = \bar{\mathbf{r}}_1 = \begin{bmatrix} \frac{l}{2}C_1 \\ \frac{l}{2}S_1 \\ 0 \end{bmatrix} \qquad \bar{\mathbf{c}}_2 = \begin{bmatrix} \frac{l}{2}C_{12} \\ \frac{l}{2}S_{12} \\ 0 \end{bmatrix} \qquad \bar{\mathbf{r}}_2 = \begin{bmatrix} lC_1 + \frac{l}{2}C_{12} \\ lS_1 + \frac{l}{2}S_{12} \\ 0 \end{bmatrix}$$

Using Eq. (3.4-25), we obtain the elements of the **D** matrix, as follows:

$$D_{11} = ({}^1\mathbf{R}_0\mathbf{z}_0)^T\mathbf{I}_1({}^1\mathbf{R}_0\mathbf{z}_0) + ({}^2\mathbf{R}_0\mathbf{z}_0)^T\mathbf{I}_2({}^2\mathbf{R}_0\mathbf{z}_0) + m_1(\mathbf{z}_0 \times \bar{\mathbf{c}}_1) \cdot (\mathbf{z}_0 \times \bar{\mathbf{r}}_1)$$

$$+ m_2[\mathbf{z}_0 \times (\mathbf{p}_1^* + \bar{\mathbf{c}}_2)] \cdot (\mathbf{z}_0 \times \bar{\mathbf{r}}_2)$$

$$= (0, 0, 1)\,\mathbf{I}_1 \begin{bmatrix} 0 \\ 0 \\ 1 \end{bmatrix} + (0, 0, 1)\,\mathbf{I}_2 \begin{bmatrix} 0 \\ 0 \\ 1 \end{bmatrix}$$

$$+ m_1 \left[\begin{bmatrix} 0 \\ 0 \\ 1 \end{bmatrix} \times \begin{bmatrix} \dfrac{l}{2}C_1 \\ \dfrac{l}{2}S_1 \\ 0 \end{bmatrix} \right] \cdot \left[\begin{bmatrix} 0 \\ 0 \\ 1 \end{bmatrix} \times \begin{bmatrix} \dfrac{l}{2}C_1 \\ \dfrac{l}{2}S_1 \\ 0 \end{bmatrix} \right]$$

$$+ m_2 \left[\begin{bmatrix} 0 \\ 0 \\ 1 \end{bmatrix} \times \begin{bmatrix} lC_1 + \dfrac{l}{2}C_{12} \\ lS_1 + \dfrac{l}{2}S_{12} \\ 0 \end{bmatrix} \right] \cdot \left[\begin{bmatrix} 0 \\ 0 \\ 1 \end{bmatrix} \times \begin{bmatrix} lC_1 + \dfrac{l}{2}C_{12} \\ lS_1 + \dfrac{l}{2}S_{12} \\ 0 \end{bmatrix} \right]$$

$$= \tfrac{1}{3}m_1 l^2 + \tfrac{4}{3}m_2 l^2 + C_2 m_2 l^2$$

$$D_{12} = D_{21} = ({}^2\mathbf{R}_0\mathbf{z}_0)^T\mathbf{I}_2({}^2\mathbf{R}_0\mathbf{z}_1) + m_2(\mathbf{z}_1 \times \bar{\mathbf{c}}_2) \cdot (\mathbf{z}_0 \times \bar{\mathbf{r}}_2) = \tfrac{1}{3}m_2 l^2 + \tfrac{1}{2}m_2 C_2 l^2$$

$$D_{22} = ({}^2\mathbf{R}_0\mathbf{z}_1)^T\mathbf{I}_2({}^2\mathbf{R}_0\mathbf{z}_1) + m_2(\mathbf{z}_1 \times \bar{\mathbf{c}}_2) \cdot [\mathbf{z}_1 \times (\bar{\mathbf{r}}_2 - \mathbf{p}_1)]$$

$$= \tfrac{1}{12}m_2 l^2 + \tfrac{1}{4}m_2 l^2 = \tfrac{1}{3}m_2 l^2$$

Thus,

$$[D_{ij}] = \begin{bmatrix} D_{11} & D_{12} \\ D_{21} & D_{22} \end{bmatrix} = \begin{bmatrix} \tfrac{1}{3}m_1 l^2 + \tfrac{4}{3}m_2 l^2 + m_2 l^2 C_2 & \tfrac{1}{3}m_2 l^2 + \tfrac{1}{2}m_2 C_2 l^2 \\ \tfrac{1}{3}m_2 l^2 + \tfrac{1}{2}m_2 C_2 l^2 & \tfrac{1}{3}m_2 l^2 \end{bmatrix}$$

To derive the $h_i^{\text{tran}}(\boldsymbol{\theta}, \dot{\boldsymbol{\theta}})$ and $h_i^{\text{rot}}(\boldsymbol{\theta}, \dot{\boldsymbol{\theta}})$ components, we need to consider only the following terms in Eqs. (3.4-26) and (3.4-27) in our example because the other terms are zero.

$$h_1^{\text{tran}} = m_2 [\dot{\theta}_1 \mathbf{z}_0 \times (\dot{\theta}_1 \mathbf{z}_0 \times \mathbf{p}_1^*)] \cdot (\mathbf{z}_0 \times \bar{\mathbf{r}}_2) + m_1 [\dot{\theta}_1 \mathbf{z}_0 \times (\dot{\theta}_1 \mathbf{z}_0 \times \bar{\mathbf{c}}_1)]$$

$$\cdot (\mathbf{z}_0 \times \bar{\mathbf{r}}_1) + m_2 [(\dot{\theta}_1 \mathbf{z}_0 + \dot{\theta}_2 \mathbf{z}_1) \times [(\dot{\theta}_1 \mathbf{z}_0 + \dot{\theta}_2 \mathbf{z}_1) \times \bar{\mathbf{c}}_2)$$

$$+ (\dot{\theta}_1 \mathbf{z}_0 \times \dot{\theta}_2 \mathbf{z}_1) \times \bar{\mathbf{c}}_2] \cdot (\mathbf{z}_0 \times \bar{\mathbf{r}}_2)$$

$$= \tfrac{1}{2} m_2 l^2 S_2 \dot{\theta}_1^2 - \tfrac{1}{2} m_2 l^2 S_2 \dot{\theta}_1^2 - \tfrac{1}{2} m_2 l^2 S_2 \dot{\theta}_2^2 - m_2 l^2 S_2 \dot{\theta}_1 \dot{\theta}_2$$

$$h_1^{\text{rot}} = ({}^{1}\mathbf{R}_0 \mathbf{z}_0 \times \dot{\theta}_1 \,{}^{1}\mathbf{R}_0 \mathbf{z}_0)^T \mathbf{I}_1 (\dot{\theta}_1 \,{}^{1}\mathbf{R}_0 \mathbf{z}_0) + ({}^{2}\mathbf{R}_0 \mathbf{z}_0)^T \mathbf{I}_2 (\dot{\theta}_1 \,{}^{2}\mathbf{R}_0 \mathbf{z}_0 \times \dot{\theta}_2 \,{}^{2}\mathbf{R}_0 \mathbf{z}_1)$$

$$+ [{}^{2}\mathbf{R}_0 \mathbf{z}_0 \times (\dot{\theta}_1 \,{}^{2}\mathbf{R}_0 \mathbf{z}_0 + \dot{\theta}_2 \,{}^{2}\mathbf{R}_0 \mathbf{z}_1)]^T \mathbf{I}_2 (\dot{\theta}_1 \,{}^{2}\mathbf{R}_0 \mathbf{z}_0 + \dot{\theta}_2 \,{}^{2}\mathbf{R}_0 \mathbf{z}_1)$$

$$= 0$$

Thus,

$$h_1 = h_1^{\text{tran}} + h_1^{\text{rot}} = -\tfrac{1}{2} m_2 l^2 S_2 \dot{\theta}_2^2 - m_2 l^2 S_2 \dot{\theta}_1 \dot{\theta}_2$$

Similarly, we can find

$$h_2^{\text{tran}} = m_2 [\dot{\theta}_1 \mathbf{z}_0 \times (\dot{\theta}_1 \mathbf{z}_0 \times \mathbf{p}_1^*)] \cdot [\mathbf{z}_1 \times (\bar{\mathbf{r}}_2 - \mathbf{p}_1)] + m_2 \{(\dot{\theta}_1 \mathbf{z}_0 + \dot{\theta}_2 \mathbf{z}_1)$$

$$\times [(\dot{\theta}_1 \mathbf{z}_0 + \dot{\theta}_2 \mathbf{z}_1) \times \bar{\mathbf{c}}_2] + (\dot{\theta}_1 \mathbf{z}_0 \times \dot{\theta}_2 \mathbf{z}_1) \times \bar{\mathbf{c}}_2\} \cdot [\mathbf{z}_1 \times (\bar{\mathbf{r}}_2 - \mathbf{p}_1)]$$

$$= \tfrac{1}{2} m_2 l^2 S_2 \dot{\theta}_1^2$$

$$h_2^{\text{rot}} = ({}^{2}\mathbf{R}_0 \mathbf{z}_1)^T \mathbf{I}_2 (\dot{\theta}_1 \,{}^{2}\mathbf{R}_0 \mathbf{z}_0 \times \dot{\theta}_2 \,{}^{2}\mathbf{R}_0 \mathbf{z}_1)$$

$$+ [{}^{2}\mathbf{R}_0 \mathbf{z}_1 \times (\dot{\theta}_1 \,{}^{2}\mathbf{R}_0 \mathbf{z}_0 + \dot{\theta}_2 \,{}^{2}\mathbf{R}_0 \mathbf{z}_1)]^T \mathbf{I}_2 (\dot{\theta}_1 \,{}^{2}\mathbf{R}_0 \mathbf{z}_0 + \dot{\theta}_2 \,{}^{2}\mathbf{R}_0 \mathbf{z}_1)$$

$$= 0$$

We note that $h_1^{\text{rot}} = h_2^{\text{rot}} = 0$ which simplifies the design of feedback control law. Thus,

$$h_2 = h_2^{\text{tran}} + h_2^{\text{rot}} = \tfrac{1}{2} m_2 l^2 S_2 \dot{\theta}_1^2$$

Therefore,

$$\mathbf{h} = \begin{bmatrix} h_1 \\ h_2 \end{bmatrix} = \begin{bmatrix} -\tfrac{1}{2} m_2 S_2 l^2 \dot{\theta}_2^2 - m_2 S_2 l^2 \dot{\theta}_1 \dot{\theta}_2 \\ \tfrac{1}{2} m_2 S_2 l^2 \dot{\theta}_1^2 \end{bmatrix}$$

To derive the elements of the $\mathbf{c}$ vector, we use Eq. (3.4-28):

$$c_1 = -\mathbf{g} \cdot [\mathbf{z}_0 \times (m_1 \bar{\mathbf{r}}_1 + m_2 \bar{\mathbf{r}}_2)] = (\tfrac{1}{2} m_1 + m_2) glC_1 + \tfrac{1}{2} m_2 glC_{12}$$

$$c_2 = -\mathbf{g} \cdot [\mathbf{z}_1 \times m_2 (\bar{\mathbf{r}}_2 - \mathbf{p}_1)] = \tfrac{1}{2} m_2 glC_{12}$$

where $\mathbf{g} = (0, -g, 0)^T$. Thus, the gravity loading vector $\mathbf{c}$ becomes

$$\mathbf{c} = \begin{bmatrix} c_1 \\ c_2 \end{bmatrix} = \begin{bmatrix} (\tfrac{1}{2}m_1 + m_2)glC_1 + \tfrac{1}{2}m_2 glC_{12} \\ \tfrac{1}{2}m_2 glC_{12} \end{bmatrix}$$

where $g = 9.8062$ m/s^2. Based on the above results, it follows that the equations of motion of the two-link robot arm by the generalized d'Alembert method are:

$$\begin{bmatrix} \tau_1(t) \\ \tau_2(t) \end{bmatrix} = \begin{bmatrix} \tfrac{1}{3}m_1 l^2 + \tfrac{4}{3}m_2 l^2 + m_2 C_2 l^2 & \tfrac{1}{3}m_2 l^2 + \tfrac{1}{2}m_2 C_2 l^2 \\ \tfrac{1}{3}m_2 l^2 + \tfrac{1}{2}m_2 C_2 l^2 & \tfrac{1}{3}m_2 l^2 \end{bmatrix} \begin{bmatrix} \ddot{\theta}_1(t) \\ \ddot{\theta}_2(t) \end{bmatrix}$$

$$+ \begin{bmatrix} -\tfrac{1}{2}m_2 S_2 l^2 \dot{\theta}_2^2 - m_2 S_2 l^2 \dot{\theta}_1 \dot{\theta}_2 \\ \tfrac{1}{2}m_2 S_2 l^2 \dot{\theta}_1^2 \end{bmatrix}$$

$$+ \begin{bmatrix} (\tfrac{1}{2}m_1 + m_2)glC_1 + \tfrac{1}{2}m_2 glC_{12} \\ \tfrac{1}{2}m_2 glC_{12} \end{bmatrix}$$

3.5 CONCLUDING REMARKS

Three different formulations for robot arm dynamics have been presented and discussed. The L-E equations of motion can be expressed in a well structured form, but they are computationally difficult to utilize for real-time control purposes unless they are simplified. The N-E formulation results in a very efficient set of recursive equations, but they are difficult to use for deriving advanced control laws. The G-D equations of motion give fairly well "structured" equations at the expense of higher computational cost. In addition to having faster computation time than the L-E equations of motion, the G-D equations of motion explicitly indicate the contributions of the translational and rotational effects of the links. Such information is useful for control analysis in obtaining an appropriate approximate model of a manipulator. Furthermore, the G-D equations of motion can be used in manipulator design. To briefly summarize the results, a user is able to choose between a formulation which is highly structured but computationally inefficient (L-E), a formulation which has efficient computations at the expense of the "structure" of the equations of motion (N-E), and a formulation which retains the "structure" of the problem with only a moderate computational penalty (G-D).

REFERENCES

Further reading on general concepts on dynamics can be found in several excellent mechanics books (Symon [1971] and Crandall et al. [1968]). The derivation of

Lagrange-Euler equations of motion using the 4×4 homogeneous transformation matrix was first carried out by Uicker [1965]. The report by Lewis [1974] contains a more detailed derivation of Lagrange-Euler equations of motion for a six-joint manipulator. An excellent report written by Bejczy [1974] reviews the details of the dynamics and control of an extended Stanford robot arm (the JPL arm). The report also discusses a scheme for obtaining simplified equations of motion. Exploiting the recursive nature of the lagrangian formulation, Hollerbach [1980] further improved the computation time of the generalized torques based on the lagrangian formulation.

Simplification of L-E equations of motion can be achieved via a differential transformation (Paul [1981]), a model reduction method (Bejczy and Lee [1983]), and an equivalent-composite approach (Luh and Lin [1981*b*]). The differential transformation technique converts the partial derivative of the homogeneous transformation matrices into a matrix product of the transformation and a differential matrix, thus reducing the acceleration-related matrix D_{ik} to a much simpler form. However, the Coriolis and centrifugal term, h_{ikm}, which contains the second-order partial derivative was not simplified by Paul [1981]. Bejczy and Lee [1983] developed the model reduction method which is based on the homogeneous transformation and on the lagrangian dynamics and utilized matrix numeric analysis technique to simplify the Coriolis and centrifugal term. Luh and Lin [1981*b*] utilized the N-E equations of motion and compared their terms in a computer to eliminate various terms and then rearranged the remaining terms to form the equations of motion in a symbolic form.

As an alternative to deriving more efficient equations of motion is to develop efficient algorithms for computing the generalized forces/torques based on the N-E equations of motion. Armstrong [1979], and Orin et al. [1979] were among the first to exploit the recursive nature of the Newton-Euler equations of motion. Luh et al. [1980*a*] improved the computations by referencing all velocities, accelerations, inertial matrices, location of the center of mass of each link, and forces/moments, to their own link coordinate frames. Walker and Orin [1982] extended the N-E formulation to computing the joint accelerations for computer simulation of robot motion.

Though the structure of the L-E and the N-E equations of motion are different, Turney et al. [1980] explicitly verified that one can obtain the L-E motion equations from the N-E equations, while Silver [1982] investigated the equivalence of the L-E and the N-E equations of motion through tensor analysis. Huston and Kelly [1982] developed an algorithmic approach for deriving the equations of motion suitable for computer implementation. Lee et al. [1983], based on the generalized d'Alembert principle, derived equations of motion which are expressed explicitly in vector-matrix form suitable for control analysis.

Neuman and Tourassis [1983] and Murray and Neuman [1984] developed computer software for obtaining the equations of motion of manipulators in symbolic form. Neuman and Tourassis [1985] developed a discrete dynamic model of a manipulator.

PROBLEMS

3.1 (*a*) What is the meaning of the generalized coordinates for a robot arm? (*b*) Give two *different* sets of generalized coordinates for the robot arm shown in the figure below. Draw two separate figures of the arm indicating the generalized coordinates that you chose.

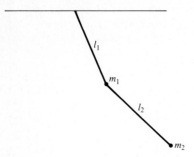

3.2 As shown in the figure below, a particle fixed in an intermediate coordinate frame (x_1, y_1, x_1) is located at $(-1, 1, 2)$ in that coordinate frame. The intermediate coordinate frame is moving translationally with a velocity of $3t\mathbf{i} + 2t\mathbf{j} + 4\mathbf{k}$ with respect to the reference frame (x_0, y_0, x_0) where $\mathbf{i}$, $\mathbf{j}$, and $\mathbf{k}$ are unit vectors along the x_0, y_0, and z_0 axes, respectively. Find the acceleration of the particle with respect to the reference frame.

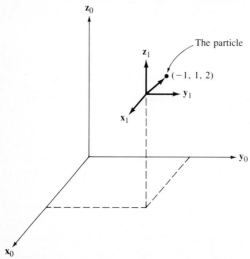

3.3 With reference to Secs. 3.3.1 and 3.3.2, a particle at rest in the starred coordinate system is located by a vector $\mathbf{r}(t) = 3t\mathbf{i} + 2t\mathbf{j} + 4\mathbf{k}$ with respect to the unstarred coordinate system (reference frame), where $(\mathbf{i}, \mathbf{j}, \mathbf{k})$ are unit vectors along the principal axes of the reference frame. If the starred coordinate frame is *only* rotating with respect to the reference frame with $\omega = (0, 0, 1)^T$, find the Coriolis and centripetal accelerations.

3.4 Discuss the differences between Eq. (3.3-13) and Eq. (3.3-17) when (*a*) $\mathbf{h} = \mathbf{0}$ and (*b*) $d\mathbf{h}/dt = \mathbf{0}$ (that is, $\mathbf{h}$ is a constant vector).

3.5 With references to the cube of mass M and side $2a$ shown in the figure below, (x_0, y_0, z_0) is the reference coordinate frame, $(\mathbf{u}, \mathbf{v}, \mathbf{w})$ is the body-attached coordinate frame, and (x_{cm}, y_{cm}, z_{cm}) is another body-attached coordinate frame at the center of mass of the cube. (*a*) Find the inertia tensor in the ($\mathbf{x}_0$, $\mathbf{y}_0$, $\mathbf{z}_0$) coordinate system. (*b*) Find the inertia tensor at the center of mass in the ($\mathbf{x}_{cm}$, $\mathbf{y}_{cm}$, $\mathbf{z}_{cm}$) coordinate system.

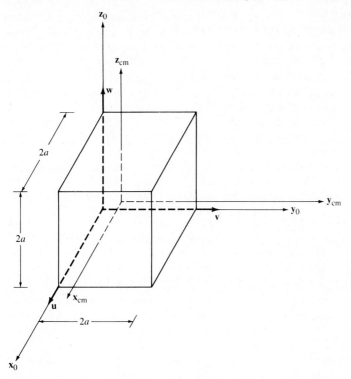

3.6 Repeat Prob. 3.5 for this rectangular block of mass M and sides $2a$, $2b$, and $2c$:

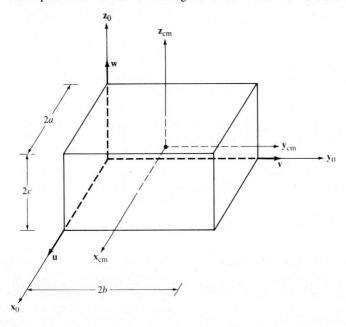

3.7 Assume that the cube in Prob. 3.5 is being rotated through an angle of α about the $\mathbf{z}_0$ axis and then rotated through an angle of θ about the $\mathbf{u}$ axis. Determine the inertia tensor in the $(\mathbf{x}_0, \mathbf{y}_0, \mathbf{z}_0)$ coordinate system.

3.8 Repeat Prob. 3.7 for the rectangular block in Prob. 3.6.

3.9 We learned that the Newton-Euler formulation of the dynamic model of a manipulator is computationally more efficient than the Lagrange-Euler formulation. However, most researchers still use the Lagrange-Euler formulation. Why is this so? (Give two reasons.)

3.10 A robotics researcher argues that if a robot arm is always moving at a very slow speed, then its Coriolis and centrifugal forces/torques can be omitted from the equations of motion formulated by the Lagrange-Euler approach. Will these "approximate" equations of motion be computationally more efficient than the Newton-Euler equations of motion? Explain and justify your answer.

3.11 We discussed two formulations for robot arm dynamics in this chapter, namely, the Lagrange-Euler formulation and the Newton-Euler formulation. Since they describe the same physical system, their equations of motion should be "equivalent." Given a set point on a preplanned trajectory at time t_1, $(\mathbf{q}^d(t_1), \dot{\mathbf{q}}^d(t_1), \ddot{\mathbf{q}}^d(t_1))$, one should be able to find the $\mathbf{D}(\mathbf{q}^d(t_1))$, the $\mathbf{h}(\mathbf{q}^d(t_1))$, $\dot{\mathbf{q}}^d(t_1))$, and the $\mathbf{c}(\mathbf{q}^d(t_1))$ matrices from the L-E equations of motion. Instead of finding them from the L-E equations of motion, can you state a *procedure* indicating how you can obtain the above matrices from the N-E equations of motion using the same set point from the trajectory?

3.12 The dynamic coefficients of the equations of motion of a manipulator can be obtained from the N-E equations of motion using the technique of *probing* as discussed in Prob. 3.11. Assume that N multiplications and M additions are required to compute the torques applied to the joint motors for a particular robot. What is the *smallest* number of multiplications and additions in terms of N, M, and n needed to find all the elements in the $\mathbf{D}(\mathbf{q})$ matrix in the L-E equations of motion, where n is the number of degrees of freedom of the robot?

3.13 In the Lagrange-Euler derivation of equations of motion, the gravity vector $\mathbf{g}$ given in Eq. (3.3-22) is a row vector of the form $(0, 0, -|g|, 0)$, where there is a negative sign for a level system. In the Newton-Euler formulation, the gravity effect as given in Table 3.2 is $(0, 0, |g|)^T$ for a level system, and there is no negative sign. Explain the discrepancy.

3.14 In the recursive Newton-Euler equations of motion referred to its own link coordinate frame, the matrix $(^i\mathbf{R}_0 \mathbf{I}_i \, ^0\mathbf{R}_i)$ is the inertial tensor of link i about the ith coordinate frame. Derive the relationship between this matrix and the pseudo-inertia matrix $\mathbf{J}_i$ of the Lagrange-Euler equations of motion.

3.15 Compare the differences between the representation of angular velocity and kinetic energy of the Lagrange-Euler and Newton-Euler equations of motion in the following table (fill in the blanks):

	Lagrange-Euler	Newton-Euler
Angular velocity		
Kinetic energy		

3.16 The two-link robot arm shown in the figure below is attached to the ceiling and under the influence of the gravitational acceleration $g = 9.8062$ m/sec^2; $(\mathbf{x}_0, \mathbf{y}_0, \mathbf{z}_0)$ is the reference frame; θ_1, θ_2 are the generalized coordinates; d_1, d_2 are the lengths of the links; and m_1, m_2 are the respective masses. Under the assumption of lumped equivalent masses, the mass of each link is lumped at the end of the link. (*a*) Find the link transformation matrices $^{i-1}\mathbf{A}_i$, $i = 1, 2$. (*b*) Find the pseudo-inertia matrix $\mathbf{J}_i$ for each link. (*c*) Derive the Lagrange-Euler equations of motion by first finding the elements in the $\mathbf{D}(\boldsymbol{\theta})$, $\mathbf{h}(\boldsymbol{\theta}, \dot{\boldsymbol{\theta}})$, and $\mathbf{c}(\boldsymbol{\theta})$ matrices.

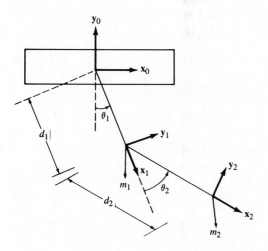

3.17 Given the same two-link robot arm as in Prob. 3.16, do the following steps to derive the Newton-Euler equations of motion and then compare them with the Lagrange-Euler equations of motion. (*a*) What are the initial conditions for the recursive Newton-Euler equations of motion? (*b*) Find the inertia tensor $^i\mathbf{R}_0\mathbf{I}_i\,^0\mathbf{R}_i$ for each link. (*c*) Find the other constants that will be needed for the recursive Newton-Euler equations of motion, such as $^i\mathbf{R}_0\bar{\mathbf{s}}_i$ and $^i\mathbf{R}_0\,\mathbf{p}_i^*$. (*d*) Derive the Newton-Euler equations of motion for this robot arm, assuming that $\mathbf{f}_{n+1}$ and $\mathbf{n}_{n+1}$ have zero reaction force/torque.

3.18 Use the Lagrange-Euler formulation to derive the equations of motion for the two-link $\theta - d$ robot arm shown below, where $(\mathbf{x}_0, \mathbf{y}_0, \mathbf{z}_0)$ is the reference frame, θ and d are the generalized coordinates, and m_1, m_2 are the link masses. Mass m_1 of link 1 is assumed to be located at a constant distance r_1 from the axis of rotation of joint 1, and mass m_2 of link 2 is assumed to be located at the end point of link 2.

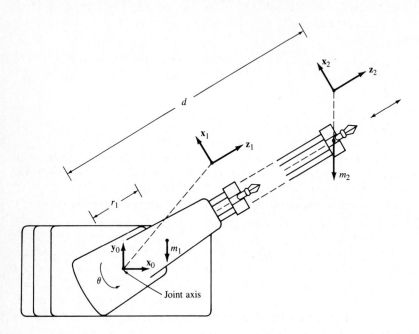

FOUR

PLANNING OF MANIPULATOR TRAJECTORIES

A mighty maze! but not without a plan.
Alexander Pope

4.1 INTRODUCTION

With the discussion of kinematics and dynamics of a serial link manipulator in the previous chapters as background, we now turn to the problem of controlling the manipulator so that it follows a preplanned path. Before moving a robot arm, it is of considerable interest to know whether there are any obstacles present in its path (obstacle constraint) and whether the manipulator hand must traverse a specified path (path constraint). These two constraints combined give rise to four possible control modes, as tabulated in Table 4.1. From this table, it is noted that the control problem of a manipulator can be conveniently divided into two coherent subproblems—motion (or trajectory) planning and motion control. This chapter focuses attention on various trajectory planning schemes for obstacle-free motion. It also deals with the formalism of describing the desired manipulator motion as sequences of points in space (position and orientation of the manipulator) through which the manipulator must pass, as well as the space curve that it traverses. The space curve that the manipulator hand moves along from the initial location (position and orientation) to the final location is called the *path*. We are interested in developing suitable formalisms for defining and describing the desired motions of the manipulator hand between the path endpoints.

Trajectory planning schemes generally "interpolate" or "approximate" the desired path by a class of polynomial functions and generates a sequence of time-based "control set points" for the control of the manipulator from the initial location to its destination. Path endpoints can be specified either in joint coordinates or in cartesian coordinates. However, they are usually specified in cartesian coordinates because it is easier to visualize the correct end-effector configurations in cartesian coordinates than in joint coordinates. Furthermore, joint coordinates are not suitable as a working coordinate system because the joint axes of most manipulators are not orthogonal and they do not separate position from orientation. If joint coordinates are desired at these locations, then the inverse kinematics solution routine can be called upon to make the necessary conversion.

Quite frequently, there exists a number of possible trajectories between the two given endpoints. For example, one may want to move the manipulator along

149

Table 4.1 Control modes of a manipulator

		Obstacle constraint	
		Yes	No
Path constraint	Yes	Off-line collision-free path planning plus on-line path tracking	Off-line path planning plus on-line path tracking
	No	Positional control plus on-line obstacle detection and avoidance	Positional control

a straight-line path that connects the endpoints (straight-line trajectory); or to move the manipulator along a smooth, polynomial trajectory that satisfies the position and orientation constraints at both endpoints (joint-interpolated trajectory). In this chapter, we discuss the formalisms for planning both joint-interpolated and straight-line path trajectories. We shall first discuss simple trajectory planning that satisfies path constraints and then extend the concept to include manipulator dynamics constraints.

A systematic approach to the trajectory planning problem is to view the trajectory planner as a black box, as shown in Fig. 4.1. The trajectory planner accepts input variables which indicate the constraints of the path and outputs a sequence of time-based intermediate configurations of the manipulator hand (position and orientation, velocity, and acceleration), expressed either in joint or cartesian coordinates, from the initial location to the final location. Two common approaches are used to plan manipulator trajectories. The first approach requires the user to explicitly specify a set of constraints (e.g., continuity and smoothness) on position, velocity, and acceleration of the manipulator's generalized coordinates at selected locations (called *knot points* or *interpolation points*) along the trajectory. The trajectory planner then selects a parameterized trajectory from a class of functions (usually the class of polynomial functions of degree n or less, for some n, in the time interval $[t_0, t_f]$) that "interpolates" and satisfies the constraints at the interpolation points. In the second approach, the user explicitly specifies the path that the manipulator must traverse by an analytical function, such as a straight-line path in cartesian coordinates, and the trajectory planner determines a desired trajectory either in joint coordinates or cartesian coordinates that approximates the desired path. In the first approach, the constraint specification and the planning of the manipulator trajectory are performed in joint coordinates. Since no constraints are imposed on the manipulator hand, it is difficult for the user to trace the path that

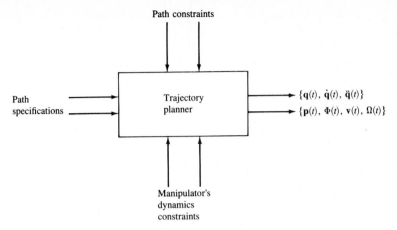

Figure 4.1 Trajectory planner block diagram.

the manipulator hand traverses. Hence, the manipulator hand may hit obstacles with no prior warning. In the second approach, the path constraints are specified in cartesian coordinates, and the joint actuators are servoed in joint coordinates. Hence, to find a trajectory that approximates the desired path closely, one must convert the cartesian path constraints to joint path constraints by some functional approximations and then find a parameterized trajectory that satisfies the joint path constraints.

The above two approaches for planning manipulator trajectories should result in simple trajectories that are meant to be efficient, smooth, and accurate with a fast computation time (near real time) for generating the sequence of control set points along the desired trajectory of the manipulator. However, the sequences of the time-based joint-variable space vectors $\{q(t), \dot{q}(t), \ddot{q}(t)\}$ are generated without taking the dynamics of the manipulator into consideration. Thus, large tracking errors may result in the servo control of the manipulator. We shall discuss this problem in Sec. 4.4.3. This chapter begins with a discussion of general issues that arise in trajectory planning in Sec. 4.2; joint-interpolated trajectory in Sec. 4.3; straight-line trajectory planning in Sec. 4.4; and a cubic polynomial trajectory along a straight-line path in joint coordinates with manipulator dynamics taken into consideration in Sec. 4.4.3. Section 4.5 summarizes the results.

4.2 GENERAL CONSIDERATIONS ON TRAJECTORY PLANNING

Trajectory planning can be conducted either in the joint-variable space or in the cartesian space. For joint-variable space planning, the time history of all joint variables and their first two time derivatives are planned to describe the desired motion of the manipulator. For cartesian space planning, the time history of the manipulator hand's position, velocity, and acceleration are planned, and the corresponding joint positions, velocities, and accelerations are derived from the hand information. Planning in the joint-variable space has three advantages: (1) the trajectory is planned directly in terms of the controlled variables during motion, (2) the trajectory planning can be done in near real time, and (3) the joint trajectories are easier to plan. The associated disadvantage is the difficulty in determining the locations of the various links and the hand during motion, a task that is usually required to guarantee obstacle avoidance along the trajectory.

In general, the basic algorithm for generating joint trajectory set points is quite simple:

$$t = t_0;$$
loop: Wait for next control interval;
$$t = t + \Delta t;$$
$\mathbf{h}(t)$ = where the manipulator joint position should be at time t;
If $t = t_f$, then exit;
go to *loop*;

where Δt is the control sampling period for the manipulator.

From the above algorithm, we see that the computation consists of a trajectory function (or trajectory planner) $\mathbf{h}(t)$ which must be updated in every control interval. Thus, four constraints are imposed on the planned trajectory. First, the trajectory set points must be readily calculable in a noniterative manner. Second, intermediate positions must be determined and specified deterministically. Third, the continuity of the joint position and its first two time derivatives must be guaranteed so that the planned joint trajectory is smooth. Finally, extraneous motions, such as "wandering," must be minimized.

The above four constraints on the planned trajectory will be satisfied if the time histories of the joint variables can be specified by polynomial sequences. If the joint trajectory for a given joint (say joint i) uses p polynomials, then $3(p + 1)$ coefficients are required to specify initial and terminal conditions (joint position, velocity, and acceleration) and guarantee continuity of these variables at the polynomial boundaries. If an additional intermediate condition such as position is specified, then an additional coefficient is required for each intermediate condition. In general, two intermediate positions may be specified: one near the initial position for departure and the other near the final position for arrival which will guarantee safe departure and approach directions, in addition to a better controlled motion. Thus, one seventh-degree polynomial for each joint variable connecting the initial and final positions would suffice, as would two quartic and one cubic (4-3-4) trajectory segments, two cubics and one quintic (3-5-3) trajectory segments, or five cubic (3-3-3-3-3) trajectory segments. This will be discussed further in the next section.

For cartesian path control, the above algorithm can be modified to:

$t = t_0$;

loop: Wait for next control interval;

$t = t + \Delta t$;

$\mathbf{H}(t) = $ where the manipulator hand should be at time t;

$\mathbf{Q}[\mathbf{H}(t)] = $ joint solution corresponding to $\mathbf{H}(t)$;

If $t = t_f$, then exit;

go to *loop*;

Here, in addition to the computation of the manipulator hand trajectory function $\mathbf{H}(t)$ at every control interval, we need to convert the cartesian positions into their corresponding joint solutions, $\mathbf{Q}[\mathbf{H}(t)]$. The matrix function $\mathbf{H}(t)$ indicates the desired location of the manipulator hand at time t and can be easily realized by a 4×4 transformation matrix, as discussed in Sec. 4.4.

Generally, cartesian path planning can be realized in two coherent steps: (1) generating or selecting a set of knot points or interpolation points in cartesian coordinates according to some rules along the cartesian path and then (2) specifying a class of functions to link these knot points (or to approximate these path segments) according to some criteria. For the latter step, the criteria chosen are quite often dictated by the following control algorithms to ensure the desired path tracking. There are two major approaches for achieving it: (1) The cartesian space-oriented method in which most of the computation and optimization is performed in cartesian coordinates and the subsequent control is performed at the hand level.† The servo sample points on the desired straight-line path are selected at a fixed servo interval and are converted into their corresponding joint solutions in real time while controlling the manipulator. The resultant trajectory is a piecewise straight line. Paul [1979], Taylor [1979], and Luh and Lin [1981] all reported methods for using a straight line to link adjacent cartesian knot points. (2) The joint space-oriented method in which a low-degree polynomial function in the joint-variable space is used to approximate the path segment bounded by two adjacent knot points on the straight-line path and the resultant control is done at the joint level.‡ The resultant cartesian path is a nonpiecewise straight line. Taylor's bounded deviation joint path (Taylor [1979]) and Lin's cubic polynomial trajectory method (Lin et al. [1983]) all used low-degree polynomials in the joint-variable space to approximate the straight-line path.

The cartesian space-oriented method has the advantage of being a straightforward concept, and a certain degree of accuracy is assured along the desired straight-line path. However, since all the available control algorithms are invariably based on joint coordinates because, at this time, there are no sensors capable

---

† The error actuating signal to the joint actuators is computed based on the error between the target cartesian position and the actual cartesian position of the manipulator hand.

‡ The error actuating signal to the joint actuators is computed based on the error between the target joint position and the actual joint position of the manipulator hand.

of measuring the manipulator hand in cartesian coordinates, cartesian space path planning requires transformations between the cartesian and joint coordinates in real time—a task that is computationally intensive and quite often leads to longer control intervals. Furthermore, the transformation from cartesian coordinates to joint coordinates is ill-defined because it is not a one-to-one mapping. In addition, if manipulator dynamics are included in the trajectory planning stage, then path constraints are specified in cartesian coordinates while physical constraints, such as torque and force, velocity, and acceleration limits of each joint motor, are bounded in joint coordinates. Thus, the resulting optimization problem will have mixed constraints in two different coordinate systems.

Because of the various disadvantages mentioned above, the joint space-oriented method, which converts the cartesian knot points into their corresponding joint coordinates and uses low-degree polynomials to interpolate these joint knot points, is widely used. This approach has the advantages of being computationally faster and makes it easier to deal with the manipulator dynamics constraints. However, it loses accuracy along the cartesian path when the sampling points fall on the fitted, smooth polynomials. We shall examine several planning schemes in these approaches in Sec. 4.4.

4.3 JOINT-INTERPOLATED TRAJECTORIES

To servo a manipulator, it is required that its robot arm's configuration at both the initial and final locations must be specified before the motion trajectory is planned. In planning a joint-interpolated motion trajectory for a robot arm, Paul [1972] showed that the following considerations are of interest:

1. When picking up an object, the motion of the hand must be directed away from an object; otherwise the hand may crash into the supporting surface of the object.
2. If we specify a departure position (lift-off point) along the normal vector to the surface out from the initial position and if we require the hand (i.e., the origin of the hand coordinate frame) to pass through this position, we then have an admissible departure motion. If we further specify the time required to reach this position, we could then control the speed at which the object is to be lifted.
3. The same set of lift-off requirements for the arm motion is also true for the set-down point of the final position motion (i.e., we must move to a normal point out from the surface and then slow down to the final position) so that the correct approach direction can be obtained and controlled.
4. From the above, we have four positions for each arm motion: initial, lift-off, set-down, and final (see Fig. 4.2).
5. Position constraints
   (a) Initial position: velocity and acceleration are given (normally zero).
   (b) Lift-off position: continuous motion for intermediate points.
   (c) Set-down position: same as lift-off position.
   (d) Final position: velocity and acceleration are given (normally zero).

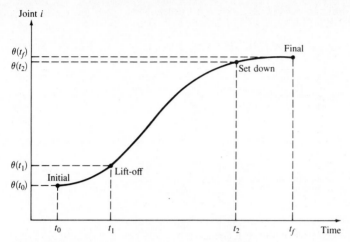

Figure 4.2 Position conditions for a joint trajectory.

6. In addition to these constraints, the extrema of all the joint trajectories must be within the physical and geometric limits of each joint.
7. Time considerations
    (*a*) Initial and final trajectory segments: time is based on the rate of approach of the hand to and from the surface and is some fixed constant based on the characteristics of the joint motors.
    (*b*) Intermediate points or midtrajectory segment: time is based on maximum velocity and acceleration of the joints, and the maximum of these times is used (i.e., the maximum time of the slowest joint is used for normalization).

The constraints of a typical joint trajectory are listed in Table 4.2. Based on these constraints, we are concerned with selecting a class of polynomial functions of degree n or less such that the required joint position, velocity, and acceleration at these knot points (initial, lift-off, set-down, and final positions) are satisfied, and the joint position, velocity, and acceleration are continuous on the entire time interval $[t_0, t_f]$. One approach is to specify a seventh-degree polynomial for each joint i,

$$q_i(t) = a_7 t^7 + a_6 t^6 + a_5 t^5 + a_4 t^4 + a_3 t^3 + a_2 t^2 + a_1 t + a_0 \qquad (4.3\text{-}1)$$

where the unknown coefficients a_j can be determined from the known positions and continuity conditions. However, the use of such a high-degree polynomial to interpolate the given knot points may not be satisfactory. It is difficult to find its extrema and it tends to have extraneous motion. An alternative approach is to split the entire joint trajectory into several trajectory segments so that different interpolating polynomials of a lower degree can be used to interpolate in each trajectory

Table 4.2 Constraints for planning joint-interpolated trajectory

Initial position:

1. Position (given)
2. Velocity (given, normally zero)
3. Acceleration (given, normally zero)

Intermediate positions:

4. Lift-off position (given)
5. Lift-off position (continuous with previous trajectory segment)
6. Velocity (continuous with previous trajectory segment)
7. Acceleration (continuous with previous trajectory segment)
8. Set-down position (given)
9. Set-down position (continuous with next trajectory segment)
10. Velocity (continuous with next trajectory segment)
11. Acceleration (continuous with next trajectory segment)

Final position:

12. Position (given)
13. Velocity (given, normally zero)
14. Acceleration (given, normally zero)

---

segment. There are different ways a joint trajectory can be split, and each method possesses different properties. The most common methods are the following:

4-3-4 *Trajectory*. Each joint has the following three trajectory segments: the first segment is a fourth-degree polynomial specifying the trajectory from the initial position to the lift-off position. The second trajectory segment (or midtrajectory segment) is a third-degree polynomial specifying the trajectory from the lift-off position to the set-down position. The last trajectory segment is a fourth-degree polynomial specifying the trajectory from the set-down position to the final position.

3-5-3 *Trajectory*. Same as 4-3-4 trajectory, but uses polynomials of different degrees for each segment: a third-degree polynomial for the first segment, a fifth-degree polynomial for the second segment, and a third-degree polynomial for the last segment.

5-Cubic *Trajectory*. Cubic spline functions of third-degree polynomials for five trajectory segments are used.

Note that the foregoing discussion is valid for each joint trajectory; that is, each joint trajectory is split into either a three-segment or a five-segment trajectory. The number of polynomials for a 4-3-4 trajectory of an N-joint manipulator will have N joint trajectories or $N \times 3 = 3N$ trajectory segments and $7N$ polynomial coefficients to evaluate plus the extrema of the $3N$ trajectory segments. We shall discuss the planning of a 4-3-4 joint trajectory and a 5-cubic joint trajectory in the next section.

4.3.1 Calculation of a 4-3-4 Joint Trajectory

Since we are determining N joint trajectories in each trajectory segment, it is convenient to introduce a normalized time variable, $t \in [0, 1]$, which allows us to treat the equations of each trajectory segment for each joint angle in the same way, with time varying from $t = 0$ (initial time for all trajectory segments) to $t = 1$ (final time for all trajectory segments). Let us define the following variables:

t : normalized time variable, $t \in [0, 1]$

τ : real time in seconds

τ_i : real time at the end of the ith trajectory segment

$t_i = \tau_i - \tau_{i-1}$: real time required to travel through the ith segment

$$t = \frac{\tau - \tau_{i-1}}{\tau_i - \tau_{i-1}}; \quad \tau \in [\tau_{i-1}, \tau_i]; \quad t \in [0, 1]$$

The trajectory consists of the polynomial sequences, $h_i(t)$, which together form the trajectory for joint j. The polynomial equations for each joint variable in each trajectory segment expressed in normalized time are:

$$h_1(t) = a_{14}t^4 + a_{13}t^3 + a_{12}t^2 + a_{11}t + a_{10} \qquad \text{(1st segment)} \qquad (4.3\text{-}2)$$

$$h_2(t) = a_{23}t^3 + a_{22}t^2 + a_{21}t + a_{20} \qquad \text{(2nd segment)} \qquad (4.3\text{-}3)$$

and $\quad h_n(t) = a_{n4}t^4 + a_{n3}t^3 + a_{n2}t^2 + a_{n1}t + a_{n0} \qquad \text{(last segment)} \qquad (4.3\text{-}4)$

The subscript of each polynomial equation indicates the segment number, and n indicates the last trajectory segment. The unknown coefficient a_{ji} indicates the ith coefficient for the j trajectory segment of a joint trajectory. The boundary conditions that this set of joint trajectory segment polynomials must satisfy are:

1. Initial position $= \theta_0 = \theta(t_0)$
2. Magnitude of initial velocity $= v_0$ (normally zero)
3. Magnitude of initial acceleration $= a_0$ (normally zero)
4. Lift-off position $= \theta_1 = \theta(t_1)$
5. Continuity in position at t_1 [that is, $\theta(t_1^-) = \theta(t_1^+)$]
6. Continuity in velocity at t_1 [that is, $v(t_1^-) = v(t_1^+)$]
7. Continuity in acceleration at t_1 [that is, $a(t_1^-) = a(t_1^+)$]
8. Set-down position $= \theta_2 = \theta(t_2)$
9. Continuity in position at t_2 [that is, $\theta(t_2^-) = \theta(t_2^+)$]
10. Continuity in velocity at t_2 [that is, $v(t_2^-) = v(t_2^+)$]
11. Continuity in acceleration at t_2 [that is, $a(t_2^-) = a(t_2^+)$]
12. Final position $= \theta_f = \theta(t_f)$

13. Magnitude of final velocity = v_f (normally zero)
14. Magnitude of final acceleration = a_f (normally zero)

The boundary conditions for the 4-3-4 joint trajectory are shown in Fig. 4.3. The first and second derivatives of these polynomial equations with respect to real time τ can be written as:

$$v_i(t) = \frac{dh_i(t)}{d\tau} = \frac{dh_i(t)}{dt}\frac{dt}{d\tau} = \frac{1}{\tau_i - \tau_{i-1}}\frac{dh_i(t)}{dt}$$

$$= \frac{1}{t_i}\frac{dh_i(t)}{dt} = \frac{1}{t_i}\dot{h}_i(t) \qquad i = 1, 2, n \qquad (4.3\text{-}5)$$

and

$$a_i(t) = \frac{d^2h_i(t)}{d\tau^2} = \frac{1}{(\tau_i - \tau_{i-1})^2}\frac{d^2h_i(t)}{dt^2}$$

$$= \frac{1}{t_i^2}\frac{d^2h_i(t)}{dt^2} = \frac{1}{t_i^2}\ddot{h}_i(t) \qquad i = 1, 2, n \qquad (4.3\text{-}6)$$

For the first trajectory segment, the governing polynomial equation is of the fourth degree:

$$h_1(t) = a_{14}t^4 + a_{13}t^3 + a_{12}t^2 + a_{11}t + a_{10} \qquad t \in [0, 1] \qquad (4.3\text{-}7)$$

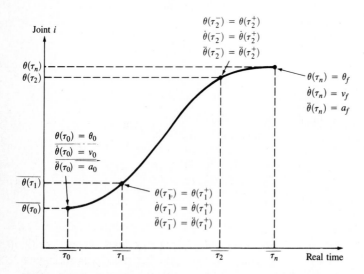

Figure 4.3 Boundary conditions for a 4-3-4 joint trajectory.

From Eqs. (4.3-5) and (4.3-6), its first two time derivatives with respect to real time are

$$v_1(t) = \frac{\dot{h}_1(t)}{t_1} = \frac{4a_{14}t^3 + 3a_{13}t^2 + 2a_{12}t + a_{11}}{t_1} \tag{4.3-8}$$

and

$$a_1(t) = \frac{\ddot{h}_1(t)}{t_1^2} = \frac{12a_{14}t^2 + 6a_{13}t + 2a_{12}}{t_1^2} \tag{4.3-9}$$

1. For $t = 0$ (at the initial position of this trajectory segment). Satisfying the boundary conditions at this position leads to

$$a_{10} = h_1(0) = \theta_0 \quad \text{(given)} \tag{4.3-10}$$

$$v_0 = \frac{\dot{h}_1(0)}{t_1} = \left[\frac{4a_{14}t^3 + 3a_{13}t^2 + 2a_{12}t + a_{11}}{t_1} \right]_{t=0} = \frac{a_{11}}{t_1} \tag{4.3-11}$$

which gives

$$a_{11} = v_0 t_1$$

and

$$a_0 = \frac{\ddot{h}_1(0)}{t_1^2} = \left[\frac{12a_{14}t^2 + 6a_{13}t + 2a_{12}}{t_1^2} \right]_{t=0} = \frac{2a_{12}}{t_1^2} \tag{4.3-12}$$

which yields

$$a_{12} = \frac{a_0 t_1^2}{2}$$

With these unknowns determined, Eq. (4.3-7) can be rewritten as:

$$h_1(t) = a_{14}t^4 + a_{13}t^3 + \left[\frac{a_0 t_1^2}{2} \right] t^2 + (v_0 t_1)t + \theta_0 \qquad t \in [0, 1] \tag{4.3-13}$$

2. For $t = 1$ (at the final position of this trajectory segment). At this position, we relax the requirement that the interpolating polynomial must pass through the position exactly. We only require that the velocity and acceleration at this position

have to be continuous with the velocity and acceleration, respectively, at the beginning of the next trajectory segment. The velocity and acceleration at this position are:

$$v_1(1) \triangleq v_1 = \frac{\dot{h}_1(1)}{t_1} = \frac{4a_{14} + 3a_{13} + a_0 t_1^2 + v_0 t_1}{t_1} \qquad (4.3\text{-}14)$$

$$a_1(1) \triangleq a_1 = \frac{\ddot{h}_1(1)}{t_1^2} = \frac{12a_{14} + 6a_{13} + a_0 t_1^2}{t_1^2} \qquad (4.3\text{-}15)$$

For the second trajectory segment, the governing polynomial equation is of the third degree:

$$h_2(t) = a_{23}t^3 + a_{22}t^2 + a_{21}t + a_{20} \qquad t \in [0, 1] \qquad (4.3\text{-}16)$$

1. For $t = 0$ (at the lift-off position). Using Eqs. (4.3-5) and (4.3-6), the velocity and acceleration at this position are, respectively,

$$h_2(0) = a_{20} = \theta_2(0) \qquad (4.3\text{-}17)$$

$$v_1 = \frac{\dot{h}_2(0)}{t_2} = \left[\frac{3a_{23}t^2 + 2a_{22}t + a_{21}}{t_2} \right]_{t=0} = \frac{a_{21}}{t_2} \qquad (4.3\text{-}18)$$

which gives

$$a_{21} = v_1 t_2$$

and

$$a_1 = \frac{\ddot{h}_2(0)}{t_2^2} = \left[\frac{6a_{23}t + 2a_{22}}{t_2^2} \right]_{t=0} = \frac{2a_{22}}{t_2^2} \qquad (4.3\text{-}19)$$

which yields

$$a_{22} = \frac{a_1 t_2^2}{2}$$

Since the velocity and acceleration at this position must be continuous with the velocity and acceleration at the end of the previous trajectory segment respectively, we have

$$\frac{\dot{h}_2(0)}{t_2} = \frac{\dot{h}_1(1)}{t_1} \qquad \text{and} \qquad \frac{\ddot{h}_2(0)}{t_2^2} = \frac{\ddot{h}_1(1)}{t_1^2} \qquad (4.3\text{-}20)$$

which, respectively, leads to

$$\left[\frac{3a_{23}t^2 + 2a_{22}t + a_{21}}{t_2} \right]_{t=0} = \left[\frac{4a_{14}t^3 + 3a_{13}t^2 + 2a_{12}t + a_{11}}{t_1} \right]_{t=1}$$

(4.3-21)

or

$$\frac{-a_{21}}{t_2} + \frac{4a_{14}}{t_1} + \frac{3a_{13}}{t_1} + \frac{a_0 t_1^2}{t_1} + \frac{v_0 t_1}{t_1} = 0 \qquad (4.3\text{-}22)$$

and

$$\left[\frac{6a_{23}t + 2a_{22}}{t_2^2} \right]_{t=0} = \left[\frac{12a_{14}t^2 + 6a_{13}t + 2a_{12}}{t_1^2} \right]_{t=1}$$

(4.3-23)

or

$$\frac{-2a_{22}}{t_2^2} + \frac{12a_{14}}{t_1^2} + \frac{6a_{13}}{t_1^2} + \frac{a_0 t_1^2}{t_1^2} = 0 \qquad (4.3\text{-}24)$$

2. For $t = 1$ (at the set-down position). Again the velocity and acceleration at this position must be continuous with the velocity and acceleration at the beginning of the next trajectory segment. The velocity and acceleration at this position are obtained, respectively, as:

$$h_2(1) = a_{23} + a_{22} + a_{21} + a_{20} \qquad (4.3\text{-}25)$$

$$v_2(1) = \frac{\dot{h}_2(1)}{t_2} = \left[\frac{3a_{23}t^2 + 2a_{22}t + a_{21}}{t_2} \right]_{t=1} \qquad (4.3\text{-}26)$$

$$= \frac{3a_{23} + 2a_{22} + a_{21}}{t_2}$$

and

$$a_2(1) = \frac{\ddot{h}_2(1)}{t_2^2} = \left[\frac{6a_{23}t + 2a_{22}}{t_2^2} \right]_{t=1} = \frac{6a_{23} + 2a_{22}}{t_2^2} \qquad (4.3\text{-}27)$$

For the last trajectory segment, the governing polynomial equation is of the fourth degree:

$$h_n(t) = a_{n4}t^4 + a_{n3}t^3 + a_{n2}t^2 + a_{n1}t + a_{n0} \qquad t \in [0, 1] \qquad (4.3\text{-}28)$$

If we substitute $\bar{t} = t - 1$ into t in the above equation, we have shifted the normalized time t from $t \in [0, 1]$ to $\bar{t} \in [-1, 0]$. Then Eq. (4.3-28) becomes

$$h_n(\bar{t}) = a_{n4}\bar{t}^4 + a_{n3}\bar{t}^3 + a_{n2}\bar{t}^2 + a_{n1}\bar{t} + a_{n0} \qquad \bar{t} \in [-1, 0] \qquad (4.3\text{-}29)$$

Using Eqs. (4.3-5) and (4.3-6), its first and second derivatives with respect to real time are

$$v_n(\bar{t}) = \frac{\dot{h}_n(\bar{t})}{t_n} = \frac{4a_{n4}\bar{t}^3 + 3a_{n3}\bar{t}^2 + 2a_{n2}\bar{t} + a_{n1}}{t_n} \qquad (4.3\text{-}30)$$

and

$$a_n(\bar{t}) = \frac{\ddot{h}_n(\bar{t})}{t_n^2} = \frac{12a_{n4}\bar{t}^2 + 6a_{n3}\bar{t} + 2a_{n2}}{t_n^2} \qquad (4.3\text{-}31)$$

1. For $\bar{t} = 0$ (at the final position of this segment). Satisfying the boundary conditions at this final position of the trajectory, we have

$$h_n(0) = a_{n0} = \theta_f \qquad (4.3\text{-}32)$$

$$v_f = \frac{\dot{h}_n(0)}{t_n} = \frac{a_{n1}}{t_n} \qquad (4.3\text{-}33)$$

which gives

$$a_{n1} = v_f t_n$$

and

$$a_f = \frac{\ddot{h}_n(0)}{t_n^2} = \frac{2a_{n2}}{t_n^2} \qquad (4.3\text{-}34)$$

which yields

$$a_{n2} = \frac{a_f t_n^2}{2}$$

2. For $\bar{t} = -1$ (at the starting position of this trajectory segment). Satisfying the boundary conditions at this position, we have, at the set-down position,

$$h_n(-1) = a_{n4} - a_{n3} + \frac{a_f t_n^2}{2} - v_f t_n + \theta_f = \theta_2(1) \qquad (4.3\text{-}35)$$

and

$$\frac{\dot{h}_n(-1)}{t_n} = \left[\frac{4a_{n4}\bar{t}^3 + 3a_{n3}\bar{t}^2 + 2a_{n2}\bar{t} + a_{n1}}{t_n}\right]_{\bar{t}=-1}$$

$$= \frac{-4a_{n4} + 3a_{n3} - a_f t_n^2 + v_f t_n}{t_n} \qquad (4.3\text{-}36)$$

and

$$\frac{\ddot{h}_n(-1)}{t_n^2} = \left[\frac{12a_{n4}\bar{t}^2 + 6a_{n3}\bar{t} + 2a_{n2}}{t_n^2}\right]_{\bar{t}=-1}$$

$$= \frac{12a_{n4} - 6a_{n3} + a_f t_n^2}{t_n^2} \qquad (4.3\text{-}37)$$

The velocity and acceleration continuity conditions at this set-down point are

$$\frac{\dot{h}_2(1)}{t_2} = \frac{\dot{h}_n(-1)}{t_n} \quad \text{and} \quad \frac{\ddot{h}_2(1)}{t_2^2} = \frac{\ddot{h}_n(-1)}{t_n^2} \qquad (4.3\text{-}38)$$

or

$$\frac{4a_{n4} - 3a_{n3} + a_f t_n^2 - v_f t_n}{t_n} + \frac{3a_{23}}{t_2} + \frac{2a_{22}}{t_2} + \frac{a_{21}}{t_2} = 0 \qquad (4.3\text{-}39)$$

and

$$\frac{-12a_{n4} + 6a_{n3} - a_f t_n^2}{t_n^2} + \frac{6a_{23}}{t_2^2} + \frac{2a_{22}}{t_2^2} = 0 \qquad (4.3\text{-}40)$$

The difference of joint angles between successive trajectory segments can be found to be

$$\delta_1 = \theta_1 - \theta_0 = h_1(1) - h_1(0) = a_{14} + a_{13} + \frac{a_0 t_1^2}{2} + v_0 t_1 \qquad (4.3\text{-}41)$$

$$\delta_2 = \theta_2 - \theta_1 = h_2(1) - h_2(0) = a_{23} + a_{22} + a_{21} \qquad (4.3\text{-}42)$$

and

$$\delta_n = \theta_f - \theta_2 = h_n(0) - h_n(-1) = -a_{n4} + a_{n3} - \frac{a_f t_n^2}{2} + v_f t_n \qquad (4.3\text{-}43)$$

All the unknown coefficients of the trajectory polynomial equations can be determined by simultaneously solving Eqs. (4.3-41), (4.3-22), (4.3-24), (4.3-42),

(4.3-39), (4.3-40), and (4.3-43). Rewriting them in matrix vector notation, we have

$$\mathbf{y} = \mathbf{C}\mathbf{x} \tag{4.3-44}$$

where

$$\mathbf{y} = \left[\delta_1 - \frac{a_0 t_1^2}{2} - v_0 t_1 , \ -a_0 t_1 - v_0 , \ -a_0 , \delta_2 , \right. \tag{4.3-45}$$

$$\left. -a_f t_n + v_f, a_f, \delta_n + \frac{a_f t_n^2}{2} - v_f t_n \right]^T$$

$$\mathbf{C} = \begin{bmatrix} 1 & 1 & 0 & 0 & 0 & 0 & 0 \\ 3/t_1 & 4/t_1 & -1/t_2 & 0 & 0 & 0 & 0 \\ 6/t_1^2 & 12/t_1^2 & 0 & -2/t_2^2 & 0 & 0 & 0 \\ 0 & 0 & 1 & 1 & 1 & 0 & 0 \\ 0 & 0 & 1/t_2 & 2/t_2 & 3/t_2 & -3/t_n & 4/t_n \\ 0 & 0 & 0 & 2/t_2^2 & 6/t_2^2 & 6/t_n^2 & -12/t_n^2 \\ 0 & 0 & 0 & 0 & 0 & 1 & -1 \end{bmatrix} \tag{4.3-46}$$

and

$$\mathbf{x} = (a_{13} , a_{14} , a_{21} , a_{22} , a_{23} , a_{n3} , a_{n4})^T \tag{4.3-47}$$

Then the planning of the joint trajectory (for each joint) reduces to solving the matrix vector equation in Eq. (4.3-44):

$$y_i = \sum_{j=1}^{7} c_{ij} x_j \tag{4.3-48}$$

or

$$\mathbf{x} = \mathbf{C}^{-1}\mathbf{y} \tag{4.3-49}$$

The structure of matrix $\mathbf{C}$ makes it easy to compute the unknown coefficients and the inverse of $\mathbf{C}$ always exists if the time intervals t_i, $i = 1, 2, n$ are positive values. Solving Eq. (4.3-49), we obtain all the coefficients for the polynomial equations for the joint trajectory segments for joint j.

Since we made a change in normalized time to run from [0, 1] to [−1, 0] for the last trajectory segment, after obtaining the coefficients a_{ni} from the above matrix equation, we need to reconvert the normalized time back to $[0, 1]$. This can be accomplished by substituting $t = \bar{t} + 1$ into $\bar{t}$ in Eq. (4.3-29). Thus we obtain

$$h_n(t) = a_{n4}t^4 + (-4a_{n4} + a_{n3})t^3 + (6a_{n4} - 3a_{n3} + a_{n2})t^2$$

$$+ (-4a_{n4} + 3a_{n3} - 2a_{n2} + a_{n1})t$$

$$+ (a_{n4} - a_{n3} + a_{n2} - a_{n1} + a_{n0}) \qquad t \in [0, 1] \qquad (4.3\text{-}50)$$

The resulting polynomial equations for the 4-3-4 trajectory, obtained by solving the above matrix equation, are listed in Table 4.3. Similarly, we can apply this technique to compute a 3-5-3 joint trajectory. This is left as an exercise to the reader. The polynomial equations for a 3-5-3 joint trajectory are listed in Table 4.4.

4.3.2 Cubic Spline Trajectory (Five Cubics)

The interpolation of a given function by a set of cubic polynomials, preserving continuity in the first and second derivatives at the interpolation points is known as cubic spline functions. The degree of approximation and smoothness that can be achieved is relatively good. In general, a spline curve is a polynomial of degree k with continuity of derivative of order $k - 1$, at the interpolation points. In the case of cubic splines, the first derivative represents continuity in the velocity and the second derivative represents continuity in the acceleration. Cubic splines offer several advantages. First, it is the lowest degree polynomial function that allows continuity in velocity and acceleration. Second, low-degree polynomials reduce the effort of computations and the possibility of numerical instabilities.

The general equation of five-cubic polynomials for each joint trajectory segment is:

$$h_j(t) = a_{j3}t^3 + a_{j2}t^2 + a_{j1}t + a_{j0} \qquad j = 1, 2, 3, 4, n \qquad (4.3\text{-}51)$$

with $\tau_{j-1} \leqslant \tau \leqslant \tau_j$ and $t \in [0, 1]$. The unknown coefficient a_{ji} indicates the ith coefficient for joint j trajectory segment and n indicates the last trajectory segment.

In using five-cubic polynomial interpolation, we need to have five trajectory segments and six interpolation points. However, from our previous discussion, we only have four positions for interpolation, namely, initial, lift-off, set-down, and final positions. Thus, two extra interpolation points must be selected to provide enough boundary conditions for solving the unknown coefficients in the polynomial sequences. We can select these two extra knot points between the lift-off and set-down positions. It is not necessary to know these two locations exactly; we only require that the time intervals be known and that continuity of velocity and acceleration be satisfied at these two locations. Thus, the boundary conditions that this set of joint trajectory segment polynomials must satisfy are (1) position constraints at the initial, lift-off, set-down, and final positions and (2) continuity of velocity and acceleration at all the interpolation points. The boundary conditions

Table 4.3 Polynomial equations for 4-3-4 joint trajectory

First trajectory segment:

$$h_1(t) = \left[\delta_1 - v_0 t_1 - \frac{a_0 t_1^2}{2} - \sigma \right] t^4 + \sigma t^3 + \left[\frac{a_0 t_1^2}{2} \right] t^2 + (v_0 t_1)t + \theta_0$$

$$v_1 = \frac{\dot{h}_1(1)}{t_1} = \frac{4\delta_1}{t_1} - 3v_0 - a_0 t_1 - \frac{\sigma}{t_1}$$

$$a_1 = \frac{\ddot{h}_1(1)}{t_1^2} = \frac{12\delta_1}{t_1^2} - \frac{12v_0}{t_1} - 5a_0 - \frac{6\sigma}{t_1^2}$$

Second trajectory segment:

$$h_2(t) = \left[\delta_2 - v_1 t_2 - \frac{a_1 t_2^2}{2} \right] t^3 + \left[\frac{a_1 t_2^2}{2} \right] t^2 + (v_1 t_2)t + \theta_1$$

$$v_2 = \frac{\dot{h}_2(1)}{t_2} = \frac{3\delta_2}{t_2} - 2v_1 - \frac{a_1 t_2}{2}$$

$$a_2 = \frac{\ddot{h}_2(1)}{t_2^2} = \frac{6\delta_2}{t_2^2} - \frac{6v_1}{t_2} - 2a_1 t_2$$

Last trajectory segment:

$$h_n(t) = \left[9\delta_n - 4v_2 t_n - \frac{a_2 t_n^2}{2} - 5v_f t_n + \frac{a_f t_n^2}{2} \right] t^4$$

$$+ \left[-8\delta_n + 5v_f t_n - \frac{a_f t_n^2}{2} + 3v_2 t_n \right] t^3 + \left[\frac{a_2 t_n^2}{2} \right] t^2 + (v_2 t_n)t + \theta_2$$

where $\sigma = f/g$ and

$$f = 2\delta_1 \left[4 + \frac{2t_n}{t_2} + \frac{2t_n}{t_1} + \frac{3t_2}{t_1} \right] - \frac{\delta_2 t_1}{t_2} \left[3 + \frac{t_n}{t_2} \right] + \frac{2\delta_n t_1}{t_n}$$

$$- v_0 t_1 \left[6 + \frac{6t_2}{t_1} + \frac{4t_n}{t_1} + \frac{3t_n}{t_2} \right] - v_f t_1 - a_0 t_1 t_n \left[\frac{5}{3} + \frac{t_1}{t_2} + \frac{2t_1}{t_n} + \frac{5t_2}{2t_n} \right] + a_f t_1 t_n$$

$$g = \frac{t_n}{t_2} + \frac{2t_n}{t_1} + 2 + \frac{3t_2}{t_1}$$

Table 4.4 Polynomial equations for a 3-5-3 joint trajectory

First trajectory segment:

$$h_1(t) = \left[\delta_1 - v_0 t_1 - \frac{a_0 t_1^2}{2} \right] t^3 + \left[\frac{a_0 t_1^2}{2} \right] t^2 + (v_0 t_1)t + \theta_0$$

$$v_1 = \frac{\dot{h}_1(1)}{t_1} = \frac{3\delta_1}{t_1} - 2v_0 - \frac{a_0 t_1}{2}$$

$$a_1 = \frac{\ddot{h}_1(1)}{t_1^2} = \frac{6\delta_1}{t_1^2} - \frac{6v_0}{t_1} - 2a_0$$

Second trajectory segment:

$$h_2(t) = \left[6\delta_2 - 3v_1 t_2 - 3v_2 t_2 - \frac{a_1 t_2^2}{2} + \frac{a_2 t_2^2}{2} \right] t^5$$

$$+ \left[-15\delta_2 + 8v_1 t_2 + 7v_2 t_2 + \frac{3a_1 t_2^2}{2} - a_2 t_2^2 \right] t^4$$

$$+ \left[10\delta_2 - 6v_1 t_2 - 4v_2 t_2 - \frac{3a_1 t_2^2}{2} + \frac{a_2 t_2^2}{2} \right] t^3 + \left[\frac{a_1 t_2^2}{2} \right] t^2$$

$$+ (v_1 t_2)t + \theta_1$$

$$v_2 = \frac{\dot{h}_2(1)}{t_2} = \frac{3\delta_n}{t_n} - 2v_f + \frac{a_f t_n}{2}$$

$$a_2 = \frac{\ddot{h}_2(1)}{t_2^2} = \frac{-6\delta_n}{t_n^2} + \frac{6v_f}{t_n} - 2a_f$$

Last trajectory segment:

$$h_n(t) = \left[\delta_n - v_f t_n + \frac{a_f t_n^2}{2} \right] t^3 + (-3\delta_n + 3v_f t_n - a_f t_n^2)t^2$$

$$+ \left[3\delta_n - 2v_f t_n + \frac{a_f t_n^2}{2} \right] t + \theta_2$$

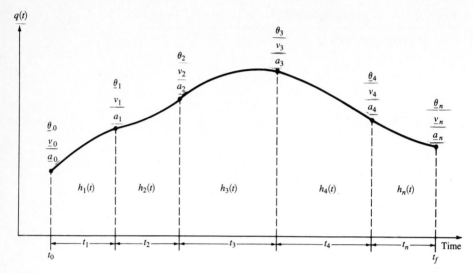

Figure 4.4 Boundary conditions for a 5-cubic joint trajectory.

for a five-cubic joint trajectory are shown in Fig. 4.4, where the underlined variables represent the known values before calculating the five-cubic polynomials.

The first and second derivatives of the polynomials with respect to real time are:

$$v_j(t) = \frac{\dot{h}_j(t)}{t_j} = \frac{3a_{j3}t^2 + 2a_{j2}t + a_{j1}}{t_j} \qquad j = 1, 2, 3, 4, n \qquad (4.3\text{-}52)$$

and

$$a_j(t) = \frac{\ddot{h}_j(t)}{t_j^2} = \frac{6a_{j3}t + 2a_{j2}}{t_j^2} \qquad j = 1, 2, 3, 4, n \qquad (4.3\text{-}53)$$

where t_j is the real time required to travel through the jth trajectory segment. Given the positions, velocities, and accelerations at the initial and final positions, the polynomial equations for the initial and final trajectory segments [$h_1(t)$ and $h_n(t)$] are completely determined. Once these polynomial equations are calculated, $h_2(t)$, $h_3(t)$, and $h_4(t)$ can be determined using the position constraints and continuity conditions.

For the first trajectory segment, the governing polynomial equation is

$$h_1(t) = a_{13}t^3 + a_{12}t^2 + a_{11}t + a_{10} \qquad (4.3\text{-}54)$$

At $t = 0$, satisfying the boundary conditions at this position, we have

$$h_1(0) = a_{10} = \theta_0 \qquad \text{(given)} \qquad (4.3\text{-}55)$$

$$v_0 \triangleq \frac{\dot{h}_1(0)}{t_1} = \frac{a_{11}}{t_1} \qquad (4.3\text{-}56)$$

from which

$$a_{11} = v_0 t_1$$

and $$a_0 \triangleq \frac{\ddot{h}_1(0)}{t_1^2} = \frac{2a_{12}}{t_1^2}$$ (4.3-57)

which yields

$$a_{12} = \frac{a_0 t_1^2}{2}$$

At $t = 1$, satisfying the position constraint at this position, we have

$$h_1(1) = a_{13} + \frac{a_0 t_1^2}{2} + v_0 t_1 + \theta_0 = \theta_1$$ (4.3-58)

from which a_{13} is found to be

$$a_{13} = \delta_1 - v_0 t_1 - \frac{a_0 t_1^2}{2}$$ (4.3-59)

where $\delta_i = \theta_i - \theta_{i-1}$. Thus, the first trajectory segment polynomial is completely determined:

$$h_1(t) = \left[\delta_1 - v_0 t_1 - \frac{a_0 t_1^2}{2} \right] t^3 + \left[\frac{a_0 t_1^2}{2} \right] t^2 + (v_0 t_1)t + \theta_0$$ (4.3-60)

With this polynomial equation, the velocity and acceleration at $t = 1$ are found to be

$$\frac{\dot{h}_1(1)}{t_1} \triangleq v_1 = \frac{3\delta_1 - (a_0 t_1^2)/2 - 2v_0 t_1}{t_1} = \frac{3\delta_1}{t_1} - 2v_0 - \frac{a_0 t_1}{2}$$ (4.3-61)

and $$\frac{\ddot{h}_1(1)}{t_1^2} \triangleq a_1 = \frac{6\delta_1 - 2a_0 t_1^2 - 6v_0 t_1}{t_1^2} = \frac{6\delta_1}{t_1^2} - \frac{6v_0}{t_1} - 2a_0$$ (4.3-62)

The velocity and acceleration must be continuous with the velocity and acceleration at the beginning of the next trajectory segment.

For the last trajectory segment, the polynomial equation is

$$h_n(t) = a_{n3}t^3 + a_{n2}t^2 + a_{n1}t + a_{n0}$$ (4.3-63)

At $t = 0$ and $t = 1$, satisfying the boundary conditions, we have

$$h_n(0) = a_{n0} = \theta_4 \quad \text{(given)} \tag{4.3-64}$$

$$h_n(1) = a_{n3} + a_{n2} + a_{n1} + \theta_4 = \theta_f \tag{4.3-65}$$

$$\frac{\dot{h}_n(1)}{t_n} \triangleq v_f = \frac{3a_{n3} + 2a_{n2} + a_{n1}}{t_n} \tag{4.3-66}$$

and

$$\frac{\ddot{h}_n(1)}{t_n^2} \triangleq a_f = \frac{6a_{n3} + 2a_{n2}}{t_n^2} \tag{4.3-67}$$

Solving the above three equations for the unknown coefficients a_{n3}, a_{n2}, a_{n1}, we obtain

$$h_n(t) = \left[\delta_n - v_f t_n + \frac{a_f t_n^2}{2} \right] t^3 + (-3\delta_n + 3v_f t_n - a_f t_n^2)t^2$$

$$+ \left[3\delta_n - 2v_f t_n + \frac{a_f t_n^2}{2} \right] t + \theta_4 \tag{4.3-68}$$

where $\delta_n = \theta_f - \theta_4$.

For the second trajectory segment, the equation is

$$h_2(t) = a_{23}t^3 + a_{22}t^2 + a_{21}t + a_{20} \tag{4.3-69}$$

At $t = 0$, satisfying the position constraint and the continuity of velocity and acceleration with the previous trajectory segment, we have

$$h_2(0) = a_{20} = \theta_1 \quad \text{(given)} \tag{4.3-70}$$

$$v_1 = \frac{\dot{h}_2(0)}{t_2} = \frac{a_{21}}{t_2} = \frac{\dot{h}_1(1)}{t_1} \tag{4.3-71}$$

so that

$$a_{21} = v_1 t_2$$

and

$$a_1 = \frac{\ddot{h}_2(0)}{t_2^2} = \frac{2a_{22}}{t_2^2} = \frac{\ddot{h}_1(1)}{t_1^2} \tag{4.3-72}$$

which gives

$$a_{22} = \frac{a_1 t_2^2}{2}$$

With these unknowns determined, the polynomial equation becomes

$$h_2(t) = a_{23}t^3 + \left[\frac{a_1 t_2^2}{2}\right] t^2 + (v_1 t_2)t + \theta_1 \tag{4.3-73}$$

where $\quad v_1 = \dfrac{3\delta_1}{t_1} - 2v_0 - \dfrac{a_0 t_1}{2} \qquad a_1 = \dfrac{6\delta_1}{t_1^2} - \dfrac{6v_0}{t_1} - 2a_0$

and a_{23} remains to be found. With this polynomial equation, at $t = 1$, we obtain the velocity and acceleration which must be continuous with the velocity and acceleration at the beginning of the next trajectory segment.

$$h_2(1) = \theta_2 = a_{23} + \frac{a_1 t_2^2}{2} + v_1 t_2 + \theta_1 \tag{4.3-74}$$

$$\frac{\dot{h}_2(1)}{t_2} = v_2 = \frac{3a_{23} + a_1 t_2^2 + v_1 t_2}{t_2} = v_1 + a_1 t_2 + \frac{3a_{23}}{t_2} \tag{4.3-75}$$

and $\quad \dfrac{\ddot{h}_2(1)}{t_2^2} = a_2 = \dfrac{6a_{23} + a_1 t_2^2}{t_2^2} = a_1 + \dfrac{6a_{23}}{t_2^2} \tag{4.3-76}$

Note that θ_2, v_2, and a_2 all depend on the value of a_{23}.

For the third trajectory segment, the equation is

$$h_3(t) = a_{33}t^3 + a_{32}t^2 + a_{31}t + a_{30} \tag{4.3-77}$$

At $t = 0$, satisfying the continuity of velocity and acceleration with the previous trajectory segment, we have

$$h_3(0) = a_{30} = \theta_2 = a_{23} + \frac{a_1 t_2^2}{2} + v_1 t_2 + \theta_1 \tag{4.3-78}$$

$$v_2 \triangleq \frac{\dot{h}_3(0)}{t_3} = \frac{a_{31}}{t_3} = \frac{\dot{h}_2(1)}{t_2} \tag{4.3-79}$$

so that

$$a_{31} = v_2 t_3$$

and $\quad a_2 \triangleq \dfrac{\ddot{h}_3(0)}{t_3^2} = \dfrac{2a_{32}}{t_3^2} = \dfrac{\ddot{h}_2(1)}{t_2^2} \tag{4.3-80}$

which yields

$$a_{32} = \frac{a_2 t_3^2}{2}$$

With these undetermined unknowns, the polynomial equation can be written as

$$h_3(t) = a_{33}t^3 + \left(\frac{a_2 t_3^2}{2}\right)t^2 + v_2 t_3 t + \theta_2 \qquad (4.3\text{-}81)$$

At $t = 1$, we obtain the velocity and acceleration which are continuous with the velocity and acceleration at the beginning of the next trajectory segment.

$$h_3(1) = \theta_3 = \theta_2 + v_2 t_3 + \frac{a_2 t_3^2}{2} + a_{33} \qquad (4.3\text{-}82)$$

$$\frac{\dot{h}_3(1)}{t_3} = v_3 = \frac{3a_{33} + a_2 t_3^2 + v_2 t_3}{t_3} = v_2 + a_2 t_3 + \frac{3a_{33}}{t_3} \qquad (4.3\text{-}83)$$

and $\quad \dfrac{\ddot{h}_3(1)}{t_3^2} = a_3 = \dfrac{6a_{33} + a_2 t_3^2}{t_3^2} = a_2 + \dfrac{6a_{33}}{t_3^2} \qquad (4.3\text{-}84)$

Note that θ_3, v_3, and a_3 all depend on a_{33} and implicitly depend on a_{23}.

For the fourth trajectory segment, the equation is

$$h_4(t) = a_{43}t^3 + a_{42}t^2 + a_{41}t + a_{40} \qquad (4.3\text{-}85)$$

At $t = 0$, satisfying the position constraint and the continuity of velocity and acceleration with the previous trajectory segment, we have

$$h_4(0) = a_{40} = \theta_3 = \theta_2 + v_2 t_3 + \frac{a_2 t_3^2}{2} + a_{33} \qquad (4.3\text{-}86)$$

$$v_3 = \frac{\dot{h}_4(0)}{t_4} = \frac{a_{41}}{t_4} = \frac{\dot{h}_3(1)}{t_3} \qquad (4.3\text{-}87)$$

which gives

$$a_{41} = v_3 t_4$$

and $\quad a_3 = \dfrac{\ddot{h}_4(0)}{t_4^2} = \dfrac{2a_{42}}{t_4^2} = \dfrac{\ddot{h}_3(1)}{t_3^2} \qquad (4.3\text{-}88)$

which yields

$$a_{42} = \frac{a_3 t_4^2}{2}$$

With these unknowns determined, the polynomial equation becomes

$$h_4(t) = a_{43}t^3 + \left[\frac{a_3 t_4^2}{2}\right]t^2 + (v_3 t_4)t + \theta_3 \qquad (4.3\text{-}89)$$

where θ_3, v_3, and a_3 are given in Eqs. (4.3-82), (4.3-83), and (4.3-84), respectively, and a_{23}, a_{33}, a_{43} remain to be found. In order to completely determine the polynomial equations for the three middle trajectory segments, we need to determine the coefficients a_{23}, a_{33}, and a_{43}. This can be done by matching the condition at the endpoint of trajectory $h_4(t)$ with the initial point of $h_5(t)$:

$$h_4(1) = a_{43} + \frac{a_3 t_4^2}{2} + v_3 t_4 + \theta_3 = \theta_4 \qquad (4.3\text{-}90)$$

$$\frac{\dot{h}_4(1)}{t_4} = \frac{3a_{43}}{t_4} + a_3 t_4 + v_3 = v_4 = \frac{3\delta_n}{t_n} - 2v_f + \frac{a_f t_n}{2} \qquad (4.3\text{-}91)$$

and $\quad \dfrac{\ddot{h}_4(1)}{t_4^2} = \dfrac{6a_{43}}{t_4^2} + a_3 = a_4 = \dfrac{-6\delta_n}{t_n^2} + \dfrac{6v_f}{t_n} - 2a_f \qquad (4.3\text{-}92)$

These three equations can be solved to determine the unknown coefficients a_{23}, a_{33}, and a_{43}. Solving for a_{23}, a_{33}, and a_{43}, the five-cubic polynomial equations are completely determined and they are listed below.

$$h_1(t) = \left[\delta_1 - v_0 t_1 - \frac{a_0 t_1^2}{2}\right]t^3 + \left[\frac{a_0 t_1^2}{2}\right]t^2 + (v_0 t_1)t + \theta_0 \qquad (4.3\text{-}93)$$

$$v_1 = \frac{3\delta_1}{t_1} - 2v_0 - \frac{a_0 t_1}{2} \qquad a_1 = \frac{6\delta_1}{t_1^2} - \frac{6v_0}{t_1} - 2a_0 \qquad (4.3\text{-}94)$$

$$h_2(t) = a_{23}t^3 + \left[\frac{a_1 t_2^2}{2}\right]t^2 + (v_1 t_2)t + \theta_1 \qquad (4.3\text{-}95)$$

$$\theta_2 = a_{23} + \frac{a_1 t_2^2}{2} + v_1 t_2 + \theta_1 \qquad (4.3\text{-}96)$$

$$v_2 = v_1 + a_1 t_2 + \frac{3a_{23}}{t_2} \qquad a_2 = a_1 + \frac{6a_{23}}{t_2^2} \qquad (4.3\text{-}97)$$

$$h_3(t) = a_{33}t^3 + \left[\frac{a_2 t_3^2}{2}\right]t^2 + v_2 t_3 t + \theta_2 \qquad (4.3\text{-}98)$$

$$\theta_3 = \theta_2 + v_2 t_3 + \frac{a_2 t_3^2}{2} + a_{33} \qquad (4.3\text{-}99)$$

$$v_3 = v_2 + a_2 t_3 + \frac{3a_{33}}{t_3} \qquad a_3 = a_2 + \frac{6a_{33}}{t_3^2} \tag{4.3-100}$$

$$h_4(t) = a_{43} t^3 + \left[\frac{a_3 t_4^2}{2} \right] t^2 + (v_3 t_4) t + \theta_3 \tag{4.3-101}$$

$$h_n(t) = \left[\delta_n - v_f t_n + \frac{a_f t_n^2}{2} \right] t^3 + (-3\delta_n + 3v_f t_n - a_f t_n^2) t^2 \tag{4.3-102}$$

$$+ \left[3\delta_n - 2v_f t_n + \frac{a_f t_n^2}{2} \right] t + \theta_4$$

$$v_4 = \frac{3\delta_n}{t_n} - 2v_f + \frac{a_f t_n}{2} \qquad a_4 = \frac{-6\delta_n}{t_n^2} + \frac{6v_f}{t_n} - 2a_f \tag{4.3-103}$$

$$a_{23} = t_2^2 \frac{x_1}{D} \qquad a_{33} = t_3^2 \frac{x_2}{D} \qquad a_{43} = t_4^2 \frac{x_3}{D} \tag{4.3-104}$$

with

$$x_1 = k_1(u - t_2) + k_2(t_4^2 - d) - k_3[(u - t_4)d + t_4^2(t_4 - t_2)] \tag{4.3-105}$$

$$x_2 = -k_1(u + t_3) + k_2(c - t_4^2) + k_3[(u - t_4)c + t_4^2(u - t_2)] \tag{4.3-106}$$

$$x_3 = k_1(u - t_4) + k_2(d - c) + k_3[(t_4 - t_2)c - d(u - t_2)] \tag{4.3-107}$$

$$D = u(u - t_2)(u - t_4) \tag{4.3-108}$$

$$u = t_2 + t_3 + t_4 \tag{4.3-109}$$

$$k_1 = \theta_4 - \theta_1 - v_1 u - a_1 \frac{u^2}{2} \tag{4.3-110}$$

$$k_2 = \frac{v_4 - v_1 - a_1 u - (a_4 - a_1)u/2}{3} \tag{4.3-111}$$

$$k_3 = \frac{a_4 - a_1}{6} \tag{4.3-112}$$

$$c = 3u^2 - 3ut_2 + t_2^2 \tag{4.3-113}$$

$$d = 3t_4^2 + 3t_3 t_4 + t_3^2 \tag{4.3-114}$$

So, it has been demonstrated that given the initial, the lift-off, the set-down, and the final positions, as well as the time to travel each trajectory (t_i), five-cubic polynomial equations can be uniquely determined to satisfy all the position constraints and continuity conditions. What we have just discussed is using a five-cubic polynomial to spline a joint trajectory with six interpolation points. A more general approach to finding the cubic polynomial for n interpolation points will be discussed in Sec. 4.4.3.

4.4 PLANNING OF CARTESIAN PATH TRAJECTORIES

In the last section, we described low-degree polynomial functions for generating joint-interpolated trajectory set points for the control of a manipulator. Although the manipulator joint coordinates fully specify the position and orientation of the manipulator hand, they are not suitable for specifying a goal task because most of the manipulator joint coordinates are not orthogonal and they do not separate position from orientation. For a more sophisticated robot system, programming languages are developed for controlling a manipulator to accomplish a task. In such systems, a task is usually specified as sequences of cartesian knot points through which the manipulator hand or end effector must pass. Thus, in describing the motions of the manipulator in a task, we are more concerned with the formalism of describing the target positions to which the manipulator hand has to move, as well as the space curve (or path) that it traverses.

Paul [1979] describes the design of manipulator cartesian paths made up of straight-line segments for the hand motion. The velocity and acceleration of the hand between these segments are controlled by converting them into the joint coordinates and smoothed by a quadratic interpolation routine. Taylor [1979] extended and refined Paul's method by using the dual-number quaternion representation to describe the location of the hand. Because of the properties of quaternions, transitions between the hand locations due to rotational operations require less computation, while the translational operations yield no advantage. We shall examine their approaches in designing straight-line cartesian paths in the next two sections.

4.4.1 Homogeneous Transformation Matrix Approach

In a programmable robotic system, the desired motion can be specified as sequences of cartesian knot points, each of which can be described in terms of homogeneous transformations relating the manipulator hand coordinate system to the workspace coordinate system. The corresponding joint coordinates at these cartesian knot points can be computed from the inverse kinematics solution routine and a quadratic polynominal can be used to smooth the two consecutive joint knot points in joint coordinates for control purposes. Thus, the manipulator hand is controlled to move along a straight line connected by these knot points. This technique has the advantage of enabling us to control the manipulator hand to track moving objects. Although the target positions are described by transforms, they do not specify how the manipulator hand is to be moved from one transform to another. Paul [1979] used a straight-line translation and two rotations to achieve the motion between two consecutive cartesian knot points. The first rotation is about a unit vector $\mathbf{k}$ and serves to align the tool or end-effector along the desired approach angle and the second rotation aligns the orientation of the tool about the tool axis.

In general, the manipulator target positions can be expressed in the following fundamental matrix equation:

$$\mathbf{^{0}T_{6}}\,\mathbf{^{6}T_{tool}} = \mathbf{^{0}C_{base}}(t)\,\mathbf{^{base}P_{obj}} \tag{4.4-1}$$

where

$\mathbf{^{0}T_{6}}$ = 4×4 homogeneous transformation matrix describing the manipulator hand position and orientation with respect to the base coordinate frame.

$\mathbf{^{6}T_{tool}}$ = 4×4 homogeneous transformation matrix describing the tool position and orientation with respect to the hand coordinate frame. It describes the tool endpoint whose motion is to be controlled.

$\mathbf{^{0}C_{base}}(t)$ = 4×4 homogeneous transformation matrix function of time describing the working coordinate frame of the object with respect to the base coordinate frame.

$\mathbf{^{base}P_{obj}}$ = 4×4 homogeneous transformation matrix describing the desired gripping position and orientation of the object for the end-effector with respect to the working coordinate frame.

If the $\mathbf{^{6}T_{tool}}$ is combined with $\mathbf{^{0}T_{6}}$ to form the arm matrix, then $\mathbf{^{6}T_{tool}}$ is a 4×4 identity matrix and can be omitted. If the working coordinate system is the same as the base coordinate system of the manipulator, then $\mathbf{^{0}C_{base}}(t)$ is a 4×4 identity matrix at all times.

Looking at Eq. (4.4-1), one can see that the left-hand-side matrices describe the gripping position and orientation of the manipulator, while the right-hand-side matrices describe the position and orientation of the feature of the object where we would like the manipulator's tool to grasp. Thus, we can solve for $\mathbf{^{0}T_{6}}$ which describes the configuration of the manipulator for grasping the object in a correct and desired manner:

$$\mathbf{^{0}T_{6}} = \mathbf{^{0}C_{base}}(t)\,\mathbf{^{base}P_{obj}}\,[\mathbf{^{6}T_{tool}}]^{-1} \tag{4.4-2}$$

If $\mathbf{^{0}T_{6}}$ were evaluated at a sufficiently high rate and converted into corresponding joint angles, the manipulator could be servoed to follow the trajectory.

Utilizing Eq. (4.4-1), a sequence of N target positions defining a task can be expressed as

$$\mathbf{^{0}T_{6}}\,(\mathbf{^{6}T_{tool}})_{1} = [\mathbf{^{0}C_{base}}(t)]_{1}\,(\mathbf{^{base}P_{obj}})_{1}$$

$$\mathbf{^{0}T_{6}}\,(\mathbf{^{6}T_{tool}})_{2} = [\mathbf{^{0}C_{base}}(t)]_{2}\,(\mathbf{^{base}P_{obj}})_{2} \tag{4.4-3}$$

$$\vdots$$

$$\mathbf{^{0}T_{6}}\,(\mathbf{^{6}T_{tool}})_{N} = [\mathbf{^{0}C_{base}}(t)]_{N}\,(\mathbf{^{base}P_{obj}})_{N}$$

Simplifying the notation of superscript and subscript in the above equation, we have

$$\mathbf{T}_6 {}^{\text{tool}}\mathbf{T}_1 = \mathbf{C}_1(t)\,\mathbf{P}_1$$

$$\mathbf{T}_6 {}^{\text{tool}}\mathbf{T}_2 = \mathbf{C}_2(t)\,\mathbf{P}_2 \tag{4.4-4}$$

$$\vdots$$

$$\mathbf{T}_6 {}^{\text{tool}}\mathbf{T}_N = \mathbf{C}_N(t)\,\mathbf{P}_N$$

From the positions defined by $\mathbf{C}_i(t)\,\mathbf{P}_i$ we can obtain the distance between consecutive points, and if we are further given linear and angular velocities, we can obtain the time requested T_i to move from position i to position $i+1$. Since tools and moving coordinate systems are specified at positions with respect to the base coordinate system, moving from one position to the next is best done by specifying both positions and tools with respect to the destination position. This has the advantage that the tool appears to be at rest from the moving coordinate system. In order to do this, we need to redefine the present position and tools with respect to the subsequent coordinate system. This can easily be done by redefining the $\mathbf{P}_i$ transform using a two subscript notation as $\mathbf{P}_{ij}$ which indicates the position $\mathbf{P}_i$ expressed with respect to the jth coordinate system. Thus, if the manipulator needs to be controlled from position 1 to position 2, then at position 1, expressing it with respect to its own coordinate system, we have

$$\mathbf{T}_6 {}^{\text{tool}}\mathbf{T}_1 = \mathbf{C}_1(t)\,\mathbf{P}_{11} \tag{4.4-5}$$

and expressing it with respect to the position 2 coordinate system, we have

$$\mathbf{T}_6 {}^{\text{tool}}\mathbf{T}_2 = \mathbf{C}_2(t)\,\mathbf{P}_{12} \tag{4.4-6}$$

We can now obtain $\mathbf{P}_{12}$ from these equations:

$$\mathbf{P}_{12} = \mathbf{C}_2^{-1}(t)\,\mathbf{C}_1(t)\,\mathbf{P}_{11}\,({}^{\text{tool}}\mathbf{T}_1)^{-1}\,{}^{\text{tool}}\mathbf{T}_2 \tag{4.4-7}$$

The purpose of the above equation is to find $\mathbf{P}_{12}$ given $\mathbf{P}_{11}$. Thus, the motion between any two consecutive positions i and $i+1$ can be stated as a motion from

$$\mathbf{T}_6 = \mathbf{C}_{i+1}(t)\,\mathbf{P}_{i,i+1}\,({}^{\text{tool}}\mathbf{T}_{i+1})^{-1} \tag{4.4-8}$$

to

$$\mathbf{T}_6 = \mathbf{C}_{i+1}(t)\,\mathbf{P}_{i+1,i+1}\,({}^{\text{tool}}\mathbf{T}_{i+1})^{-1} \tag{4.4-9}$$

where $\mathbf{P}_{i,i+1}$ and $\mathbf{P}_{i+1,i+1}$ represent transforms, as discussed above. Paul [1979] used a simple way to control the manipulator hand moving from one transform to the other. The scheme involves a translation and a rotation about a fixed axis in space coupled with a second rotation about the tool axis to produce controlled linear and angular velocity motion of the manipulator hand. The first rotation serves to align the tool in the required approach direction and the second rotation serves to align the orientation vector of the tool about the tool axis.

The motion from position i to position $i+1$ can be expressed in terms of a "drive" transform, $\mathbf{D}(\lambda)$, which is a function of a normalized time λ, as

$$\mathbf{T}_6(\lambda) = \mathbf{C}_{i+1}(\lambda)\,\mathbf{P}_{i,\,i+1}\,\mathbf{D}(\lambda)\,(^{\text{tool}}\mathbf{T}_{i+1})^{-1} \tag{4.4-10}$$

where

$\lambda = \dfrac{t}{T}$, $\lambda \in [0, 1]$

t = real time since the beginning of the motion

T = total time for the traversal of this segment

At position i, the real time is zero, λ is zero, $\mathbf{D}(0)$ is a 4×4 identity matrix, and

$$\mathbf{P}_{i+1,\,i+1} = \mathbf{P}_{i,\,i+1}\,\mathbf{D}(1) \tag{4.4-11}$$

which gives

$$\mathbf{D}(1) = (\mathbf{P}_{i,\,i+1})^{-1}\,\mathbf{P}_{i+1,\,i+1} \tag{4.4-12}$$

Expressing the positions i and $i+1$ in their respective homogeneous transform matrices, we have

$$\mathbf{P}_{i,\,i+1} \triangleq \mathbf{A} = \begin{bmatrix} \mathbf{n}_A & \mathbf{s}_A & \mathbf{a}_A & \mathbf{p}_A \\ 0 & 0 & 0 & 1 \end{bmatrix} = \begin{bmatrix} n_x^A & s_x^A & a_x^A & p_x^A \\ n_y^A & s_y^A & a_y^A & p_y^A \\ n_z^A & s_z^A & a_z^A & p_z^A \\ 0 & 0 & 0 & 1 \end{bmatrix} \tag{4.4-13}$$

and $\quad \mathbf{P}_{i+1,\,i+1} \triangleq \mathbf{B} = \begin{bmatrix} \mathbf{n}_B & \mathbf{s}_B & \mathbf{a}_B & \mathbf{p}_B \\ 0 & 0 & 0 & 1 \end{bmatrix} = \begin{bmatrix} n_x^B & s_x^B & a_x^B & p_x^B \\ n_y^B & s_y^B & a_y^B & p_y^B \\ n_z^B & s_z^B & a_z^B & p_z^B \\ 0 & 0 & 0 & 1 \end{bmatrix} \tag{4.4-14}$

Using Eq. (2.2-27) to invert $\mathbf{P}_{i,\,i+1}$ and multiply with $\mathbf{P}_{i+1,\,i+1}$, we obtain

$$\mathbf{D}(1) = \begin{bmatrix} \mathbf{n}_A \cdot \mathbf{n}_B & \mathbf{n}_A \cdot \mathbf{s}_B & \mathbf{n}_A \cdot \mathbf{a}_B & \mathbf{n}_A \cdot (\mathbf{p}_B - \mathbf{p}_A) \\ \mathbf{s}_A \cdot \mathbf{n}_B & \mathbf{s}_A \cdot \mathbf{s}_B & \mathbf{s}_A \cdot \mathbf{a}_B & \mathbf{s}_A \cdot (\mathbf{p}_B - \mathbf{p}_A) \\ \mathbf{a}_A \cdot \mathbf{n}_B & \mathbf{a}_A \cdot \mathbf{s}_B & \mathbf{a}_A \cdot \mathbf{a}_B & \mathbf{a}_A \cdot (\mathbf{p}_B - \mathbf{p}_A) \\ 0 & 0 & 0 & 1 \end{bmatrix} \tag{4.4-15}$$

where the dot indicates the scalar product of two vectors.

If the drive function consists of a translational motion and two rotational motions, then both the translation and the rotations will be directly proportional to λ. If λ varies linearly with the time, then the resultant motion represented by

$\mathbf{D}(\lambda)$ will correspond to a constant linear velocity and two angular velocities. The translational motion can be represented by a homogeneous transformation matrix $\mathbf{L}(\lambda)$ and the motion will be along the straight line joining $\mathbf{P}_i$ and $\mathbf{P}_{i+1}$. The first rotational motion can be represented by a homogeneous transformation matrix $\mathbf{R}_A(\lambda)$ and it serves to rotate the approach vector from $\mathbf{P}_i$ to the approach vector at $\mathbf{P}_{i+1}$. The second rotational motion represented by $\mathbf{R}_B(\lambda)$ serves to rotate the orientation vector from $\mathbf{P}_i$ into the orientation vector at $\mathbf{P}_{i+1}$ about the tool axis. Thus, the drive function can be represented as

$$\mathbf{D}(\lambda) = \mathbf{L}(\lambda)\,\mathbf{R}_A(\lambda)\,\mathbf{R}_B(\lambda) \qquad (4.4\text{-}16)$$

where

$$\mathbf{L}(\lambda) = \begin{bmatrix} 1 & 0 & 0 & \lambda x \\ 0 & 1 & 0 & \lambda y \\ 0 & 0 & 1 & \lambda z \\ 0 & 0 & 0 & 1 \end{bmatrix} \qquad (4.4\text{-}17)$$

$$\mathbf{R}_A(\lambda) = \begin{bmatrix} S^2\psi V(\lambda\theta) + C(\lambda\theta) & -S\psi C\psi V(\lambda\theta) & C\psi S(\lambda\theta) & 0 \\ -S\psi C\psi V(\lambda\theta) & C^2\psi V(\lambda\theta) + C(\lambda\theta) & S\psi S(\lambda\theta) & 0 \\ -C\psi S(\lambda\theta) & -S\psi S(\lambda\theta) & C(\lambda\theta) & 0 \\ 0 & 0 & 0 & 1 \end{bmatrix}$$

$$(4.4\text{-}18)$$

$$\mathbf{R}_B(\lambda) = \begin{bmatrix} C(\lambda\phi) & -S(\lambda\phi) & 0 & 0 \\ S(\lambda\phi) & C(\lambda\phi) & 0 & 0 \\ 0 & 0 & 1 & 0 \\ 0 & 0 & 0 & 1 \end{bmatrix} \qquad (4.4\text{-}19)$$

where

$$V(\lambda\theta) = \text{Versine}(\lambda\theta) = 1 - \cos(\lambda\theta) \qquad (4.4\text{-}20)$$

$$C(\lambda\theta) = \cos(\lambda\theta) \qquad S(\lambda\theta) = \sin(\lambda\theta)$$

$$C(\lambda\phi) = \cos(\lambda\phi) \qquad S(\lambda\phi) = \sin(\lambda\phi)$$

and $\lambda \in [0, 1]$. The rotation matrix $\mathbf{R}_A(\lambda)$ indicates a rotation of an angle θ about the orientation vector of $\mathbf{P}_i$ which is rotated an angle of ψ about the approach vector. $\mathbf{R}_B(\lambda)$ represents a rotation of ϕ about the approach vector of the tool at $\mathbf{P}_{i+1}$.

Multiplying the matrices in Eqs. (4.4-17) to (4.4-19) together, we have

$$\mathbf{D}(\lambda) = \begin{bmatrix} \mathbf{dn} & \mathbf{do} & \mathbf{da} & \mathbf{dp} \\ 0 & 0 & 0 & 1 \end{bmatrix} \tag{4.4-21}$$

where

$$\mathbf{do} = \begin{bmatrix} -S(\lambda\phi)[S^2\psi V(\lambda\theta) + C(\lambda\theta)] + C(\lambda\phi)[-S\psi C\psi V(\lambda\theta)] \\ -S(\lambda\phi)[-S\psi C\psi V(\lambda\theta)] + C(\lambda\theta)[C^2\psi V(\lambda\theta) + C(\lambda\theta)] \\ -S(\lambda\phi)[-C\psi S(\lambda\theta)] + C(\lambda\theta)[-S\psi S(\lambda\theta)] \end{bmatrix}$$

$$\mathbf{da} = \begin{bmatrix} C\psi S(\lambda\theta) \\ S\psi S(\lambda\theta) \\ C(\lambda\theta) \end{bmatrix} \qquad \mathbf{dp} = \begin{bmatrix} \lambda x \\ \lambda y \\ \lambda z \end{bmatrix}$$

and

$$\mathbf{dn} = \mathbf{do} \times \mathbf{da}$$

Using the inverse transform technique on Eq. (4.4-16), we may solve for x, y, z by postmultiplying Eq. (4.4-16) by $\mathbf{R}_B^{-1}(\lambda)\,\mathbf{R}_A^{-1}(\lambda)$ and equating the elements of the position vector,

$$x = \mathbf{n}_A \cdot (\mathbf{p}_B - \mathbf{p}_A)$$

$$y = \mathbf{s}_A \cdot (\mathbf{p}_B - \mathbf{p}_A) \tag{4.4-22}$$

$$z = \mathbf{a}_A \cdot (\mathbf{p}_B - \mathbf{p}_A)$$

By postmultiplying both sides of Eq. (4.4-16) by $\mathbf{R}_B^{-1}(\lambda)$ and then premultiplying by $\mathbf{L}^{-1}(\lambda)$, we can solve for θ and ψ by equating the elements of the third column with the elements of the third column from Eq. (4.4-16),

$$\psi = \tan^{-1}\left[\frac{\mathbf{s}_A \cdot \mathbf{a}_B}{\mathbf{n}_A \cdot \mathbf{a}_B}\right] \qquad -\pi \leqslant \psi < \pi \tag{4.4-23}$$

$$\theta = \tan^{-1}\left\{\frac{[(\mathbf{n}_A \cdot \mathbf{a}_B)^2 + (\mathbf{s}_A \cdot \mathbf{a}_B)^2]^{\frac{1}{2}}}{\mathbf{a}_A \cdot \mathbf{a}_B}\right\} \qquad 0 \leqslant \theta \leqslant \pi \tag{4.4-24}$$

To find ϕ, we premultiply both sides of Eq. (4.4-16) by $\mathbf{L}^{-1}(\lambda)$ and then $\mathbf{R}_A^{-1}(\lambda)$ and equate the elements to obtain

$$S\phi = -S\psi C\psi V(\lambda\theta)(\mathbf{n}_A \cdot \mathbf{n}_B) + [C^2\psi V(\lambda\theta) + C(\lambda\theta)](\mathbf{s}_A \cdot \mathbf{n}_B)$$

$$- S\psi S(\lambda\theta)(\mathbf{a}_A \cdot \mathbf{n}_B) \tag{4.4-25}$$

and $\quad C\phi = -S\psi C\psi V(\lambda\theta)(\mathbf{n}_A \cdot \mathbf{s}_B) + [C^2\psi V(\lambda\theta) + C(\lambda\theta)](\mathbf{s}_A \cdot \mathbf{s}_B)$

$$- S\psi S(\lambda\theta)(\mathbf{a}_A \cdot \mathbf{s}_B) \tag{4.4-26}$$

then

$$\phi = \tan^{-1} \left[\frac{S\phi}{C\phi} \right] \qquad -\pi \leqslant \phi < \pi \qquad (4.4\text{-}27)$$

Transition Between Two Path Segments. Quite often, a manipulator has to move on connected straight-line segments to satisfy a task motion specification or to avoid obstacles. In order to avoid discontinuity of velocity at the endpoint of each segment, we must accelerate or decelerate the motion from one segment to another segment. This can be done by initiating a change in velocity τ unit of time before the manipulator reaches an endpoint and maintaining the acceleration constant until τ unit of time into the new motion segment (see Fig. 4.5). If the acceleration for each variable is maintained at a constant value from time $-\tau$ to τ, then the acceleration necessary to change both the position and velocity is

$$\ddot{\mathbf{q}}(t) = \frac{1}{2\tau^2} \left[\Delta \mathbf{C} \frac{\tau}{T} + \Delta \mathbf{B} \right] \qquad (4.4\text{-}28)$$

where $-\tau < t < T$ and

$$\ddot{\mathbf{q}}(t) = \begin{bmatrix} \ddot{x} \\ \ddot{y} \\ \ddot{z} \\ \ddot{\theta} \\ \ddot{\phi} \end{bmatrix} \qquad \Delta \mathbf{C} = \begin{bmatrix} x_{BC} \\ y_{BC} \\ z_{BC} \\ \theta_{BC} \\ \phi_{BC} \end{bmatrix} \qquad \Delta \mathbf{B} = \begin{bmatrix} x_{BA} \\ y_{BA} \\ z_{BA} \\ \theta_{BA} \\ \phi_{BA} \end{bmatrix}$$

where $\Delta \mathbf{C}$ and $\Delta \mathbf{B}$ are vectors whose elements are cartesian distances and angles from points B to C and from points B to A, respectively.

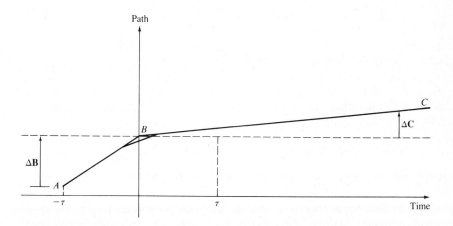

Figure 4.5 Straight line transition between two segments.

From Eq. (4.4-28), the velocity and position for $-\tau < t < \tau$ are given by

$$\dot{\mathbf{q}}(t) = \frac{1}{\tau} \left[\Delta\mathbf{C}\frac{\tau}{T} + \Delta\mathbf{B} \right] \lambda - \frac{\Delta\mathbf{B}}{\tau} \tag{4.4-29}$$

$$\mathbf{q}(t) = \left[\left(\Delta\mathbf{C}\frac{\tau}{T} + \Delta\mathbf{B} \right) \lambda - 2\,\Delta\mathbf{B} \right] \lambda + \Delta\mathbf{B} \tag{4.4-30}$$

where

$$\mathbf{q}(t) = \begin{bmatrix} x \\ y \\ z \\ \theta \\ \phi \end{bmatrix} \qquad \dot{\mathbf{q}}(t) = \begin{bmatrix} \dot{x} \\ \dot{y} \\ \dot{z} \\ \dot{\theta} \\ \dot{\phi} \end{bmatrix} \qquad \lambda \triangleq \frac{t + \tau}{2\,\tau} \tag{4.4-31}$$

For $\tau < t < T$, the motion is described by

$$\mathbf{q} = \Delta\mathbf{C}\lambda \qquad \dot{\mathbf{q}} = 0 \tag{4.4-32}$$

where

$$\lambda \triangleq \frac{t}{T}$$

It is noted that, as before, λ represents normalized time in the range $[0, 1]$. The reader should bear in mind, however, that the normalization factors are usually different for different time intervals.

For the motion from A to B and to C, we define a ψ as a linear interpolation between the motions for $-\tau < t < \tau$ as

$$\psi = (\psi_{BC} - \psi_{AB})\lambda + \psi_{AB} \tag{4.4-33}$$

where ψ_{AB} and ψ_{BC} are defined for the motion from A to B and from B to C, respectively, as in Eq. (4.4-23). Thus, ψ will change from ψ_{AB} to ψ_{BC}.

In summary, to move from a position $\mathbf{P}_i$ to a position $\mathbf{P}_{i+1}$, the drive function $\mathbf{D}(\lambda)$ is computed using Eqs. (4.4-16) to (4.4-27); then $\mathbf{T}_6(\lambda)$ can be evaluated by Eq. (4.4-10) and the corresponding joint values can be calculated from the inverse kinematics routine. If necessary, quadratic polynominal functions can be used to interpolate between the points obtained from the inverse kinematics routine.

Example: A robot is commanded to move in straight line motion to place a bolt into one the holes in the bracket shown in Fig. 4.6. Write down all the necessary matrix equations relationships as discussed above so that the robot can move along the dotted lines and complete the task. You may use symbols to indicate intermediate positions along the straight line paths.

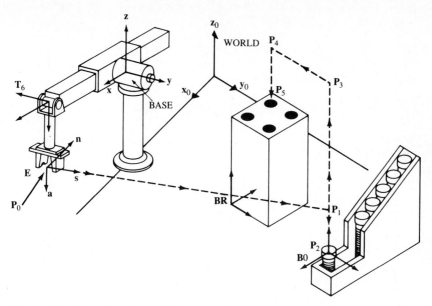

Figure 4.6 Figure for the example.

SOLUTION: Let $\mathbf{P}_i$ be the cartesian knot points that the manipulator hand must traverse ($i = 0, 1, 2, 3, 4, 5$). Then the governing matrix equations are:

At $\mathbf{P}_0$: $\quad$ **[BASE] [T6] [E] = [INIT] [P$_0$]** $\qquad\qquad$ (4.4-34)

At $\mathbf{P}_1$: $\quad$ **[BASE] [T6] [E] = [B0] [P$_1$]** $\qquad\qquad\quad$ (4.4-35)

At $\mathbf{P}_2$: $\quad$ **[BASE] [T6] [E] = [B0] [P$_2$]** $\qquad\qquad\quad$ (4.4-36)

At $\mathbf{P}_3$: $\quad$ **[BASE] [T6] [E] = [B0] [P$_3$]** $\qquad\qquad\quad$ (4.4-37)

At $\mathbf{P}_4$: $\quad$ **[BASE] [T6] [E] = [BR] [P$_4$]** $\qquad\qquad\quad$ (4.4-38)

At $\mathbf{P}_5$: $\quad$ **[BASE] [T6] [E] = [BR] [P$_5$]** $\qquad\qquad\quad$ (4.4-39)

where **[WORLD]**, **[BASE]**, **[INIT]**, **[B0]**, **[BR]**, **[T6]**, **[E]**, **[P$_0$]**, **[P$_1$]**, **[P$_2$]**, **[P$_3$]**, **[P$_4$]**, and **[P$_5$]** are 4×4 coordinate matrices. **[BASE]**, **[INIT]**, **[B0]**, and **[BR]** are expressed with respect to **[WORLD]**; **[T6]** is expressed with respect to **[BASE]**; **[E]** is expressed with respect to **[T6]**; **[P$_0$]** is expressed with respect to **[INIT]**; **[P$_1$]**, **[P$_2$]**, and **[P$_3$]** are expressed with respect to **[B0]**; and **[P$_4$]** and **[P$_5$]** are expressed with respect to **[BR]**.

To move from location $\mathbf{P}_0$ to location $\mathbf{P}_1$, we use the double subscript to describe Eq. (4.4-34) with respect to $\mathbf{P}_1$ coordinate frame. From Eq. (4.4-34), we have

$$[\mathbf{T6}] = [\mathbf{BASE}]^{-1} [\mathbf{INIT}] [\mathbf{P}_{00}] [\mathbf{E}]^{-1} \qquad (4.4\text{-}40)$$

and expressing it with respect to $\mathbf{P}_1$, we have

$$[\mathbf{T6}] = [\mathbf{BASE}]^{-1} [\mathbf{B0}] [\mathbf{P}_{01}] [\mathbf{E}]^{-1} \qquad (4.4\text{-}41)$$

Equating Eqs. (4.4-40) and (4.4-41), we have

$$[\mathbf{P}_{01}] = [\mathbf{B0}]^{-1} [\mathbf{INIT}] [\mathbf{P}_{00}] \qquad (4.4\text{-}42)$$

Thus, moving from location $\mathbf{P}_0$ to location $\mathbf{P}_1$ in a straight-line motion means that the manipulator hand must change configuration from

$$[\mathbf{T6}] = [\mathbf{BASE}]^{-1} [\mathbf{B0}] [\mathbf{P}_{01}] [\mathbf{E}]^{-1} \qquad (4.4\text{-}43a)$$

to $\qquad\qquad [\mathbf{T6}] = [\mathbf{BASE}]^{-1} [\mathbf{B0}] [\mathbf{P}_{11}] [\mathbf{E}]^{-1} \qquad (4.4\text{-}43b)$

Moving from locations $\mathbf{P}_i$ to $\mathbf{P}_{i+1}$, $i = 1, 2, 3, 4$, can be solved in the same manner. □

4.4.2 Planning Straight-Line Trajectories Using Quaternions

Paul's straight-line motion trajectory scheme uses the homogeneous transformation matrix approach to represent target position. This representation is easy to understand and use. However, the matrices are moderately expensive to store and computations on them require more operations than for some other representations. Furthermore, matrix representation for rotations is highly redundant, and this may lead to numerical inconsistencies. Taylor [1979] noted that using a quaternion to represent rotation will make the motion more uniform and efficient. He proposed two approaches to the problem of planning straight-line motion between knot points. The first approach, called *cartesian path control*, is a refinement of Paul's technique but using a quaternion representation for rotations. This method is simple and provides more uniform rotational motion. However, it requires considerable real time computation and is vulnerable to degenerate manipulator configurations. The second approach, called *bounded deviation joint path*, requires a motion planning phase which selects enough knot points so that the manipulator can be controlled by linear interpolation of joint values without allowing the manipulator hand to deviate more than a prespecified amount from the straight-line path. This approach greatly reduces the amount of computation that must be done at every sample interval.

Quaternion Representation. The quaternion concept has been successfully applied to the analysis of spatial mechanisms for the last several decades. We shall use quaternions to facilitate the representation of the orientation of a manipulator hand for planning a straight-line trajectory. A quaternion is a quadruple of ordered real numbers, s, a, b, c, associated, respectively, with four units: the real number $+1$, and three other units **i**, **j**, **k**, having cyclical permutation:

$$\mathbf{i}^2 = \mathbf{j}^2 = \mathbf{k}^2 = -1$$

$$\mathbf{ij} = \mathbf{k} \qquad \mathbf{jk} = \mathbf{i} \qquad \mathbf{ki} = \mathbf{j}$$

$$\mathbf{ji} = -\mathbf{k} \qquad \mathbf{kj} = -\mathbf{i} \qquad \mathbf{ik} = -\mathbf{j}$$

The units **i**, **j**, **k** of a quaternion may be interpreted as the three basis vectors of a cartesian set of axes. Thus, a quaternion Q may be written as a scalar part s and a vector part v:

$$Q = [s + \mathbf{v}] = s + a\mathbf{i} + b\mathbf{j} + c\mathbf{k} = (s, a, b, c) \qquad (4.4\text{-}44)$$

The following properties of quaternion algebra are basic:

Scalar part of Q: s

Vector part of Q: $a\mathbf{i} + b\mathbf{j} + c\mathbf{k}$

Conjugate of Q: $s - (a\mathbf{i} + b\mathbf{j} + c\mathbf{k})$

Norm of Q: $s^2 + a^2 + b^2 + c^2$

Reciprocal of Q: $\dfrac{s - (a\mathbf{i} + b\mathbf{j} + c\mathbf{k})}{s^2 + a^2 + b^2 + c^2}$

Unit quaternion: $s + a\mathbf{i} + b\mathbf{j} + c\mathbf{k}$, where $s^2 + a^2 + b^2 + c^2 = 1$

It is important to note that quaternions include the real numbers $(s, 0, 0, 0)$ with a single unit 1, the complex numbers $(s, a, 0, 0)$ with two units 1 and **i**, and the vectors $(0, a, b, c)$ in a three-dimensional space. The addition (subtraction) of two quaternions equals adding (subtracting) corresponding elements in the quadruples. The multiplication of two quaternions can be written as

$$Q_1 Q_2 = (s_1 + a_1\mathbf{i} + b_1\mathbf{j} + c_1\mathbf{k})(s_2 + a_2\mathbf{i} + b_2\mathbf{j} + c_2\mathbf{k})$$

$$= (s_1 s_2 - \mathbf{v}_1 \cdot \mathbf{v}_2 + s_2 \mathbf{v}_1 + s_1 \mathbf{v}_2 + \mathbf{v}_1 \times \mathbf{v}_2) \qquad (4.4\text{-}45)$$

and is obtained by distributing the terms on the right as in ordinary algebra, except that the order of the units must be preserved. In general, the product of two vectors in three-dimensional space, expressed as quaternions, is not a vector but a quaternion. That is, $Q_1 = [0 + \mathbf{v}_1] = (0, a_1, b_1, c_1)$ and $Q_2 = [0 + \mathbf{v}_2] = (0, a_2, b_2, c_2)$ and from Eq. (4.4-45),

$$Q_1 Q_2 = -\mathbf{v}_1 \cdot \mathbf{v}_2 + \mathbf{v}_1 \times \mathbf{v}_2$$

With the aid of quaternion algebra, finite rotations in space may be dealt with in a simple and efficient manner. If we use the notation

$$S = \sin\left(\frac{\theta}{2}\right) \qquad \text{and} \qquad C = \cos\left(\frac{\theta}{2}\right)$$

then we can represent a rotation Rot $(\mathbf{n}, \theta)$ of angle θ about an axis $\mathbf{n}$ by a quaternion,

$$\text{Rot}(\mathbf{n}, \theta) = \left[\cos\left(\frac{\theta}{2}\right) + \sin\left(\frac{\theta}{2}\right)\mathbf{n} \right] \tag{4.4-46}$$

Example: A rotation of $90°$ about $\mathbf{k}$ followed by a rotation of $90°$ about $\mathbf{j}$ is represented by the quaternion product

$$(\cos 45° + \mathbf{j}\sin 45°)(\cos 45° + \mathbf{k}\sin 45°) = (\tfrac{1}{2} + \mathbf{j}\tfrac{1}{2} + \mathbf{k}\tfrac{1}{2} + \mathbf{i}\tfrac{1}{2})$$

$$= \left[\tfrac{1}{2} + \frac{\mathbf{i} + \mathbf{j} + \mathbf{k}}{\sqrt{3}}\frac{\sqrt{3}}{2} \right]$$

$$= \left[\cos 60° + \sin 60°\frac{\mathbf{i} + \mathbf{j} + \mathbf{k}}{\sqrt{3}} \right]$$

$$= \text{Rot}\left[\frac{\mathbf{i} + \mathbf{j} + \mathbf{k}}{\sqrt{3}}, 120° \right]$$

The resultant rotation is a rotation of $120°$ about an axis equally inclined to the $\mathbf{i}, \mathbf{j}, \mathbf{k}$ axes. Note that we could represent the rotations about the $\mathbf{j}$ and $\mathbf{k}$ axes using the rotation matrices discussed in Chap. 2. However, the quaternion gives a much simpler representation. Thus, one can change the representation for rotations from quaternion to matrix or vice versa. □

For the remainder of this section, finite rotations will be represented in quaternion as Rot $(\mathbf{n}, \theta) = [\cos(\theta/2) + \sin(\theta/2)\mathbf{n}]$ for a rotation of angle θ about an axis $\mathbf{n}$. Table 4.5 lists the computational requirements of some common rotation operations, using quaternion and matrix representations.

Table 4.5 Computational requirements using quaternions and matrices

Operation	Quaternion representation	Matrix representation
$\mathbf{R}_1\mathbf{R}_2$	9 adds, 16 multiplies	15 adds, 24 multiplies
$\mathbf{R}v$	12 adds, 22 multiplies	6 adds, 9 multiplies
$\mathbf{R} \rightarrow \text{Rot}(\mathbf{n}, \theta)$	4 multiplies, 1 square root, 1 arctangent	8 adds, 10 multiplies, 2 square roots, 1 arctangent

Cartesian Path Control Scheme. It is required to move the manipulator's hand coordinate frame along a straight-line path between two knot points specified by $\mathbf{F}_0$ and $\mathbf{F}_1$ in time T, where each coordinate frame is represented by a homogeneous transformation matrix,

$$\mathbf{F}_i = \begin{bmatrix} \mathbf{R}_i & \mathbf{p}_i \\ 0 & 1 \end{bmatrix} .$$

The motion along the path consists of translation of the tool frame's origin from $\mathbf{p}_0$ to $\mathbf{p}_1$ coupled with rotation of the tool frame orientation part from $\mathbf{R}_0$ to $\mathbf{R}_1$. Let $\lambda(t)$ be the remaining fraction of the motion still to be traversed at time t. Then for uniform motion, we have

$$\lambda(t) = \frac{T - t}{T} \qquad (4.4\text{-}47)$$

where T is the total time needed to traverse the segment and t is time starting from the beginning of the segment traversal. The tool frame's position and orientation at time t are given, respectively, by

$$\mathbf{p}(t) = \mathbf{p}_1 - \lambda(t)(\mathbf{p}_1 - \mathbf{p}_0) \qquad (4.4\text{-}48)$$

$$\mathbf{R}(t) = \mathbf{R}_1 \, \mathrm{Rot}\,[\mathbf{n},\, -\theta \lambda(t)] \qquad (4.4\text{-}49)$$

where $\mathrm{Rot}(\mathbf{n}, \theta)$ is a rotation by θ about an axis $\mathbf{n}$ to reorient $\mathbf{R}_0$ into $\mathbf{R}_1$,

$$\mathrm{Rot}\,(\mathbf{n}, \theta) = \mathbf{R}_0^{-1}\mathbf{R}_1 \qquad (4.4\text{-}50)$$

where $\mathrm{Rot}\,(\mathbf{n}, \theta)$ represents the resultant rotation of $\mathbf{R}_0^{-1}\mathbf{R}_1$ in quaternion form. It is worth noting that $\mathbf{p}_1 - \mathbf{p}_0$ in Eq. (4.4-48) and $\mathbf{n}$ and θ in Eq. (4.4-49) need to be evaluated only once per segment if the frame $\mathbf{F}_1$ is fixed. On the other hand, if the destination point is changing, then $\mathbf{F}_1$ will be changing too. In this case, $\mathbf{p}_1 - \mathbf{p}_0$, $\mathbf{n}$, and θ should be evaluated per step. This can be accomplished by the pursuit formulation as described by Taylor [1979].

If the manipulator hand is required to move from one segment to another while maintaining constant acceleration, then it must accelerate or decelerate from one segment to the next. In order to accomplish this, the transition must start τ time before the manipulator reaches the intersection of the two segments and complete the transition to the new segment at time τ after the intersection with the new segment has been reached. From this requirement, the boundary conditions for the segment transition are

$$\mathbf{p}(T_1 - \tau) = \mathbf{p}_1 - \frac{\tau \Delta \mathbf{p}_1}{T_1} \qquad (4.4\text{-}51)$$

$$\mathbf{p}(T_1 + \tau) = \mathbf{p}_1 + \frac{\tau \Delta \mathbf{p}_2}{T_2} \tag{4.4-52}$$

$$\frac{d}{dt}\mathbf{p}(t)\Big|_{t = T_1 - \tau} = \frac{\Delta \mathbf{p}_1}{T_1} \tag{4.4-53}$$

$$\frac{d}{dt}\mathbf{p}(t)\Big|_{t = T_1 + \tau} = \frac{\Delta \mathbf{p}_2}{T_2} \tag{4.4-54}$$

where $\Delta \mathbf{p}_1 = \mathbf{p}_1 - \mathbf{p}_2$, $\Delta \mathbf{p}_2 = \mathbf{p}_2 - \mathbf{p}_1$, and T_1 and T_2 are the traversal times for the two segments. If we apply a constant acceleration to the transition,

$$\frac{d^2}{dt^2}\mathbf{p}(t) = \mathbf{a}_p \tag{4.4-55}$$

then integrating the above equation twice and applying the boundary conditions gives the position equation of the tool frame,

$$\mathbf{p}(t') = \mathbf{p}_1 - \frac{(\tau - t')^2}{4\tau T_1}\Delta \mathbf{p}_1 + \frac{(\tau + t')^2}{4\tau T_2}\Delta \mathbf{p}_2 \tag{4.4-56}$$

where $t' = T_1 - t$ is the time from the intersection of two segments. Similarly, the orientation equation of the tool frame is obtained as

$$\mathbf{R}(t) = \mathbf{R}_1 \text{Rot}\left[\mathbf{n}_1, \ -\frac{(\tau - t')^2}{4\tau T_1}\theta_1 \right] \text{Rot}\left[\mathbf{n}_2, \ -\frac{(\tau + t')^2}{4\tau T_2}\theta_2 \right] \tag{4.4-57}$$

where

$$\text{Rot}(\mathbf{n}_1, \theta_1) = \mathbf{R}_0^{-1}\mathbf{R}_1 \quad \text{and} \quad \text{Rot}(\mathbf{n}_2, \theta_2) = \mathbf{R}_1^{-1}\mathbf{R}_2$$

The last two terms represent the respective rotation matrix in quaternion form. The above equations for the position and orientation of the tool frame along the straight-line path produce a smooth transition between the two segments. It is worth pointing out that the angular acceleration will not be constant unless the axes $\mathbf{n}_1$ and $\mathbf{n}_2$ are parallel or unless one of the spin rates

$$\phi_1 = \frac{\theta_1}{T_1} \quad \text{or} \quad \phi_2 = \frac{\theta_2}{T_2}$$

is zero.

Bounded Deviation Joint Path. The cartesian path control scheme described above requires a considerable amount of computation time, and it is difficult to deal with the constraints on the joint-variable space behavior of the manipulator in real time. Several possible ways are available to deal with this problem. One could precompute and store the joint solution by simulating the real time algorithm before the execution of the motion. Then motion execution would be trivial as the servo set points could be read readily from memory. Another possible way is to precompute the joint solution for every nth sample interval and then perform joint interpolation using low-degree polynominals to fit through these intermediate points to generate the servo set points. The difficulty of this method is that the number of intermediate points required to keep the manipulator hand acceptably close to the cartesian straight-line path depends on the particular motion being made. Any predetermined interval small enough to guarantee small deviations will require a wasteful amount of precomputation time and memory storage. In view of this, Taylor [1979] proposed a joint variable space motion strategy called *bounded deviation joint path*, which selects enough intermediate points during the preplanning phase to guarantee that the manipulator hand's deviation from the cartesian straight-line path on each motion segment stays within prespecified error bounds.

The scheme starts with a precomputation of all the joint solution vectors q_i corresponding to the knot points F_i on the desired cartesian straight-line path. The joint-space vectors q_i are then used as knot points for a joint-variable space interpolation strategy analogous to that used for the position equation of the cartesian control path. That is, for motion from the knot point q_0 to q_1, we have

$$\mathbf{q}(t) = \mathbf{q}_1 - \frac{T_1 - t}{T_1} \Delta\mathbf{q}_1 \tag{4.4-58}$$

and, for transition between q_0 to q_1 and q_1 to q_2, we have

$$\mathbf{q}(t') = \mathbf{q}_1 - \frac{(\tau - t')^2}{4\tau T_1}\Delta\mathbf{q}_1 + \frac{(\tau + t')^2}{4\tau T_2}\Delta\mathbf{q}_2 \tag{4.4-59}$$

where $\Delta\mathbf{q}_1 = \mathbf{q}_1 - \mathbf{q}_2$, $\Delta\mathbf{q}_2 = \mathbf{q}_2 - \mathbf{q}_1$, and T_1, T_2, τ, and t' have the same meaning as discussed before. The above equations achieve uniform velocity between the joint knot points and make smooth transitions with constant acceleration between segments. However, the tool frame may deviate substantially from the desired straight-line path. The deviation error can be seen from the difference between the $F_j(t)$, which corresponds to the manipulator hand frame at the joint knot point $q_j(t)$, and $F_d(t)$, which corresponds to the the manipulator hand frame at the cartesian knot point $F_j(t)$. Defining the displacement and rotation deviations respectively as

$$\delta_p = |\mathbf{p}_j(t) - \mathbf{p}_d(t)| \tag{4.4-60}$$

$$\delta_R = |\text{angle part of Rot }(\mathbf{n}, \phi) = \mathbf{R}_d^{-1}(t)\,\mathbf{R}_j(t)| \tag{4.4-61}$$

$$= |\phi|$$

and specifying the maximum deviations, $\delta_p^{\max}$ and $\delta_R^{\max}$, for the displacement and orientation parts, respectively, we would like to bound the deviation errors as

$$\delta_p \leqslant \delta_p^{\max} \qquad \text{and} \qquad \delta_R \leqslant \delta_R^{\max} \qquad (4.4\text{-}62)$$

With this deviation error bounds, we need to select enough intermediate points between two consecutive joint knot points such that Eq. (4.4-62) is satisfied. Taylor [1979] presented a bounded deviation joint path which is basically a recursive bisector method for finding the intermediate points such that Eq. (4.4-62) is satisfied. The algorithm converges quite rapidly to produce a good set of intermediate points, though they are not a minimal set. His algorithm is as follows.

Algorithm BDJP: Given the maximum deviation error bounds $\delta_p^{\max}$ and $\delta_R^{\max}$ for the position and orientation of the tool frame, respectively, and the cartesian knot points $\mathbf{F}_i$ along the desired straight-line path, this algorithm selects enough joint knot points such that the manipulator hand frame will not deviate more than the prespecified error bounds along the desired straight-line path.

S1. *Compute joint solution.* Compute the joint solution vectors $\mathbf{q}_0$ and $\mathbf{q}_1$ corresponding to $\mathbf{F}_0$ and $\mathbf{F}_1$, respectively.

S2. *Find joint space midpoint.* Compute the joint-variable space midpoint

$$\mathbf{q}_m = \mathbf{q}_1 - \tfrac{1}{2} \Delta \mathbf{q}_1$$

where $\Delta \mathbf{q}_1 = \mathbf{q}_1 - \mathbf{q}_0$, and use $\mathbf{q}_m$ to compute the hand frame $\mathbf{F}_m$ corresponding to the joint values $\mathbf{q}_m$.

S3. *Find cartesian space midpoint.* Compute the corresponding cartesian path midpoint $\mathbf{F}_c$:

$$\mathbf{p}_c = \frac{\mathbf{p}_0 + \mathbf{p}_1}{2} \qquad \text{and} \qquad \mathbf{R}_c = \mathbf{R}_1 \operatorname{Rot}\left(\mathbf{n}_1, -\frac{\theta}{2}\right)$$

where $\operatorname{Rot}(\mathbf{n}, \theta) = \mathbf{R}_0^{-1} \mathbf{R}_1$.

S4. *Find the deviation errors.* Compute the deviation error between $\mathbf{F}_m$ and $\mathbf{F}_c$,

$$\delta_p = |\mathbf{p}_m - \mathbf{p}_c| \qquad \delta_R = |\text{angle part of } \operatorname{Rot}(\mathbf{n}, \phi) = \mathbf{R}_c^{-1} \mathbf{R}_m|$$

$$= |\phi|$$

S5. *Check error bounds.* If $\delta_p \leqslant \delta_p^{\max}$ and $\delta_R \leqslant \delta_R^{\max}$, then stop. Otherwise, compute the joint solution vector $\mathbf{q}_c$ corresponding to the cartesian space midpoint $\mathbf{F}_c$, and apply steps S2 to S5 recursively for the two subsegments by replacing $\mathbf{F}_1$ with $\mathbf{F}_c$ and $\mathbf{F}_c$ with $\mathbf{F}_0$.

Convergence of the above algorithm is quite rapid. The maximum deviation error is usually reduced by approximately a factor of 4 for each recursive iteration.

Taylor [1979] investigated the rate of convergence of the above algorithm for a cylindrical robot (two prismatic joints coupled with a rotary joint) and found that it ranges from a factor of 2 to a factor of 4 depending on the positions of the manipulator hand.

In summary, the bounded deviation joint path scheme relies on a preplanning phase to interpolate enough intermediate points in the joint-variable space so that the manipulator may be driven in the joint-variable space without deviating more than a prespecified error from the desired straight-line path.

4.4.3 Cubic Polynomial Joint Trajectories with Torque Constraint

Taylor's straight-line trajectory planning schemes generate the joint-space vectors $\{\mathbf{q}(t), \dot{\mathbf{q}}(t), \ddot{\mathbf{q}}(t)\}$ along the desired cartesian path without taking the dynamics of the manipulator into consideration. However, the actuator of each joint is subject to saturation and cannot furnish an unlimited amount of torque and force. Thus, torque and force constraints must be considered in the planning of straight-line trajectory. This suggests that the control of the manipulator should be considered in two coherent phases of execution: off-line optimum trajectory planning, followed by on-line path tracking control.

In planning a cartesian straight-line trajectory, the path is constrained in cartesian coordinates while the actuator torques and forces at each joint is bounded in joint coordinates. Hence, it becomes an optimization problem with mixed constraints (path and torque constraints) in two different coordinate systems. One must either convert the cartesian path into joint paths by some low-degree polynomial function approximation and optimize the joint paths and control the robot at the joint level (Lee and Chung [1984]); or convert the joint torque and force bounds into their corresponding cartesian bounds and optimize the cartesian path and control the robot at the hand level (Lee and Lee [1984]).

Although it involves numerous nonlinear transformations between the cartesian and joint coordinates, it is easier to approach the trajectory planning problem in the joint-variable space. Lin et al. [1983] proposed a set of joint spline functions to fit the segments among the selected knot points along the given cartesian path. This approach involves the conversion of the desired cartesian path into its functional representation of N joint trajectories, one for each joint. Since no transformation is known to map the straight-line path into its equivalent representation in the joint-variable space, the curve fitting methods must be used to approximate the cartesian path. Thus, to approximate a desired cartesian path in the joint-variable space, one can select enough knot points along the path, and each path segment specified by two adjacent knot points can then be interpolated by N joint polynomial functions, one function for each joint trajectory. These functions must pass through the selected knot points. Since cubic polynomial trajectories are smooth and have small overshoot of angular displacement between two adjacent knot points, Lin et al. [1983] adopted the idea of using cubic spline polynomials to fit the segment between two adjacent knots. Joint displacements for the $n - 2$

selected knot points are interpolated by piecewise cubic polynomials. In order to satisfy the continuity conditions for the joint displacement, velocity, and acceleration on the entire trajectory for the cartesian path, two extra knot points with unspecified joint displacements must be added to provide enough degrees of freedom for solving the cubic polynomials under continuity conditions. Thus, the total number of knot points becomes n and each joint trajectory consists of $n - 1$ piecewise cubic polynomials. Using the continuity conditions, the two extra knot points are then expressed as a combination of unknown variables and known constants. Then, only $n - 2$ equations need to be solved. The resultant matrix equation has a banded structure which facilitates computation. After solving the matrix equation, the resulting spline functions are expressed in terms of time intervals between adjacent knots. To minimize the total traversal time along the path, these time intervals must be adjusted subject to joint constraints within each of the $n - 1$ cubic polynomials. Thus, the problem reduces to an optimization of minimizing the total traveling time by adjusting the time intervals.

Let $\mathbf{H}(t)$ be the hand coordinate system expressed by a 4×4 homogeneous transformation matrix. The hand is required to pass through a sequence of n cartesian knot points, $[\mathbf{H}(t_1), \mathbf{H}(t_2), \ldots, \mathbf{H}(t_n)]$. The corresponding joint position vectors, $(q_{11}, q_{21}, \ldots, q_{N1})$, $(q_{12}, q_{22}, \ldots, q_{N2})$, $\ldots$, $(q_{1n}, q_{21}, \ldots, q_{Nn})$, at these n cartesian knot points can be solved using the inverse kinematics routine, where q_{ji} is the angular displacement of joint j at the ith knot point corresponding to $\mathbf{H}_i(t)$. Thus, the objective is to find a cubic polynomial trajectory for each joint j which fits the joint positions $[q_{j1}(t_1), q_{j2}(t_2), \ldots, q_{jn}(t_n)]$, where $t_1 < t_2 < \cdots < t_n$ is an ordered time sequence indicating when the hand should pass through these joint knot points. At the initial time $t = t_1$ and the final time $t = t_n$, the joint displacement, velocity, and acceleration are specified, respectively, as q_{j1}, v_{j1}, a_{j1} and q_{jn}, v_{jn}, a_{jn}. In addition, joint displacements q_{jk} at $t = t_k$ for $k = 3, 4, \ldots, n - 2$ are also specified for the joint trajectory to pass through. However, q_2 and q_{n-1} are not specified; these are the two extra knot points required to provide the freedom for solving the cubic polynomials.

Let $Q_{ji}(t)$ be the piecewise cubic polynomial function for joint j between the knot points $\mathbf{H}_i$ and $\mathbf{H}_{i+1}$, defined on the time interval $[t_i, t_{i+1}]$. Then the problem is to spline $Q_{ji}(t)$, for $i = 1, 2, \ldots, n-1$, together such that the required displacement, velocity, and acceleration are satisfied and are continuous on the entire time interval $[t_1, t_n]$. Since the polynomial $Q_{ji}(t)$ is cubic, its second-time derivative $\ddot{Q}_{ji}(t)$ must be a linear function of time t, $i = 1, \ldots, n-1$

$$Q_{ji}(t) = \frac{t_{i+1} - t_i}{u_i}\ddot{Q}_{ji}(t_i) + \frac{(t - t_i)}{u_i}\ddot{Q}_{ji}(t_{i+1}) \qquad j = 1, \ldots, N \qquad (4.4\text{-}63)$$

where $u_i = t_{i+1} - t_i$ is the time spent in traveling segment i. Integrating $\ddot{Q}_{ji}(t)$ twice and satisfying the boundary conditions of $Q_{ji}(t_i) = q_{ji}$ and

$Q_{ji}(t_{i+1}) = q_{j,i+1}$ leads to the following interpolating functions:

$$Q_{ji}(t) = \frac{\ddot{Q}_{ji}(t_i)}{6u_i}(t_{i+1} - t)^3 + \frac{\ddot{Q}_{ji}(t_{i+1})}{6u_i}(t - t_i)^3$$

$$+ \left[\frac{q_{j,i+1}}{u_i} - \frac{u_i \ddot{Q}_{ji}(t_{i+1})}{6} \right] (t-t_i)$$

$$+ \left[\frac{q_{j,i}}{u_i} - \frac{u_i \ddot{Q}_{ji}(t_i)}{6} \right] (t_{i+1} - t)$$

$$i = 1, \dots, n - 1$$

$$j = 1, \dots, N \tag{4.4-64}$$

Thus, for $i = 1, 2, \dots, n - 1$, $Q_{ji}(t)$ is determined if $\ddot{Q}_{ji}(t_i)$ and $\ddot{Q}_{ji}(t_{i+1})$ are known. This leads to a system of $n - 2$ linear equations with unknowns $\ddot{Q}_{ji}(t_i)$ for $i = 2, \dots, n - 1$ and knowns u_i for $i = 1, 2, \dots, n - 1$,

$$\mathbf{A\ddot{Q} = b} \tag{4.4-65}$$

where

$$\mathbf{\ddot{Q}} = \begin{bmatrix} \ddot{Q}_{j2}(t_2) \\ \ddot{Q}_{j3}(t_3) \\ \cdot \\ \cdot \\ \cdot \\ \ddot{Q}_{j,n-1}(t_{n-1}) \end{bmatrix}$$

$$\mathbf{A} = \begin{bmatrix} 3u_1 + 2u_2 + \dfrac{u_1^2}{u_2} & u_2 & 0 & 0 & 0 & 0 & 0 & 0 \\[2mm] u_2 - \dfrac{u_1^2}{u_2} & 2(u_2 + u_3) & u_3 & \cdot & \cdot & \cdot & \cdot & \cdot \\[2mm] 0 & u_3 & 2(u_3 + u_4) & \cdot & \cdot & \cdot & \cdot & \\[2mm] \cdot & 0 & u_4 & 2(u_4 + u_{n-3}) & \cdot & u_{n-3} & & \\[2mm] \cdot & \cdot & 0 & \cdot & \cdot & \cdot & 2(u_{n-3} + u_{n-2}) & u_{n-2} - \dfrac{u_{n-1}^2}{u_{n-2}} \\[2mm] 0 & 0 & 0 & 0 & 0 & 0 & u_{n-2} & 3u_{n-1} + 2u_{n-2} + \dfrac{u_{n-1}^2}{u_{n-2}} \end{bmatrix}$$

and

$$
\mathbf{b} =
\begin{bmatrix}
6\left[\dfrac{q_{j3}}{u_2} + \dfrac{q_{j1}}{u_1}\right] - 6\left[\dfrac{1}{u_1} + \dfrac{1}{u_2}\right]\left[q_{j1} + u_1 v_{j1} + \dfrac{u_1^2}{3}a_{j1}\right] - u_1 a_{j1} \\[2ex]
\dfrac{6}{u_2}\left[q_{j1} + u_1 v_{j1} + \dfrac{u_1^2}{3}a_{j1}\right] + \dfrac{6q_{j4}}{u_3} - 6\left[\dfrac{1}{u_2} + \dfrac{1}{u_3}\right]q_{j3} \\[2ex]
6\left[\dfrac{q_{j5} - q_{j4}}{u_4} - \dfrac{q_4 - q_{j3}}{u_3}\right] \\[2ex]
6\left[\dfrac{q_{j5} - q_{j4}}{u_4} - \dfrac{q_4 - q_{j3}}{u_3}\right] \\[2ex]
\cdot \\
\cdot \\
\cdot \\[1ex]
\dfrac{6}{u_{n-2}}\left[q_{jn} - v_{jn}u_{n-1} + \dfrac{u_{n-1}^2}{3}a_{jn}\right] - 6\left[\dfrac{1}{u_{n-2}} + \dfrac{1}{u_{n-3}}\right]q_{j,n-2} + \dfrac{6}{u_{n-3}}q_{j,n-3} \\[2ex]
-6\left[\dfrac{1}{u_{n-1}} + \dfrac{1}{u_{n-2}}\right]\left[q_{jn} - v_{jn}u_{n-1} + \dfrac{u_{n-1}^2}{3}a_{jn}\right] + \dfrac{6q_{jn}}{u_{n-1}} + \dfrac{6q_{n-2}}{u_{n-1}u_{n-2}} - u_{n-1}a_{jn}
\end{bmatrix}
$$

The banded structure of the matrix $\mathbf{A}$ makes it easy to solve for $\ddot{\mathbf{Q}}$ which is substituted into Eq. (4.4-63) to obtain the resulting solution $Q_{ji}(t)$. The resulting solution $Q_{ji}(t)$ is given in terms of time intervals u_i and the given values of joint displacements, velocities, and accelerations. The above banded matrix $\mathbf{A}$ of Eq. (4.4-64) is always nonsingular if the time intervals u_i are positive. Thus, the cubic polynomial joint trajectory always has a unique solution.

Since the actuator of each joint motor is subject to saturation and cannot furnish an unlimited amount of torque and force, the total time spent on traveling the specified path approximated by the cubic polynomials is constrainted by the maximum values of each joint velocity, acceleration, and jerk which is the rate of change of acceleration. In order to maximize the speed of traversing the path, the total traveling time for the manipulator must be minimized. This can be achieved by adjusting the time intervals u_i between two adjacent knot points subject to the velocity, acceleration, jerk, and torque constraints. The problem can then be stated as:

Minimize the objective function

$$
T = \sum_{i=1}^{n-1} u_i \tag{4.4-66}
$$

subject to the following constraints:

Velocity constraint: $|\dot{Q}_{ji}(t)| \leqslant V_j \qquad \begin{aligned} j &= 1, \ldots, N \\ i &= 1, \ldots, n-1 \end{aligned}$

Acceleration constraint: $|\ddot{Q}_{ji}(t)| \leqslant A_j$

$j = 1, \ldots, N$

$i = 1, \ldots, n-1$

Jerk constraint: $\left| \dfrac{d^3}{dt^3} Q_{ji}(t) \right| \leqslant J_j$

$j = 1, \ldots, N$

$i = 1, \ldots, n-1$

Torque constraint: $|\tau_j(t)| \leqslant \Gamma_j$ $\qquad j = 1, \ldots, N$

where T is the total traveling time, V_j, A_j, J_j, and Γ_j are, respectively, the velocity, acceleration, jerk, and torque limits of joint j.

The above constraints can be expressed in explicit forms as follows.

Velocity Constraints. Differentiating Eq. (4.4-64), and replacing $\ddot{Q}_{ji}(t_i)$ and $\ddot{Q}_{ji}(t_{i+1})$ by ω_{ji} and $\omega_{j,i+1}$, respectively, leads to the following expressions for $\dot{Q}_{ji}(t)$ and $\ddot{Q}_{ji}(t)$:

$$\dot{Q}_{ji}(t) = \frac{\omega_{ji}}{2u_i}(t_{i+1} - t)^2 + \frac{\omega_{j,i+1}}{2u_i}(t - t_i)^2 + \left[\frac{q_{j,i+1}}{u_i} - \frac{u_i \omega_{j,i+1}}{6} \right]$$

$$- \left[\frac{q_{ji}}{u_i} - \frac{u_i \omega_{ji}}{6} \right] \tag{4.4-67}$$

and $\quad \ddot{Q}_{ji}(t) = \dfrac{\omega_{j,i+1}}{u_i}(t - t_i) - \dfrac{\omega_{ji}}{u_i}(t - t_{i+1}) \tag{4.4-68}$

where ω_{ji} is the acceleration at $\mathbf{H}_i$ and equal to $\ddot{Q}_{ji}(t_i)$ if the time instant at which $Q_{ji}(t)$ passes through $\mathbf{H}_i$ is t_i.

The maximum absolute value of velocity exists at t_i, t_{i+1}, or $\bar{t}_i$, where $\bar{t}_i \in [t_i, t_{i+1}]$ and satisfies $\ddot{Q}_{ji}(\bar{t}_i) = 0$. The velocity constraints then become

$$\max_{t \in [t_i, t_{i+1}]} |\dot{Q}_{ji}| = \max \left[|\dot{Q}_{ji}(t_i)|, |\dot{Q}_{ji}(t_{i+1})|, |\dot{Q}_{ji}(\bar{t}_i)| \right] \leqslant V_j$$

$$i = 1, 2, \ldots, n-1 \tag{4.4-69}$$

$$j = 1, 2, \ldots, N$$

where

$$|\dot{Q}_{ji}(t_i)| = \left| \frac{\omega_{ji}}{2}u_i + \frac{q_{j,i+1} - q_{ji}}{u_i} + \frac{(\omega_{ji} - \omega_{j,i+1})u_i}{6} \right|$$

$$|\dot{Q}_{ji}(t_{i+1})| = \left| \frac{\omega_{j,i+1}}{2}u_i + \frac{q_{j,i+1} - q_{ji}}{u_i} + \frac{(\omega_{ji} - \omega_{j,i+1})u_i}{6} \right|$$

and

$$
|\dot{Q}_{ji}(\bar{t}_i)| = \begin{cases} \left| \dfrac{\omega_{ji}\omega_{j,\,i+1}u_i}{2(\omega_{ji} - \omega_{j,\,i+1})} + \dfrac{(\omega_{ji} - \omega_{j,\,i+1})u_i}{6} + \dfrac{q_{j,\,i+1} - q_{ji}}{u_i} \right| \\ \qquad \text{if } \omega_{ji} \neq \omega_{j,\,i+1} \text{ and } \bar{t}_i \in [t_i,\, t_{i+1}] \\[2mm] 0 \qquad \text{if } \omega_{ji} = \omega_{j,\,i+1} \text{ or } \bar{t}_i \notin [t_i,\, t_{i+1}] \end{cases}
$$

Acceleration Constraints. The acceleration is a linear function of time between two adjacent knot points. Thus, the maximum absolute value of acceleration occurs at either t_i or t_{i+1} and equals the maximum of $\{|\omega_{ji}|, |\omega_{j,\,i+1}|\}$. Thus, the acceleration constraints become

$$
\max \{|\omega_{j1}|, |\omega_{j2}|, \ldots, |\omega_{jn}|\} \leq A_j \qquad j = 1, 2, \ldots, N \quad (4.4\text{-}70)
$$

Jerk Constraints. The jerk is the rate of change of acceleration. Thus the constraints are represented by

$$
\left| \frac{\omega_{j,\,i+1} - \omega_{ji}}{u_i} \right| \leq J_j \qquad \begin{aligned} &j = 1, 2, \ldots, N \\ &i = 1, 2, \ldots, n - 1 \end{aligned} \quad (4.4\text{-}71)
$$

Torque Constraints. The torque $\tau(t)$ can be computed from the dynamic equations of motion [Eq. (3.2-25)]

$$
\tau_j(t) = \sum_{k=1}^{N} D_{jk}(\mathbf{Q}_i(t))\ddot{Q}_{ji}(t) + \sum_{k=1}^{N}\sum_{m=1}^{N} h_{jkm}(\mathbf{Q}_i(t))\dot{Q}_{ki}(t)\dot{Q}_{mi}(t) + c_j(\mathbf{Q}_i(t))
$$

$$
(4.4\text{-}72)
$$

where

$$
\mathbf{Q}_i(t) = (Q_{1i}(t), Q_{2i}(t), \ldots, Q_{Ni}(t))^T \qquad \begin{aligned} &j = 1, 2, \ldots, N \\ &i = 1, 2, \ldots, n - 1 \end{aligned}
$$

If the torque constraints are not satisfied, then dynamic time scaling of the trajectory must be performed to ensure the satisfaction of the torque constraints (Lin and Chang [1985], Hollerbach [1984]).

With this formulation, the objective is to find an appropriate optimization algorithm that will minimize the total traveling time subject to the velocity, acceleration, jerk, and torque constraints. There are several optimization algorithms available, and Lin et al. [1983] utilized Nelder and Mead's flexible polyhedron search to obtain an iterative algorithm which minimizes the total traveling time subject to the constraints on joint velocities, accelerations, jerks, and torques. Results using this optimization technique can be found in Lin et al. [1983].

4.5 CONCLUDING REMARKS

Two major approaches for trajectory planning have been discussed: the joint-interpolated approach and the cartesian space approach. The joint-interpolated approach plans polynomial sequences that yield smooth joint trajectory. In order to yield faster computation and less extraneous motion, lower-degree polynomial sequences are preferred. The joint trajectory is split into several trajectory segments and each trajectory segment is splined by a low-degree polynomial. In particular, 4-3-4 and five-cubic polynomial sequences have been discussed.

Several methods have been discussed in the cartesian space planning. Because servoing is done in the joint-variable space while a path is specified in cartesian coordinates, the most common approach is to plan the straight-line path in the joint-variable space using low-degree polynomials to approximate the path. Paul [1979] used a translation and two rotations to accomplish the straight-line motion of the manipulator hand. Taylor [1979] improved the technique by using a quaternion approach to represent the rotational operation. He also developed a bounded deviation joint control scheme which involved selecting more intermediate interpolation points when the joint polynomial approximation deviated too much from the desired straight-line path. Lin et al. [1983] used cubic joint polynomials to spline n interpolation points selected by the user on the desired straight-line path. Then, the total traveling time along the knot points was minimized subject to joint velocity, acceleration, jerk, and torque constraints. These techniques represent a shift away from the real-time planning objective to an off-line planning phase. In essence, this decomposes the control of robot manipulators into off-line motion planning followed by on-line tracking control, a topic that is discussed in detail in Chap. 5.

REFERENCES

Further reading on joint-interpolated trajectories can be found in Paul [1972], Lewis [1973, 1974], Brady et al. [1982], and Lee et al. [1986]. Most of these joint-interpolated trajectories seldom include the physical manipulator dynamics and actuator torque limit into the planning schemes. They focused on the requirement that the joint trajectories must be smooth and continuous by specifying velocity and acceleration bounds along the trajectory. In addition to the continuity constraints, Hollerbach [1984] developed a time-scaling scheme to determine whether a planned trajectory is realizable within the dynamics and torque limits which depend on instantaneous joint position and velocity.

The design of a manipulator path made up of straight line segments in the cartesian space is discussed by Paul [1979], using the homogeneous transformation matrix to represent target positions for the manipulator hand to traverse. Movement between two consecutive target positions is accomplished by two sequential operations: a translation and a rotation to align the approach vector of the manipulator hand and a final rotation about the tool axis to align the gripper orientation.

A quadratic polynomial interpolation routine in the joint-variable space is then used to guarantee smooth transition between two connected path segments. Taylor [1979], using the quaternion representation, extended Paul's method for a better and uniform motion. In order to achieve real-time trajectory planning objective, both approaches neglect the physical manipulator torque constraint.

Other existing cartesian planning schemes are designed to satisfy the continuity and the torque constraints simultaneously. To include the torque constraint in the trajectory planning stage, one usually assumes that the maximum allowable torque is constant at every position and velocity. For example, instead of using varying torque constraint, Lin et al. [1983] and Luh and Lin [1984] used the velocity, acceleration, and jerk bounds which are assumed constant for each joint. They selected several knot points on the desired cartesian path, solved the inverse kinematics, and found appropriate smooth, lower-degree polynomial functions which guaranteed the continuity conditions to fit through these knot points in the joint-variable space. Then, by relaxing the normalized time to the servo time, the dynamic constraint with the constant torque bound assumption was included along the trajectory. Due to the joint-interpolated functions, the location of the manipulator hand at each servo instant may not be exactly on the desired path, but rather on the joint-interpolated polynomial functions.

Lee [1985] developed a discrete time trajectory planning scheme to determine the trajectory set points exactly on a given straight-line path which satisfies both the smoothness and torque constraints. The trajectory planning problem is formulated as a maximization of the distance between two consecutive cartesian set points on a given straight-line path subject to the smoothness and torque constraints. Due to the discrete time approximations of joint velocity, acceleration, and jerk, the optimization solution involves intensive computations which prevent useful applications. Thus, to reduce the computational cost, the optimization is realized by iterative search algorithms.

PROBLEMS

4.1 A single-link rotary robot is required to move from $\theta(0) = 30°$ to $\theta(2) = 100°$ in 2 s. The joint velocity and acceleration are both zero at the initial and final positions. (*a*) What is the highest degree polynomial that can be used to accomplish the motion? (*b*) What is the lowest degree polynomial that can be used to accomplish the motion?

4.2 With reference to Prob. 4.1, (*a*) determine the coefficients of a cubic polynomial that accomplishes the motion; (*b*) determine the coefficients of a quartic polynomial that accomplishes the motion; and (*c*) determine the coefficients of a quintic polynomial that accomplishes the motion. You may split the joint trajectory into several trajectory segments.

4.3 Consider the two-link robot arm discussed in Sec. 3.2.6, and assume that each link is 1 m long. The robot arm is required to move from an initial position $(x_0, y_0) = (1.96, 0.50)$ to a final position $(x_f, y_f) = (1.00, 0.75)$. The initial and final velocity and acceleration are zero. Determine the coefficients of a cubic polynomial for each joint to accomplish the motion. You may split the joint trajectory into several trajectory segments.

4.4 In planning a 4-3-4 trajectory one needs to solve a matrix equation, as in Eq. (4.3-46). Does the matrix inversion of Eq. (4.3-46) always exist? Justify your answer.

4.5 Given a PUMA 560 series robot arm whose joint coordinate frames have been established as in Fig. 2.11, you are asked to design a 4-3-4 trajectory for the following conditions: The initial position of the robot arm is expressed by the homogeneous transformation matrix $\mathbf{T}_{\text{initial}}$:

$$\mathbf{T}_{\text{initial}} = \begin{bmatrix} -0.660 & -0.436 & -0.612 & -184.099 \\ -0.750 & 0.433 & 0.500 & 892.250 \\ 0.047 & 0.789 & -0.612 & -34.599 \\ 0 & 0 & 0 & 1 \end{bmatrix}$$

The final position of the robot arm is expressed by the homogeneous transformation matrix $\mathbf{T}_{\text{final}}$

$$\mathbf{T}_{\text{final}} = \begin{bmatrix} -0.933 & -0.064 & 0.355 & 412.876 \\ -0.122 & 0.982 & -0.145 & 596.051 \\ -0.339 & -0.179 & -0.924 & -545.869 \\ 0 & 0 & 0 & 1 \end{bmatrix}$$

The lift-off and set-down positions of the robot arm are obtained from a rule of thumb by taking 25 percent of d_6 (the value of d_6 is 56.25 mm). What are the homogeneous transformation matrices at the lift-off and set-down positions (that is, $\mathbf{T}_{\text{lift-off}}$ and $\mathbf{T}_{\text{set-down}}$)?

4.6 Given a PUMA 560 series robot arm whose joint coordinate frames have been established as in Fig. 2.11, you are asked to design a 4-3-4 trajectory for the following conditions: The initial position of the robot arm is expressed by the homogeneous transformation matrix $\mathbf{T}_{\text{initial}}$:

$$\mathbf{T}_{\text{initial}} = \begin{bmatrix} -1 & 0 & 0 & 0 \\ 0 & 1 & 0 & 600.0 \\ 0 & 0 & -1 & -100.0 \\ 0 & 0 & 0 & 1 \end{bmatrix}$$

The set-down position of the robot arm is expressed by the homogeneous transformation matrix $\mathbf{T}_{\text{set-down}}$:

$$\mathbf{T}_{\text{set-down}} = \begin{bmatrix} 0 & 1 & 0 & 100.0 \\ 1 & 0 & 0 & 400.0 \\ 0 & 0 & -1 & -50.0 \\ 0 & 0 & 0 & 1 \end{bmatrix}$$

(*a*) The lift-off and set-down positions of the robot arm are obtained from a rule of thumb by taking 25 percent of d_6 (the value of d_6 is 56.25 mm) plus any required rotations. What is the homogeneous transformation matrix at the lift-off (that is, $\mathbf{T}_{\text{lift-off}}$) if the hand is rotated 60° about the **s** axis at the initial point to arrive at the lift-off point? (*b*) What is the homogeneous transformation matrix at the final position (that is, $\mathbf{T}_{\text{final}}$) if the hand is rotated −60° about the **s** axis at the set-down point to arrive at the final position?

4.7 A manipulator is required to move along a straight line from point **A** to point **B**, where **A** and **B** are respectively described by

$$
A = \begin{bmatrix} -1 & 0 & 0 & 5 \\ 0 & 1 & 0 & 10 \\ 0 & 0 & -1 & 15 \\ 0 & 0 & 0 & 1 \end{bmatrix} \quad \text{and} \quad B = \begin{bmatrix} 0 & -1 & 0 & 20 \\ 0 & 0 & 1 & 30 \\ -1 & 0 & 0 & 5 \\ 0 & 0 & 0 & 1 \end{bmatrix}
$$

The motion from **A** to **B** consists of a translation and two rotations, as described in Sec. 4.4.1. Determine θ, ψ, ϕ and x, y, z for the drive transform. Aslo find three intermediate transforms between **A** and **B**.

4.8 A manipulator is required to move along a straight line from point **A** to point **B** rotating at constant angular velocity about a vector **k** and at an angle θ. The points **A** and **B** are given by a 4×4 homogeneous transformation matrices as

$$
A = \begin{bmatrix} -1 & 0 & 0 & 10 \\ 0 & 1 & 0 & 10 \\ 0 & 0 & -1 & 10 \\ 0 & 0 & 0 & 1 \end{bmatrix} \quad B = \begin{bmatrix} 0 & -1 & 0 & 10 \\ 0 & 0 & 1 & 30 \\ -1 & 0 & 0 & 10 \\ 0 & 0 & 0 & 1 \end{bmatrix}
$$

Find the vector **k** and the angle θ. Also find three intermediate transforms between **A** and **B**.

4.9 Express the rotation results of Prob. 4.8 in quaternion form.

4.10 Give a quaternion representation for the following rotations: a rotation of $60°$ about **j** followed by a rotation of $120°$ about **i**. Find the resultant rotation in quaternion representation.

4.11 Show that the inverse of the banded structure matrix **A** in Eq. (4.4-65) always exists.

FIVE

CONTROL OF ROBOT MANIPULATORS

> Let us realize that what happens around us
> is largely outside our control, but that
> the way we choose to react to it
> is inside our control.
> *Quoted by J. Petty in "Apples of Gold"*

5.1 INTRODUCTION

Given the dynamic equations of motion of a manipulator, the purpose of robot arm control is to maintain the dynamic response of the manipulator in accordance with some prespecified performance criterion. Although the control problem can be stated in such a simple manner, its solution is complicated by inertial forces, coupling reaction forces, and gravity loading on the links. In general, the control problem consists of (1) obtaining dynamic models of the manipulator, and (2) using these models to determine control laws or strategies to achieve the desired system response and performance. The first part of the control problem has been discussed extensively in Chap. 3. This chapter concentrates on the latter part of the control problem.

From the control analysis point of view, the movement of a robot arm is usually accomplished in two distinct control phases. The first is the gross motion control in which the arm moves from an initial position/orientation to the vicinity of the desired target position/orientation along a planned trajectory. The second is the fine motion control in which the end-effector of the arm dynamically interacts with the object using sensory feedback information to complete the task.

Current industrial approaches to robot arm control system design treat each joint of the robot arm as a simple joint servomechanism. The servomechanism approach models the varying dynamics of a manipulator inadequately because it neglects the motion and configuration of the whole arm mechanism. These changes in the parameters of the controlled system are significant enough to render conventional feedback control strategies ineffective. The result is reduced servo response speed and damping, limiting the precision and speed of the end-effector and making it appropriate only for limited-precision tasks. As a result, manipulators controlled this way move at slow speeds with unnecessary vibrations. Any significant performance gain in this and other areas of robot arm control require the consideration of more efficient dynamic models, sophisticated control techniques, and the use of computer architectures. This chapter focuses on deriving

201

strategies which utilize the dynamic models discussed in Chap. 3 to efficiently control a manipulator.

Considering the robot arm control as a path-trajectory tracking problem (see Fig. 5.1), motion control can be classified into three major categories for the purpose of discussion:

1. Joint motion controls

   Joint servomechanism (PUMA robot arm control scheme)
   Computed torque technique
   Minimum-time control
   Variable structure control
   Nonlinear decoupled control

2. Resolved motion controls (cartesian space control)

   Resolved motion rate control
   Resolved motion acceleration control
   Resolved motion force control

3. Adaptive controls

   Model-referenced adaptive control
   Self-tuning adaptive control
   Adaptive perturbation control with feedforward compensation
   Resolved motion adaptive control

For these control methods, we assume that the desired motion is specified by a time-based path/trajectory of the manipulator either in joint or cartesian coordinates. Each of the above control methods will be described in the following sections.

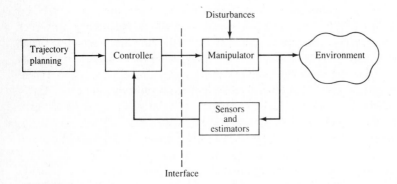

Figure 5.1 Basic control block diagram for robot manipulators.

5.2 CONTROL OF THE PUMA ROBOT ARM

Current industrial practice treats each joint of the robot arm as a simple servomechanism. For the PUMA 560 series robot arm, the controller consists of a DEC LSI-11/02 computer and six Rockwell 6503 microprocessors, each with a joint encoder, a digital-to-analog converter (DAC), and a current amplifier. The control structure is hierarchically arranged. At the top of the system hierarchy is the LSI-11/02 microcomputer which serves as a supervisory computer. At the lower level are the six 6503 microprocessors—one for each degree of freedom (see Fig. 5.2). The LSI-11/02 computer performs two major functions: (1) on-line user interaction and subtask scheduling from the user's VAL† commands, and (2) subtask coordination with the six 6503 microprocessors to carry out the command. The on-line interaction with the user includes parsing, interpreting, and decoding the VAL commands, in addition to reporting appropriate error messages to the user. Once a VAL command has been decoded, various internal routines are called to perform scheduling and coordination functions. These functions, which reside in the EPROM memory of the LSI-11/02 computer, include:

1. Coordinate systems transformations (e.g., from world to joint coordinates or vice versa).
2. Joint-interpolated trajectory planning; this involves sending incremental location updates corresponding to each set point to each joint every 28 ms.
3. Acknowledging from the 6503 microprocessors that each axis of motion has completed its required incremental motion.
4. Looking ahead two instructions to perform continuous path interpolation if the robot is in a continuous path mode.

At the lower level in the system hierarchy are the joint controllers, each of which consists of a digital servo board, an analog servo board, and a power amplifier for each joint. The 6503 microprocessor is an integral part of the joint controller which directly controls each axis of motion. Each microprocessor resides on a digital servo board with its EPROM and DAC. It communicates with the LSI-11/02 computer through an interface board which functions as a demultiplexer that routes trajectory set points information to each joint controller. The interface board is in turn connected to a 16-bit DEC parallel interface board (DRV-11) which transmits the data to and from the Q-bus of the LSI-11/02 (see Fig. 5.2). The microprocessor computes the joint error signal and sends it to the analog servo board which has a current feedback designed for each joint motor.

There are two servo loops for each joint control (see Fig. 5.2). The outer loop provides position error information and is updated by the 6503 microprocessor about every 0.875 ms. The inner loop consists of analog devices and a com-

---

† VAL is a software package from Unimation Inc. for control of the PUMA robot arm.

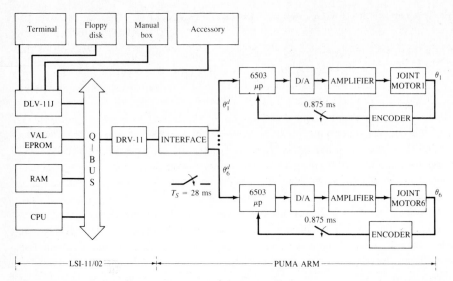

Figure 5.2 PUMA robot arm servo control architecture.

pensator with derivative feedback to dampen the velocity variable. Both servo loop gains are constant and tuned to perform as a "critically damped joint system" at a speed determined by the VAL program. The main functions of the microprocessors include:

1. Every 28 ms, receive and acknowledge trajectory set points from the LSI-11/02 computer and perform interpolation between the current joint value and the desired joint value.
2. Every 0.875 ms, read the register value which stores the incremental values from the encoder mounted at each axis of rotation.
3. Update the error actuating signals derived from the joint-interpolated set points and the values from the axis encoders.
4. Convert the error actuating signal to current using the DACs, and send the current to the analog servo board which moves the joint.

It can be seen that the PUMA robot control scheme is basically a proportional plus integral plus derivative control method (PID controller). One of the main disadvantages of this control scheme is that the feedback gains are constant and prespecified. It does not have the capability of updating the feedback gains under varying payloads. Since an industrial robot is a highly nonlinear system, the inertial loading, the coupling between joints and the gravity effects are all either position-dependent or position- and velocity-dependent terms. Furthermore, at high speeds the inertial loading term can change drastically. Thus, the above control scheme using constant feedback gains to control a nonlinear system does not perform well under varying speeds and payloads. In fact, the PUMA arm moves with noticeable vibration at reduced speeds. One solution to the problem is the use of digital control in which the applied torques to the robot arm are obtained by a computer based on an appropriate dynamic model of the arm. A version of this method is discussed in Sec. 5.3.

5.3 COMPUTED TORQUE TECHNIQUE

Given the Lagrange-Euler or Newton-Euler equations of motion of a manipulator, the control problem is to find appropriate torques/forces to servo all the joints of the manipulator in real time in order to track a desired time-based trajectory as closely as possible. The drive motor torque required to servo the manipulator is based on a dynamic model of the manipulator (L-E or N-E formulations). The motor-voltage (or motor-current) characteristics are also modeled in the computation scheme and the computed torque is converted to the applied motor voltage (or current). This applied voltage is computed at such a high rate that sampling effects generally can be ignored in the analysis.

Because of modeling errors and parameter variations in the model, position and derivative feedback signals will be used to compute the correction torques which, when added to the torques computed based on the manipulator model, provide the corrective drive signal for the joint motors.

5.3.1 Transfer Function of a Single Joint

This section deals with the derivation of the transfer function of a single joint robot from which a proportional plus derivative controller (PD controller) will be obtained. This will be followed by a discussion of controller design for multijoint manipulators based on the Lagrange-Euler and/or Newton-Euler equations of motion. The analysis here treats the "single-joint" robot arm as a continuous time system, and the Laplace transform technique is used to simplify the analysis.

Most industrial robots are either electrically, hydraulically, or pneumatically actuated. Electrically driven manipulators are constructed with a dc permanent magnet torque motor for each joint. Basically, the dc torque motor is a permanent magnet, armature excited, continuous rotation motor incorporating such features as high torque-power ratios, smooth, low-speed operation, linear torque-speed characteristics, and short time constants. Use of a permanent magnet field and dc power provide maximum torque with minimum input power and minimum weight. These features also reduce the motor inductance and hence the electrical time constant. In Fig. 5.3, an equivalent circuit of an armature-controlled dc permanent magnet torque motor for a joint is shown based on the following variables:

V_a Armature voltage, volts

V_f Field voltage, volts

L_a Armature inductance, Henry

L_f Field inductance, Henry

R_a Armature resistance, ohms

R_f Field resistance, ohms

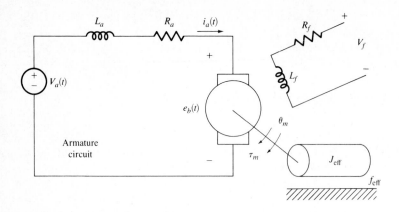

Figure 5.3 Equivalent circuit of an armature-controlled dc motor.

i_a Armature current, amperes

i_f Field current, amperes

e_b Back electromotive force (emf), volts

τ Torque delivered by the motor, oz·in

θ_m Angular displacement of the motor shaft, radians

θ_L Angular displacement of the load shaft, radians

J_m Moment of inertia of the motor referred to the motor shaft, oz · in · s²/ rad

f_m Viscous-friction coefficient of the motor referred to the motor shaft, oz · in · s/rad

J_L Moment of inertia of the load referred to the load shaft, oz · in · s²/rad

f_L Viscous-friction coefficient of the load referred to the load shaft, oz · in · s/rad

N_m Number of teeth of the input gear (motor gear)

N_L Number of teeth of the output gear (load gear)

The motor shaft is coupled to a gear train to the load of the link. With reference to the gear train shown in Fig. 5.4, the total linear distance traveled on each gear is the same. That is,

$$d_m = d_L \quad \text{and} \quad r_m\theta_m = r_L\theta_L \tag{5.3-1}$$

where r_m and r_L are, respectively, the radii of the input gear and the output gear. Since the radius of the gear is proportional to the number of teeth it has, then

$$N_m\theta_m = N_L\theta_L \tag{5.3-2}$$

or
$$\frac{N_m}{N_L} = \frac{\theta_L}{\theta_m} = n < 1 \tag{5.3-3}$$

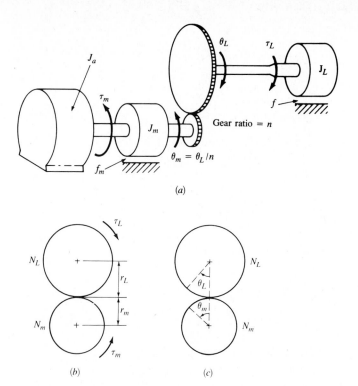

(a)

(b)   (c)

Figure 5.4 Analysis of a gear train.

where n is the gear ratio and it relates θ_L to θ_m by

$$\theta_L(t) = n\theta_m(t) \tag{5.3-4}$$

Taking the first two time derivatives, we have

$$\dot{\theta}_L(t) = n\dot{\theta}_m(t) \tag{5.3-5}$$

and

$$\ddot{\theta}_L(t) = n\ddot{\theta}_m(t) \tag{5.3-6}$$

If a load is attached to the output gear, then the torque developed at the motor shaft is equal to the sum of the torques dissipated by the motor and its load. That is,

$$\begin{bmatrix} \text{Torque from} \\ \text{motor} \\ \text{shaft} \end{bmatrix} = \begin{bmatrix} \text{torque} \\ \text{on} \\ \text{motor} \end{bmatrix} + \begin{bmatrix} \text{torque on load} \\ \text{referred to} \\ \text{the motor shaft} \end{bmatrix} \tag{5.3-7}$$

or, in equation form,

$$\tau(t) = \tau_m(t) + \tau_L^*(t) \tag{5.3-8}$$

The load torque referred to the load shaft is

$$\tau_L(t) = J_L\ddot{\theta}_L(t) + f_L\dot{\theta}_L(t) \tag{5.3-9}$$

and the motor torque referred to the motor shaft is

$$\tau_m(t) = J_m\ddot{\theta}_m(t) + f_m\dot{\theta}_m(t) \tag{5.3-10}$$

Recalling that conservation of work requires that the work done by the load referred to the load shaft, $\tau_L\theta_L$, be equal to the work done by the load referred to the motor shaft, $\tau_L^*\theta_m$, leads to

$$\tau_L^*(t) = \frac{\tau_L(t)\theta_L(t)}{\theta_m(t)} = n\tau_L(t) \tag{5.3-11}$$

Using Eqs. (5.3-9), (5.3-5), and (5.3-6), we have

$$\tau_L^*(t) = n^2[J_L\ddot{\theta}_m(t) + f_L\dot{\theta}_m(t)] \tag{5.3-12}$$

Using Eqs. (5.3-10) and (5.3-12), the torque developed at the motor shaft [Eq. (5.3-8)] is

$$\tau(t) = \tau_m(t) + \tau_L^*(t) = (J_m + n^2J_L)\ddot{\theta}_m(t) + (f_m + n^2f_L)\dot{\theta}_m(t)$$

$$= J_{\text{eff}}\,\ddot{\theta}_m(t) + f_{\text{eff}}\,\dot{\theta}_m(t) \tag{5.3-13}$$

where $J_{\text{eff}} = J_m + n^2J_L$ is the effective moment of inertia of the combined motor and load referred to the motor shaft and $f_{\text{eff}} = f_m + n^2f_L$ is the effective viscous friction coefficient of the combined motor and load referred to the motor shaft.

Based on the above results, we can now derive the transfer function of this single joint manipulator system. Since the torque developed at the motor shaft increases linearly with the armature current, independent of speed and angular position, we have

$$\tau(t) = K_a i_a(t) \tag{5.3-14}$$

where K_a is known as the motor-torque proportional constant in oz $\cdot$ in/A. Applying Kirchhoff's voltage law to the armature circuit, we have

$$V_a(t) = R_a i_a(t) + L_a\frac{di_a(t)}{dt} + e_b(t) \tag{5.3-15}$$

where e_b is the back electromotive force (emf) which is proportional to the angular velocity of the motor,

$$e_b(t) = K_b\dot{\theta}_m(t) \tag{5.3-16}$$

and K_b is a proportionality constant in V · s/rad. Taking the Laplace transform of the above equations and solving for $I_a(s)$, we have

$$I_a(s) = \frac{V_a(s) - sK_b\Theta_m(s)}{R_a + sL_a} \tag{5.3-17}$$

Taking the Laplace transform of Eq. (5.3-13), we have

$$T(s) = s^2 J_{\text{eff}}\Theta_m(s) + s f_{\text{eff}}\Theta_m(s) \tag{5.3-18}$$

Taking the Laplace transform of Eq. (5.3-14), and substituting $I_a(s)$ from Eq. (5.3-17), we have

$$T(s) = K_a I_a(s) = K_a \left[\frac{V_a(s) - sK_b\Theta_m(s)}{R_a + sL_a} \right] \tag{5.3-19}$$

Equating Eqs. (5.3-18) and (5.3-19) and rearranging the terms, we obtain the transfer function from the armature voltage to the angular displacement of the motor shaft,

$$\frac{\Theta_m(s)}{V_a(s)} = \frac{K_a}{s[s^2 J_{\text{eff}}L_a + (L_a f_{\text{eff}} + R_a J_{\text{eff}})s + R_a f_{\text{eff}} + K_a K_b]} \tag{5.3-20}$$

Since the electrical time constant of the motor is much smaller than the mechanical time constant, we can neglect the armature inductance effect, L_a. This allows us to simplify the above equation to

$$\frac{\Theta_m(s)}{V_a(s)} = \frac{K_a}{s(sR_a J_{\text{eff}} + R_a f_{\text{eff}} + K_a K_b)} = \frac{K}{s(T_m s + 1)} \tag{5.3-21}$$

where

$$K \triangleq \frac{K_a}{R_a f_{\text{eff}} + K_a K_b} \qquad \text{motor gain constant}$$

and

$$T_m \triangleq \frac{R_a J_{\text{eff}}}{R_a f_{\text{eff}} + K_a K_b} \qquad \text{motor time constant}$$

Since the output of the control system is the angular displacement of the joint $[\Theta_L(s)]$, using Eq. (5.3-4) and its Laplace transformed equivalence, we can relate the angular position of the joint $\Theta_L(s)$ to the armature voltage $V_a(s)$,

$$\frac{\Theta_L(s)}{V_a(s)} = \frac{nK_a}{s(sR_a J_{\text{eff}} + R_a f_{\text{eff}} + K_a K_b)} \tag{5.3-22}$$

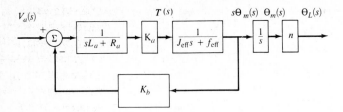

Figure 5.5 Open-loop transfer function of a single-joint robot arm.

Eq. (5.3-22) is the transfer function of the "single-joint" manipulator relating the applied voltage to the angular displacement of the joint. The block diagram of the system is shown in Fig. 5.5.

5.3.2 Positional Controller for a Single Joint

The purpose of a positional controller is to servo the motor so that the actual angular displacement of the joint will track a desired angular displacement specified by a preplanned trajectory, as discussed in Chap. 4. The technique is based on using the error signal between the desired and actual angular positions of the joint to actuate an appropriate voltage. In other words, the applied voltage to the motor is linearly proportional to the error between the desired and actual angular displacement of the joint,

$$
V_a(t) = \frac{K_p e(t)}{n} = \frac{K_p[\theta_L^d(t) - \theta_L(t)]}{n}
\tag{5.3-23}
$$

where K_p is the position feedback gain in volts per radian, $e(t) = \theta_L^d(t) - \theta_L(t)$ is the system error, and the gear ratio n is included to compute the applied voltage referred to the motor shaft. Equation (5.3-23) indicates that the actual angular displacement of the joint is fed back to obtain the error which is amplified by the position feedback gain K_p to obtain the applied voltage. In reality, we have changed the single-joint robot system from an open-loop control system [Eq. (5.3-22)] to a closed-loop control system with unity negative feedback. This closed-loop control system is shown in Fig. 5.6. The actual angular position of the joint can be measured either by an optical encoder or by a potentiometer.

Taking the Laplace transform of Eq. (5.3-23),

$$
V_a(s) = \frac{K_p[\Theta_L^d(s) - \Theta_L(s)]}{n} = \frac{K_p E(s)}{n}
\tag{5.3-24}
$$

and substituting $V_a(s)$ into Eq. (5.3-22), yields the open-loop transfer function relating the error actuating signal $[E(s)]$ to the actual displacement of the joint:

$$
\frac{\Theta_L(s)}{E(s)} \triangleq G(s) = \frac{K_a K_p}{s(s R_a J_{\text{eff}} + R_a f_{\text{eff}} + K_a K_b)}
\tag{5.3-25}
$$

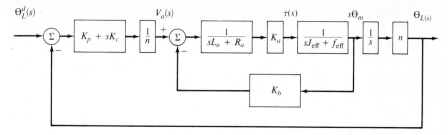

Figure 5.6 Feedback control of a single-joint manipulator.

After some simple algebraic manipulation, we can obtain the closed-loop transfer function relating the actual angular displacement $\Theta_L(s)$ to the desired angular displacement $\Theta_L^d(s)$:

$$\frac{\Theta_L(s)}{\Theta_L^d(s)} = \frac{G(s)}{1 + G(s)} = \frac{K_a K_p}{s^2 R_a J_{\text{eff}} + s(R_a f_{\text{eff}} + K_a K_b) + K_a K_p}$$

$$= \frac{K_a K_p / R_a J_{\text{eff}}}{s^2 + [(R_a f_{\text{eff}} + K_a K_b)/R_a J_{\text{eff}}] \, s + K_a K_p / R_a J_{\text{eff}}} \tag{5.3-26}$$

Equation (5.3-26) shows that the proportional controller for the single-joint robot is a second-order system which is always stable if all the system parameters are positive. In order to increase the system response time and reduce the steady-state error, one can increase the positional feedback gain K_p and incorporate some damping into the system by adding a derivative of the positional error. The angular velocity of the joint can be measured by a tachometer or approximated from the position data between two consecutive sampling periods. With this added feedback term, the applied voltage to the joint motor is linearly proportional to the position error and its derivative; that is,

$$V_a(t) = \frac{K_p[\theta_L^d(t) - \theta_L(t)] + K_v[\dot{\theta}_L^d(t) - \dot{\theta}_L(t)]}{n}$$

$$= \frac{K_p e(t) + K_v \dot{e}(t)}{n} \tag{5.3-27}$$

where K_v is the error derivative feedback gain, and the gear ratio n is included to compute the applied voltage referred to the motor shaft. Equation (5.3-27) indicates that, in addition to the positional error feedback, the velocity of the motor is measured or computed and fed back to obtain the velocity error which is multiplied by the velocity feedback gain K_v. Since, as discussed in Chap. 4, the desired joint trajectory can be described by smooth polynomial functions whose first two time derivatives exist within $[t_0, t_f]$, the desired velocity can be computed from the polynomial function and utilized to obtain the velocity error for feedback purposes. The summation of these voltages is then applied to the joint motor. This closed-loop control system is shown in Fig. 5.6.

Taking the Laplace transform of Eq. (5.3-27) and substituting $V_a(s)$ into Eq. (5.3-22) yields the transfer function relating the error actuating signal $[E(s)]$ to the actual displacement of the joint:

$$\frac{\Theta_L(s)}{E(s)} \triangleq G_{PD}(s) = \frac{K_a(K_p + sK_v)}{s(sR_a J_{eff} + R_a f_{eff} + K_a K_b)}$$

$$= \frac{K_a K_v s + K_a K_p}{s(sR_a J_{eff} + R_a f_{eff} + K_a K_b)} \tag{5.3-28}$$

Some simple algebraic manipulation yields the closed-loop transfer function relating the actual angular displacement $[\Theta_L(s)]$ to the desired angular displacement $[\Theta_L^d(s)]$:

$$\frac{\Theta_L(s)}{\Theta_L^d(s)} = \frac{G_{PD}(s)}{1 + G_{PD}(s)}$$

$$= \frac{K_a K_v s + K_a K_p}{s^2 R_a J_{eff} + s(R_a f_{eff} + K_a K_b + K_a K_v) + K_a K_p} \tag{5.3-29}$$

Note that if K_v is equal to zero, Eq. (5.3-29) reduces to Eq. (5.3-26).

Equation (5.3-29) is a second-order system with a finite zero located at $-K_p/K_v$ in the left half plane of the s plane. Depending on the location of this zero, the system could have a large overshoot and a long settling time. From Fig. 5.7, we notice that the manipulator system is also under the influence of disturbances $[D(s)]$ which are due to gravity loading and centrifugal effects of the link. Because of this disturbance, the torque generated at the motor shaft has to compensate for the torques dissipated by the motor, the load, and also the disturbances. Thus, from Eq. (5.3-18),

$$T(s) = [s^2 J_{eff} + s f_{eff}]\Theta_m(s) + D(s) \tag{5.3-30}$$

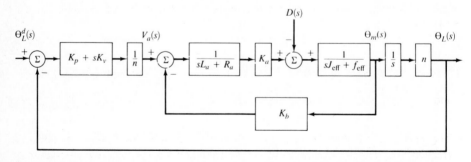

Figure 5.7 Feedback control block diagram of a manipulator with disturbances.

where $D(s)$ is the Laplace transform equivalent of the disturbances. The transfer function relating the disturbance inputs to the actual joint displacement is given by

$$\left. \frac{\Theta_L(s)}{D(s)} \right|_{\Theta_L^d(s)=0} = \frac{-nR_a}{s^2 R_a J_{\text{eff}} + s(R_a f_{\text{eff}} + K_a K_b + K_a K_v) + K_a K_p} \tag{5.3-31}$$

From Eqs. (5.3-29) and (5.3-31) and using the superposition principle, we can obtain the actual displacement of the joint from these two inputs, as follows:

$$\Theta_L(s) = \frac{K_a(K_p + sK_v)\Theta_L^d(s) - nR_a D(s)}{s^2 R_a J_{\text{eff}} + s(R_a f_{\text{eff}} + K_a K_b + K_a K_v) + K_a K_p} \tag{5.3-32}$$

We are interested in looking at the performance of the above closed-loop system, with particular emphasis on the steady state error of the system due to step and ramp inputs and the bounds of the position and velocity feedback gains. This is covered in the following section.

5.3.3 Performance and Stability Criteria

The performance of a closed-loop second-order control system is based on several criteria, such as fast rise time, small or zero steady-state error, and fast settling time. We shall first investigate the bounds for the position and velocity feedback gains. Assuming for a moment that the disturbances are zero, we see that from Eqs. (5.3-29) and (5.3-31) that the system is basically a second-order system with a finite zero, as indicated in the previous section. The effect of this finite zero usually causes a second-order system to peak early and to have a larger overshoot (than the second-order system without a finite zero). We shall temporarily ignore the effect of this finite zero and try to determine the values of K_p and K_v to have a critically damped or overdamped system.

The reader will recall that the characteristic equation of a second-order system can be expressed in the following standard form:

$$s^2 + 2\zeta\omega_n s + \omega_n^2 = 0 \tag{5.3-33}$$

where ζ and ω_n are, respectively, the damping ratio and the undamped natural frequency of the system. Relating the closed-loop poles of Eq. (5.3-29) to Eq. (5.3-33), we see that

$$\omega_n^2 = \frac{K_a K_p}{J_{\text{eff}} R_a} \tag{5.3-34}$$

and

$$2\zeta\omega_n = \frac{R_a f_{\text{eff}} + K_a K_b + K_a K_v}{J_{\text{eff}} R_a} \tag{5.3-35}$$

The performance of the second-order system is dictated by its natural undamped frequency ω_n and the damping ratio ζ. For reasons of safety, the manipulator system cannot have an underdamped response for a step input. In order to have good performance (as outlined above), we would like to have a critically damped or an overdamped system, which requires that the system damping ratio be greater than or equal to unity. From Eq. (5.3-34), the position feedback gain is found from the natural frequency of the system:

$$K_p = \frac{\omega_n^2 J_{\text{eff}} R_a}{K_a} > 0 \qquad (5.3\text{-}36)$$

Substituting ω_n from Eq. (5.3-34) into Eq. (5.3-35), we find that

$$\zeta = \frac{R_a f_{\text{eff}} + K_a K_b + K_a K_v}{2\sqrt{K_a K_p J_{\text{eff}} R_a}} \geqslant 1 \qquad (5.3\text{-}37)$$

where the equality of the above equation gives a critically damped system response and the inequality gives an overdamped system response. From Eq. (5.3-37), the velocity feedback gain K_v can be found to be

$$K_v \geqslant \frac{2\sqrt{K_a K_p J_{\text{eff}} R_a} - R_a f_{\text{eff}} - K_a K_b}{K_a} \qquad (5.3\text{-}38)$$

In order not to excite the structural oscillation and resonance of the joint, Paul [1981] suggested that the undamped natural frequency ω_n may be set to no more than one-half of the structural resonant frequency of the joint, that is,

$$\omega_n \leqslant 0.5\omega_r \qquad (5.3\text{-}39)$$

where ω_r is the structural resonant frequency in radians per second.

The structural resonant frequency is a property of the material used in constructing the manipulator. If the effective stiffness of the joint is k_{stiff}, then the restoring torque $k_{\text{stiff}}\theta_m(t)$ opposes the inertial torque of the motor,

$$J_{\text{eff}}\ddot{\theta}_m(t) + k_{\text{stiff}}\theta_m(t) = 0 \qquad (5.3\text{-}40)$$

Taking the Laplace transform, the characteristic equation of Eq. (5.3-40) is

$$J_{\text{eff}}s^2 + k_{\text{stiff}} = 0 \qquad (5.3\text{-}41)$$

and solving the above characteristic equation gives the structural resonant frequency of the system

$$\omega_r = \left(\frac{k_{\text{stiff}}}{J_{\text{eff}}}\right)^{1/2} \qquad (5.3\text{-}42)$$

Although the stiffness of the joint is fixed, if a load is added to the manipulator's end-effector, the effective moment of inertia will increase which, in effect, reduces the structural resonant frequency. If a structural resonant frequency ω_0 is measured at a known moment of inertia J_0, then the structural resonant frequency at the other moment of inertia J_{eff} is given by

$$\omega_r = \omega_0 \left[\frac{J_0}{J_{\text{eff}}} \right]^{1/2} \tag{5.3-43}$$

Using the condition of Eq. (5.3-39), K_p from Eq. (5.3-36) is bounded by

$$0 < K_p \leqslant \frac{\omega_r^2 J_{\text{eff}} R_a}{4K_a} \tag{5.3-44}$$

which, using Eq. (5.3-43), reduces to

$$0 < K_p \leqslant \frac{\omega_0^2 J_0 R_a}{4K_a} \tag{5.3-45}$$

After finding K_p, the velocity feedback gain K_v can be found from Eq. (5.3-38):

$$K_v \geqslant \frac{R_a \omega_0 \sqrt{J_0 J_{\text{eff}}} - R_a f_{\text{eff}} - K_a K_b}{K_a} \tag{5.3-46}$$

Next we investigate the steady-state errors of the above system for step and ramp inputs. The system error is defined as $e(t) = \theta_L^d(t) - \theta_L(t)$. Using Eq. (5.3-32), the error in the Laplace transform domain can be expressed as

$$
\begin{aligned}
E(s) &= \Theta_L^d(s) - \Theta_L(s) \\
&= \frac{[s^2 J_{\text{eff}} R_a + s(R_a f_{\text{eff}} + K_a K_b)]\Theta_L^d(s) + nR_a D(s)}{s^2 R_a J_{\text{eff}} + s(R_a f_{\text{eff}} + K_a K_b + K_a K_v) + K_a K_p}
\end{aligned}
\tag{5.3-47}
$$

For a step input of magnitude A, that is, $\theta_L^d(t) = A$, and if the disturbance input is unknown, then the steady-state error of the system due to a step input can be found from the final value theorem, provided the limits exist; that is,

$$
\begin{aligned}
e_{ss}(\text{step}) &\triangleq e_{ssp} = \lim_{t \to \infty} e(t) = \lim_{s \to 0} sE(s) \\
&= \lim_{s \to 0} s \frac{[(s^2 J_{\text{eff}} R_a + s(R_a f_{\text{eff}} + K_a K_b)]A/s + nR_a D(s)}{s^2 R_a J_{\text{eff}} + s(R_a f_{\text{eff}} + K_a K_b + K_a K_v) + K_a K_p} \\
&= \lim_{s \to 0} s \left[\frac{nR_a D(s)}{s^2 R_a J_{\text{eff}} + s(R_a f_{\text{eff}} + K_a K_b + K_a K_v) + K_a K_p} \right]
\end{aligned}
$$

$$\tag{5.3-48}$$

which is a function of the disturbances. Fortunately, we do know some of the disturbances, such as gravity loading and centrifugal torque due to the velocity of the joint. Other disturbances that we generally do not know are the frictional torque due to the gears and the system noise. Thus, we can identify each of these torques separately as

$$\tau_D(t) = \tau_G(t) + \tau_C(t) + \tau_e \qquad (5.3-49)$$

where $\tau_G(t)$ and $\tau_C(t)$ are, respectively, torques due to gravity and centrifugal effects of the link, and τ_e are disturbances other than the gravity and centrifugal torques and can be assumed to be a very small constant value. The corresponding Laplace transform of Eq. (5.3-49) is

$$D(s) = T_G(s) + T_C(s) + \frac{T_e}{s} \qquad (5.3-50)$$

To compensate for gravity loading and centrifugal effects, we can precompute these torque values and feed the computed torques forward into the controller to minimize their effects as shown in Fig. 5.8. This is called *feedforward compensation*.

Let us denote the computed torques as $\tau_{comp}(t)$ whose Laplace transform is $T_{comp}(s)$. With this computed torque and using Eq. (5.3-50), the error equation of Eq. (5.3-47) is modified to

$$E(s) = \frac{[s^2 J_{eff} R_a + s(R_a f_{eff} + K_a K_b)]\Theta_L^d(s) + \dfrac{n R_a [T_G(s) + T_C(s) + T_e/s - T_{comp}(s)]}{s^2 R_a J_{eff} + s(R_a f_{eff} + K_a K_b + K_a K_v) + K_a K_p}} \qquad (5.3-51)$$

For a step input, $\Theta_L^d(s) = A/s$, the steady-state position error of the system is given by

$$e_{ssp} = \lim_{s \to 0} s \left[\frac{n R_a [T_G(s) + T_C(s) + T_e/s - T_{comp}(s)]}{s^2 R_a J_{eff} + s(R_a f_{eff} + K_a K_b + K_a K_v) + K_a K_p} \right] \qquad (5.3-52)$$

For the steady-state position error, the contribution from the disturbances due to the centrifugal effect is zero as time approaches infinity. The reason for this is that the centrifugal effect is a function of $\dot\theta_L^2(t)$ and, as time approaches infinity, $\dot\theta_L(\infty)$ approaches zero. Hence, its contribution to the steady-state position error is zero. If the computed torque $\tau_{comp}(t)$ is equivalent to the gravity loading of the link, then the steady-state position error reduces to

$$e_{ssp} = \frac{n R_a T_e}{K_a K_p} \qquad (5.3-53)$$

Since K_p is bounded by Eq. (5.3-45), the above steady-state position error reduces to

$$e_{ssp} = \frac{4n T_e}{\omega_0^2 J_0} \qquad (5.3-54)$$

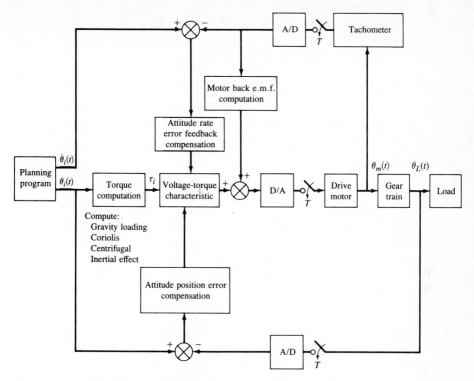

Figure 5.8 Compensation of disturbances.

which is small because τ_e is assumed to be small. The computation of $\tau_G(t)$ will be discussed later using the dynamic model of the manipulator.

If the input to the system is a ramp function, then $\Theta_L^d(s) = A/s^2$, and if we again assume that the disturbances are known as in Eq. (5.3-50), then the steady-state error of the system due to a ramp input is

$$
e_{ss}(\text{ramp}) \triangleq e_{ssv} = \lim_{s \to 0} s \frac{[s^2 J_{\text{eff}} R_a + s(R_a f_{\text{eff}} + K_a K_b)]A/s^2}{s^2 R_a J_{\text{eff}} + s(R_a f_{\text{eff}} + K_a K_b + K_a K_v) + K_a K_p}
$$

$$
+ \lim_{s \to 0} s \frac{n R_a [T_G(s) + T_C(s) + T_e/s - T_{\text{comp}}(s)]}{s^2 R_a J_{\text{eff}} + s(R_a f_{\text{eff}} + K_a K_b + K_a K_v) + K_a K_p}
$$

$$
= \frac{(R_a f_{\text{eff}} + K_a K_b)A}{K_a K_p}
$$

$$
+ \lim_{s \to 0} s \frac{n R_a [T_G(s) + T_C(s) + T_e/s - T_{\text{comp}}(s)]}{s^2 R_a J_{\text{eff}} + s(R_a f_{\text{eff}} + K_a K_b + K_a K_v) + K_a K_p} \qquad (5.3\text{-}55)
$$

Again, in order to reduce the steady-state velocity error, the computed torque $[\tau_{\text{comp}}(t)]$ needs to be equivalent to the gravity and centrifugal effects. Thus, the steady-state velocity error reduces to

$$e_{\text{ssv}} = \frac{(R_a f_{\text{eff}} + K_a K_b)A}{K_a K_p} + e_{\text{ssp}} \tag{5.3-56}$$

which has a finite steady-state error. The computation of $\tau_{\text{comp}}(t)$ depends on the dynamic model of the manipulator. In general, as discussed in Chap. 3, the Lagrange-Euler equations of motion of a six-joint manipulator, excluding the dynamics of the electronic control device, gear friction, and backlash, can be written as [Eq. (3.2-24)]

$$\tau_i(t) = \sum_{k=i}^{6} \sum_{j=1}^{k} \text{Tr} \left[\frac{\partial \, ^0\mathbf{T}_k}{\partial q_j} \mathbf{J}_k \left(\frac{\partial \, ^0\mathbf{T}_k}{\partial q_i} \right)^T \right] \ddot{q}_j(t)$$

$$+ \sum_{r=i}^{6} \sum_{j=1}^{r} \sum_{k=1}^{r} \text{Tr} \left[\frac{\partial^2 \, ^0\mathbf{T}_r}{\partial q_j \partial q_k} \mathbf{J}_r \left(\frac{\partial \, ^0\mathbf{T}_r}{\partial q_i} \right)^T \right] \dot{q}_j(t)\dot{q}_k(t)$$

$$- \sum_{j=i}^{6} m_j \mathbf{g} \left(\frac{\partial \, ^0\mathbf{T}_j}{\partial q_i} \right) \bar{\mathbf{r}}_j \quad \text{for } i = 1, 2, \ldots, 6 \tag{5.3-57}$$

where $\tau_i(t)$ is the generalized applied torque for joint i to drive the ith link, $\dot{q}_i(t)$ and $\ddot{q}_i(t)$ are the angular velocity and angular acceleration of joint i, respectively, and q_i is the generalized coordinate of the manipulator and indicates its angular position. $^0\mathbf{T}_i$ is a 4×4 homogeneous link transformation matrix which relates the spatial relationship between two coordinate frames (the ith and the base coordinate frames), $\bar{\mathbf{r}}_i$ is the position of the center of mass of link i with respect to the ith coordinate system, $\mathbf{g} = (g_x, g_y, g_z, 0)$ is the gravity row vector and $|\mathbf{g}| = 9.8062 \text{ m/s}^2$, and $\mathbf{J}_i$ is the pseudo-inertia matrix of link i about the ith coordinate frame and can be written as in Eq. (3.2-18).

Equation (5.3-57) can be expressed in matrix form explicitly as

$$\sum_{k=1}^{6} D_{ik} \ddot{q}_k(t) + \sum_{k=1}^{6} \sum_{m=1}^{6} h_{ikm} \dot{q}_k(t)\dot{q}_m(t) + c_i = \tau_i(t) \quad i = 1, 2, \ldots, 6 \tag{5.3-58}$$

where

$$D_{ik} = \sum_{j=\max(i,k)}^{6} \text{Tr} \left[\frac{\partial \, ^0\mathbf{T}_j}{\partial q_k} \mathbf{J}_j \left(\frac{\partial \, ^0\mathbf{T}_j}{\partial q_i} \right)^T \right] \quad i, k = 1, 2, \ldots, 6 \tag{5.3-59}$$

$$h_{ikm} = \sum_{j=\max(i,k,m)}^{6} \text{Tr} \left[\frac{\partial^2 \, ^0\mathbf{T}_j}{\partial q_k \partial q_m} \mathbf{J}_j \left(\frac{\partial \, ^0\mathbf{T}_j}{\partial q_i} \right)^T \right] \quad i, k, m = 1, 2, \ldots, 6$$

$$\tag{5.3-60}$$

$$c_i = \sum_{j=i}^{6} \left[-m_j \mathbf{g} \left(\frac{\partial \,^0\mathbf{T}_j}{\partial q_i} \right) \bar{\mathbf{r}}_j \right] \qquad i = 1, 2, \ldots, 6 \qquad (5.3\text{-}61)$$

Eq. (5.3-57) can be rewritten in a matrix notation as

$$\tau_i(t) = [D_{i1}, D_{i2}, D_{i3}, D_{i4}, D_{i5}, D_{i6}] \begin{bmatrix} \ddot{q}_1(t) \\ \ddot{q}_2(t) \\ \ddot{q}_3(t) \\ \ddot{q}_4(t) \\ \ddot{q}_5(t) \\ \ddot{q}_6(t) \end{bmatrix}$$

$$+ [\dot{q}_1(t), \dot{q}_2(t), \dot{q}_3(t), \dot{q}_4(t), \dot{q}_5(t), \dot{q}_6(t)] \qquad (5.3\text{-}62)$$

$$\times \begin{bmatrix} h_{i11} & h_{i12} & h_{i13} & h_{i14} & h_{i15} & h_{i16} \\ h_{i21} & h_{i22} & h_{i23} & h_{i24} & h_{i25} & h_{i26} \\ h_{i31} & h_{i32} & h_{i33} & h_{i34} & h_{i35} & h_{i36} \\ \cdot & \cdot & \cdot & \cdot & \cdot & \cdot \\ h_{i61} & h_{i62} & h_{i63} & h_{i64} & h_{i65} & h_{i66} \end{bmatrix} \begin{bmatrix} \dot{q}_1(t) \\ \dot{q}_2(t) \\ \dot{q}_3(t) \\ \dot{q}_4(t) \\ \dot{q}_5(t) \\ \dot{q}_6(t) \end{bmatrix} + c_i$$

Using the Lagrange-Euler equations of motion as formulated above, the computed torque for the gravity loading and centrifugal and Coriolis effects for joint i can be found, respectively, as

$$\tau_G(t) = c_i \qquad i = 1, 2, \ldots, 6 \qquad (5.3\text{-}63)$$

and

$$\tau_C(t) = [\dot{q}_1(t), \dot{q}_2(t), \dot{q}_3(t), \dot{q}_4(t), \dot{q}_5(t), \dot{q}_6(t)]$$

$$\times \begin{bmatrix} h_{i11} & h_{i12} & h_{i13} & h_{i14} & h_{i15} & h_{i16} \\ h_{i21} & h_{i22} & h_{i23} & h_{i24} & h_{i25} & h_{i26} \\ h_{i31} & h_{i32} & h_{i33} & h_{i34} & h_{i35} & h_{i36} \\ \cdot & \cdot & \cdot & \cdot & \cdot & \cdot \\ h_{i61} & h_{i62} & h_{i63} & h_{i64} & h_{i65} & h_{i66} \end{bmatrix} \begin{bmatrix} \dot{q}_1(t) \\ \dot{q}_2(t) \\ \dot{q}_3(t) \\ \dot{q}_4(t) \\ \dot{q}_5(t) \\ \dot{q}_6(t) \end{bmatrix} \qquad i = 1, 2, \ldots, 6$$

$$(5.3\text{-}64)$$

This compensation leads to what is usually known as the "inverse dynamics problem" or "computed torque" technique. This is covered in the next section.

5.3.4 Controller for Multijoint Robots

For a manipulator with multiple joints, one of the basic control schemes is the computed torque technique based on the L-E or the N-E equations of motion. Basically the computed torque technique is a feedforward control and has feedforward and feedback components. The control components compensate for the interaction forces among all the various joints and the feedback component computes the necessary correction torques to compensate for any deviations from the desired trajectory. It assumes that one can accurately compute the counterparts of $D(q)$, $h(q, \dot{q})$, and $c(q)$ in the L-E equations of motion [Eq. (3.2-26)] to minimize their nonlinear effects, and use a proportional plus derivative control to servo the joint motors. Thus, the structure of the control law has the form of

$$\tau(t) = D_a(q)\{\ddot{q}^d(t) + K_v[\dot{q}^d(t) - \dot{q}(t)] + K_p[q^d(t) - q(t)]\} + h_a(q, \dot{q}) + c_a(q)$$
(5.3-65)

where K_v and K_p are 6×6 derivative and position feedback gain matrices, respectively, and the manipulator has 6 degrees of freedom.

Substituting $\tau(t)$ from Eq. (5.3-65) into Eq. (3.2-26), we have

$$D(q)\ddot{q}(t) + h(q, \dot{q}) + c(q) = D_a(q)\{\ddot{q}^d(t) + K_v[\dot{q}^d(t) - \dot{q}(t)] + K_p[q^d(t) - q(t)]\}$$

$$+ h_a(q, \dot{q}) + c_a(q)$$
(5.3-66)

If $D_a(q)$, $h_a(q, \dot{q})$, $c_a(q)$ are equal to $D(q)$, $h(q, \dot{q})$, and $c(q)$, respectively, then Eq. (5.3-66) reduces to

$$D(q)[\ddot{e}(t) + K_v\dot{e}(t) + K_p e(t)] = 0$$
(5.3-67)

where $e(t) \triangleq q^d(t) - q(t)$ and $\dot{e}(t) \triangleq \dot{q}^d(t) - \dot{q}(t)$.

Since $D(q)$ is always nonsingular, K_p and K_v can be chosen appropriately so the characteristic roots of Eq. (5.3-67) have negative real parts, then the position error vector $e(t)$ approaches zero asymptotically.

The computation of the joint torques based on the complete L-E equations of motion [Eq. (5.3-65)] is very inefficient. As a result, Paul [1972] concluded that real-time closed-loop digital control is impossible or very difficult. Because of this reason, it is customary to simplify Eq. (5.3-65) by neglecting the velocity-related coupling term $h_a(q, \dot{q})$ and the off-diagonal elements of the acceleration-related matrix $D_a(q)$. In this case, the structure of the control law has the form

$$\tau(t) = \text{diag}[D_a(q)]\{\ddot{q}^d(t) + K_v[\dot{q}^d(t) - \dot{q}(t)] + K_p[q^d(t) - q(t)]\}$$

$$+ c_a(q)$$
(5.3-68)

A computer simulation study had been conducted to which showed that these terms cannot be neglected when the robot arm is moving at high speeds (Paul [1972]).

An analogous control law in the joint-variable space can be derived from the N-E equations of motion to servo a robot arm. The control law is computed recursively using the N-E equations of motion. The recursive control law can be obtained by substituting $\ddot{q}_i(t)$ into the N-E equations of motion to obtain the necessary joint torque for each actuator:

$$\ddot{q}_i(t) = \ddot{q}_i^d(t) + \sum_{j=1}^{n} K_v^{ij}[\dot{q}_j^d(t) - \dot{q}_j(t)] + \sum_{j=1}^{n} K_p^{ij}[q_j^d(t) - q_j(t)] \quad (5.3\text{-}69)$$

where K_v^{ij} and K_p^{ij} are the derivative and position feedback gains for joint i respectively and $e_j(t) = q_j^d(t) - q_j(t)$ is the position error for joint j. The physical interpretation of putting Eq. (5.3-69) into the N-E recursive equations can be viewed as follows:

1. The first term will generate the desired torque for each joint if there is no modeling error and the physical system parameters are known. However, there are errors due to backlash, gear friction, uncertainty about the inertia parameters, and time delay in the servo loop so that deviation from the desired joint trajectory will be inevitable.
2. The remaining terms in the N-E equations of motion will generate the correction torque to compensate for small deviations from the desired joint trajectory.

The above recursive control law is a proportional plus derivative control and has the effect of compensating for inertial loading, coupling effects, and gravity loading of the links. In order to achieve a critically damped system for each joint subsystem (which in turn loosely implies that the whole system behaves as a critically damped system), the feedback gain matrices $\mathbf{K}_p$ and $\mathbf{K}_v$ (diagonal matrices) can be chosen as discussed in Sec. 5.3.3, or as in Paul [1981] or Luh [1983b].

In summary, the computed torque technique is a feedforward compensation control. Based on complete L-E equations of motion, the joint torques can be computed in $O(n^4)$ time. The analogous control law derived from the N-E equations of motion can be computed in $O(n)$ time. One of the main drawbacks of this control technique is that the convergence of the position error vector depends on the dynamic coefficients of $\mathbf{D}(\mathbf{q})$, $\mathbf{h}(\mathbf{q}, \dot{\mathbf{q}})$, and $\mathbf{c}(\mathbf{q})$ in the equations of motion.

5.3.5 Compensation of Digitally Controlled Systems

In a sampled-data control system, time is normalized to the sampling period Δt; i.e., velocity is expressed as radians per Δt rather than radians per second. This has the effect of scaling the link equivalent inertia up by f_s^2, where f_s is the sampling frequency ($f_s = 1/\Delta t$).

It is typical to use 60-Hz sampling frequency (16-msec sampling period) because of its general availability and because the mechanical resonant frequency of most manipulators is around 5 to 10 Hz. Although the Nyquist sampling theorem indicates that, if the sampling rate is at least twice the cutoff frequency of the system, one should be able to recover the signal, the sampling rate for a continuous time system is more stringent than that. To minimize any deterioration of the controller due to sampling, the rate of sampling must be much greater than the natural frequency of the arm (inversely, the sampling period must be much less than the smallest time constant of the arm). Thus, to minimize the effect of sampling, usually 20 times the cutoff frequency is chosen. That is,

$$\Delta t = \frac{1}{20\,\omega_n/2\pi} = \frac{1}{20 f_n} \tag{5.3-70}$$

5.3.6 Voltage-Torque Conversion

Torque in an armature-controlled dc motor is theoretically a linear function of the armature voltage. However, due to bearing friction at low torques and saturation characteristics at high torques, the actual voltage-torque curves are not linear. For these reasons, a computer conversion of computed torque to required input voltage is usually accomplished via lookup tables or calculation from piecewise linear approximation formulas. The output voltage is usually a constant value and the voltage pulse width varies. A typical voltage-torque curve is shown in Fig. 5.9, where V_ϕ is the motor drive at which the joint will move at constant velocity exerting zero force in the direction of motion, and F_ϕ is the force/torque that the joint will exert at drive level V_ϕ with a negative velocity. The slopes and slope differences are obtained from the experimental curves.

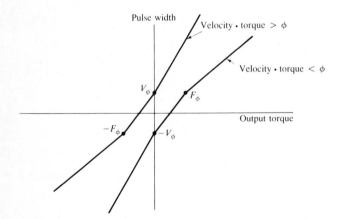

Figure 5.9 Voltage-torque conversion curve.

5.4 NEAR-MINIMUM-TIME CONTROL

For most manufacturing tasks, it is desirable to move a manipulator at its highest speed to minimize the task cycle time. This prompted Kahn and Roth [1971] to investigate the time-optimal control problem for mechanical manipulators. The objective of minimum-time control is to transfer the end-effector of a manipulator from an initial position to a specified desired position in minimum time.

Let us briefly discuss the basics of time-optimal control for a six-link manipulator. The state space representation of the equations of motion of a six-link robot can be formulated from the L-E equations of motion. Let us define a $2n$-dimensional state vector of a manipulator as

$$\mathbf{x}^T(t) = [\mathbf{q}^T(t), \dot{\mathbf{q}}^T(t)] = [q_1(t), \ldots, q_n(t), \dot{q}_1(t), \ldots, \dot{q}_n(t)]$$

$$\triangleq [\mathbf{x}_1^T(t), \mathbf{x}_2^T(t)] \triangleq [x_1(t), x_2(t), \ldots, x_{2n}(t)] \tag{5.4-1}$$

and an n-dimensional input vector as

$$\mathbf{u}^T(t) = [\tau_1(t), \tau_2(t), \ldots, \tau_n(t)] \tag{5.4-2}$$

The L-E equations of motion can be expressed in state space representation as

$$\dot{\mathbf{x}}(t) = \mathbf{f}[\mathbf{x}(t), \mathbf{u}(t)] \tag{5.4-3}$$

where $\mathbf{f}(\cdot)$ is a $2n \times 1$ continuously differentiable vector-valued function. Since $\mathbf{D}(\mathbf{q})$ is always nonsingular, the above equation can be expressed as

$$\dot{\mathbf{x}}_1(t) = \mathbf{x}_2(t)$$

and

$$\dot{\mathbf{x}}_2(t) = \mathbf{f}_2[\mathbf{x}(t)] + \mathbf{b}[\mathbf{x}_1(t)]\mathbf{u}(t) \tag{5.4-4}$$

where $\mathbf{f}_2(\mathbf{x})$ is an $n \times 1$ vector-valued function,

$$\mathbf{f}_2(\mathbf{x}) \equiv -\mathbf{D}^{-1}(\mathbf{x}_1)[\mathbf{h}(\mathbf{x}_1, \mathbf{x}_2) + \mathbf{c}(\mathbf{x}_1)] \tag{5.4-5}$$

and it can be shown that $\mathbf{b}(\mathbf{x}_1)$ is equivalent to the matrix $\mathbf{D}^{-1}(\mathbf{x}_1)$.

At the initial time $t = t_0$, the system is assumed to be in the initial state $\mathbf{x}(t_0) = \mathbf{x}_0$, and at the final minimum time $t = t_f$ the system is required to be in the desired final state $\mathbf{x}(t_f) = \mathbf{x}_f$. Furthermore, the admissible controls of the system are assumed to be bounded and satisfy the constraints,

$$|u_i| \leqslant (u_i)_{\max} \qquad \text{for all } t \tag{5.4-6}$$

Then the time-optimal control problem is to find an admissible control which transfers the system from the initial state $\mathbf{x}_0$ to the final state $\mathbf{x}_f$, while minimizing the performance index in Eq. (5.4-7) and subject to the constraints of Eq. (5.4-3),

$$J = \int_{t_0}^{t_f} dt = t_f - t_0 \tag{5.4-7}$$

Using the Pontryagin minimum principle (Kirk [1970]), an optimal control which minimizes the above functional J must minimize the hamiltonian. In terms of the optimal state vector $\mathbf{x}^*(t)$, the optimal control vector $\mathbf{v}^*(t)$, the optimal adjoint variables $\mathbf{p}^*(t)$, and the hamiltonian function,

$$H(\mathbf{x}, \mathbf{p}, \mathbf{v}) = \mathbf{p}^T \mathbf{f}(\mathbf{x}, \mathbf{v}) + 1 \tag{5.4-8}$$

the necessary conditions for $\mathbf{v}^*(t)$ to be an optimal control are

$$\dot{\mathbf{x}}^*(t) = \frac{\partial H(\mathbf{x}^*, \mathbf{p}^*, \mathbf{u}^*)}{\partial \mathbf{p}} \qquad \text{for all } t \in [t_0, t_f] \tag{5.4-9}$$

$$\dot{\mathbf{p}}^*(t) = -\frac{\partial H(\mathbf{x}^*, \mathbf{p}^*, \mathbf{u}^*)}{\partial \mathbf{x}} \qquad \text{for all } t \in [t_0, t_f] \tag{5.4-10}$$

and $\quad H(\mathbf{x}^*, \mathbf{p}^*, \mathbf{u}^*) \leqslant H(\mathbf{x}^*, \mathbf{p}^*, \mathbf{u}) \qquad \text{for all } t \in [t_0, t_f] \tag{5.4-11}$

and for all admissible controls. Obtaining $\mathbf{v}^*(t)$ from Eqs. (5.4-8) to (5.4-11), the optimization problem reduces to a two point boundary value problem with boundary conditions on the state $\mathbf{x}(t)$ at the initial and final times. Due to the non-linearity of the equations of motion, a numerical solution is usually the only approach to this problem. However, the numerical solution only computes the control function (open-loop control) and does not accommodate any system disturbances. In addition, the solution is optimal for the special initial and final conditions. Hence, the computations of the optimal control have to be performed for each manipulator motion. Furthermore, in practice, the numerical procedures do not provide an acceptable solution for the control of mechanical manipulators. Therefore, as an alternative to the numerical solution, Kahn and Roth [1971] proposed an approximation to the optimal control which results in a near-minimum-time control.

The suboptimal feedback control is obtained by approximating the nonlinear system [Eq. (5.4-4)] by a linear system and analytically finding an optimal control for the linear system. The linear system is obtained by a change of variables followed by linearization of the equations of motion. A transformation is used to decouple the controls in the linearized system. Defining a new set of dependent variables, $\xi_i(t)$, $i = 1, 2, \ldots, 2n$, the equations of motion can be transformed, using the new state variables, to

$$\xi_i(t) = x_i(t) - x_i(t_f) \qquad i = 1, 2, \ldots, n$$

and $\qquad \xi_i(t) = x_i(t) \qquad\qquad i = n + 1, \ldots, 2n \tag{5.4-12}$

The first $n\xi_i(t)$ is the error of the angular position, and the second $n\xi_i(t)$ is the error of the rate of the angular position. Because of this change of variables, the control problem becomes one of moving the system from an initial state $\xi(t_0)$ to the origin of the ξ space.

In order to obtain the linearized system, Eq. (5.4-12) is substituted into Eq. (5.4-4) and a Taylor series expansion is used to linearize the system about the origin of the ξ space. In addition, all sine and cosine functions of ξ_i are replaced by their series representations. As a result, the linearized equations of motion are

$$\dot{\xi}(t) = \mathbf{A}\xi(t) + \mathbf{B}\mathbf{v}(t) \tag{5.4-13}$$

where $\xi^T(t) = (\xi_1, \xi_2, \ldots, \xi_n)$ and $\mathbf{v}(t)$ is related to $\mathbf{u}(t)$ by $\mathbf{v}(t) = \mathbf{u}(t) + \mathbf{c}$, where the vector $\mathbf{c}$ contains the steady-state torques due to gravity at the final state. Although Eq. (5.4-13) is linear, the control functions $\mathbf{v}(t)$ are coupled. By properly selecting a set of basis vectors from the linearly independent columns of the controllability matrices of $\mathbf{A}$ and $\mathbf{B}$ to decouple the control function, a new set of equations with no coupling in control variables can be obtained:

$$\dot{\zeta}(t) = \overline{\mathbf{A}}\zeta(t) + \overline{\mathbf{B}}\mathbf{v}(t) \tag{5.4-14}$$

Using a three-link manipulator as an example and applying the above equations to it, we can obtain a three double-integrator system with unsymmetric bounds on controls:

$$\dot{\zeta}_{2i-1}(t) = v_i \tag{5.4-15}$$

$$\dot{\zeta}_{2i}(t) = \zeta_{2i-1} \qquad i = 1, 2, 3$$

where $v_i^- \leqslant v_i \leqslant v_i^+$ and

$$v_i^+ = (u_i)_{\max} + c_i \tag{5.4-16}$$

$$v_i^- = -(u_i)_{\max} + c_i$$

where c_i is the ith element of vector $\mathbf{c}$.

From this point on, a solution to the time-optimal control and switching surfaces† problem can be obtained by the usual procedures. The linearized and decoupled suboptimal control [Eqs. (5.4-15) and (5.4-16)] generally results in response times and trajectories which are reasonably close to the time-optimal solutions. However, this control method is usually too complex to be used for manipulators with 4 or more degrees of freedom and it neglects the effect of unknown external loads.

---

† Recall that time-optimal controls are piecewise constant functions of time, and thus we are interested in regions of the state space over which the control is constant. These regions are separated by curves in two-dimensional space, by surfaces in three-dimensional space, and by hypersurfaces in n-dimensional space. These separating surfaces are called switching curves, switching surfaces, and switching hypersurfaces, respectively.

5.5 VARIABLE STRUCTURE CONTROL

In 1978, Young [1978] proposed to use the theory of variable structure systems for the control of manipulators. Variable structure systems (VSS) are a class of systems with discontinuous feedback control. For the last 20 years, the theory of variable structure systems has found numerous applications in control of various processes in the steel, chemical, and aerospace industries. The main feature of VSS is that it has the so-called *sliding mode* on the switching surface. Within the sliding mode, the system remains insensitive to parameter variations and disturbances and its trajectories lie in the switching surface. It is this insensitivity property of VSS that enables us to eliminate the interactions among the joints of a manipulator. The sliding phenomena do not depend on the system parameters and have a stable property. Hence, the theory of VSS can be used to design a variable structure controller (VSC) which induces the sliding mode and in which lie the robot arm's trajectories. Such design of the variable structure controller does not require accurate dynamic modeling of the manipulator; the bounds of the model parameters are sufficient to construct the controller. Variable structure control differs from time-optimal control in the sense that the variable structure controller induces the sliding mode in which the trajectories of the system lie. Furthermore, the system is insensitive to system parameter variations in the sliding mode.

Let us consider the variable structure control for a six-link manipulator. From Eq. (5.4-1), defining the state vector $\mathbf{x}^T(t)$ as

$$
\begin{aligned}
\mathbf{x}^T &= (q_1, \ldots, q_6, \dot{q}_1, \ldots, \dot{q}_6) \\
&\triangleq (p_1, \ldots, p_6, v_1, \ldots, v_6) \\
&= (\mathbf{p}^T, \mathbf{v}^T)
\end{aligned}
\tag{5.5-1}
$$

and introducing the position error vector $\mathbf{e}_1(t) = \mathbf{p}(t) - \mathbf{p}^d$ and the velocity error vector $\mathbf{e}_2(t) = \mathbf{v}(t)$ (with $\mathbf{v}^d = 0$), we have changed the tracking problem to a regulator problem. The error equations of the system become

$$
\dot{\mathbf{e}}_1(t) = \mathbf{v}(t)
$$

and
$$
\dot{\mathbf{v}}(t) = \mathbf{f}_2(\mathbf{e}_1 + \mathbf{p}^d, \mathbf{v}) + \mathbf{b}(\mathbf{e}_1 + \mathbf{p}^d)\mathbf{u}(t)
\tag{5.5-2}
$$

where $\mathbf{f}_2(\cdot)$ and $\mathbf{b}(\cdot)$ are defined in Eq. (5.4-5). For the regulator system problem in Eq. (5.5-2), a variable structure control $\mathbf{u}(\mathbf{p}, \mathbf{v})$ can be constructed as

$$
u_i(\mathbf{p}, \mathbf{v}) =
\begin{cases}
u_i^+ (\mathbf{p}, \mathbf{v}) & \text{if } s_i(e_i, v_i) > 0 \\
\\
u_i^- (\mathbf{p}, \mathbf{v}) & \text{if } s_i(e_i, v_i) < 0
\end{cases}
\qquad i = 1, \ldots, 6
\tag{5.5-3}
$$

where $s_i(e_i, v_i)$ are the switching surfaces found to be

$$s_i(e_i, v_i) = c_i e_i + v_i \qquad c_i > 0 \qquad i = 1, \ldots, 6 \tag{5.5-4}$$

and the synthesis of the control reduces to choosing the feedback controls as in Eq. (5.5-3) so that the sliding mode occurs on the intersection of the switching planes. By solving the algebraic equations of the switching planes,

$$\dot{s}_i(e_i, v_i) = 0 \qquad i = 1, \ldots, 6 \tag{5.5-5}$$

a unique control exists and is found to be

$$\mathbf{u}_{eq} = -\mathbf{D}(\mathbf{p})(\mathbf{f}_2(\mathbf{p}, \mathbf{v}) + \mathbf{C}\mathbf{v}) \tag{5.5-6}$$

where $\mathbf{C} \equiv \text{diag}\, [c_1, c_2, \ldots, c_6]$. Then, the sliding mode is obtained from Eq. (5.5-4) as

$$\dot{e}_i = -c_i e_i \qquad i = 1, \ldots, 6 \tag{5.5-7}$$

The above equation represents six uncoupled first-order linear systems, each representing 1 degree of freedom of the manipulator when the system is in the sliding mode. As we can see, the controller [Eq. (5.5-3)] forces the manipulator into the sliding mode and the interactions among the joints are completely eliminated. When in the sliding mode, the controller [Eq. (5.5-6)] is used to control the manipulator. The dynamics of the manipulator in the sliding mode depend only on the design parameters c_i. With the choice of $c_i > 0$, we can obtain the asymptotic stability of the system in the sliding mode and make a speed adjustment of the motion in sliding mode by varying the parameters c_i.

In summary, the variable structure control eliminates the nonlinear interactions among the joints by forcing the system into the sliding mode. However, the controller produces a discontinuous feedback control signal that change signs rapidly. The effects of such control signals on the physical control device of the manipulator (i.e., chattering) should be taken into consideration for any applications to robot arm control. A more detailed discussion of designing a multi-input controller for a VSS can be found in Young [1978].

5.6 NONLINEAR DECOUPLED FEEDBACK CONTROL

There is a substantial body of nonlinear control theory which allows one to design a near-optimal control strategy for mechanical manipulators. Most of the existing robot control algorithms emphasize nonlinear compensations of the interactions among the links, e.g., the computed torque technique. Hemami and Camana [1976] applied the nonlinear feedback control technique to a simple locomotion system which has a particular class of nonlinearity (sine, cosine, and polynomial) and obtained decoupled subsystems, postural stability, and desired periodic trajectories. Their approach is different from the method of linear system decoupling where the system to be decoupled must be linear. Saridis and Lee [1979] proposed an iterative algorithm for sequential improvement of a nonlinear suboptimal control law. It provides an approximate optimal control for a manipulator. To achieve such a high quality of control, this method also requires a considerable amount of computational time. In this section, we shall briefly describe the general nonlinear decoupling theory (Falb and Wolovich [1967], Freund [1982]) which will be utilized together with the Newton-Euler equations of motion to compute a nonlinear decoupled controller for robot manipulators.

Given a general nonlinear system as in

$$\dot{\mathbf{x}}(t) = \mathbf{A}(\mathbf{x}) + \mathbf{B}(\mathbf{x})\mathbf{u}(t)$$

and

$$\mathbf{y}(t) = \mathbf{C}(\mathbf{x}) \tag{5.6-1}$$

where $\mathbf{x}(t)$ is an n-dimensional vector, $\mathbf{u}(t)$ and $\mathbf{y}(t)$ are m-dimensional vectors, and $\mathbf{A}(\mathbf{x})$, $\mathbf{B}(\mathbf{x})$, and $\mathbf{C}(\mathbf{x})$ are matrices of compatible order. Let us define a nonlinear operator N_A^K as

$$N_A^K C_i(\mathbf{x}) = \left[\frac{\partial}{\partial \mathbf{x}} N_A^{K-1} C_i(\mathbf{x}) \right] \mathbf{A}(\mathbf{x}) \qquad \begin{aligned} K &= 1, 2, \ldots, n \\ i &= 1, 2, \ldots, m \end{aligned} \tag{5.6-2}$$

where $C_i(\mathbf{x})$ is the ith component of $\mathbf{C}(\mathbf{x})$ and $N_A^0 C_i(\mathbf{x}) = C_i(\mathbf{x})$. Also, let us define the differential order d_i of the nonlinear system as

$$d_i = \min \left\{ j: \left[\frac{\partial}{\partial \mathbf{x}} N_A^{j-1} C_i(\mathbf{x}) \right] \mathbf{B}(\mathbf{x}) \neq 0, \ j = 1, 2, \ldots, n \right\} \tag{5.6-3}$$

Then, the control objective is to find a feedback decoupled controller $\mathbf{u}(t)$:

$$\mathbf{u}(t) = \mathbf{F}(\mathbf{x}) + \mathbf{G}(\mathbf{x}) \, \mathbf{w}(t) \tag{5.6-4}$$

where $\mathbf{w}(t)$ is an m-dimensional reference input vector, $\mathbf{F}(\mathbf{x})$ is an $m \times 1$ feedback vector for decoupling and pole assignment, and $\mathbf{G}(\mathbf{x})$ is an $m \times m$ input gain matrix so that the overall system has a decoupled input-output relationship.

Substituting $\mathbf{u}(t)$ from Eq. (5.6-4) into the system equation of Eq. (5.6-1) results in the following expressions:

$$\dot{\mathbf{x}}(t) = \mathbf{A}(\mathbf{x}) + \mathbf{B}(\mathbf{x})\mathbf{F}(\mathbf{x}) + \mathbf{B}(\mathbf{x})\mathbf{G}(\mathbf{x})\mathbf{w}(t)$$

and $\qquad\qquad \mathbf{y}(t) = \mathbf{C}(\mathbf{x}) \tag{5.6-5}$

In order to obtain the decoupled input-output relationships in the above system, $\mathbf{F}(\mathbf{x})$ and $\mathbf{G}(\mathbf{x})$ are chosen, respectively, as follows:

$$\mathbf{F}(\mathbf{x}) = \mathbf{F}_1^*(\mathbf{x}) + \mathbf{F}_2^*(\mathbf{x}) \tag{5.6-6}$$

where

$$\mathbf{F}_1^*(\mathbf{x}) = -\mathbf{D}^{*-1}(\mathbf{x})\mathbf{C}^*(\mathbf{x})$$

$$\mathbf{F}_2^*(\mathbf{x}) = -\mathbf{D}^{*-1}(\mathbf{x})\mathbf{M}^*(\mathbf{x})$$

and

$$\mathbf{G}(\mathbf{x}) = \mathbf{D}^{*-1}(\mathbf{x})\,\mathbf{\Lambda}$$

$\mathbf{F}_1^*(\mathbf{x})$ represents the state feedback that yields decoupling, while $\mathbf{F}_2^*(\mathbf{x})$ performs the control part with arbitrary pole assignment. The input gain of the decoupled part can be chosen by $\mathbf{G}(\mathbf{x})$, and $\mathbf{D}^*(\mathbf{x})$ is an $m \times m$ matrix whose ith row is given by

$$\mathbf{D}_i^*(\mathbf{x}) = \left[\frac{\partial}{\partial \mathbf{x}} N_A^{d_i-1} C_i(\mathbf{x}) \right] \mathbf{B}(\mathbf{x}) \qquad \text{for } d_i \neq 0 \qquad (5.6\text{-}7)$$

$\mathbf{C}^*(\mathbf{x})$ is an m-dimensional vector whose ith component is given by

$$C_i^*(\mathbf{x}) = N_A^{d_i} C_i(\mathbf{x}) \qquad (5.6\text{-}8)$$

$\mathbf{M}^*(\mathbf{x})$ is an m-dimensional vector whose ith component is given by

$$M_i^*(\mathbf{x}) = \sum_{K=0}^{d_i-1} \alpha_{K,i} N_A^K C_i(\mathbf{x}) \qquad \text{for } d_i \neq 0 \qquad (5.6\text{-}9)$$

and $\mathbf{\Lambda}$ is a diagonal matrix whose elements are constant values λ_i for $i = 1, 2, \ldots, m$. Then, the system in Eq. (5.6-1) can be represented in terms of

$$\mathbf{y}^*(t) = \mathbf{C}^*(\mathbf{x}) + \mathbf{D}^*(\mathbf{x})\,\mathbf{u}(t) \qquad (5.6\text{-}10)$$

where $\mathbf{y}^*(t)$ is an output vector whose ith component is $y_i^{(d_i)}(t)$. That is,

$$y_i^{(d_i)}(t) = C_i^*(\mathbf{x}) + \mathbf{D}_i^*\mathbf{u}(t) \qquad (5.6\text{-}11)$$

Utilizing Eq. (5.6-4) and Eqs. (5.6-6) to (5.6-11), we obtain

$$y_i^{(d_i)}(t) + \alpha_{d_i-1,i}\, y_i^{(d_i-1)}(t) + \cdots + \alpha_{0,i} y_i(t) = \lambda_i \omega_i(t) \qquad (5.6\text{-}12)$$

where $\alpha_{K,i}$ and λ_i are arbitrary scalars.

To show that the ith component of $\mathbf{y}^*(t)$ has the form of Eq. (5.6-11), let us assume that $d_i = 1$. Then, $y_i(t) = C_i(\mathbf{x})$ and, by differentiating it successively, we have

$$y_i^{(1)}(t) = \dot{y}_i(t) = \frac{\partial C_i(\mathbf{x})}{\partial \mathbf{x}} \dot{\mathbf{x}}(t)$$

$$= \frac{\partial C_i(\mathbf{x})}{\partial \mathbf{x}} [\mathbf{A}(\mathbf{x}) + \mathbf{B}(\mathbf{x})\mathbf{F}(\mathbf{x}) + \mathbf{B}(\mathbf{x})\mathbf{G}(\mathbf{x})\mathbf{w}(t)]$$

$$= N_{A+BF}^1 C_i(\mathbf{x}) + \frac{\partial C_i(\mathbf{x})}{\partial \mathbf{x}} [\mathbf{B}(\mathbf{x})\mathbf{G}(\mathbf{x})\mathbf{w}(t)]$$

Using the identity, $N_{A+BF}^{d_i} C_i(\mathbf{x}) = N_A^{d_i} C_i(\mathbf{x}) + [\partial/\partial\mathbf{x} N_A^{d_i-1} C_i(\mathbf{x})]\mathbf{B}(\mathbf{x})\mathbf{F}(\mathbf{x})$, $y_i^{(1)}(t)$ can be written as

$$y_i^{(1)}(t) = N_A^1 C_i(\mathbf{x}) + \left[\frac{\partial}{\partial \mathbf{x}} C_i(\mathbf{x})\right] \mathbf{B}(\mathbf{x})\mathbf{F}(\mathbf{x}) + \frac{\partial C_i(\mathbf{x})}{\partial \mathbf{x}}[\mathbf{B}(\mathbf{x})\mathbf{G}(\mathbf{x})\mathbf{w}(t)]$$

Using Eqs. (5.6-4) and (5.6-7), it becomes

$$y_i^{(1)}(t) = C_i^*(\mathbf{x}) + D_i^*(\mathbf{x})\mathbf{u}(t)$$

Similar comments hold for $d_i = 2, 3, \ldots$ to yield Eq. (5.6-11). Thus, the resultant system has decoupled input-output relations and becomes a time-invariant second-order system which can be used to model each joint of the robot arm.

As discussed in Chap. 3, the Lagrange-Euler equations of motion of a six-link robot can be written as

$$\begin{bmatrix} D_{11} & \cdots & D_{16} \\ \vdots & & \vdots \\ D_{16} & \cdots & D_{66} \end{bmatrix} \begin{bmatrix} \ddot{\theta}_1(t) \\ \vdots \\ \ddot{\theta}_6(t) \end{bmatrix} + \begin{bmatrix} h_1(\theta, \dot{\theta}) \\ \vdots \\ h_6(\theta, \dot{\theta}) \end{bmatrix} + \begin{bmatrix} c_1(\theta) \\ \vdots \\ c_6(\theta) \end{bmatrix} = \begin{bmatrix} u_1(t) \\ \vdots \\ u_6(t) \end{bmatrix} \tag{5.6-13}$$

which can be rewritten in vector-matrix notation as

$$\mathbf{D}(\theta)\ddot{\theta} + \mathbf{h}(\theta, \dot{\theta}) + \mathbf{c}(\theta) = \mathbf{u}(t) \tag{5.6-14}$$

where $\mathbf{u}(t)$ is a 6×1 applied torque vector for joint actuators, $\theta(t)$ is the angular positions, $\dot{\theta}(t)$ is the angular velocities, $\ddot{\theta}(t)$ is a 6×1 acceleration vector, $\mathbf{c}(\theta)$ is a 6×1 gravitational force vector, $\mathbf{h}(\theta, \dot{\theta})$ is a 6×1 Coriolis and centrifugal force vector, and $\mathbf{D}(\theta)$ is a 6×6 acceleration-related matrix. Since $\mathbf{D}(\theta)$ is always nonsingular, the above equation can be rewritten as

$$\ddot{\theta}(t) = -\mathbf{D}^{-1}(\theta)[\mathbf{h}(\theta, \dot{\theta}) + \mathbf{c}(\theta)] + \mathbf{D}^{-1}(\theta)\mathbf{u}(t) \tag{5.6-15}$$

or, explicitly,

$$\ddot{\theta}(t) = - \begin{bmatrix} D_{11} & \cdots & D_{16} \\ \cdot & & \cdot \\ \cdot & & \cdot \\ D_{16} & \cdots & D_{66} \end{bmatrix}^{-1} \begin{bmatrix} h_1(\theta, \dot{\theta}) + c_1(\theta) \\ \cdot \\ \cdot \\ h_6(\theta, \dot{\theta}) + c_6(\theta) \end{bmatrix}$$ (5.6-16)

$$+ \begin{bmatrix} D_{11} & & D_{16} \\ \cdot & \cdots & \cdots & \cdots & \cdot \\ \cdot & & \cdot \\ \cdot & \cdots & \cdots & \cdots & \cdot \\ D_{16} & & D_{66} \end{bmatrix}^{-1} \begin{bmatrix} u_1(t) \\ \cdot \\ \cdot \\ u_6(t) \end{bmatrix}$$

The above dynamic model consists of second-order differential equations for each joint variable; hence, $d_i = 2$. Treating each joint variable $\theta_i(t)$ as an output variable, the above equation can be related to Eq. (5.6-11) as

$$y_i^{(2)}(t) = \ddot{y}_i(t) = -[\mathbf{D}^{-1}(\theta)]_i[\mathbf{h}(\theta, \dot{\theta}) + \mathbf{c}(\theta)] + [\mathbf{D}^{-1}(\theta)]_i\mathbf{u}(t)$$

$$= C_i^*(\mathbf{x}) + \mathbf{D}_i^*(\mathbf{x})\mathbf{u}(t)$$ (5.6-17)

where

$$C_i^*(\mathbf{x}) = -[\mathbf{D}^{-1}(\theta)]_i[\mathbf{h}(\theta, \dot{\theta}) + \mathbf{c}(\theta)]$$ (5.6-18)

$$\mathbf{x}^T(t) = [\theta^T(t), \dot{\theta}^T(t)]$$

and $$\mathbf{D}_i^*(\mathbf{x}) = [\mathbf{D}^{-1}(\theta)]_i$$ (5.6-19)

and $[\mathbf{D}^{-1}(\theta)]_i$ is the ith row of the $\mathbf{D}^{-1}(\theta)$ matrix. Thus, the controller $\mathbf{u}(t)$ for the decoupled system [Eq. (5.6-5)] must be

$$\mathbf{u}(t) = -\mathbf{D}^{*-1}(\mathbf{x})[\mathbf{C}^*(\mathbf{x}) + \mathbf{M}^*(\mathbf{x}) - \mathbf{\Lambda}\,\mathbf{w}(t)]$$

$$= -\mathbf{D}(\theta)\{-\mathbf{D}^{-1}(\theta)[\mathbf{h}(\theta, \dot{\theta}) + \mathbf{c}(\theta)] + \mathbf{M}^*(\mathbf{x}) - \mathbf{\Lambda}\,\mathbf{w}(t)\}$$

$$= \mathbf{h}(\theta, \dot{\theta}) + \mathbf{c}(\theta) - \mathbf{D}(\theta)[\mathbf{M}^*(\mathbf{x}) - \mathbf{\Lambda}\,\mathbf{w}(t)]$$ (5.6-20)

Explicitly, for joint i,

$$u_i(t) = h_i(\theta, \dot{\theta}) + c_i(\theta) - [D_{i1} \cdots D_{i6}] \begin{bmatrix} \alpha_{11}\dot{\theta}_1(t) + \alpha_{01}\theta_1(t) - \lambda_1 w_1(t) \\ \cdot \\ \cdot \\ \alpha_{16}\dot{\theta}_6(t) + \alpha_{06}\theta_6(t) - \lambda_6 w_6(t) \end{bmatrix}$$

(5.6-21)

From the above equation, we note that the controller $u_i(t)$ for joint i depends only on the current dynamic variables and the input $\mathbf{w}(t)$. Substituting $\mathbf{u}(t)$ from Eq. (5.6-20) into Eq. (5.6-14), we have

$$\mathbf{D}(\boldsymbol{\theta})\ddot{\boldsymbol{\theta}}(t) + \mathbf{h}(\boldsymbol{\theta}, \dot{\boldsymbol{\theta}}) + \mathbf{c}(\boldsymbol{\theta})$$

$$= \mathbf{h}(\boldsymbol{\theta}, \dot{\boldsymbol{\theta}}) + \mathbf{c}(\boldsymbol{\theta}) - \mathbf{D}(\boldsymbol{\theta}) \begin{bmatrix} \alpha_{11}\dot{\theta}_1(t) + \alpha_{01}\theta_1(t) - \lambda_1 w_1(t) \\ \vdots \\ \alpha_{16}\dot{\theta}_6(t) + \alpha_{06}\theta_6(t) - \lambda_6 w_6(t) \end{bmatrix} \quad (5.6\text{-}22)$$

which leads to

$$\mathbf{D}(\boldsymbol{\theta}) \begin{bmatrix} \ddot{\theta}_1(t) + \alpha_{11}\dot{\theta}_1(t) + \alpha_{01}\theta_1(t) - \lambda_1 w_1(t) \\ \vdots \\ \ddot{\theta}_6(t) + \alpha_{16}\dot{\theta}_6(t) + \alpha_{06}\theta_6(t) - \lambda_6 w_6(t) \end{bmatrix} = 0 \quad (5.6\text{-}23)$$

Since $\mathbf{D}(\boldsymbol{\theta})$ is always nonsingular, the above equation becomes

$$\ddot{\theta}_i(t) + \alpha_{1i}\dot{\theta}_i(t) + \alpha_{0i}\theta_i(t) = \lambda_i \omega_i(t) \qquad i = 1, 2, \ldots, 6 \quad (5.6\text{-}24)$$

which indicates the final decoupled input-output relationships of the system. It is interesting to note that the parameters α_{1i}, α_{0i}, and λ_i can be selected arbitrarily, provided that the stability criterion is maintained. Hence, the manipulator can be considered as six independent, decoupled, second-order, time-invariant systems and the controller $\mathbf{u}(t)$ [Eq. (5.6-20)] can be computed efficiently based on the manipulator dynamics. An efficient way of computing the controller $\mathbf{u}(t)$ is through the use of the Newton-Euler equations of motion. Hence, to compute the controller $u_i(t)$ for joint i, $\ddot{\theta}_i(t)$ is substituted with $\lambda_i w_i(t) - \alpha_{1i}\dot{\theta}_i(t) - \alpha_{0i}\theta_i(t)$ in the Newton-Euler equations of motion.

5.7 RESOLVED MOTION CONTROL

In the last section, several methods were discussed for controlling a mechanical manipulator in the joint-variable space to follow a joint-interpolated trajectory. In many applications, resolved motion control, which commands the manipulator hand to move in a desired cartesian direction in a coordinated position and rate control, is more appropriate. *Resolved motion* means that the motions of the various joint motors are combined and resolved into separately controllable hand motions along the world coordinate axes. This implies that several joint motors must run simultaneously at different time-varying rates in order to achieve desired coordinated hand motion along any world coordinate axis. This enables the user to specify the

direction and speed along any arbitrarily oriented path for the manipulator to follow. This motion control greatly simplifies the specification of the sequence of
motions for completing a task because users are usually more adapted to the cartesian coordinate system than the manipulator's joint angle coordinates.

In general, the desired motion of a manipulator is specified in terms of a
time-based hand trajectory in cartesian coordinates, while the servo control system
requires that the reference inputs be specified in joint coordinates. The mathematical relationship between these two coordinate systems is important in designing
efficient control in the cartesian space. We shall briefly describe the basic kinematics theory relating these two coordinate systems for a six-link robot arm that will
lead us to understand various important resolved motion control methods.

The location of the manipulator hand with respect to a fixed reference coordinate system can be realized by establishing an orthonormal coordinate frame at the
hand (the hand coordinate frame), as shown in Fig. 5.10. The problem of finding
the location of the hand is reduced to finding the position and orientation of the
hand coordinate frame with respect to the inertial frame of the manipulator. This
can be conveniently achieved by a 4 × 4 homogeneous transformation matrix:

$$
{}^{\text{base}}\mathbf{T}_{\text{hand}}(t) = \begin{bmatrix} n_x(t) & s_x(t) & a_x(t) & p_x(t) \\ n_y(t) & s_y(t) & a_y(t) & p_y(t) \\ n_z(t) & s_z(t) & a_z(t) & p_z(t) \\ 0 & 0 & 0 & 1 \end{bmatrix} = \begin{bmatrix} \mathbf{n}(t) & \mathbf{s}(t) & \mathbf{a}(t) & \mathbf{p}(t) \\ 0 & 0 & 0 & 1 \end{bmatrix}
$$

$$(5.7\text{-}1)$$

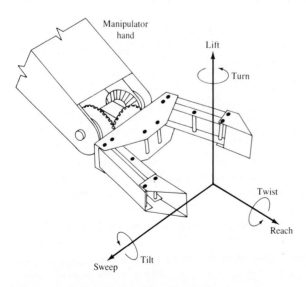

Figure 5.10 The hand coordinate system.

where **p** is the position vector of the hand, and **n**, **s**, **a** are the unit vectors along the principal axes of the coordinate frame describing the orientation of the hand. Instead of using the rotation submatrix [**n**, **s**, **a**] to describe the orientation, we can use three Euler angles, yaw $\alpha(t)$, pitch $\beta(t)$, and roll $\gamma(t)$, which are defined as rotations of the hand coordinate frame about the $\mathbf{x}_0$, $\mathbf{y}_0$, and $\mathbf{z}_0$ of the reference frame, respectively. One can obtain the elements of [**n**, **s**, **a**] from the Euler rotation matrix resulting from a rotation of the α angle about the $\mathbf{x}_0$ axis, then a rotation of the β angle about the $\mathbf{y}_0$ axis, and a rotation of the γ angle about the $\mathbf{z}_0$ axis of the reference frame [Eq. (2.2-19)]. Thus:

$$
{}^{\text{base}}\mathbf{R}_{\text{hand}}(t) = \begin{bmatrix} n_x(t) & s_x(t) & a_x(t) \\ n_y(t) & s_y(t) & a_y(t) \\ n_z(t) & s_z(t) & a_z(t) \end{bmatrix}
$$

$$
= \begin{bmatrix} C\gamma & -S\gamma & 0 \\ S\gamma & C\gamma & 0 \\ 0 & 0 & 1 \end{bmatrix} \begin{bmatrix} C\beta & 0 & S\beta \\ 0 & 1 & 0 \\ -S\beta & 0 & C\beta \end{bmatrix} \begin{bmatrix} 1 & 0 & 0 \\ 0 & C\alpha & -S\alpha \\ 0 & S\alpha & C\alpha \end{bmatrix}
$$

$$
= \begin{bmatrix} C\gamma C\beta & -S\gamma C\alpha + C\gamma S\beta S\alpha & S\gamma S\alpha + C\gamma S\beta C\alpha \\ S\gamma C\beta & C\gamma C\alpha + S\gamma S\beta S\alpha & -C\gamma S\alpha + S\gamma S\beta C\alpha \\ -S\beta & C\beta S\alpha & C\beta C\alpha \end{bmatrix} \quad (5.7\text{-}2)
$$

where $\sin\alpha \equiv S\alpha$, $\cos\alpha \equiv C\alpha$, $\sin\beta \equiv S\beta$, $\cos\beta \equiv C\beta$, $\sin\gamma \equiv S\gamma$, and $\cos\gamma \equiv C\gamma$.

Let us define the position $\mathbf{p}(t)$, Euler angles $\boldsymbol{\Phi}(t)$, linear velocity $\mathbf{v}(t)$, and angular velocity $\boldsymbol{\Omega}(t)$ vectors of the manipulator hand with respect to the reference frame, respectively:

$$
\mathbf{p}(t) \triangleq [p_x(t), p_y(t), p_z(t)]^T \qquad \boldsymbol{\Phi}(t) \triangleq [\alpha(t), \beta(t), \gamma(t)]^T
$$

$$
\mathbf{v}(t) \triangleq [v_x(t), v_y(t), v_z(t)]^T \qquad \boldsymbol{\Omega}(t) \triangleq [\omega_x(t), \omega_y(t), \omega_z(t)]^T \quad (5.7\text{-}3)
$$

The linear velocity of the hand with respect to the reference frame is equal to the time derivative of the position of the hand:

$$
\mathbf{v}(t) = \frac{d\mathbf{p}(t)}{dt} = \dot{\mathbf{p}}(t) \quad (5.7\text{-}4)
$$

Since the inverse of a direction cosine matrix is equivalent to its transpose, the instantaneous angular velocities of the hand coordinate frame about the principal axes of the reference frame can be obtained from Eq. (5.7-2):

$$\mathbf{R}\frac{d\mathbf{R}^T}{dt} = -\frac{d\mathbf{R}}{dt}\mathbf{R}^T = -\begin{bmatrix} 0 & -\omega_z & \omega_y \\ \omega_z & 0 & -\omega_x \\ -\omega_y & \omega_x & 0 \end{bmatrix}$$

$$= \begin{bmatrix} 0 & -S\beta\dot{\alpha} + \dot{\gamma} & -S\gamma C\beta\dot{\alpha} - C\gamma\dot{\beta} \\ S\beta\dot{\alpha} - \dot{\gamma} & 0 & C\gamma C\beta\dot{\alpha} - S\gamma\dot{\beta} \\ S\gamma C\beta\dot{\alpha} + C\gamma\dot{\beta} & -C\gamma C\beta\dot{\alpha} + S\gamma\dot{\beta} & 0 \end{bmatrix} \quad (5.7\text{-}5)$$

From the above equation, the relation between the $[\omega_x(t), \omega_y(t), \omega_z(t)]^T$ and $[\dot{\alpha}(t), \dot{\beta}(t), \dot{\gamma}(t)]^T$ can be found by equating the nonzero elements in the matrices:

$$\begin{bmatrix} \omega_x(t) \\ \omega_y(t) \\ \omega_z(t) \end{bmatrix} = \begin{bmatrix} C\gamma C\beta & -S\gamma & 0 \\ S\gamma C\beta & C\gamma & 0 \\ -S\beta & 0 & 1 \end{bmatrix} \begin{bmatrix} \dot{\alpha}(t) \\ \dot{\beta}(t) \\ \dot{\gamma}(t) \end{bmatrix} \quad (5.7\text{-}6)$$

Its inverse relation can be found easily:

$$\begin{bmatrix} \dot{\alpha}(t) \\ \dot{\beta}(t) \\ \dot{\gamma}(t) \end{bmatrix} = \sec\beta \begin{bmatrix} C\gamma & S\gamma & 0 \\ -S\gamma C\beta & C\gamma C\beta & 0 \\ C\gamma S\beta & S\gamma S\beta & C\beta \end{bmatrix} \begin{bmatrix} \omega_x(t) \\ \omega_y(t) \\ \omega_z(t) \end{bmatrix} \quad (5.7\text{-}7)$$

or expressed in matrix-vector form,

$$\dot{\Phi}(t) \triangleq [S(\Phi)]\,\Omega(t) \quad (5.7\text{-}8)$$

Based on the moving coordinate frame concept, the linear and angular velocities of the hand can be obtained from the velocities of the lower joints:

$$\begin{bmatrix} \mathbf{v}(t) \\ \Omega(t) \end{bmatrix} = [\mathbf{N}(\mathbf{q})]\dot{\mathbf{q}}(t) = [\mathbf{N}_1(\mathbf{q}), \mathbf{N}_2(\mathbf{q}), \ldots, \mathbf{N}_6(\mathbf{q})]\,\dot{\mathbf{q}}(t) \quad (5.7\text{-}9)$$

where $\dot{\mathbf{q}}(t) = (\dot{q}_1, \ldots, \dot{q}_6)^T$ is the joint velocity vector of the manipulator, and $\mathbf{N}(\mathbf{q})$ is a 6×6 jacobian matrix whose ith column vector $\mathbf{N}_i(\mathbf{q})$ can be found to be (Whitney [1972]):

$$
\mathbf{N}_i(\mathbf{q}) = \begin{cases} \begin{bmatrix} \mathbf{z}_{i-1} \times (\mathbf{p} - \mathbf{p}_{i-1}) \\ \mathbf{z}_{i-1} \end{bmatrix} & \text{if joint } i \text{ is rotational} \\[2em] \begin{bmatrix} \mathbf{z}_{i-1} \\ \mathbf{0} \end{bmatrix} & \text{if joint } i \text{ is translational} \end{cases} \tag{5.7-10}
$$

where $\times$ indicates the vector cross product, $\mathbf{p}_{i-1}$ is the position of the origin of the $(i-1)$th coordinate frame with respect to the reference frame, $\mathbf{z}_{i-1}$ is the unit vector along the axis of motion of joint i, and $\mathbf{p}$ is the position of the hand with respect to the reference coordinate frame.

If the inverse jacobian matrix exists at $\mathbf{q}(t)$, then the joint velocities $\dot{\mathbf{q}}(t)$ of the manipulator can be computed from the hand velocities using Eq. (5.7-9):

$$
\dot{\mathbf{q}}(t) = \mathbf{N}^{-1}(\mathbf{q}) \begin{bmatrix} \mathbf{v}(t) \\ \mathbf{\Omega}(t) \end{bmatrix} \tag{5.7-11}
$$

Given the desired linear and angular velocities of the hand, this equation computes the joint velocities and indicates the rates at which the joint motors must be maintained in order to achieve a steady hand motion along the desired cartesian direction.

The accelerations of the hand can be obtained by taking the time derivative of the velocity vector in Eq. (5.7-9):

$$
\begin{bmatrix} \dot{\mathbf{v}}(t) \\ \dot{\mathbf{\Omega}}(t) \end{bmatrix} = \dot{\mathbf{N}}(\mathbf{q}, \dot{\mathbf{q}})\dot{\mathbf{q}}(t) + \mathbf{N}(\mathbf{q})\ddot{\mathbf{q}}(t) \tag{5.7-12}
$$

where $\ddot{\mathbf{q}}(t) = [\ddot{q}_1(t), \ldots, \ddot{q}_6(t)]^T$ is the joint acceleration vector of the manipulator. Substituting $\dot{\mathbf{q}}(t)$ from Eq. (5.7-11) into Eq. (5.7-12) gives

$$
\begin{bmatrix} \dot{\mathbf{v}}(t) \\ \dot{\mathbf{\Omega}}(t) \end{bmatrix} = \dot{\mathbf{N}}(\mathbf{q}, \dot{\mathbf{q}})\mathbf{N}^{-1}(\mathbf{q}) \begin{bmatrix} \mathbf{v}(t) \\ \mathbf{\Omega}(t) \end{bmatrix} + \mathbf{N}(\mathbf{q})\ddot{\mathbf{q}}(t) \tag{5.7-13}
$$

and the joint accelerations $\ddot{\mathbf{q}}(t)$ can be computed from the hand velocities and accelerations as

$$
\ddot{\mathbf{q}}(t) = \mathbf{N}^{-1}(\mathbf{q}) \begin{bmatrix} \dot{\mathbf{v}}(t) \\ \dot{\mathbf{\Omega}}(t) \end{bmatrix} - \mathbf{N}^{-1}(\mathbf{q})\dot{\mathbf{N}}(\mathbf{q}, \dot{\mathbf{q}})\mathbf{N}^{-1}(\mathbf{q}) \begin{bmatrix} \mathbf{v}(t) \\ \mathbf{\Omega}(t) \end{bmatrix} \tag{5.7-14}
$$

The above kinematic relations between the joint coordinates and the cartesian coordinates will be used in Sec. 5.7.1 for various resolved motion control methods and in deriving the resolved motion equations of motion of the manipulator hand in cartesian coordinates.

5.7.1 Resolved Motion Rate Control

Resolved motion rate control (RMRC) means that the motions of the various joint motors are combined and run simultaneously at different time-varying rates in order to achieve steady hand motion along any world coordinate axis. The mathematics that relate the world coordinates, such as lift p_x, sweep p_y, reach p_z, yaw α, pitch β, and roll γ to the joint angle coordinate of a six-link manipulator is inherently nonlinear and can be expressed by a nonlinear vector-valued function as

$$\mathbf{x}(t) = \mathbf{f}[\mathbf{q}(t)] \tag{5.7-15}$$

where $\mathbf{f}(\mathbf{q})$ is a 6×1 vector-valued function, and

$$\mathbf{x}(t) = \text{world coordinates} = (p_x, p_y, p_z, \alpha, \beta, \gamma)^T$$

and
$$\mathbf{q}(t) = \text{generalized coordinates} = (q_1, q_2, \ldots, q_n)^T$$

The relationship between the linear and angular velocities and the joint velocities of a six-link manipulator is given by Eq. (5.7-9).

For a more general discussion, if we assume that the manipulator has m degrees of freedom while the world coordinates of interest are of dimension n, then the joint angles and the world coordinates are related by a nonlinear function, as in Eq. (5.7-15).

If we differentiate Eq. (5.7-15) with respect to time, we have

$$\frac{d\mathbf{x}(t)}{dt} = \dot{\mathbf{x}}(t) = \mathbf{N}(\mathbf{q})\dot{\mathbf{q}}(t) \tag{5.7-16}$$

where $\mathbf{N}(\mathbf{q})$ is the jacobian matrix with respect to $\mathbf{q}(t)$, that is,

$$N_{ij} = \frac{\partial f_i}{\partial q_j} \qquad 1 \leqslant i \leqslant n, \ 1 \leqslant j \leqslant m \tag{5.7-17}$$

We see that if we work with rate control, the relationship is linear, as indicated by Eq. (5.7-16). When $\mathbf{x}(t)$ and $\mathbf{q}(t)$ are of the same dimension, that is, $m = n$, then the manipulator is nonredundant and the jacobian matrix can be inverted at a particular nonsingular position $\mathbf{q}(t)$:

$$\dot{\mathbf{q}}(t) = \mathbf{N}^{-1}(\mathbf{q})\dot{\mathbf{x}}(t) \tag{5.7-18}$$

From Eq. (5.7-18), given the desired rate along the world coordinates, one can easily find the combination of joint motor rates to achieve the desired hand motion. Various methods of computing the inverse jacobian matrix can be used. A resolved motion rate control block diagram is shown in Fig. 5.11.

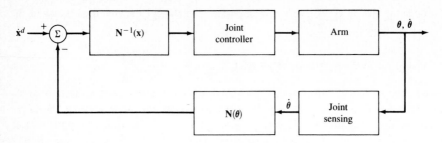

Figure 5.11 The resolved motion rate control block diagram.

If $m > n$, then the manipulator is redundant and the inverse jacobian matrix does not exist. This reduces the problem to finding the generalized inverse of the jacobian matrix. In this case, if the rank of $\mathbf{N(q)}$ is n, then $\dot{\mathbf{q}}(t)$ can be found by minimizing an error criterion formed by adjoining Eq. (5.7-16) with a Lagrange multiplier to a cost criterion, that is,

$$C = \frac{1}{2}\dot{\mathbf{q}}^T\mathbf{A}\dot{\mathbf{q}} + \lambda^T[\dot{\mathbf{x}} - \mathbf{N(q)}\dot{\mathbf{q}}] \tag{5.7-19}$$

where λ is a Lagrange multiplier vector, and $\mathbf{A}$ is an $m \times m$ symmetric, positive definite matrix.

Minimizing the cost criterion C with respect to $\dot{\mathbf{q}}(t)$ and λ, we have

$$\dot{\mathbf{q}}(t) = \mathbf{A}^{-1}\mathbf{N}^T(\mathbf{q})\lambda \tag{5.7-20}$$

and $$\dot{\mathbf{x}}(t) = \mathbf{N(q)}\dot{\mathbf{q}}(t) \tag{5.7-21}$$

Substituting $\dot{\mathbf{q}}(t)$ from Eq. (5.7-20) into Eq. (5.7-21), and solving for λ, yields

$$\lambda = [\mathbf{N(q)}\mathbf{A}^{-1}\mathbf{N}^T(\mathbf{q})]^{-1}\dot{\mathbf{x}}(t) \tag{5.7-22}$$

Substituting λ into Eq. (5.7-20), we obtain

$$\dot{\mathbf{q}}(t) = \mathbf{A}^{-1}\mathbf{N}^T(\mathbf{q})[\mathbf{N(q)}\mathbf{A}^{-1}\mathbf{N}^T(\mathbf{q})]^{-1}\dot{\mathbf{x}}(t) \tag{5.7-23}$$

If the matrix $\mathbf{A}$ is an identity matrix, then Eq. (5.7-23) reduces to Eq. (5.7-18).

Quite often, it is of interest to command the hand motion along the hand coordinate system rather than the world coordinate system (see Fig. 5.10). In this case, the desired hand rate motion $\dot{\mathbf{h}}(t)$ along the hand coordinate system is related to the world coordinate motion by

$$\dot{\mathbf{x}}(t) = {}^0\mathbf{R}_h\dot{\mathbf{h}}(t) \tag{5.7-24}$$

where ${}^0\mathbf{R}_h$ is an $n \times 6$ matrix that relates the orientation of the hand coordinate system to the world coordinate system. Given the desired hand rate motion $\dot{\mathbf{h}}(t)$

with respect to the hand coordinate system, and using Eqs. (5.7-23) to (5.7-24), the joint rate $\dot{\mathbf{q}}(t)$ can be computed by:

$$\dot{\mathbf{q}}(t) = \mathbf{A}^{-1}\mathbf{N}^{T}(\mathbf{q})[\mathbf{N}(\mathbf{q})\mathbf{A}^{-1}\mathbf{N}^{T}(\mathbf{q})]^{-1}{}^{0}\mathbf{R}_{h}\dot{\mathbf{h}}(t) \qquad (5.7\text{-}25)$$

In Eqs. (5.7-23) and (5.7-25), the angular position $\mathbf{q}(t)$ depends on time t, so we need to evaluate $\mathbf{N}^{-1}(\mathbf{q})$ at each sampling time t for the calculation of $\dot{\mathbf{q}}(t)$. The added computation in obtaining the inverse jacobian matrix at each sampling time and the singularity problem associated with the matrix inversion are important issues in using this control method.

5.7.2 Resolved Motion Acceleration Control

The resolved motion acceleration control (RMAC) (Luh et al. [1980b]) extends the concept of resolved motion rate control to include acceleration control. It presents an alternative position control which deals directly with the position and orientation of the hand of a manipulator. All the feedback control is done at the hand level, and it assumes that the desired accelerations of a preplanned hand motion are specified by the user.

The actual and desired position and orientation of the hand of a manipulator can be represented by 4×4 homogeneous transformation matrices, respectively, as

$$\mathbf{H}(t) = \begin{bmatrix} \mathbf{n}(t) & \mathbf{s}(t) & \mathbf{a}(t) & \mathbf{p}(t) \\ 0 & 0 & 0 & 1 \end{bmatrix}$$

and
$$\mathbf{H}^{d}(t) = \begin{bmatrix} \mathbf{n}^{d}(t) & \mathbf{s}^{d}(t) & \mathbf{a}^{d}(t) & \mathbf{p}^{d}(t) \\ 0 & 0 & 0 & 1 \end{bmatrix} \qquad (5.7\text{-}26)$$

where $\mathbf{n}$, $\mathbf{s}$, $\mathbf{a}$ are the unit vectors along the principal axes $\mathbf{x}$, $\mathbf{y}$, $\mathbf{z}$ of the hand coordinate system, respectively, and $\mathbf{p}(t)$ is the position vector of the hand with respect to the base coordinate system. The orientation submatrix $[\mathbf{n}, \mathbf{s}, \mathbf{a}]$ can be defined in terms of Euler angles of rotation (α, β, γ) with respect to the base coordinate system as in Eq. (5.7-2).

The position error of the hand is defined as the difference between the desired and the actual position of the hand and can be expressed as

$$\mathbf{e}_{p}(t) = \mathbf{p}^{d}(t) - \mathbf{p}(t) = \begin{bmatrix} p_{x}^{d}(t) - p_{x}(t) \\ p_{y}^{d}(t) - p_{y}(t) \\ p_{z}^{d}(t) - p_{z}(t) \end{bmatrix} \qquad (5.7\text{-}27)$$

Similarly, the orientation error is defined by the discrepancies between the desired and actual orientation axes of the hand and can be represented by

$$\mathbf{e}_{0}(t) = \tfrac{1}{2}[\mathbf{n}(t) \times \mathbf{n}^{d} + \mathbf{s}(t) \times \mathbf{s}^{d} + \mathbf{a}(t) \times \mathbf{a}^{d}] \qquad (5.7\text{-}28)$$

Thus, control of the manipulator is achieved by reducing these errors of the hand to zero.

Considering a six-link manipulator, we can combine the linear velocities $\mathbf{v}(t)$ and the angular velocities $\boldsymbol{\omega}(t)$ of the hand into a six-dimensional vector as $\dot{\mathbf{x}}(t)$,

$$\dot{\mathbf{x}}(t) = \begin{bmatrix} \mathbf{v}(t) \\ \boldsymbol{\omega}(t) \end{bmatrix} = \mathbf{N}(\mathbf{q})\dot{\mathbf{q}}(t) \tag{5.7-29}$$

where $\mathbf{N}(\mathbf{q})$ is a 6×6 matrix as given in Eq. (5.7-10). Equation (5.7-29) is the basis for resolved motion rate control where joint velocities are solved from the hand velocities. If this idea is extended further to solve for the joint accelerations from the hand acceleration $\ddot{\mathbf{x}}(t)$, then the time derivative of $\dot{\mathbf{x}}(t)$ is the hand acceleration

$$\ddot{\mathbf{x}}(t) = \mathbf{N}(\mathbf{q})\ddot{\mathbf{q}}(t) + \dot{\mathbf{N}}(\mathbf{q}, \dot{\mathbf{q}})\dot{\mathbf{q}}(t) \tag{5.7-30}$$

The closed-loop resolved motion acceleration control is based on the idea of reducing the position and orientation errors of the hand to zero. If the cartesian path for a manipulator is preplanned, then the desired position $\mathbf{p}^d(t)$, the desired velocity $\mathbf{v}^d(t)$, and the desired acceleration $\dot{\mathbf{v}}^d(t)$ of the hand are known with respect to the base coordinate system. In order to reduce the position error, one may apply joint torques and forces to each joint actuator of the manipulator. This essentially makes the actual linear acceleration of the hand, $\dot{\mathbf{v}}(t)$, satisfy the equation

$$\dot{\mathbf{v}}(t) = \dot{\mathbf{v}}^d(t) + k_1[\mathbf{v}^d(t) - \mathbf{v}(t)] + k_2[\mathbf{p}^d(t) - \mathbf{p}(t)] \tag{5.7-31}$$

where k_1 and k_2 are scalar constants. Equation (5.7-31) can be rewritten as

$$\ddot{\mathbf{e}}_p(t) + k_1\dot{\mathbf{e}}_p(t) + k_2\mathbf{e}_p(t) = 0 \tag{5.7-32}$$

where $\mathbf{e}_p(t) = \mathbf{p}^d(t) - \mathbf{p}(t)$. The input torques and forces must be chosen so as to guarantee the asymptotic convergence of the position error of the hand. This requires that k_1 and k_2 be chosen such that the characteristic roots of Eq. (5.7-32) have negative real parts.

Similarly, to reduce the orientation error of the hand, one has to choose the input torques and forces to the manipulator so that the angular acceleration of the hand satisfies the expression

$$\dot{\boldsymbol{\omega}}(t) = \dot{\boldsymbol{\omega}}^d(t) + k_1[\boldsymbol{\omega}^d(t) - \boldsymbol{\omega}(t)] + k_2\mathbf{e}_0 \tag{5.7-33}$$

Let us group $\mathbf{v}^d$ and $\boldsymbol{\omega}^d$ into a six-dimensional vector and the position and orientation errors into an error vector:

$$\dot{\mathbf{x}}^d(t) = \begin{bmatrix} \mathbf{v}^d(t) \\ \boldsymbol{\omega}^d(t) \end{bmatrix} \quad \text{and} \quad \mathbf{e}(t) = \begin{bmatrix} \mathbf{e}_p(t) \\ \mathbf{e}_0(t) \end{bmatrix} \tag{5.7-34}$$

Combining Eqs. (5.7-31) and (5.7-33), we have

$$\ddot{\mathbf{x}}(t) = \ddot{\mathbf{x}}^d(t) + k_1[\dot{\mathbf{x}}^d(t) - \dot{\mathbf{x}}(t)] + k_2\mathbf{e}(t) \qquad (5.7\text{-}35)$$

Substituting Eqs. (5.7-29) and (5.7-30) into Eq. (5.7-35) and solving for $\ddot{\mathbf{q}}(t)$ gives

$$
\begin{aligned}
\ddot{\mathbf{q}}(t) &= \mathbf{N}^{-1}(\mathbf{q})[\ddot{\mathbf{x}}^d(t) + k_1(\dot{\mathbf{x}}^d(t) - \dot{\mathbf{x}}(t)) + k_2\mathbf{e}(t) - \dot{\mathbf{N}}(\mathbf{q}, \dot{\mathbf{q}})\dot{\mathbf{q}}(t)] \\
&= -k_1\dot{\mathbf{q}}(t) + \mathbf{N}^{-1}(\mathbf{q})[\ddot{\mathbf{x}}^d(t) + k_1\dot{\mathbf{x}}^d(t) + k_2\mathbf{e}(t) - \dot{\mathbf{N}}(\mathbf{q}, \dot{\mathbf{q}})\dot{\mathbf{q}}(t)]
\end{aligned}
$$
$$(5.7\text{-}36)$$

Equation (5.7-36) is the basis for the closed-loop resolved acceleration control for manipulators. In order to compute the applied joint torques and forces to each joint actuator of the manipulator, the recursive Newton-Euler equations of motion are used. The joint position $\mathbf{q}(t)$, and joint velocity $\dot{\mathbf{q}}(t)$ are measured from the potentiometers, or optical encoders, of the manipulator. The quantities $\mathbf{v}$, ω, $\mathbf{N}$, $\mathbf{N}^{-1}$, $\dot{\mathbf{N}}$, and $\mathbf{H}(t)$ can be computed from the above equations. These values together with the desired position $\mathbf{p}^d(t)$, desired velocity $\mathbf{v}^d(t)$, and the desired acceleration $\mathbf{v}^d(t)$ of the hand obtained from a planned trajectory can be used to compute the joint acceleration using Eq. (5.7-36). Finally the applied joint torques and forces can be computed recursively from the Newton-Euler equations of motion. As in the case of RMRC, this control method is characterized by extensive computational requirements, singularities associated with the jacobian matrix, and the need to plan a manipulator hand trajectory with acceleration information.

5.7.3 Resolved Motion Force Control

The basic concept of resolved motion force control (RMFC) is to determine the applied torques to the joint actuators in order to perform the cartesian position control of the robot arm. An advantage of RMFC is that the control is not based on the complicated dynamic equations of motion of the manipulator and still has the ability to compensate for changing arm configurations, gravity loading forces on the links, and internal friction. Similar to RMAC, all the control of RMFC is done at the hand level.

The RMFC is based on the relationship between the resolved force vector $\mathbf{F}$ obtained from a wrist force sensor and the joint torques at the joint actuators. The control technique consists of the cartesian position control and the force convergent control. The position control calculates the desired forces and moments to be applied to the end-effector in order to track a desired cartesian trajectory. The force convergent control determines the necessary joint torques to each actuator so that the end-effector can maintain the desired forces and moments obtained from the position control. A control block diagram of the RMFC is shown in Fig. 5.12.

We shall briefly discuss the mathematics that governs this control technique. A more detailed discussion can be found in Wu and Paul [1982]. The basic control concept of the RMFC is based on the relationship between the resolved

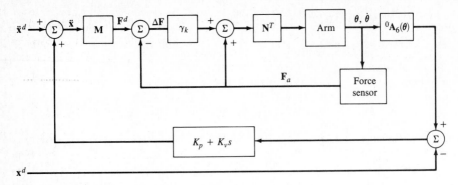

Figure 5.12 Resolved motion force control.

force vector, $\mathbf{F} = (F_x, F_y, F_z, M_x, M_y, M_z)^T$, and the joint torques, $\tau = (\tau_1, \tau_2, \ldots, \tau_n)^T$, which are applied to each joint actuator in order to counterbalance the forces felt at the hand, where $(F_x, F_y, F_z)^T$ and $(M_x, M_y, M_z)^T$ are the cartesian forces and moments in the hand coordinate system, respectively. The underlying relationship between these quantities is

$$\tau(t) = \mathbf{N}^T(\mathbf{q})\mathbf{F}(t) \tag{5.7-37}$$

where $\mathbf{N}$ is the jacobian matrix, as in Eq. (5.7-10).

Since the objective of RMFC is to track the cartesian position of the end-effector, an appropriate time-based position trajectory has to be specified as functions of the arm transformation matrix ${}^0\mathbf{A}_6(t)$, the velocity $(v_x, v_y, v_z)^T$, and the angular velocity $(\omega_x, \omega_y, \omega_z)^T$ about the hand coordinate system. That is, the desired time-varying arm transformation matrix, ${}^0\mathbf{A}_6(t + \Delta t)$, can be represented as

$$
{}^0\mathbf{A}_6(t + \Delta t) = {}^0\mathbf{A}_6(t)
\begin{bmatrix}
1 & -\omega_z(t) & \omega_y(t) & v_x(t) \\
\omega_z(t) & 1 & -\omega_x(t) & v_y(t) \\
-\omega_y(t) & \omega_x(t) & 1 & v_z(t) \\
0 & 0 & 0 & 1
\end{bmatrix}
\Delta t \tag{5.7-38}
$$

then, the desired cartesian velocity $\dot{\mathbf{x}}^d(t) = (v_x, v_y, v_z, \omega_x, \omega_y, \omega_z)^T$ can be obtained from the element of the following equation

$$
\begin{bmatrix}
1 & -\omega_z(t) & \omega_y(t) & v_x(t) \\
\omega_z(t) & 1 & -\omega_x(t) & v_y(t) \\
-\omega_y(t) & \omega_x(t) & 1 & v_z(t) \\
0 & 0 & 0 & 1
\end{bmatrix}
= \frac{1}{\Delta t}[({}^0\mathbf{A}_6)^{-1}(t) \, {}^0\mathbf{A}_6(t + \Delta t)] \tag{5.7-39}
$$

The cartesian velocity error $\dot{\mathbf{x}}^d - \dot{\mathbf{x}}$ can be obtained using the above equation. The velocity error $\dot{\mathbf{x}}^d - \dot{\mathbf{x}}$ used in Eq. (5.7-31) is different from the above velocity error because the above error equation uses the homogeneous transformation matrix method. In Eq. (5.7-31), the velocity error is obtained simply by differentiating $\mathbf{p}^d(t) - \mathbf{p}(t)$.

Similarly, the desired cartesian acceleration $\ddot{\mathbf{x}}^d(t)$ can be obtained as:

$$\ddot{\mathbf{x}}^d(t) = \frac{\dot{\mathbf{x}}^d(t + \Delta t) - \dot{\mathbf{x}}^d(t)}{\Delta t} \qquad (5.7\text{-}40)$$

Based on the proportional plus derivative control approach, if there is no error in position and velocity of the hand, then we want the actual cartesian acceleration $\ddot{\mathbf{x}}(t)$ to track the desired cartesian acceleration as closely as possible. This can be done by setting the actual cartesian acceleration as

$$\ddot{\mathbf{x}}(t) = \ddot{\mathbf{x}}^d(t) + K_v[\dot{\mathbf{x}}^d(t) - \dot{\mathbf{x}}(t)] + K_p[\mathbf{x}^d(t) - \mathbf{x}(t)] \qquad (5.7\text{-}41)$$

or
$$\ddot{\mathbf{x}}_e(t) + K_v\dot{\mathbf{x}}_e(t) + K_p\mathbf{x}_e(t) = 0 \qquad (5.7\text{-}42)$$

By choosing the values of K_v and K_p so that the characteristic roots of Eq. (5.7-42) have negative real parts, $\mathbf{x}(t)$ will converge to $\mathbf{x}^d(t)$ asymptotically.

Based on the above control technique, the desired cartesian forces and moments to correct the position errors can be obtained using Newton's second law:

$$\mathbf{F}^d(t) = \mathbf{M}\ddot{\mathbf{x}}(t) \qquad (5.7\text{-}43)$$

where $\mathbf{M}$ is the mass matrix with diagonal elements of total mass of the load m and the moments of inertia I_{xx}, I_{yy}, I_{zz} at the principal axes of the load. Then, using the Eq. (5.7-37), the desired cartesian forces $\mathbf{F}^d$ can be resolved into the joint torques:

$$\tau(t) = \mathbf{N}^T(\mathbf{q})\mathbf{F}^d = \mathbf{N}^T(\mathbf{q})\,\mathbf{M}\ddot{\mathbf{x}}(t) \qquad (5.7\text{-}44)$$

In general, the above RMFC works well when the mass and the load are negligible, as compared with the mass of the manipulator. But, if the mass and the load approaches the mass of the manipulator, the position of the hand usually does not converge to the desired position. This is due to the fact that some of the joint torques are spent to accelerate the links. In order to compensate for these loading and acceleration effects, a force convergence control is incorporated as a second part of the RMFC.

The force convergent control method is based on the Robbins-Monro stochastic approximation method to determine the actual cartesian force $\mathbf{F}_a$ so that the observed cartesian force $\mathbf{F}_0$ (measured by a wrist force sensor) at the hand will converge to the desired cartesian force $\mathbf{F}^d$ obtained from the above position control technique. If the error between the measured force vector $\mathbf{F}_0$ and the desired cartesian force is greater than a user-designed threshold $\Delta\mathbf{F}(k) = \mathbf{F}^d(k) - \mathbf{F}_0(k)$, then the actual cartesian force is updated by

$$\mathbf{F}_a(k + 1) = \mathbf{F}_a(k) + \gamma_k\Delta\mathbf{F}(k) \qquad (5.7\text{-}45)$$

where $\gamma_k = 1/(k + 1)$ for $k = 0, 1, \ldots, N$. Theoretically, the value of N must be large. However, in practice, the value of N can be chosen based on the force convergence. Based on a computer simulation study (Wu and Paul [1982]), a value of $N = 1$ or 2 gives a fairly good convergence of the force vector.

In summary, the RMFC with force convergent control has the advantage that the control method can be extended to various loading conditions and to a manipulator with any number of degrees of freedom without increasing the computational complexity.

5.8 ADAPTIVE CONTROL

Most of the schemes discussed in the previous sections control the arm at the hand or joint level and emphasize nonlinear compensations of the interaction forces between the various joints. These control algorithms sometimes are inadequate because they require accurate modeling of the arm dynamics and neglect the changes of the load in a task cycle. These changes in the payload of the controlled system often are significant enough to render the above feedback control strategies ineffective. The result is reduced servo response speed and damping, which limits the precision and speed of the end-effector. Any significant gain in performance for tracking the desired time-based trajectory as closely as possible over a wide range of manipulator motion and payloads require the consideration of adaptive control techniques.

5.8.1 Model-Referenced Adaptive Control

Among various adaptive control methods, the model-referenced adaptive control (MRAC) is the most widely used and it is also relatively easy to implement. The concept of model-referenced adaptive control is based on selecting an appropriate reference model and adaptation algorithm which modifies the feedback gains to the actuators of the actual system. The adaptation algorithm is driven by the errors between the reference model outputs and the actual system outputs. A general control block diagram of the model-referenced adaptive control system is shown in Fig. 5.13.

Dubowsky and DesForges [1979] proposed a simple model-referenced adaptive control for the control of mechanical manipulators. In their analysis, the payload is taken into consideration by combining it to the final link, and the end-effector dimension is assumed to be small compared with the length of other links. Then, the selected reference model provides an effective and flexible means of specifying desired closed-loop performance of the controlled system. A linear second-order time invariant differential equation is selected as the reference model for each degree of freedom of the robot arm. The manipulator is controlled by adjusting the position and velocity feedback gains to follow the model so that its closed-loop performance characteristics closely match the set of desired performance characteristics in the reference model. As a result, this adaptive control scheme only

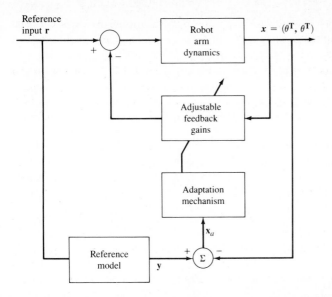

Figure 5.13 A general control block diagram for model-referenced adaptive control.

requires moderate computations which can be implemented with a low-cost microprocessor. Such a model-referenced, adaptive control algorithm does not require complex mathematical models of the system dynamics nor the a priori knowledge of the environment (loads, etc). The resulting model-referenced adaptive system is capable of maintaining uniformly good performance over a wide range of motions and payloads.

Defining the vector $\mathbf{y}(t)$ to represent the reference model response and the vector $\mathbf{x}(t)$ to represent the manipulator response, the joint i of the reference model can be described by

$$a_i \ddot{y}_i(t) + b_i \dot{y}_i(t) + y_i(t) = r_i(t) \tag{5.8-1}$$

In terms of natural frequency ω_{ni} and damping ratio ζ_i of a second-order linear system, a_i and b_i correspond to

$$a_i = \frac{1}{\omega_{ni}^2} \quad \text{and} \quad b_i = \frac{2\zeta_i}{\omega_{ni}} \tag{5.8-2}$$

If we assume that the manipulator is controlled by position and velocity feedback gains, and that the coupling terms are negligible, then the manipulator dynamic equation for joint i can be written as

$$\alpha_i(t)\ddot{x}_i(t) + \beta_i(t)\dot{x}_i(t) + x_i(t) = r_i(t) \tag{5.8-3}$$

where the system parameters $\alpha_i(t)$ and $\beta_i(t)$ are assumed to vary slowly with time.

Several techniques are available to adjust the feedback gains of the controlled system. Due to its simplicity, a steepest descent method is used to minimize a quadratic function of the system error, which is the difference between the response of the actual system [Eq. (5.8-3)] and the response of the reference model [Eq. (5.8-1)]:

$$J_i(e_i) = \frac{1}{2}(k_2^i \ddot{e}_i + k_1^i \dot{e}_i + k_0^i e_i)^2 \qquad i = 1, 2, \ldots, n \qquad (5.8\text{-}4)$$

where $e_i = y_i - x_i$, and the values of the weighting factors, k_j^i, are selected from stability considerations to obtain stable system behavior.

Using a steepest descent method, the system parameters adjustment mechanism which will minimize the system error is governed by

$$\dot{\alpha}_i(t) = [k_2^i \ddot{e}_i(t) + k_1^i \dot{e}_i(t) + k_0^i e_i(t)][k_2^i \ddot{u}_i(t) + k_1^i \dot{u}_i(t) + k_0^i u_i(t)] \qquad (5.8\text{-}5)$$

$$\dot{\beta}_i(t) = [k_2^i \ddot{e}_i(t) + k_1^i \dot{e}_i(t) + k_0^i e_i(t)][k_2^i \ddot{w}_i(t) + k_1^i \dot{w}_i(t) + k_0^i w_i(t)] \qquad (5.8\text{-}6)$$

where $u_i(t)$ and $w_i(t)$ and their derivatives are obtained from the solutions of the following differential equations:

$$a_i \ddot{u}_i(t) + b_i \dot{u}_i(t) + u_i(t) = -\ddot{y}_i(t) \qquad (5.8\text{-}7)$$

$$a_i \ddot{w}_i(t) + b_i \dot{w}_i(t) + w_i(t) = -\dot{y}_i(t) \qquad (5.8\text{-}8)$$

and $\dot{y}_i(t)$ and $\ddot{y}_i(t)$ are the first two time derivatives of response of the reference model. The closed-loop adaptive system involves solving the reference model equations for a given desired input; then the differential equations in Eqs. (5.8-7) and (5.8-8) are solved to yield $u_i(t)$ and $w_i(t)$ and their derivatives for Eqs. (5.8-5) and (5.8-6). Finally, solving the differential equations in Eqs. (5.8-5) and (5.8-6), yields $\alpha_i(t)$ and $\beta_i(t)$.

The fact that this control approach is not dependent on a complex mathematical model is one of its major advantages, but stability considerations of the closed-loop adaptive system are critical. A stability analysis is difficult, and Dubowsky and DesForges [1979] carried out an investigation of this adaptive system using a linearized model. However, the adaptability of the controller can become questionable if the interaction forces among the various joints are severe.

5.8.2 Adaptive Control Using an Autoregressive Model

Koivo and Guo [1983] proposed an adaptive, self-tuning controller using an autoregressive model to fit the input-output data from the manipulator. The control algorithm assumes that the interaction forces among the joints are negligible. A block diagram of the control system is shown in Fig. 5.14. Let the input torque to joint i be u_i, and the output angular position of the manipulator be y_i. The input-output

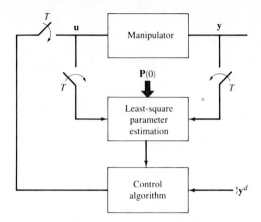

Figure 5.14 Adaptive control with autoregressive model.

pairs (u_i, y_i) may be described by an autoregressive model which match these pairs as closely as possible:

$$y_i(k) = \sum_{m=1}^{n} [a_i^m y_i(k - m) + b_i^m u_i(k - m)] + a_i^0 + e_i(k) \quad (5.8\text{-}9)$$

where a_i^0 is a constant forcing term, $e_i(k)$ is the modeling error which is assumed to be white gaussian noise with zero mean and independent of u_i and $y_i(k - m)$, $m \geqslant 1$. The parameters a_i^m and b_i^m are determined so as to obtain the best least-squares fit of the measured input-output data pairs. These parameters can be obtained by minimizing the following criterion:

$$E_N^i(\alpha_i) = \frac{1}{N+1} \sum_{k=0}^{N} e_i^2(k) \quad (5.8\text{-}10)$$

where N is the number of measurements. Let α_i be the ith parameter vector:

$$\alpha_i = (a_i^0, a_i^1, \ldots, a_i^n, b_i^0, b_i^1, \ldots, b_i^n)^T \quad (5.8\text{-}11)$$

and let $\psi_i(k - 1)$ be the vector of the input-output pairs:

$$\psi_i(k-1) = [1, y_i(k-1), \ldots, y_i(k-n), u_i(k-1), \ldots, u_i(k-n)]^T \quad (5.8\text{-}12)$$

Then, a recursive least-squares estimation of α_i can be found as

$$\hat{\alpha}_i(N) = \hat{\alpha}_i(N - 1) + \mathbf{P}_i(N)\psi_i(N - 1)[y_i(N) - \hat{\alpha}_i^T(N - 1)\psi_i(N - 1)] \quad (5.8\text{-}13)$$

with

$$\mathbf{P}_i(N) = \frac{1}{\mu_i} \left[\frac{\mathbf{P}_i(N-1)\boldsymbol{\psi}_i(N-1)\boldsymbol{\psi}_i^T(N-1)\mathbf{P}_i(N-1)}{\mu_i + \boldsymbol{\psi}_i^T(N-1)\mathbf{P}_i(N-1)\boldsymbol{\psi}_i(N-1)} \right] \qquad (5.8\text{-}14)$$

where $0 < \mu_i \leqslant 1$ is a "forgetting" factor which provides an exponential weighting of past data in the estimation algorithm and by which the algorithm allows a slow drift of the parameters; $\mathbf{P}_i$ is a $(2n + 1) \times (2n + 1)$ symmetric matrix, and the hat notation is used to indicate an estimate of the parameters.

Using the above equations to compute the estimates of the autoregressive model, the model can be represented by:

$$y_i(k) = \hat{\boldsymbol{\alpha}}_i^T \boldsymbol{\psi}_i(k-1) + e_i(k) \qquad (5.8\text{-}15)$$

In order to track the trajectory set points, a performance criterion for joint i is defined as

$$J_i^k(\mathbf{u}) = E\{[y_i(k+2) - y_i^d(k+2)]^2 + \gamma_i u_i^2(k+1)|\boldsymbol{\psi}_i(k)\} \qquad (5.8\text{-}16)$$

where $E[\,\cdot\,]$ represents an expectation operation conditioned on $\boldsymbol{\psi}_i(k)$ and γ_i is a user-defined nonnegative weighting factor.

The optimal control that minimizes the above performance criterion is found to be:

$$u_i(k+1)$$

$$= \frac{-\hat{b}_i^1(k)}{[\hat{b}_i^1(k)]^2 + \gamma_i} \left\{ \hat{a}_i^0(k) + \hat{a}_i^1(k)[\hat{\boldsymbol{\alpha}}_i^T\boldsymbol{\psi}_i(k)] + \sum_{m=2}^n \hat{a}_i^m(k)y_i(k+2-m) \right.$$

$$\left. + \sum_{m=2}^n \hat{b}_i^m(k)u_i(k+2-m) - y_i^d(k+2) \right\} \qquad (5.8\text{-}17)$$

where $\hat{a}_i^m$, $\hat{b}_i^m$, and $\hat{\alpha}_i^m$ are the estimates of the parameters from Eqs. (5.8-13) and (5.8-14).

In summary, this adaptive control uses an autoregressive model [Eq. (5.8-9)] to fit the input-output data from the manipulator. The recursive least-squares identification scheme [Eqs. (5.8-13) and (5.8-14)] is used to estimate the parameters which are used in the optimal control [Eq. (5.8-17)] to servo the manipulator.

5.8.3 Adaptive Perturbation Control

Based on perturbation theory, Lee and Chung [1984, 1985] proposed an adaptive control strategy which tracks a desired time-based manipulator trajectory as closely as possible for all times over a wide range of manipulator motion and payloads. Adaptive perturbation control differs from the above adaptive schemes in the sense that it takes all the interactions among the various joints into consideration. The adaptive control discussed in this section is based on linearized perturbation equations in the vicinity of a nominal trajectory. The nominal trajectory is specified by an interpolated joint trajectory whose angular position, angular velocity, and angular acceleration are known at every sampling instant. The highly coupled nonlinear dynamic equations of a manipulator are then linearized about the planned manipulator trajectory to obtain the linearized perturbation system. The controlled system is characterized by feedforward and feedback components which can be computed separately and simultaneously. Using the Newton-Euler equations of motion as inverse dynamics of the manipulator, the feedforward component computes the nominal torques which compensate all the interaction forces between the various joints along the nominal trajectory. The feedback component computes the perturbation torques which reduce the position and velocity errors of the manipulator to zero along the nominal trajectory. An efficient, recursive, real-time, least-squares identification scheme is used to identify the system parameters in the perturbation equations. A one-step optimal control law is designed to control the linearized perturbation system about the nominal trajectory. The parameters and the feedback gains of the linearized system are updated and adjusted in each sampling period to obtain the necessary control effort. The total torques applied to the joint actuators then consist of the nominal torques computed from the Newton-Euler equations of motion and the perturbation torques computed from the one-step optimal control law of the linearized system. This adaptive control strategy reduces the manipulator control problem from nonlinear control to controlling a linear system about a nominal trajectory.

The adaptive control is based on the linearized perturbation equations about the referenced trajectory. We need to derive appropriate linearized perturbation equations suitable for developing the feedback controller which computes perturbation joint torques to reduce position and velocity errors along the nominal trajectory. The L-E equations of motion of an n-link manipulator can be expressed in state space representation as in Eq. (5.4-4). With this formulation, the control problem is to find a feedback control law $\mathbf{u}(t) = \mathbf{g}[\mathbf{x}(t)]$ such that the closed loop control system $\dot{\mathbf{x}}(t) = \mathbf{f}\{\mathbf{x}(t), \mathbf{g}[\mathbf{x}(t)]\}$ is asymptotically stable and tracks a desired trajectory as closely as possible over a wide range of payloads for all times.

Suppose that the nominal states $\mathbf{x}_n(t)$ of the system [Eq. (5.4-4)] are known from the planned trajectory, and the corresponding nominal torques $\mathbf{u}_n(t)$ are also known from the computations of the joint torques using the N-E equations of motion. Then, both $\mathbf{x}_n(t)$ and $\mathbf{u}_n(t)$ satisfy Eq. (5.4-4):

$$\dot{\mathbf{x}}_n(t) = \mathbf{f}[\mathbf{x}_n(t), \mathbf{u}_n(t)] \tag{5.8-18}$$

Using the Taylor series expansion on Eq. (5.4-4) about the nominal trajectory, subtracting Eq. (5.8-18) from it, and assuming that the higher order terms are negligible, the associated linearized perturbation model for this control system can be expressed as

$$\delta \dot{\mathbf{x}}(t) = \nabla_x \mathbf{f}|_n \, \delta \mathbf{x}(t) + \nabla_u \mathbf{f}|_n \, \delta \mathbf{u}(t)$$

$$= \mathbf{A}(t) \, \delta \mathbf{x}(t) + \mathbf{B}(t) \, \delta \mathbf{u}(t) \tag{5.8-19}$$

where $\nabla_x \mathbf{f}|_n$ and $\nabla_u \mathbf{f}|_n$ are the jacobian matrices of $\mathbf{f}[\mathbf{x}(t), \mathbf{u}(t)]$ evaluated at $\mathbf{x}_n(t)$ and $\mathbf{u}_n(t)$, respectively, $\delta \mathbf{x}(t) = \mathbf{x}(t) - \mathbf{x}_n(t)$ and $\delta \mathbf{u}(t) = \mathbf{u}(t) - \mathbf{u}_n(t)$.

The system parameters, $\mathbf{A}(t)$ and $\mathbf{B}(t)$, of Eq. (5.8-19) depend on the instantaneous manipulator position and velocity along the nominal trajectory and thus, vary slowly with time. Because of the complexity of the manipulator equations of motion, it is extremely difficult to find the elements of $\mathbf{A}(t)$ and $\mathbf{B}(t)$ explicitly. However, the design of a feedback control law for the perturbation equations requires that the system parameters of Eq. (5.8-19) be known at all times. Thus, parameter identification techniques must be used to identify the unknown elements in $\mathbf{A}(t)$ and $\mathbf{B}(t)$.

As a result of this formulation, the manipulator control problem is reduced to determining $\delta \mathbf{u}(t)$, which drives $\delta \mathbf{x}(t)$ to zero at all times along the nominal trajectory. The overall controlled system is thus characterized by a feedforward component and a feedback component. Given the planned trajectory set points $\mathbf{q}^d(t)$, $\dot{\mathbf{q}}^d(t)$, and $\ddot{\mathbf{q}}^d(t)$, the feedforward component computes the corresponding nominal torques $\mathbf{u}_n(t)$ from the N-E equations of motion. The feedback component computes the corresponding perturbation torques $\delta \mathbf{u}(t)$ which provide control effort to compensate for small deviations from the nominal trajectory. The computation of the perturbation torques is based on a one-step optimal control law. The main advantages of this formulation are twofold. First, it reduces a nonlinear control problem to a linear control problem about a nominal trajectory; second, the computations of the nominal and perturbation torques can be performed separately and simultaneously. Because of this parallel computational structure, adaptive control techniques can be easily implemented using present day low-cost microprocessors. A control block diagram of the method is shown in Fig. 5.15.

For implementation on a digital computer, Eq. (5.8-19) needs to be discretized to obtain an appropriate discrete linear equations for parameter identification:

$$\mathbf{x}[(k + 1)T] = \mathbf{F}(kT)\mathbf{x}(kT) + \mathbf{G}(kT)\mathbf{u}(kT)$$

$$k = 0, 1, \ldots \tag{5.8-20}$$

where T is the sampling period, $\mathbf{u}(kT)$ is an n-dimensional piecewise constant control input vector of $\mathbf{u}(t)$ over the time interval between any two consecutive sam-

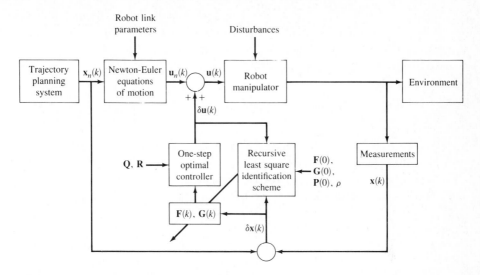

Figure 5.15 The adaptive perturbation control.

pling instants for $kT \leqslant t < (k + 1)T$, and $\mathbf{x}(kT)$ is a $2n$-dimensional perturbed state vector which is given by

$$\mathbf{x}(kT) = \boldsymbol{\Gamma}(kT, t_0)\mathbf{x}(t_0) + \int_{t_0}^{kT} \boldsymbol{\Gamma}(kT, t)\mathbf{B}(t)\mathbf{u}(t)dt \qquad (5.8\text{-}21)$$

and $\boldsymbol{\Gamma}(kT, t_0)$ is the state-transition matrix of the system. $\mathbf{F}(kT)$ and $\mathbf{G}(kT)$ are, respectively, $2n \times 2n$ and $2n \times n$ matrices and are given by

$$\mathbf{F}(kT) = \boldsymbol{\Gamma}[(k + 1)T, kT] \qquad (5.8\text{-}22)$$

and $$\mathbf{G}(kT)\mathbf{u}(kT) = \int_{kT}^{(k+1)T} \boldsymbol{\Gamma}[(k + 1)T, t]\mathbf{B}(t)\mathbf{u}(t)dt \qquad (5.8\text{-}23)$$

With this model, a total of $6n^2$ parameters in the $\mathbf{F}(kT)$ and $\mathbf{G}(kT)$ matrices need to be identified. Without confusion, we shall drop the sampling period T from the rest of the equations for clarity and simplicity.

Various identification algorithms, such as the methods of least squares, maximum likelihood, instrumental variable, cross correlation, and stochastic approximation, have been applied successfully to the parameter identification problem. Due to its simplicity and ease of application, a recursive real-time least-squares parameter identification scheme is selected here for identifying the system parameters in $\mathbf{F}(k)$ and $\mathbf{G}(k)$. In the parameter identification scheme, we make the following assumptions: (1) the parameters of the system are slowly time-varying but the variation speed is slower than the adaptation speed; (2) measurement noise is negligible; and (3) the state variables $\mathbf{x}(k)$ of Eq. (5.8-20) are measurable.

In order to apply the recursive least-squares identification algorithm to Eq. (5.8-20), we need to rearrange the system equations in a form that is suitable for parameter identification. Defining and expressing the ith row of the unknown parameters of the system at the kth instant of time in a $3n$-dimensional vector, we have

$$\theta_i^T(k) = [f_{i1}(k), \ldots, f_{ip}(k), g_{i1}(k), \ldots, g_{in}(k)] \qquad (5.8\text{-}24)$$

$$i = 1, 2, \ldots, p$$

or, expressed in matrix form, as

$$
\Theta(k) =
\begin{bmatrix}
f_{11}(k) & \cdots & f_{p1}(k) \\
\cdot & & \cdot \\
\cdot & \cdots & \cdot \\
\cdot & & \cdot \\
f_{1p}(k) & \cdots & f_{pp}(k) \\
g_{11}(k) & \cdots & g_{p1}(k) \\
\cdot & & \cdot \\
\cdot & \cdots & \cdot \\
\cdot & & \cdot \\
g_{1n}(k) & \cdots & g_{pn}(k)
\end{bmatrix}
= [\theta_1(k), \theta_2(k), \ldots, \theta_p(k)] \qquad (5.8\text{-}25)
$$

where $p = 2n$. Similarly, defining the outputs and inputs of the perturbation system [Eq. (5.8-20)] at the kth instant of time in a $3n$-dimensional vector as

$$\mathbf{z}^T(k) = [x_1(k), x_2(k), \ldots, x_p(k), u_1(k), u_2(k), \ldots, u_n(k)] \qquad (5.8\text{-}26)$$

and the states at the kth instant of time in a $2n$-dimensional vector as

$$\mathbf{x}^T(k) = [x_1(k), x_2(k), \ldots, x_p(k)] \qquad (5.8\text{-}27)$$

we have that the corresponding system equation in Eq. (5.8-20) can be written as

$$x_i(k + 1) = \mathbf{z}^T(k)\theta_i(k) \qquad i = 1, 2, \ldots, p \qquad (5.8\text{-}28)$$

With this formulation, we wish to identify the parameters in each column of $\Theta(k)$ based on the measurement vector $\mathbf{z}(k)$. In order to examine the "goodness" of the least-squares estimation algorithm, a $2n$-dimensional error vector $e(k)$, often called a *residual*, is included to account for the modeling error and noise in Eq. (5.8-20):

$$e_i(k) = x_i(k + 1) - \mathbf{z}^T(k)\hat{\theta}_i(k) \qquad i = 1, 2, \ldots, p \qquad (5.8\text{-}29)$$

Basic least-squares parameter estimation assumes that the unknown parameters are constant values and the solution is based on batch processing N sets of measurement data, which are weighted equally, to estimate the unknown parameters. Unfortunately, this algorithm cannot be applied to time-varying parameters. Furthermore, the solution requires matrix inversion which is computational intensive. In order to reduce the number of numerical computations and to track the time-varying parameters $\Theta(k)$ at each sampling period, a sequential least-squares identification scheme which updates the unknown parameters at each sampling period based on the new set of measurements at each sampling interval provides an efficient algorithmic solution to the identification problem. Such a recursive, real-time, least-squares parameter identification algorithm can be found by minimizing an exponentially weighted error criterion which has an effect of placing more weights on the squared errors of the more recent measurements; that is,

$$J_N = \sum_{j=1}^{N} \rho^{N-j} e_i^2(j) \tag{5.8-30}$$

where the error vector is weighted as

$$\mathbf{e}_i^T(N) = [\sqrt{\rho}^{N-1} e_i(1), \sqrt{\rho}^{N-2} e_i(2), \ldots, e_i(N)] \tag{5.8-31}$$

and $N > 3n$ is the number of measurements used to estimate the parameters $\theta_i(N)$. Minimizing the error criterion in Eq. (5.8-30) with respect to the unknown parameters vector θ_i and utilizing the matrix inverse lemma, a recursive real-time least-squares identification scheme can be obtained for $\theta_i(k)$ after simple algebraic manipulations:

$$\hat{\theta}_i(k+1) = \hat{\theta}_i(k) + \gamma(k)\mathbf{P}(k)\mathbf{z}(k) [x_i(k+1) - \mathbf{z}^T(k)\hat{\theta}_i(k)] \tag{5.8-32}$$

$$\mathbf{P}(k+1) = \mathbf{P}(k) - \gamma(k) \mathbf{P}(k)\mathbf{z}(k)\mathbf{z}^T(k)\mathbf{P}(k) \tag{5.8-33}$$

and $\quad \gamma(k) = [\mathbf{z}^T(k)\mathbf{P}(k)\mathbf{z}(k) + \rho]^{-1} \tag{5.8-34}$

where $0 < \rho < 1$, the hat notation is used to indicate the estimate of the parameters $\theta_i(k)$, and $\mathbf{P}(k) = \rho[\mathbf{Z}(k)\mathbf{Z}^T(k)]^{-1}$ is a $3n \times 3n$ symmetric positive definite matrix, where $\mathbf{Z}(k) = [\mathbf{z}(1), \mathbf{z}(2), \ldots, \mathbf{z}(k)]$ is the measurement matrix up to the kth sampling instant. If the errors $e_i(k)$ are identically distributed and independent with zero mean and variance σ^2, then $\mathbf{P}(k)$ can be interpreted as the covariance matrix of the estimate if ρ is chosen as σ^2.

The above recursive equations indicate that the estimate of the parameters $\hat{\theta}_i(k+1)$ at the $(k+1)$th sampling period is equal to the previous estimate $\hat{\theta}_i(k)$ corrected by the term proportional to $[x_i(k+1) - \mathbf{z}^T(k)\hat{\theta}_i(k)]$. The term $\mathbf{z}^T(k)\hat{\theta}_i(k)$ is the prediction of the value $x_i(k+1)$ based on the estimate of the parameters $\theta_i(k)$ and the measurement vector $\mathbf{z}(k)$. The components of the vector $\gamma(k)\mathbf{P}(k)\mathbf{z}(k)$ are weighting factors which indicate how the corrections and the previous estimate should be weighted to obtain the new estimate $\hat{\theta}_i(k+1)$. The parameter ρ is a weighting factor and is commonly used for tracking slowly time-varying parameters by exponentially forgetting the "aged" measurements. If $\rho << 1$, a large weighting factor is placed on the more recent sampled data by rapidly weighing out previous samples. If $\rho \approx 1$, accuracy in tracking the time-varying parameters will be lost due to the truncation of the measured data sequences. We can compromise between fast adaptation capabilities and loss of accuracy in parameter identification by adjusting the weighting factor ρ. In most applications for tracking slowly time-varying parameters, ρ is usually chosen to be $0.90 \leqslant \rho < 1.0$.

Finally, the above identification scheme [Eqs. (5.8-32) to (5.8-34)] can be started by choosing the initial values of $\mathbf{P}(0)$ to be

$$\mathbf{P}(0) = \alpha \mathbf{I}_{3n} \tag{5.8-35}$$

where α is a large positive scalar and $\mathbf{I}_{3n}$ is a $3n \times 3n$ identity matrix. The initial estimate of the unknown parameters $\mathbf{F}(k)$ and $\mathbf{G}(k)$ can be approximated by the following equations:

$$\mathbf{F}(0) \approx \mathbf{I}_{2n} + \left\{ \frac{\partial \mathbf{f}}{\partial \mathbf{x}}[\mathbf{x}_n(0), \mathbf{u}_n(0)] \right\} T + \left\{ \frac{\partial \mathbf{f}}{\partial \mathbf{x}}[\mathbf{x}_n(0), \mathbf{u}_n(0)] \right\}^2 \frac{T^2}{2} \tag{5.8-36}$$

$$\mathbf{G}(0) \approx \left\{ \frac{\partial \mathbf{f}}{\partial \mathbf{u}}[\mathbf{x}_n(0), \mathbf{u}_n(0)] \right\} T + \left\{ \frac{\partial \mathbf{f}}{\partial \mathbf{x}}[\mathbf{x}_n(0), \mathbf{u}_n(0)] \right\}$$

$$\times \left\{ \frac{\partial \mathbf{f}}{\partial \mathbf{u}}[\mathbf{x}_n(0), \mathbf{u}_n(0)] \right\} T^2$$

$$+ \left\{ \frac{\partial \mathbf{f}}{\partial \mathbf{x}}[\mathbf{x}_n(0), \mathbf{u}_n(0)] \right\}^2 \left\{ \frac{\partial \mathbf{f}}{\partial \mathbf{u}}[\mathbf{x}_n(0), \mathbf{u}_n(0)] \right\} \frac{T^3}{2} \tag{5.8-37}$$

where T is the sampling period.

With the determination of the parameters in $\mathbf{F}(k)$ and $\mathbf{G}(k)$, proper control laws can be designed to obtain the required correction torques to reduce the position and velocity errors of the manipulator along a nominal trajectory. This can be done by finding an optimal control $\mathbf{u}^*(k)$ which minimizes the performance index $J(k)$ while satisfying the constraints of Eq. (5.8-20):

$$J(k) = \tfrac{1}{2}[\mathbf{x}^T(k+1)\mathbf{Q}\mathbf{x}(k+1) + \mathbf{u}^T(k)\mathbf{R}\mathbf{u}(k)] \tag{5.8-38}$$

where $\mathbf{Q}$ is a $p \times p$ semipositive definite weighting matrix and $\mathbf{R}$ is an $n \times n$ positive definite weighting matrix. The one-step performance index in Eq. (5.8-38) indicates that the objective of the optimal control is to drive the position and velocity errors of the manipulator to zero along the nominal trajectory in a coordinated position and rate control per interval step while, at the same time, attaching a cost to the use of control effort. The optimal control solution which minimizes the functional in Eq. (5.8-38) subject to the constraints of Eq. (5.8-20) is well known and is found to be (Saridis and Lobbia [1972])

$$\mathbf{u}^*(k) = -[\mathbf{R} + \hat{\mathbf{G}}^T(k)\mathbf{Q}\hat{\mathbf{G}}(k)]^{-1}\hat{\mathbf{G}}^T(k)\,\mathbf{Q}\,\hat{\mathbf{F}}(k)\,\mathbf{x}(k) \qquad (5.8\text{-}39)$$

where $\hat{\mathbf{F}}(k)$ and $\hat{\mathbf{G}}(k)$ are the system parameters obtained from the identification algorithm [Eqs. (5.8-32) to (5.8-34)] at the kth sampling instant.

The identification and control algorithms in Eqs. (5.8-32) to (5.8-34) and Eq. (5.8-39) do not require complex computations. In Eq. (5.8-34), $[\mathbf{z}^T(k)\mathbf{P}(k)\mathbf{z}(k) + \rho]$ gives a scalar, so its inversion is trivial. Although the weighting factor ρ can be adjusted for each ith parameter vector $\boldsymbol{\theta}_i(k)$ as desired, this requires excessive computations in the $\mathbf{P}(k + 1)$ matrix. For real-time robot arm control, such adjustments are not desirable. $\mathbf{P}(k + 1)$ is computed only once at each sampling time using the same weighting factor ρ. Moreover, since $\mathbf{P}(k)$ is a symmetric positive definite matrix, only the upper diagonal matrix of $\mathbf{P}(k)$ needs to be computed. The combined identification and control algorithm can be computed in $O(n^3)$ time. The computational requirements of the adaptive perturbation control are tabulated in Table 5.1. Based on the specifications of a DEC PDP 11/45 computer, an ADDF (floating point addition) instruction requires $5.17\,\mu s$ and a MULF (floating point multiply) instruction requires $7.17\,\mu s$. If we assume that for each ADDF and MULF instruction, we need to fetch data from the core memory twice and the memory cycle time is 450 ns, then the adaptive perturbation control requires approximately 7.5 ms to compute the necessary joint torques to servo the first three joints of a PUMA robot arm for a trajectory set point.

A computer simulation study of a three-joint PUMA manipulator was conducted (Lee and Chung [1984,1985]) to evaluate and compare the performance of the adaptive controller with the controller [Eq. (5.3-65)], which is basically a proportional plus derivative control (PD controller). The study was carried out for various loading conditions along a given trajectory. The performances of the PD and adaptive controllers are compared and evaluated for three different loading

Table 5.1 Computations of the adaptive controller

Adaptive controller	Multiplications	Additions
Newton-Euler equations of motion	$117n - 24$	$103n - 21$
Least-squares identification algorithm	$30n^2 + 5n + 1$	$30n^2 + 3n - 1$
Control algorithm	$8n^3 + 2n^2 + 39$	$8n^3 - n^2 - n + 18$
Total	$8n^3 + 32n^2 + 5n + 40$	$8n^3 + 29n^2 + 2n + 17$

Table 5.2 Comparisons of the PD and adaptive controllers

Various loading conditions	Joint	PD controller Trajectory tracking			Adaptive controller Trajectory tracking		
		Max. error (degrees)	Max. error (mm)	Final position error (degrees)	Max. error (degrees)	Max. error (mm)	Final position error (degrees)
No-load and	1	0.089	1.55	0.025	0.020	0.34	0.000
10% error	2	0.098	1.71	0.039	0.020	0.36	0.004
In inertia tensor	3	0.328	2.86	0.121	0.032	0.28	0.002
½ max. load	1	0.121	2.11	0.054	0.045	0.78	0.014
and 10% error	2	0.147	2.57	0.078	0.065	1.14	0.050
In inertia tensor	3	0.480	4.19	0.245	0.096	0.83	0.077
Max. load	1	0.145	2.53	0.082	0.069	1.20	0.023
and 10% error	2	0.185	3.23	0.113	0.069	1.22	0.041
In inertia tensor	3	0.607	5.30	0.360	0.066	0.58	0.019

conditions and the results are tabulated in Table 5.2: (1) no-load and 10 percent error in inertia tensor, (2) half of maximum load and 10 percent error in inertia tensor, and (3) maximum load (5 lb) and 10 percent error in inertia tensor. In each case, a 10 percent error in inertia matrices means ±10 percent error about its measured inertial values. For all the above cases, the adaptive controller shows better performance than the PD controller with constant feedback gains both in trajectory tracking and the final position errors. Plots of angular position errors for the above cases for the adaptive control are shown in Figs. 5.16 to 5.18. Additional details of the simulation result can be found in Lee and Chung [1984, 1985].

5.8.4 Resolved Motion Adaptive Control

The adaptive control strategy of Sec. 5.8.3 in the joint variable space can be extended to control the manipulator in cartesian coordinates under various loading conditions by adopting the ideas of resolved motion rate and acceleration controls. The resolved motion adaptive control is performed at the hand level and is based on the linearized perturbation system along a desired time-based hand trajectory. The resolved motion adaptive control differs from the resolved motion acceleration control by minimizing the position/orientation and angular and linear velocities of the manipulator hand along the hand coordinate axes instead of position and orientation errors. Similar to the previous adaptive control, the controlled system is characterized by feedforward and feedback components which can be computed separately and simultaneously. The feedforward component resolves the specified

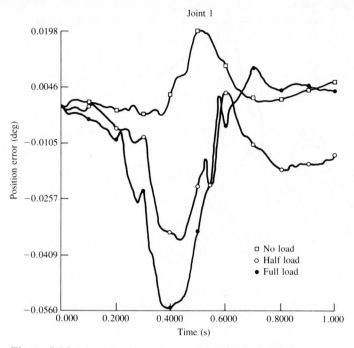

Figure 5.16 Joint 1 position error under various loads.

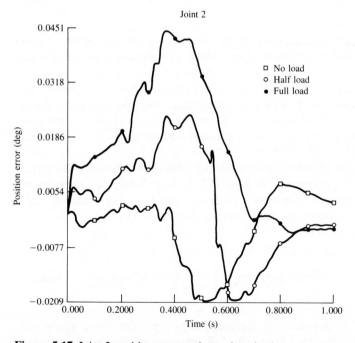

Figure 5.17 Joint 2 position error under various loads.

$$\times \begin{bmatrix} -\mathbf{h}_1(\mathbf{q}, \dot{\mathbf{q}}) - \mathbf{c}_1(\mathbf{q}) + \tau_1(t) \\ \\ -\mathbf{h}_2(\mathbf{q}, \dot{\mathbf{q}}) - \mathbf{c}_2(\mathbf{q}) + \tau_2(t) \end{bmatrix} \qquad (5.8\text{-}45)$$

where $\mathbf{0}$ is a 3×3 zero matrix. It is noted that the leftmost and middle vectors are 12×1, the center left matrix is 12×12, the right matrix is 12×6, and the rightmost vector is 6×1. Equation (5.8-45) represents the state equations of the manipulator and will be used to derive an adaptive control scheme in cartesian coordinates.

Defining the state vector for the manipulator hand as

$$\mathbf{x}(t) \triangleq (x_1, x_2, \ldots, x_{12})^T$$

$$\triangleq (p_x, p_y, p_z, \alpha, \beta, \gamma, v_x, v_y, v_z, \omega_x, \omega_y, \omega_z)^T \qquad (5.8\text{-}46)$$

$$\triangleq (\mathbf{p}^T, \mathbf{\Phi}^T, \mathbf{v}^T, \mathbf{\Omega}^T)^T$$

and the input torque vector as

$$\mathbf{u}(t) \triangleq (\tau_1, \ldots, \tau_6)^T \triangleq (u_1, \ldots, u_6)^T \qquad (5.8\text{-}47)$$

Eq. (5.8-45) can be expressed in state space representation as:

$$\dot{\mathbf{x}}(t) = \mathbf{f}[\mathbf{x}(t), \mathbf{u}(t)] \qquad (5.8\text{-}48)$$

where $\mathbf{x}(t)$ is a $2n$-dimensional vector, $\mathbf{u}(t)$ is an n-dimensional vector, $\mathbf{f}(\cdot)$ is a $2n \times 1$ continuously differentiable, nonlinear vector-valued function, and $n = 6$ is the number of degrees of freedom of the manipulator.

Equation (5.8-48) can be expressed as

$$\dot{x}_1(t) = f_1(\mathbf{x}, \mathbf{u}) = x_7(t)$$

$$\dot{x}_2(t) = f_2(\mathbf{x}, \mathbf{u}) = x_8(t)$$

$$\dot{x}_3(t) = f_3(\mathbf{x}, \mathbf{u}) = x_9(t)$$

$$\dot{x}_4(t) = f_4(\mathbf{x}, \mathbf{u}) = -\sec x_5 (x_{10} \cos x_6 + x_{11} \sin x_6) \qquad (5.8\text{-}49)$$

$$\dot{x}_5(t) = f_5(\mathbf{x}, \mathbf{u}) = \sec x_5 (x_{10} \cos x_5 \sin x_6 - x_{11} \cos x_5 \cos x_6)$$

$$\dot{x}_6(t) = f_6(\mathbf{x}, \mathbf{u}) = -\sec x_5 (x_{10} \sin x_5 \cos x_6 + x_{11} \sin x_5 \sin x_6 + x_{12} \cos x_5)$$

$$\dot{x}_{i+6}(t) = f_{i+6}(\mathbf{x}, \mathbf{u})$$

$$= g_{i+6}(\mathbf{q}, \dot{\mathbf{q}})\mathbf{x}(t) + b_{i+6}(\mathbf{q})\lambda(\mathbf{q}, \dot{\mathbf{q}}) + b_{i+6}(\mathbf{q})\mathbf{u}(t)$$

where $i = 1, \ldots, 6$ and $g_{i+6}(\mathbf{q}, \dot{\mathbf{q}})$ is the $(i + 6)$th row of the matrix:

$$
\begin{bmatrix}
\mathbf{0} & \mathbf{0} & \mathbf{I}_3 & \mathbf{0} \\
\mathbf{0} & \mathbf{0} & \mathbf{0} & \mathbf{S}(\boldsymbol{\Phi}) \\
\mathbf{0} & \mathbf{0} & \dot{\mathbf{N}}_{11}(\mathbf{q}, \dot{\mathbf{q}})\mathbf{K}_{11}(\mathbf{q}) + \dot{\mathbf{N}}_{12}(\mathbf{q}, \dot{\mathbf{q}})\mathbf{K}_{21}(\mathbf{q}) & \dot{\mathbf{N}}_{11}(\mathbf{q}, \dot{\mathbf{q}})\mathbf{K}_{12}(\mathbf{q}) + \dot{\mathbf{N}}_{12}(\mathbf{q}, \dot{\mathbf{q}})\mathbf{K}_{22}(\mathbf{q}) \\
\mathbf{0} & \mathbf{0} & \dot{\mathbf{N}}_{21}(\mathbf{q}, \dot{\mathbf{q}})\mathbf{K}_{11}(\mathbf{q}) + \dot{\mathbf{N}}_{22}(\mathbf{q}, \dot{\mathbf{q}})\mathbf{K}_{21}(\mathbf{q}) & \dot{\mathbf{N}}_{21}(\mathbf{q}, \dot{\mathbf{q}})\mathbf{K}_{12}(\mathbf{q}) + \dot{\mathbf{N}}_{22}(\mathbf{q}, \dot{\mathbf{q}})\mathbf{K}_{22}(\mathbf{q})
\end{bmatrix}
$$

and $b_{i+6}(\mathbf{q})$ is the $(i + 6)$th row of the matrix:

$$
\begin{bmatrix}
\mathbf{0} & \mathbf{0} \\
\mathbf{0} & \mathbf{0} \\
\mathbf{N}_{11}(\mathbf{q})\mathbf{E}_{11}(\mathbf{q}) + \mathbf{N}_{12}(\mathbf{q})\mathbf{E}_{21}(\mathbf{q}) & \mathbf{N}_{11}(\mathbf{q})\mathbf{E}_{12}(\mathbf{q}) + \mathbf{N}_{12}(\mathbf{q})\mathbf{E}_{22}(\mathbf{q}) \\
\mathbf{N}_{21}(\mathbf{q})\mathbf{E}_{11}(\mathbf{q}) + \mathbf{N}_{22}(\mathbf{q})\mathbf{E}_{21}(\mathbf{q}) & \mathbf{N}_{21}(\mathbf{q})\mathbf{E}_{12}(\mathbf{q}) + \mathbf{N}_{22}(\mathbf{q})\mathbf{E}_{22}(\mathbf{q})
\end{bmatrix}
$$

and

$$
\lambda(\mathbf{q}, \dot{\mathbf{q}}) =
\begin{bmatrix}
-\mathbf{h}_1(\mathbf{q}, \dot{\mathbf{q}}) - \mathbf{c}_1(\mathbf{q}) \\[1em]
-\mathbf{h}_2(\mathbf{q}, \dot{\mathbf{q}}) - \mathbf{c}_2(\mathbf{q})
\end{bmatrix}
$$

Equation (5.8-49) describes the complete manipulator dynamics in cartesian coordinates, and the control problem is to find a feedback control law $\mathbf{u}(t) = \mathbf{g}[\mathbf{x}(t)]$ to minimize the manipulator hand error along the desired hand trajectory over a wide range of payloads. Again, perturbation theory is used and Taylor series expansion is applied to Eq. (5.8-49) to obtain the associated linearized system and to design a feedback control law about the desired hand trajectory. The determination of the feedback control law for the linearized system is identical to the one in the joint coordinates [Eqs. (5.8-32) to (5.8-34) and Eq. (5.8-39)]. The resolved motion adaptive control block diagram is shown in Fig. 5.19.

The overall resolved motion adaptive control system is again characterized by a feedforward component and a feedback component. Such a formulation has the advantage of employing parallel schemes in computing these components. The feedforward component computes the desired joint torques as follows: (1) The hand trajectory set points $\mathbf{p}^d(t)$, $\boldsymbol{\Phi}^d(t)$, $\mathbf{v}^d(t)$, $\boldsymbol{\Omega}^d(t)$, $\dot{\mathbf{v}}^d(t)$, and $\dot{\boldsymbol{\Omega}}^d(t)$ are resolved into a set of values of desired joint positions, velocities, and accelerations; (2) the desired joint torques along the hand trajectory are computed from the Newton-Euler equations of motion using the computed sets of values of joint positions, velocities, and accelerations. These computed torques constitute the nominal torque values $\mathbf{u}_n(t)$. The feedback component computes the perturbation joint torques $\delta\mathbf{u}(t)$ the same way as in Eq. (5.8-39), using the recursive least-squares identification scheme in Eqs. (5.8-32) to (5.8-34).

A feasibility study of implementing the adaptive controller based on a 60-Hz sampling frequency and using present-day low-cost microprocessors can be conducted by looking at the computational requirements in terms of mathematical mul-

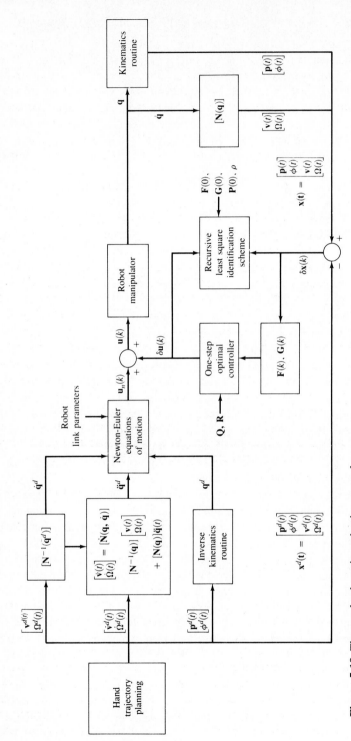

Figure 5.19 The resolved motion adaptive control.

tiplication and addition operations. We assume that multiprocessors are available for parallel computation of the controller. The feedforward component which computes the nominal joint torques along a desired hand trajectory can be computed serially in four separate stages. It requires a total of 1386 multiplications and 988 additions for a six-joint manipulator. The feedback control component which computes the perturbation joint torques can be conveniently computed serially in three separate stages. It requires about 3348 multiplications and 3118 additions for a six-joint manipulator. Since the feedforward and feedback components can be computed in parallel, the resolved motion adaptive control requires a total of 3348 multiplications and 3118 additions in each sampling period. Computational requirements in terms of multiplications and additions for the adaptive controller for a n-joint manipulator are tabulated in Table 5.3.

Based on the specification sheet of an INTEL 8087 microprocessor, an integer multiply requires 19 μs, an addition requires 17 μs, and a memory fetch or store requires 9 μs. Assuming that two memory fetches are required for each multiplication and addition operation, the proposed controller can be computed in about 233 ms which is not fast enough for closing the servo loop. (Recall from Sec. 5.3.5 that a minimum of 16 msec is required if the sampling frequency is 60 Hz). Similarly, looking at the specification sheet of a Motorola MC68000 microprocessor, an integer multiply requires 5.6 μs, an addition requires 0.96 μs, and a memory fetch or store requires 0.32 μs, the proposed controller can be computed in about 26.24 ms which is still not fast enough for closing the servo loop. Finally, looking at the specification sheet of a PDP 11/45 computer, an integer multiply requires 3.3 μs, an addition requires 300 ns, and a memory fetch or store requires 450 ns, the proposed controller can be computed in about 18 ms which translates to a sampling frequency of approximately 55 Hz. However, the PDP 11/45 is a uniprocessor machine and the parallel computation assumption is not valid. This exercise should give the reader an idea of the required processing speed for adaptive control of a manipulator. We anticipate that faster microprocessors, which will be able to compute the proposed resolved motion adaptive controller within 10 ms, will be available in a few years.

5.9 CONCLUDING REMARKS

We have reviewed various robot manipulator control methods. They vary from a simple servomechanism to advanced control schemes such as adaptive control with an identification algorithm. The control techniques are discussed in joint motion control, resolved motion control, and adaptive control. Most of the joint motion and resolved motion control methods discussed servo the arm at the hand or the joint level and emphasize nonlinear compensations of the coupling forces among the various joints. We have also discussed various adaptive control strategies. The model-referenced adaptive control is easy to implement, but suitable reference models are difficult to choose and it is difficult to establish any stability analysis of

Table 5.3 Computations of the resolved motion adaptive control†

	Adaptive controller	Number of multiplications	Number of additions
stage 1	Compute $\mathbf{q}^d$ (inverse kinematics)	(39)	(32)
stage 2	Compute $\dot{\mathbf{q}}^d$	$n^2 + 27n + 327$ (525)	$n^2 + 18n + 89$ (233)
stage 3	Compute $\ddot{\mathbf{q}}^d$	$4n^2$ (144)	$4n^2 - 3n$ (126)
stage 4	Compute τ	$117n - 24$ (678)	$103n - 21$ (597)
Total feedforward computations		$5n^2 + 144n + 342$ (1386)	$5n^2 + 118n + 100$ (988)
stage 1	Compute $(\mathbf{p}^T \Phi^T)^T$	(48)	(22)
	Compute $(\mathbf{v}^T \Omega^T)^T$	$n^2 + 27n - 21$ (177)	$n^2 + 18n - 15$ (129)
stage 2	Compute hand errors $[\mathbf{x}(k) - \mathbf{x}_n(k)]$	0 (0)	2n (12)
	Identification scheme	$33n^2 + 9n + 2$ (1244)	$34\frac{1}{2}n^2 - 1\frac{1}{2}n$ (1233)
stage 3	Compute adaptive controller	$8n^3 + 4n^2 + n + 1$ (1879)	$8n^3 - n$ (1722)
Total feedback computations		$8n^3 + 38n^2 + 37n + 30$ (3348)	$8n^3 + 35\frac{1}{2}n^2 + 17\frac{1}{2}n + 7$ (3118)
Total mathematical operations		$8n^3 + 38n^2 + 37n + 30$ (3348)	$8n^3 + 35\frac{1}{2}n^2 + 17\frac{1}{2}n + 7$ (3118)

† Number inside parentheses indicate computations for $n = 6$.

the controlled system. Self-tuning adaptive control fits the input-output data of the system with an autoregressive model. Both methods neglect the coupling forces between the joints which may be severe for manipulators with rotary joints. Adaptive control using perturbation theory may be more appropriate for various manipulators because it takes all the interaction forces between the joints into consideration. The adaptive perturbation control strategy was found suitable for controlling the manipulator in both the joint coordinates and cartesian coordinates. An adaptive perturbation control system is characterized by a feedforward component and a feedback component which can be computed separately and simultaneously in parallel. The computations of the adaptive control for a six-link robot arm may be implemented in low-cost microprocessors for controlling in the joint variable space, while the resolved motion adaptive control cannot be implemented in present-day low-cost microprocessors because they still do not have the required speed to compute the controller parameters for the "standard" 60-Hz sampling frequency.

REFERENCES

Further readings on computed torque control techniques can be found in Paul [1972], Bejczy [1974], Markiewicz [1973], Luh et al. [1980*b*], and Lee [1982]. Minimum-time control can be found in Kahn and Roth [1971], and minimum-time control with torque constraint is discussed by Bobrow and Dubowsky [1983]. Young [1978] discusses the design of a variable structure control for the control of manipulators. More general theory in variable structure control can be found in Utkin [1977] and Itkis [1976]. Various researchers have discussed nonlinear decoupled control, including Falb and Wolovich [1967], Hemami and Camana [1976], Saridis and Lee [1979], Horowitz and Tomizuka [1980], Freund [1982], Tarn et al. [1984], and Gilbert and Ha [1984].

Further readings on resolved motion control can be found in Whitney [1969, 1972] who discussed resolved motion rate control. Luh et al. [1980*b*] extended this concept to include resolved acceleration control. The disadvantage of resolved motion control lies in the fact that the inverse jacobian matrix requires intensive computations.

In order to compensate for the varying parameters of a manipulator and the changing loads that it carries, various adaptive control schemes, both in joint and cartesian coordinates, have been developed. These adaptive control schemes can be found in Dubowsky and DesForges [1979], Horowitz and Tomizuka [1980], Koivo and Guo [1983], Lee and Chung [1984, 1985], Lee and Lee [1984], and Lee et al. [1984].

An associated problem relating to control is the investigation of efficient control system architectures for computing the control laws within the required servo time. Papers written by Lee et al. [1982], Luh and Lin [1982], Orin [1984], Nigam and Lee [1985], and Lee and Chang [1986*b*], are oriented toward this goal.

PROBLEMS

5.1 Consider the development of a single-joint positional controller, as discussed in Sec. 5.3.2. If the applied voltage $V_a(t)$ is linearly proportional to the position error and to the rate of the output angular position, what is the open-loop transfer function $\Theta_L(s)/E(s)$ and the closed-loop transfer function $\Theta_L(s)/\Theta_L^d(s)$ of the system?

5.2 For the applied voltage used in Prob. 5.1, discuss the steady-state error of the system due to a step input. Repeat for a ramp input.

5.3 In the computed torque control technique, if the Newton-Euler equations of motion are used to compute the applied joint torques for a 6 degree-of-freedom manipulator with rotary joints, what is the required number of multiplications and additions per trajectory set point?

5.4 In the computed torque control technique, the analysis is performed in the continuous time, while the actual control on the robot arm is done in discrete time (i.e., by a sampled-data system) because we use a digital computer for implementing the controller. Explain the condition under which this practice is valid.

5.5 The equations of motion of the two-link robot arm in Sec. 3.2.6 can be written in a compact matrix-vector form as:

$$
\begin{bmatrix} d_{11}(\theta_2) & d_{12}(\theta_2) \\ d_{12}(\theta_2) & d_{22} \end{bmatrix} \begin{bmatrix} \ddot{\theta}_1(t) \\ \ddot{\theta}_2(t) \end{bmatrix} + \begin{bmatrix} \beta_{12}(\theta_2)\dot{\theta}_2^2 + 2\beta_{12}(\theta_2)\dot{\theta}_1\dot{\theta}_2 \\ -\beta_{12}(\theta_2)\dot{\theta}_1^2 \end{bmatrix} + \begin{bmatrix} c_1(\theta_1, \theta_2)g \\ c_2(\theta_1, \theta_2)g \end{bmatrix} = \begin{bmatrix} \tau_1(t) \\ \tau_2(t) \end{bmatrix}
$$

where g is the gravitational constant. Choose an appropriate state variable vector $\mathbf{x}(t)$ and a control vector $\mathbf{u}(t)$ for this dynamic system. Assuming that $\mathbf{D}^{-1}(\theta)$ exists, express the equations of motion of this robot arm explicitly in terms of d_{ij}'s, β_{ij}'s, and c_i's in a state-space representation with the chosen state-variable vector and control vector.

5.6 Design a variable structure controller for the robot in Prob. 5.5. (See Sec. 5.5.)

5.7 Design a nonlinear decoupled feedback controller for the robot in Prob. 5.5. (See Sec. 5.6.)

5.8 Find the jacobian matrix in the base coordinate frame for the robot in Prob. 5.5. (See Appendix B.)

5.9 Give two main disadvantages of using the resolved motion rate control.

5.10 Give two main disadvantages of using the resolved motion acceleration control.

5.11 Give two main disadvantages of using the model-referenced adaptive control.

5.12 Give two main disadvantages of using the adaptive perturbation control.

SIX

SENSING

Art thou not sensible to feeling as to sight?
William Shakespeare

6.1 INTRODUCTION

The use of external sensing mechanisms allows a robot to interact with its environment in a flexible manner. This is in contrast to preprogrammed operation in which a robot is "taught" to perform repetitive tasks via a set of programmed functions. Although the latter is by far the most predominant form of operation of present industrial robots, the use of sensing technology to endow machines with a greater degree of intelligence in dealing with their environment is, indeed, an active topic of research and development in the robotics field. A robot that can "see" and "feel" is easier to train in the performance of complex tasks while, at the same time, requires less stringent control mechanisms than preprogrammed machines. A sensory, trainable system is also adaptable to a much larger variety of tasks, thus achieving a degree of universality that ultimately translates into lower production and maintenance costs.

The function of robot sensors may be divided into two principal categories: *internal state* and *external state*. Internal state sensors deal with the detection of variables such as arm joint position, which are used for robot control, as discussed in Chap. 5. External state sensors, on the other hand, deal with the detection of variables such as range, proximity, and touch. External sensing, the topic of Chaps. 6 to 8, is used for robot guidance, as well as for object identification and handling.

External state sensors may be further classified as *contact* or *noncontact* sensors. As their name implies, the former class of sensors respond to physical contact, such as touch, slip, and torque. Noncontact sensors rely on the response of a detector to variations in acoustic or electromagnetic radiation. The most prominent examples of noncontact sensors measure range, proximity, and visual properties of an object.

The focus of this chapter is on range, proximity, touch, and force-torque sensing. Vision sensors and techniques are discussed in detail in Chaps. 7 and 8. It is of interest to note that vision and range sensing generally provide gross guidance information for a manipulator, while proximity and touch are associated with the terminal stages of object grasping. Force and torque sensors are used as feedback devices to control manipulation of an object once it has been grasped (e.g., to avoid crushing the object or to prevent it from slipping).

6.2 RANGE SENSING

A range sensor measures the distance from a reference point (usually on the sensor itself) to objects in the field of operation of the sensor. Humans estimate distance by means of stereo visual processing, as discussed in Chap. 7, while other animals, such as bats, utilize the "time of flight" concept in which distance estimates are based on the time elapsed between the transmission and return of a sonic pulse. Range sensors are used for robot navigation and obstacle avoidance, where interest lies in estimating the distance to the closest objects, to more detailed applications in which the location and general shape characteristics of objects in the work space of a robot are desired. In this section we discuss several range sensing techniques that address these problems.

6.2.1 Triangulation

One of the simplest methods for measuring range is through triangulation techniques. This approach can be easily explained with the aid of Fig. 6.1. An object is illuminated by a narrow beam of light which is swept over the surface. The sweeping motion is in the plane defined by the line from the object to the detector and the line from the detector to the source. If the detector is focused on a *small* portion of the surface then, when the detector sees the light spot, its distance D to the illuminated portion of the surface can be calculated from the geometry of Fig. 6.1 since the angle of the source with the baseline and the distance B between the source and detector are known.

The above approach yields a point measurement. If the source-detector arrangement is moved in a fixed plane (up and down and sideways on a plane perpendicular to the paper and containing the baseline in Fig. 6.1), then it is possible

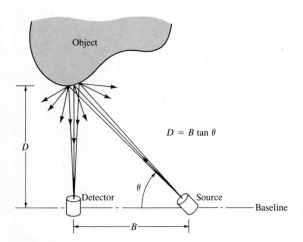

$$D = B \tan \theta$$

Figure 6.1 Range sensing by triangulation. (Adapted from Jarvis [1983*a*], © IEEE.)

Figure 6.2 (*a*) An arrangement of objects scanned by a triangulation ranging device. (*b*) Corresponding image with intensities proportional to range. (From Jarvis [1983*a*], © IEEE.)

to obtain a set of points whose distances from the detector are known. These distances are easily transformed to three-dimensional coordinates by keeping track of the location and orientation of the detector as the objects are scanned. An example is shown in Fig. 6.2. Figure 6.2*a* shows an arrangement of objects scanned in the manner just explained. Figure 6.2*b* shows the results in terms of an image whose intensity (darker is closer) is proportional to the range measured from the plane of motion of the source-detector pair.

6.2.2 Structured Lighting Approach

This approach consists of projecting a light pattern onto a set of objects and using the distortion of the pattern to calculate the range. One of the most popular light patterns in use today is a sheet of light generated through a cylindrical lens or a narrow slit.

As illustrated in Fig. 6.3, the intersection of the sheet with objects in the work space yields a light stripe which is viewed through a television camera displaced a distance *B* from the light source. The stripe pattern is easily analyzed by a computer to obtain range information. For example, an inflection indicates a change of surface, and a break corresponds to a gap between surfaces.

Specific range values are computed by first calibrating the system. One of the simplest arrangements is shown in Fig. 6.3*b*, which represents a top view of Fig. 6.3*a*. In this arrangement, the light source and camera are placed at the same height, and the sheet of light is perpendicular to the line joining the origin of the light sheet and the center of the camera lens. We call the vertical plane containing this line the *reference plane*. Clearly, the reference plane is perpendicular to the sheet of light, and any vertical flat surface that intersects the sheet will produce a vertical stripe of light (see Fig. 6.3*a*) in which every point will have the same perpendicular distance to the reference plane. The objective of the arrangement shown in Fig. 6.3*b* is to position the camera so that every such vertical stripe also appears vertical in the image plane. In this way, every point along the same column in the image will be known to have the same distance to the reference plane.

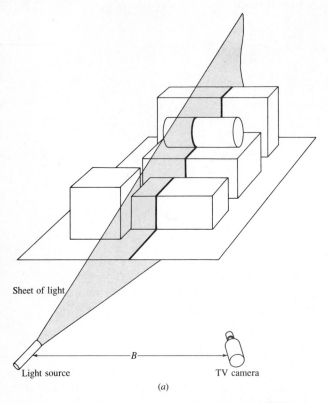

(a)

Figure 6.3 (a) Range measurement by structured lighting approach. (b) Top view of part (a) showing a specific arrangement which simplifies calibration.

Most systems based on the sheet-of-light approach use digital images. Suppose that the image seen by the camera is digitized into an $N \times M$ array (see Sec. 7.2), and let $y = 0, 1, 2, \ldots, M - 1$ be the column index of this array. As explained below, the calibration procedure consists of measuring the distance B between the light source and lens center, and then determining the angles α_c and α_0. Once these quantities are known, it follows from elementary geometry that d in Fig. 6.3b is given by

$$d = \lambda \tan \theta \qquad (6.2\text{-}1)$$

where λ is the focal length of the lens and

$$\theta = \alpha_c - \alpha_0 \qquad (6.2\text{-}2)$$

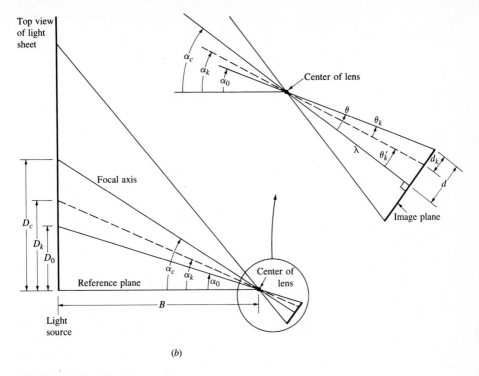

(b)

Figure 6.3 (continued)

For an M-column digital image, the distance increment d_k between columns is given by

$$d_k = k \frac{d}{M/2} = \frac{2kd}{M} \tag{6.2-3}$$

for $0 \leq k \leq M/2$. (In an image viewed on a monitor, $k = 0$ would correspond to the leftmost column and $k = M/2$ to the center column.) The angle α_k made by the projection of an arbitrary stripe is easily obtained by noting that

$$\alpha_k = \alpha_c - \theta_k' \tag{6.2-4}$$

where

$$\tan \theta_k' = \frac{d - d_k}{\lambda} \tag{6.2-5}$$

or, using Eq. (6.2-3),

$$\theta_k' = \tan^{-1} \left[\frac{d(M - 2k)}{M\lambda} \right] \tag{6.2-6}$$

where $0 \leqslant k \leqslant M/2$. For the remaining values of k (i.e., on the other side of the optical axis), we have

$$\alpha_k = \alpha_c + \theta_k'' \tag{6.2-7}$$

where

$$\theta_k'' = \tan^{-1} \left[\frac{d(2k - M)}{M\lambda} \right] \tag{6.2-8}$$

for $M/2 < k \leqslant (M - 1)$.

By comparing Eqs. (6.2-6) and (6.2-8) we note that $\theta_k'' = -\theta_k'$, so Eqs. (6.2-4) and (6.2-7) are identical for the *entire* range $0 \leqslant k \leqslant M - 1$. It then follows from Fig. 6.3b that the perpendicular distance D_k between an arbitrary light stripe and the reference plane is given by

$$D_k = B \tan \theta_k \tag{6.2-9}$$

for $0 \leqslant k \leqslant M - 1$, where α_k is given either by Eq. (6.2-4) or (6.2-7).

It is important to note that once B, α_0, α_c, M, and λ are known, the column number in the digital image completely determines the distance between the reference plane and all points in the stripe imaged on that column. Since M and λ are fixed parameters, the calibration procedure consists simply of measuring B and determining α_c and α_0, as indicated above. To determine α_c, we place a flat vertical surface so that its intersection with the sheet of light is imaged on the center of the image plane (i.e., at $y = M/2$). We then physically measure the perpendicular distance D_c between the surface and the reference plane. From the geometry of Fig. 6.3b it follows that

$$\alpha_c = \tan^{-1} \left[\frac{D_c}{B} \right] \tag{6.2-10}$$

In order to determine α_0, we move the surface closer to the reference plane until its light stripe is imaged at $y = 0$ on the image plane. We then measure D_0 and, from Fig. 6.3b,

$$\alpha_0 = \tan^{-1} \left[\frac{D_0}{B} \right] \tag{6.2-11}$$

This completes the calibration procedure.

The principal advantage of the arrangement just discussed is that it results in a relatively simple range measuring technique. Once calibration is completed, the distance associated with every column in the image is computed using Eq. (6.2-9) with $k = 0, 1, 2, \ldots, M - 1$ and the results are stored in memory. Then, during normal operation, the distance of any imaged point is obtained simply by determining its column number in the image and addressing the corresponding location in memory.

Before leaving this section, we point out that it is possible to use the concepts discussed in Sec. 7.4 to solve a more general problem in which the light source and camera are placed arbitrarily with respect to each other. The resulting expressions, however, would be considerably more complicated and difficult to handle from a computational point of view.

6.2.3 Time-of-Flight Range Finders

In this section we discuss three methods for determining range based on the time-of-flight concept introduced at the beginning of Sec. 6.2. Two of the methods utilize a laser, while the third is based on ultrasonics.

One approach for using a laser to determine range is to measure the time it takes an emitted pulse of light to return coaxially (i.e., along the same path) from a reflecting surface. The distance to the surface is given by the simple relationship $D = cT/2$, where T is the pulse transit time and c is the speed of light. It is of interest to note that, since light travels at approximately 1 ft/ns, the supporting electronic instrumentation must be capable of 50-ps time resolution in order to achieve a $\pm \frac{1}{4}$-inch accuracy in range.

A pulsed-laser system described by Jarvis [1983*a*] produces a two-dimensional array with values proportional to distance. The two-dimensional scan is accomplished by deflecting the laser light via a rotating mirror. The working range of this device is on the order of 1 to 4 m, with an accuracy of ± 0.25 cm. An example of the output of this system is shown in Fig. 6.4. Part (*a*) of this figure shows a collection of three-dimensional objects, and Fig. 6.4*b* is the corresponding sensed array displayed as an image in which the intensity at each point is proportional to the distance between the sensor and the reflecting surface at that point (darker is closer). The bright areas around the object boundaries represent discontinuity in range determined by postprocessing in a computer.

Figure 6.4 (*a*) An arrangement of objects. (*b*) Image with intensity proportional to range. (From Jarvis [1983*b*], © IEEE.)

An alternative to pulsed light is to use a continuous-beam laser and measure the delay (i.e., phase shift) between the outgoing and returning beams. We illustrate this concept with the aid of Fig. 6.5. Suppose that a beam of laser light of wavelength λ is split into two beams. One of these (called the *reference beam*) travels a distance L to a phase measuring device, and the other travels a distance D out to a reflecting surface. The total distance traveled by the reflected beam is $D' = L + 2D$. Suppose that $D = 0$. Under this condition $D' = L$ and both the reference and reflected beams arrive simultaneously at the phase measuring device. If we let D increase, the reflected beam travels a longer path and, therefore, a phase shift is introduced between the two beams at the point of measurement, as illustrated in Fig. 6.5b. In this case we have that

$$D' = L + \frac{\theta}{360} \lambda \qquad (6.2\text{-}12)$$

It is noted that if $\theta = 360°$ the two waveforms are again aligned and we cannot differentiate between $D' = L$ and $D' = L + n\lambda$, $n = 1, 2, \ldots$, based on measurements of phase shift alone. Thus, a unique solution can be obtained only if we

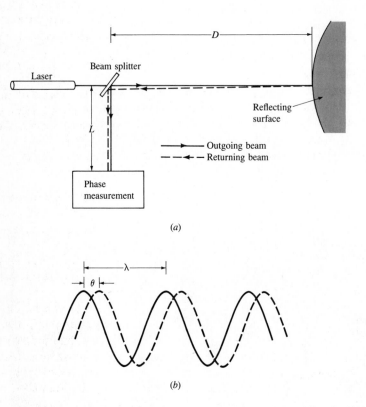

(a)

(b)

Figure 6.5 (a) Principles of range measurement by phase shift. (b) Shift between outgoing and returning light waveforms.

require that $\theta < 360°$ or, equivalently, that $2D < \lambda$. Since $D' = L + 2D$, we have by substitution into Eq. (6.2-12) that

$$D = \frac{\theta}{360} \left[\frac{\lambda}{2} \right] \qquad (6.2\text{-}13)$$

which gives distance in terms of phase shift if the wavelength is known.

Since the wavelength of laser light is small (e.g., 632.8 nm for a helium-neon laser), the method sketched in Fig. 6.5 is impractical for robotic applications. A simple solution to this problem is to modulate the amplitude of the laser light by using a waveform of much higher wavelength. (For example, recalling that $c = f\lambda$, a modulating sine wave of frequency $f = 10\,\text{MHz}$ has a wavelength of 30 m.) The approach is illustrated in Fig. 6.6. The basic technique is as before, but the reference signal is now the modulating function. The modulated laser signal is sent out to the target and the returning beam is stripped of the modulating signal, which is then compared against the reference to determine phase shift. Equation (6.2-13) still holds, but we are now working in a more practical range of wavelengths.

An important advantage of the continuous vs. the pulsed-light technique is that the former yields intensity as well as range information (Jarvis [1983a]). However, continuous systems require considerably higher power. Uncertainties in distance measurements obtained by either technique require averaging the returned signal to reduce the error. If we treat the problem as that of measurement noise being added to a true distance, and we assume that measurements are statistically independent, then it can be shown that the standard deviation of the average is equal to $1/\sqrt{N}$ times the standard deviation of the noise, where N is the number of samples averaged. In other words, the longer we average, the smaller the uncertainty will be in the distance estimate.

An example of results obtainable with a continuous, modulated laser beam scanned by a rotating mirror is shown in Fig. 6.7b. Part (a) of this figure is the range array displayed as an intensity image (brighter is closer). The true intensity information obtained with the same device is shown in part (b). Note that these two images complement each other. For example, it is difficult to count the

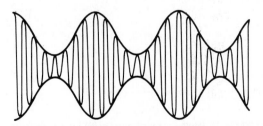

Figure 6.6 Amplitude-modulated waveform. Note the much larger wavelength of the modulating function.

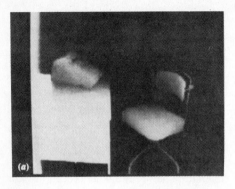 

Figure 6.7 (*a*) Range array displayed as an image. (*b*) Intensity image. (From Duda, Nitzan, and Barrett [1979], © IEEE.)

number of objects on top of the desk in Fig. 6.7*a*, a simple task in the intensity image. Conversely, it is not possible to determine the distance between the near and far edges of the desk top by examining the intensity image, while this information is readily available in the range array. Techniques for processing this type of information are discussed in Chaps. 7 and 8.

An ultrasonic range finder is another major exponent of the time-of-flight concept. The basic idea is the same as that used with a pulsed laser. An ultrasonic chirp is transmitted over a short time period and, since the speed of sound is known for a specified medium, a simple calculation involving the time interval between the outgoing pulse and the return echo yields an estimate of the distance to the reflecting surface.

In an ultrasonic ranging system manufactured by Polaroid, for example, a 1-ms chirp, consisting of 56 pulses at four frequencies, 50, 53, 57, and 60 KHz, is transmitted by a transducer 1½ inches in diameter. The signal reflected by an object is detected by the same transducer and processed by an amplifier and other circuitry capable of measuring range from approximately 0.9 to 35 ft, with an accuracy of about 1 inch. The mixed frequencies in the chirp are used to reduce signal cancellation. The beam pattern of this device is around 30°, which introduces severe limitations in resolution if one wishes to use this device to obtain a range image similar to those discussed earlier in this section. This is a common problem with ultrasonic sensors and, for this reason, they are used primarily for navigation and obstacle avoidance. The construction and operational characteristics of ultrasonic sensors are discussed in further detail in Sec. 6.3.

6.3 PROXIMITY SENSING

The range sensors discussed in the previous section yield an estimate of the distance between a sensor and a reflecting object. Proximity sensors, on the other hand, generally have a binary output which indicates the presence of an object within a specified distance interval. Typically, proximity sensors are used in robotics for near-field work in connection with object grasping or avoidance. In this section we consider several fundamental approaches to proximity sensing and discuss the basic operational characteristics of these sensors.

6.3.1 Inductive Sensors

Sensors based on a change of inductance due to the presence of a metallic object are among the most widely used industrial proximity sensors. The principle of operation of these sensors can be explained with the aid of Figs. 6.8 and 6.9. Figure 6.8a shows a schematic diagram of an inductive sensor which basically consists of a wound coil located next to a permanent magnet packaged in a simple, rugged housing.

The effect of bringing the sensor in close proximity to a ferromagnetic material causes a change in the position of the flux lines of the permanent magnet, as shown in Fig. 6.8b and c. Under static conditions there is no movement of the flux lines and, therefore, no current is induced in the coil. However, as a ferromagnetic object enters or leaves the field of the magnet, the resulting change in

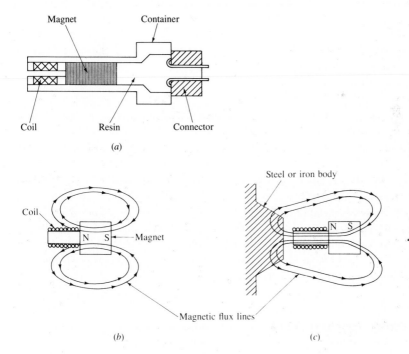

Figure 6.8 (a) An inductive sensor. (b) Shape of flux lines in the absence of a ferromagnetic body. (c) Shape of flux lines when a ferromagnetic body is brought close to the sensor. (Adapted from Canali [1981a], © Società Italiana di Fisica.)

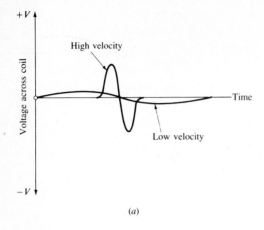

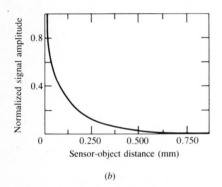

Figure 6.9 (*a*) Inductive sensor response as a function of speed. (*b*) Sensor response as a function of distance. (Adapted from Canali [1981*a*], © Società Italiana di Fisica.)

the flux lines induces a current pulse whose amplitude and shape are proportional to the rate of change in the flux.

The voltage waveform observed at the output of the coil provides an effective means for proximity sensing. Figure 6.9*a* illustrates how the voltage measured across the coil varies as a function of the speed at which a ferromagnetic material is introduced in the field of the magnet. The polarity of the voltage out of the sensor depends on whether the object is entering or leaving the field. Figure 6.9*b* illustrates the relationship between voltage amplitude and sensor-object distance. It is noted from this figure that sensitivity falls off rapidly with increasing distance, and that the sensor is effective only for fractions of a millimeter.

Since the sensor requires motion to produce an output waveform, one approach for generating a binary signal is to integrate this waveform. The binary output remains low as long as the integral value remains below a specified threshold, and then switches to high (indicating proximity of an object) when the threshold is exceeded.

6.3.2 Hall-Effect Sensors

The reader will recall from elementary physics that the Hall effect relates the voltage between two points in a conducting or semiconducting material to a magnetic field across the material. When used by themselves, Hall-effect sensors can only detect magnetized objects. However, when used in conjunction with a permanent magnet in a configuration such as the one shown in Fig. 6.10, they are capable of detecting all ferromagnetic materials. When used in this way, a Hall-effect device senses a strong magnetic field in the absence of a ferromagnetic metal in the near field (Fig. 6.10a). When such a material is brought in close proximity with the device, the magnetic field weakens at the sensor due to bending of the field lines through the material, as shown in Fig. 6.10b.

Hall-effect sensors are based on the principle of a Lorentz force which acts on a charged particle traveling through a magnetic field. This force acts on an axis perpendicular to the plane established by the direction of motion of the charged particle and the direction of the field. That is, the Lorentz force is given by $\mathbf{F} = q(\mathbf{v} \times \mathbf{B})$ where q is the charge, $\mathbf{v}$ is the velocity vector, $\mathbf{B}$ is the magnetic field vector, and "$\times$" is the vector cross product. Suppose, for example, that a current flows through a doped, n-type semiconductor which is immersed in a magnetic field, as shown in Fig. 6.11. Recalling that electrons are the majority carriers in n-type materials, and that conventional current flows opposite to electron current, we would have that the force acting on the moving, negatively charged particles would have the direction shown in Fig. 6.11. This force would act on the electrons, which would tend to collect at the bottom of the material and thus produce a voltage across it which, in this case, would be positive at the top.

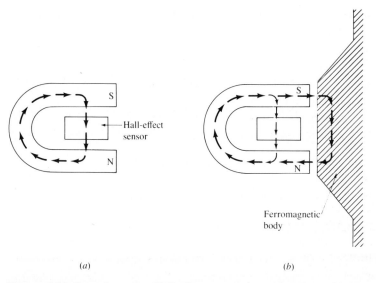

(a)   *(b)*

Figure 6.10 Operation of a Hall-effect sensor in conjunction with a permanent magnet. (Adapted from Canali [1981a], © Società Italiana di Fisica.)

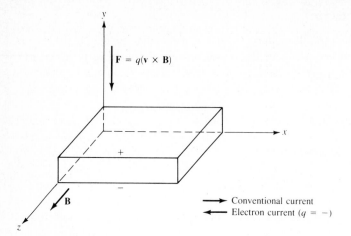

Figure 6.11 Generation of Hall voltage.

Bringing a ferromagnetic material close to the semiconductor-magnet device would decrease the strength of the magnetic field, thus reducing the Lorentz force and, ultimately, the voltage across the semiconductor. This drop in voltage is the key for sensing proximity with Hall-effect sensors. Binary decisions regarding the presence of an object are made by thresholding the voltage out of the sensor.

It is of interest to note that using a semiconductor, such as silicon, has a number of advantages in terms of size, ruggedness, and immunity to electrical interference. In addition, the use of semiconducting materials allows the construction of electronic circuitry for amplification and detection directly on the sensor itself, thus reducing sensor size and cost.

6.3.3 Capacitive Sensors

Unlike inductive and Hall-effect sensors which detect only ferromagnetic materials, capacitive sensors are potentially capable (with various degrees of sensitivity) of detecting all solid and liquid materials. As their name implies, these sensors are based on detecting a change in capacitance induced by a surface that is brought near the sensing element.

The basic components of a capacitive sensor are shown in Fig. 6.12. The sensing element is a capacitor composed of a sensitive electrode and a reference electrode. These can be, for example, a metallic disk and ring separated by a dielectric material. A cavity of dry air is usually placed behind the capacitive element to provide isolation. The rest of the sensor consists of electronic circuitry which can be included as an integral part of the unit, in which case it is normally embedded in a resin to provide sealing and mechanical support.

There are a number of electronic approaches for detecting proximity based on a change in capacitance. One of the simplest includes the capacitor as part of an

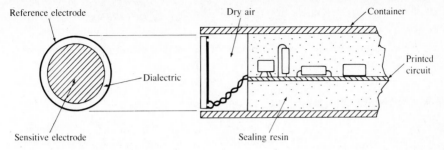

Figure 6.12 A capacitive proximity sensor. (From Canali [1981*a*], © Società Italiana di Fisica.)

oscillator circuit designed so that the oscillation starts only when the capacitance of the sensor exceeds a predefined threshold value. The start of oscillation is then translated into an output voltage which indicates the presence of an object. This method provides a binary output whose triggering sensitivity depends on the threshold value.

A more complicated approach utilizes the capacitive element as part of a circuit which is continuously driven by a reference sinusoidal waveform. A change in capacitance produces a phase shift between the reference signal and a signal derived from the capacitive element. The phase shift is proportional to the change in capacitance and can thus be used as a basic mechanism for proximity detection.

Figure 6.13 illustrates how capacitance varies as a function of distance for a proximity sensor based on the concepts just discussed. It is of interest to note that sensitivity decreases sharply past a few millimeters, and that the shape of the response curve depends on the material being sensed. Typically, these sensors are

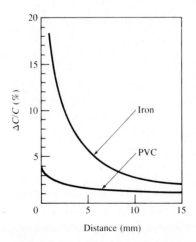

Figure 6.13 Response (percent change in capacitance) of a capacitive proximity sensor as a function of distance. (From Canali [1981*a*], © Società Italiana di Fisica.)

operated in a binary mode so that a change in the capacitance greater than a preset threshold T indicates the presence of an object, while changes below the threshold indicate the absence of an object with respect to detection limits established by the value of T.

6.3.4 Ultrasonic Sensors

The response of all the proximity sensors discussed thus far depends strongly on the material being sensed. This dependence can be reduced considerably by using ultrasonic sensors, whose operation for range detection was introduced briefly at the end of Sec. 6.2.3. In this section we discuss in more detail the construction and operation of these sensors and illustrate their use for proximity sensing.

Figure 6.14 shows the structure of a typical ultrasonic transducer used for proximity sensing. The basic element is an electroacoustic transducer, often of the piezoelectric ceramic type. The resin layer protects the transducer against humidity, dust, and other environmental factors; it also acts as an acoustical impedance matcher. Since the same transducer is generally used for both transmitting and receiving, fast damping of the acoustic energy is necessary to detect objects at close range. This is accomplished by providing acoustic absorbers, and by decoupling the transducer from its housing. The housing is designed so that it produces a narrow acoustic beam for efficient energy transfer and signal directionality.

The operation of an ultrasonic proximity sensor is best understood by analyzing the waveforms used for both transmission and detection of the acoustic energy signals. A typical set of waveforms is shown in Fig. 6.15. Waveform A is the gating signal used to control transmission. Waveform B shows the output signal as well as the resulting echo signal. The pulses shown in C result either upon transmission or reception. In order to differentiate between pulses corresponding to outgoing and returning energy, we introduce a time window (waveform D) which in essence establishes the detection capability of the sensor. That is, time interval Δt_1 is the minimum detection time, and $\Delta t_1 + \Delta t_2$ the maximum. (It is noted that these time intervals are equivalent to specifying distances since the pro-

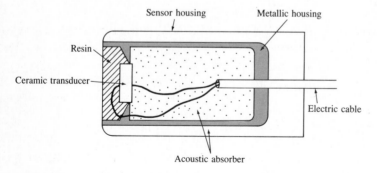

Figure 6.14 An ultrasonic proximity sensor. (Adapted from Canali [1981*b*], © Elsevier Sequoia.)

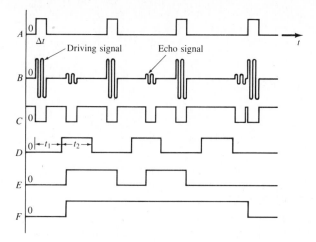

Figure 6.15 Waveforms associated with an ultrasonic proximity sensor. (Adapted from Canali [1981*b*], © Elsevier Sequoia.)

pagation velocity of an acoustic wave is known given the transmission medium.) An echo received while signal D is high produces the signal shown in E, which is reset to low at the end of a transmission pulse in signal A. Finally, signal F is set high on the positive edge of a pulse in E and is reset to low when E is low *and* a pulse occurs in A. In this manner, F will be high whenever an object is present in the distance interval specified by the parameters of waveform D. That is, F is the output of interest in an ultrasonic sensor operating in a binary mode.

6.3.5 Optical Proximity Sensors

Optical proximity sensors are similar to ultrasonic sensors in the sense that they detect proximity of an object by its influence on a propagating wave as it travels from a transmitter to a receiver. One of the most common approaches for detecting proximity by optical means is shown in Fig. 6.16. This sensor consists of a solid-state light-emitting diode (LED), which acts as a transmitter of infrared light, and a solid-state photodiode which acts as the receiver. The cones of light formed by focusing the source and detector on the same plane intersect in a long, pencil-like volume. This volume defines the field of operation of the sensor since a reflective surface which intersects the volume is illuminated by the source and simultaneously "seen" by the receiver.

Although this approach is similar in principle to the triangulation method discussed in Sec. 6.2.1, it is important to note that the detection volume shown in Fig. 6.16 does not yield a point measurement. In other words, a surface located *anywhere* in the volume will produce a reading. While it is possible to calibrate the intensity of these readings as a function of distance for known object orientations and reflective characteristics, the typical application of the arrangement

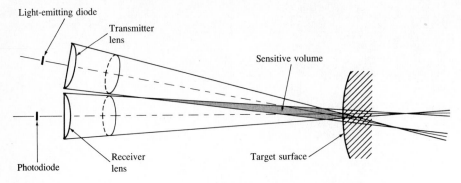

Figure 6.16 Optical proximity sensor. (From Rosen and Nitzan [1977], © IEEE.)

shown in Fig. 6.16 is in a mode where a binary signal is generated when the received light intensity exceeds a threshold value.

6.4 TOUCH SENSORS

Touch sensors are used in robotics to obtain information associated with the contact between a manipulator hand and objects in the workspace. Touch information can be used, for example, for object location and recognition, as well as to control the force exerted by a manipulator on a given object. Touch sensors can be subdivided into two principal categories: binary and analog. Binary sensors are basically switches which respond to the presence or absence of an object. Analog sensors, on the other hand, output a signal proportional to a local force. These devices are discussed in more detail in the following sections.

6.4.1 Binary Sensors

As indicated above, binary touch sensors are contact devices, such as microswitches. In the simplest arrangement, a switch is placed on the inner surface of each finger of a manipulator hand, as illustrated in Fig. 6.17. This type of sensing is useful for determining if a part is present between the fingers. By moving the hand over an object and sequentially making contact with its surface, it is also possible to center the hand over the object for grasping and manipulation.

Multiple binary touch sensors can be used on the inside surface of each finger to provide further tactile information. In addition, they are often mounted on the external surfaces of a manipulator hand to provide control signals useful for guiding the hand throughout the work space. This latter use of touch sensing is analogous to what humans do in feeling their way in a totally dark room.

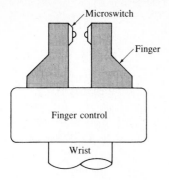

Figure 6.17 A simple robot hand equipped with binary touch sensors.

6.4.2 Analog Sensors

An analog touch sensor is a compliant device whose output is proportional to a local force. The simplest of these devices consists of a spring-loaded rod (Fig. 6.18) which is mechanically linked to a rotating shaft in such a way that the displacement of the rod due to a lateral force results in a proportional rotation of the shaft. The rotation is then measured continuously using a potentiometer or digitally using a code wheel. Knowledge of the spring constant yields the force corresponding to a given displacement.

During the past few years, considerable effort has been devoted to the development of tactile sensing arrays capable of yielding touch information over a wider area than that afforded by a single sensor. The use of these devices is illustrated in Fig. 6.19, which shows a robot hand in which the inner surface of each

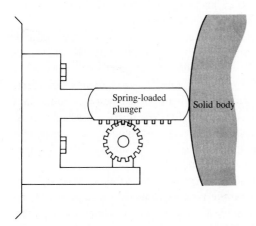

Figure 6.18 A basic analog touch sensor.

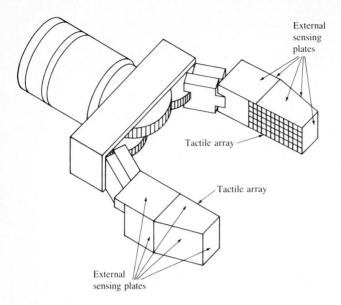

Figure 6.19 A robot hand equipped with tactile sensing arrays.

finger has been covered with a tactile sensing array. The external sensing plates are typically binary devices and have the function described at the end of Sec. 6.4.1.

Although sensing arrays can be formed by using multiple individual sensors, one of the most promising approaches to this problem consists of utilizing an array of electrodes in electrical contact with a compliant conductive material (e.g., graphite-based substances) whose resistance varies as a function of compression. In these devices, often called *artificial skins*, an object pressing against the surface causes local deformations which are measured as continuous resistance variations. The latter are easily transformed into electrical signals whose amplitude is proportional to the force being applied at any given point on the surface of the material.

Several basic approaches used in the construction of artificial skins are shown in Fig. 6.20. The scheme shown in Fig. 6.20a is based on a "window" concept, characterized by a conductive material sandwiched between a common ground and an array of electrodes etched on a fiberglass printed-circuit board. Each electrode consists of a rectangular area (and hence the name *window*) which defines one touch point. Current flows from the common ground to the individual electrodes as a function of compression of the conductive material.

In the method shown in Fig. 6.20b long, narrow electrode pairs are placed in the same substrate plane with active electronic circuits using LSI technology. The conductive material is placed above this plane and insulated from the substrate plane, except at the electrodes. Resistance changes resulting from material compression are measured and interpreted by the active circuits located between the electrode pairs.

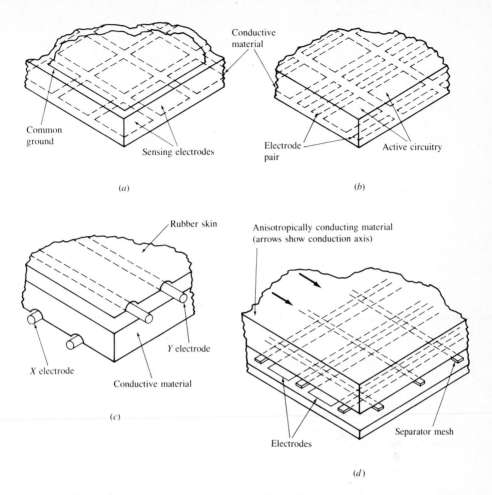

Figure 6.20 Four approaches for constructing artificial skins (see text).

Another possible technique is shown in Fig. 6.20c. In this approach the conductive material is located between two arrays of thin, flat, flexible electrodes that intersect perpendicularly. Each intersection, and the conductive material in between, constitutes one sensing point. Changes in resistance as a function of material compression are measured by electrically driving the electrodes of one array (one at a time) and measuring the current flowing in the elements of the other array. The magnitude of the current in each of these elements is proportional to the compression of the material between that element and the element being driven externally.

Finally, the arrangement shown in Fig. 6.20d requires the use of an anisotropically conductive material. Such materials have the property of being electrically conductive in only one direction. The sensor is constructed by using a linear array of thin, flat electrodes in the base. The conductive material is placed on top of this, with the conduction axis perpendicular to the electrodes and separated from them by a mesh so that there is no contact between the material and electrodes in the absence of a force. Application of sufficient force results in contact between the material and electrodes. As the force increases so does the contact area, resulting in lower resistance. As with the method in Fig. 6.20c, one array is externally driven and the resulting current is measured in the other. It is noted that touch sensitivity depends on the thickness of the separator.

The methods in Fig. 6.20c and d are based on sequentially driving the elements of one of the arrays. This often leads to difficulties in interpreting signals resulting from complex touch patterns because of "cross-point" inductions caused by alternate electrical paths. One solution is to place a diode element at each intersection to eliminate current flow through the alternate paths. Another method is to ground all paths, except the one being driven. By scanning the receiving array one path at a time, we are basically able to "look" at the contribution of the individual element intersections.

All the touch sensors discussed thus far deal with measurements of forces normal to the sensor surface. The measurement of tangential motion to determine slip is another important aspect of touch sensing. Before leaving this section, we illustrate this mode of sensing by describing briefly a method proposed by Bejczy [1980] for sensing both the direction and magnitude of slip. The device, illustrated in Fig. 6.21, consists of a free-moving dimpled ball which deflects a thin rod mounted on the axis of a conductive disk. A number of electrical contacts are

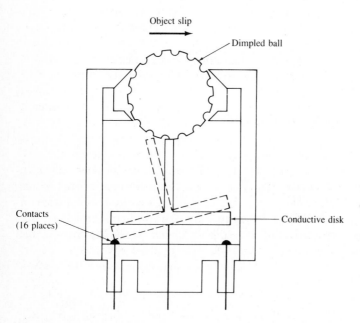

Figure 6.21 A device for sensing the magnitude and direction of slip. (Adapted from Bejczy [1980], © AAAS.)

evenly spaced under the disk. Ball rotation resulting from an object slipping past the ball causes the rod and disk to vibrate at a frequency which is proportional to the speed of the ball. The direction of ball rotation determines which of the contacts touch the disk as it vibrates, pulsing the corresponding electrical circuits and thus providing signals that can be analyzed to determine the average direction of the slip.

6.5 FORCE AND TORQUE SENSING

Force and torque sensors are used primarily for measuring the reaction forces developed at the interface between mechanical assemblies. The principal approaches for doing this are *joint* and *wrist* sensing.† A joint sensor measures the cartesian components of force and torque acting on a robot joint and adds them vectorially. For a joint driven by a dc motor, sensing is done simply by measuring the armature current. Wrist sensors, the principal topic of discussion in this section, are mounted between the tip of a robot arm and the end-effector. They consist of strain gauges that measure the deflection of the mechanical structure due to external forces. The characteristics and analysis methodology for this type of sensor are summarized in the following discussion.

6.5.1 Elements of a Wrist Sensor

Wrist sensors are small, sensitive, light in weight (about 12 oz) and relatively compact in design—on the order of 10 cm in total diameter and 3 cm in thickness, with a dynamic range of up to 200 lb. In order to reduce hysteresis and increase the accuracy in measurement, the hardware is generally constructed from one solid piece of metal, typically aluminum. As an example, the sensor shown in Fig. 6.22 uses eight pairs of semiconductor strain gauges mounted on four deflection bars—one gauge on each side of a deflection bar. The gauges on the opposite open ends of the deflection bars are wired differentially to a potentiometer circuit whose output voltage is proportional to the force component normal to the plane of the strain gauge. The differential connection of the strain gauges provides automatic compensation for variations in temperature. However, this is only a crude first-order compensation. Since the eight pairs of strain gauges are oriented normal to the **x**, **y**, and **z** axes of the force coordinate frame, the three components of force **F** and three components of moment **M** can be determined by properly adding and subtracting the output voltages, respectively. This can be done by premultiplying the sensor reading by a sensor calibration matrix, as discussed in Sec. 6.5.2.

---

† Another category is *pedestal* sensing, in which strain gauge transducers are installed between the base of a robot and its mounting surface in order to measure the components of force and torque acting on the base. In most applications, however, the base is firmly mounted on a solid surface and no provisions are made for pedestal sensing. The analysis of pedestal sensing is quite similar to that used for wrist sensing, which is discussed in detail in this section.

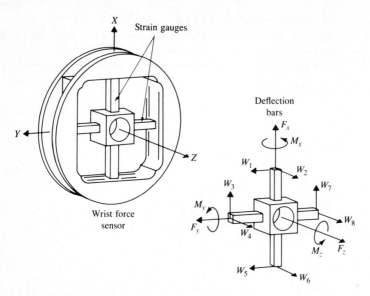

Figure 6.22 Wrist force sensor.

Most wrist force sensors function as transducers for transforming forces and moments exerted at the hand into measurable deflections or displacements at the wrist. It is important that the wrist motions generated by the force sensor do not affect the positioning accuracy of the manipulator. Thus, the required performance specifications can be summarized as follows:

1. *High stiffness.* The natural frequency of a mechanical device is related to its stiffness; thus, high stiffness ensures that disturbing forces will be quickly damped out to permit accurate readings during short time intervals. Furthermore, it reduces the magnitude of the deflections of an applied force/moment, which may add to the positioning error of the hand.
2. *Compact design.* This ensures that the device will not restrict the movement of the manipulator in a crowded workspace. It also minimizes collisions between the sensor and the other objects present in the workspace. With the compact force sensor, it is important to place the sensor as close to the tool as possible to reduce positioning error as a result of the hand rotating through small angles. In addition, it is desirable to measure as large a hand force/moment as possible; thus, minimizing the distance between the hand and the sensor reduces the lever arm for forces applied at the hand.
3. *Linearity.* Good linearity between the response of force sensing elements and the applied forces/moments permits resolving the forces and moments by simple matrix operations. Furthermore, the calibration of the force sensor is simplified. This is discussed in Sec. 6.5.2.

4. *Low hysteresis and internal friction.* Internal friction reduces the sensitivity of the force sensing elements because forces have to overcome this friction before a measurable deflection can be produced. It also produces hysteresis effects that do not restore the position measuring devices back to their original readings.

The wrist force sensor shown in Fig. 6.22 was designed with these criteria taken into consideration.

6.5.2 Resolving Forces and Moments

Assume that the coupling effects between the gauges are negligible, that the wrist force sensor is operating within the elastic range of its material, and that the strain gauges produce readings which vary linearly with respect to changes in their elongation. Then the sensor shown in Fig. 6.22 produces eight raw readings which can be resolved by computer software, using a simple force-torque balance technique, into three orthogonal force and torque components with reference to the force sensor coordinate frame. Such a transformation can be realized by specifying a 6×8 matrix, called the resolved force matrix $\mathbf{R}_F$ (or sensor calibration matrix), which is postmultiplied by the force measurements to produce the required three orthogonal force and three torque components. With reference to Fig. 6.22, the resolved force vector directed along the force sensor coordinates can be obtained mathematically as

$$\mathbf{F} = \mathbf{R}_F \mathbf{W} \tag{6.5-1}$$

where

$$\mathbf{F} \equiv (\text{forces, moments})^T = (F_x, F_y, F_z, M_x, M_y, M_z)^T$$

$$\mathbf{W} \equiv \text{raw readings} = (w_1, w_2, w_3, \ldots, w_8)^T$$

and

$$\mathbf{R}_F = \begin{bmatrix} r_{11} & \cdot & \cdot & \cdot & r_{18} \\ \cdot & \cdot & \cdot & \cdot & \cdot \\ r_{61} & \cdot & \cdot & \cdot & r_{68} \end{bmatrix} \tag{6.5-2}$$

In Eq. (6.5-2), the $r_{ij} \neq 0$ are the factors required for conversion from the raw reading $\mathbf{W}$ (in volts) to force/moment (in newton-meters). If the coupling effects between the gauges are negligible, then by looking at Fig. 6.22 and summing the forces and moments about the origin of the sensor coordinate frame located at the center of the force sensor, we can obtain the above equation with some of the r_{ij}

equal to zero. With reference to Fig. 6.22, the resolved force matrix in Eq. (6.5-2) becomes

$$
\mathbf{R}_F =
\begin{bmatrix}
0 & 0 & r_{13} & 0 & 0 & 0 & r_{17} & 0 \\
r_{21} & 0 & 0 & 0 & r_{25} & 0 & 0 & 0 \\
0 & r_{32} & 0 & r_{34} & 0 & r_{36} & 0 & r_{38} \\
0 & 0 & 0 & r_{44} & 0 & 0 & 0 & r_{48} \\
0 & r_{52} & 0 & 0 & 0 & r_{56} & 0 & 0 \\
r_{61} & 0 & r_{63} & 0 & r_{65} & 0 & r_{67} & 0
\end{bmatrix}
\tag{6.5-3}
$$

Quite often, this assumption is not valid and some coupling does exist. For some force sensors, this may produce as much as 5 percent error in the calculation of force resolution. Thus, in practice, it is usually necessary to replace the resolved force matrix $\mathbf{R}_F$ by a matrix which contains 48 nonzero elements. This "full" matrix is used to calibrate the force sensor, as discussed in Sec. 6.5.3. The resolved force vector $\mathbf{F}$ is used to generate the necessary error actuating control signal for the manipulator. The disadvantage of using a wrist force sensor is that it only provides force vectors resolved at the assembly interface for a single contact.

6.5.3 Sensor Calibration

The objective of calibrating the wrist force sensor is to determine all the 48 unknown elements in the resolved force matrix [Eq. (6.5-2)], based on experimental data. Due to the coupling effects, we need to find the full 48 nonzero elements in the $\mathbf{R}_F$ matrix. The calibration of the wrist force sensor is done by finding a pseudoinverse calibration matrix $\mathbf{R}_F^*$ which satisfies

$$
\mathbf{W} = \mathbf{R}_F^* \mathbf{F}
\tag{6.5-4}
$$

and

$$
\mathbf{R}_F^* \mathbf{R}_F \cong \mathbf{I}_{8 \times 8}
\tag{6.5-5}
$$

where $\mathbf{R}_F^*$ is an 8×6 matrix and $\mathbf{I}_{8 \times 8}$ is an 8×8 identity matrix. Then the calibration matrix $\mathbf{R}_F$ from Eq. (6.5-1) can be found from the pseudoinverse of $\mathbf{R}_F^*$ in Eq. (6.5-4) using a least-squares-fit technique. Premultiplying Eq. (6.5-4) by $(\mathbf{R}_F^*)^T$, we have

$$
(\mathbf{R}_F^*)^T \mathbf{W} = [(\mathbf{R}_F^*)^T \mathbf{R}_F^*] \mathbf{F}
\tag{6.5-6}
$$

Inverting the matrix $[(\mathbf{R}_F^*)^T \mathbf{R}_F^*]$ yields

$$
\mathbf{F} = [(\mathbf{R}_F^*)^T \mathbf{R}_F^*]^{-1} (\mathbf{R}_F^*)^T \mathbf{W}
\tag{6.5-7}
$$

Therefore comparing Eq. (6.5-1) and Eq. (6.5-7), we have

$$\mathbf{R}_F \cong [(\mathbf{R}_F^*)^T \mathbf{R}_F^*]^{-1} (\mathbf{R}_F^*)^T \tag{6.5-8}$$

The $\mathbf{R}_F^*$ matrix can be identified by placing known weights along the axes of the sensor coordinate frame. Details about the experimental procedure for calibrating the resolved force matrix can be found in a paper by Shimano and Roth [1979].

6.6 CONCLUDING REMARKS

The material presented in this chapter is representative of the state of the art in external robot sensors. It must be kept in mind, however, that the performance of these sensors is still rather primitive when compared with human capabilities.

As indicated at the beginning of this chapter, the majority of present industrial robots perform their tasks using preprogrammed techniques and without the aid of sensory feedback. The relatively recent widespread interest in flexible automation, however, has led to increased efforts in the area of sensor-driven robotic systems as a means of increasing the scope of application of these machines. Thus, sensor development is indeed a dynamic field where new techniques and applications are commonplace in the literature. For this reason, the topics included in this chapter were selected primarily for their value as fundamental material which would serve as a foundation for further study of this and related fields.

REFERENCES

Several survey articles on robotic sensing are Rosen and Nitzan [1977], Bejczy [1980], Galey and Hsia [1980], McDermott [1980], and Merritt [1982]. Further reading on laser range finders may be found in Duda et al. [1979] and Jarvis [1983a, 1983b]. For further reading on the material in Sec. 6.3 see Spencer [1980], Catros and Espiau [1980], and Canali et al. [1981a, 1981b]. Further reading for the material in Sec. 6.4 may be found in Harmon [1982], Hillis [1982], Marck [1981], and Raibert and Tanner [1982]. See also the papers by Beni et al. [1983], McDermott [1980], and Hackwood et al. [1983]. Pedestal sensors (Sec. 6.5) are discussed by Drake [1977]. Additional reading on force-torque sensing may be found in Nevins and Whitney [1978], Shimano and Roth [1979], Meindl and Wise [1979], and Wise [1982].

PROBLEMS

6.1 Show the validity of Eq. (6.2-8).

6.2 A sheet-of-light range sensor illuminating a work space with two objects produced the following output on a television screen:

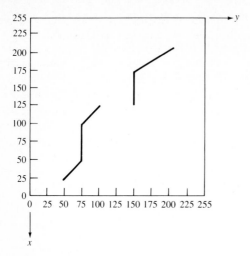

Assuming that the ranging system is set up as in Fig. 6.3*b* with $M = 256$, $D_0 = 1$ m, $D_c = 2$ m, $B = 3$ m, and $\lambda = 35$ mm, obtain the distance between the objects in the direction of the light sheet.

6.3 (*a*) A helium-neon (wavelength 632.8 nm) continuous-beam laser range finder is modulated with a 30-MHz sine wave. What is the distance to an object that produces a phase shift of 180°? (*b*) What is the upper limit on the distance for which this device would produce a unique reading?

6.4 Compute the upper limit on the frequency of a modulating sine wave to achieve a working distance of up to (but not including) 5 m using a continuous-beam laser range finder.

6.5 (*a*) Suppose that the accuracy of a laser range finder is corrupted by noise with a gaussian distribution of mean 0 and standard deviation of 100 cm. How many measurements would have to be averaged to obtain an accuracy of ± 0.5 cm with a .95 probability? (*b*) If, instead of being 0, the mean of the noise were 5 cm, how would you compensate the range measurements for this effect?

6.6 With reference to Fig. 6.15, give a set of waveforms for an ultrasonic sensor capable of measuring range instead of just yielding a binary output associated with proximity.

6.7 Suppose that an ultrasonic proximity sensor is used to detect the presence of objects within 0.5 m of the device. At time $t = 0$ the transducer is pulsed for 0.1 ms. Assume that it takes 0.4 ms for resonances to die out within the transducer and 20 ms for echoes in the environment to die out. Given that sound travels at 344 m/s: (*a*) What range of time should be used as a window? (*b*) At what time can the device be pulsed again? (*c*) What is the minimum detectable distance?

6.8 An optical proximity sensor (Fig. 6.16) has a sensitive volume formed by the intersection of two identical beams. The cone formed by each beam originates at the lens and has a vertex located 4 cm in front of the center of the opposite lens. Given that each lens has a diameter of 4 mm, and that the lens centers are 6 mm apart, over what approximate range will this sensor detect an object? Assume that an object is detected anywhere in the sensitive volume.

6.9 A 3×3 touch array is scanned by driving the rows (one at a time) with 5 V. A column is read by holding it at ground and measuring the current. Assume that the undriven rows and unread columns are left in high impedance. A given force pattern against the array results in the following resistances at each electrode intersection (row, column): 100 Ω at (1, 1), (1, 3), (3, 1), and (3, 3); and 50 Ω at (2, 2) and (3, 2). All other intersections have infinite resistance. Compute the current measured at each row-column intersection in the array, taking into account the cross-point problem.

6.10 Repeat Prob. 6.9 assuming (*a*) that all undriven rows and all columns are held at ground; and (*b*) that a diode (0.6 voltage drop) is in series with the resistance at each junction.

6.11 A wrist force sensor is mounted on a PUMA robot equipped with a parallel jaw gripper and a sensor calibration procedure has been performed to obtain the calibration matrix $\mathbf{R}_F$. Unfortunately, after you have performed all the measurements, someone remounts a different gripper on the robot. Do you need to recalibrate the wrist force sensor? Justify your answer.

SEVEN

LOW-LEVEL VISION

Where there is no vision, the people perish.
Proverbs

7.1 INTRODUCTION

As is true in humans, vision capabilities endow a robot with a sophisticated sensing mechanism that allows the machine to respond to its environment in an "intelligent" and flexible manner. The use of vision and other sensing schemes, such as those discussed in Chap. 6, is motivated by the continuing need to increase the flexibility and scope of applications of robotic systems. While proximity, touch, and force sensing play a significant role in the improvement of robot performance, vision is recognized as the most powerful of robot sensory capabilities. As might be expected, the sensors, concepts, and processing hardware associated with machine vision are considerably more complex than those associated with the sensory approaches discussed in Chap. 6.

Robot vision may be defined as the process of extracting, characterizing, and interpreting information from images of a three-dimensional world. This process, also commonly referred to as *machine* or *computer vision*, may be subdivided into six principal areas: (1) sensing, (2) preprocessing, (3) segmentation, (4) description, (5) recognition, and (6) interpretation. Sensing is the process that yields a visual image. Preprocessing deals with techniques such as noise reduction and enhancement of details. Segmentation is the process that partitions an image into objects of interest. Description deals with the computation of features (e.g., size, shape) suitable for differentiating one type of object from another. Recognition is the process that identifies these objects (e.g., wrench, bolt, engine block). Finally, interpretation assigns meaning to an ensemble of recognized objects.

It is convenient to group these various areas according to the sophistication involved in their implementation. We consider three levels of processing: low-, medium-, and high-level vision. While there are no clearcut boundaries between these subdivisions, they do provide a useful framework for categorizing the various processes that are inherent components of a machine vision system. For instance, we associate with low-level vision those processes that are primitive in the sense that they may be considered "automatic reactions" requiring no intelligence on the part of the vision system. In our discussion, we shall treat sensing and preprocessing as low-level vision functions. This will take us from the image formation process itself to compensations such as noise reduction, and finally to the extraction of primitive image features such as intensity discontinuities. This range of processes may be compared with the sensing and adaptation process a human goes through in trying to find a seat in a dark theater immediately after walking in during a bright afternoon. The intelligent process of finding an unoccupied space cannot begin until a suitable image is available.

We will associate with medium-level vision those processes that extract, characterize, and label components in an image resulting from low-level vision. In terms of our six subdivisions, we will treat segmentation, description, and recognition of individual objects as medium-level vision functions. High-level vision refers to processes that attempt to emulate cognition. While algorithms for low- and medium-level vision encompass a reasonably well-defined spectrum of activities, our knowledge and understanding of high-level vision processes is considerably more vague and speculative. As discussed in Chap. 8, these limitations lead to the formulation of constraints and idealizations intended to reduce the complexity of this task.

The categories and subdivisions discussed above are suggested to a large extent by the way machine vision systems are generally implemented. It is not implied that these subdivisions represent a model of human vision nor that they are carried out independently of each other. We know, for example, that recognition and interpretation are highly interrelated functions in a human. These relationships, however, are not yet understood to the point where they can be modeled analytically. Thus, the subdivision of functions addressed in this discussion may be viewed as a practical approach for implementing state-of-the-art machine vision systems, given our level of understanding and the analytical tools currently available in this field.

The material in this chapter deals with sensing, preprocessing, and with concepts and techniques required to implement low-level vision functions. Topics in higher-level vision are discussed in Chap. 8. Although true vision is inherently a three-dimensional activity, most of the work in machine vision is carried out using images of a three-dimensional scene, with depth information being obtained by special imaging techniques, such as the structured-lighting approach discussed in Sec. 7.3, or by the use of stereo imaging, as discussed in Sec. 7.4.

7.2 IMAGE ACQUISITION

Visual information is converted to electrical signals by visual sensors. When sampled spatially and quantized in amplitude, these signals yield a *digital image*. In this section we are interested in three main topics: (1) the principal imaging techniques used for robotic vision, (2) the effects of sampling on spatial resolution, and (3) the effects of amplitude quantization on intensity resolution. The mathematics of image formation are discussed in Sec. 7.4.

The principal devices used for robotic vision are television cameras, consisting either of a tube or solid-state imaging sensor, and associated electronics. Although an in-depth treatment of these devices is beyond the scope of the present discussion, we will consider the principles of operation of the vidicon tube, a commonly used representative of the tube family of TV cameras. Solid-state imaging sensors will be introduced via a brief discussion of charge-coupled devices (CCDs), which are one of the principal exponents of this technology. Solid-state imaging devices offer a number of advantages over tube cameras, including lighter weight, smaller size, longer life, and lower power consumption. However, the resolution of certain tubes is still beyond the capabilities of solid-state cameras.

As shown schematically in Fig. 7.1a, the vidicon camera tube is a cylindrical glass envelope containing an electron gun at one end, and a faceplate and target at the other. The beam is focused and deflected by voltages applied to the coils shown in Fig. 7.1a. The deflection circuit causes the beam to scan the inner surface of the target in order to "read" the image, as explained below. The inner surface of the glass faceplate is coated with a transparent metal film which forms an electrode from which an electrical video signal is derived. A thin photosensi-

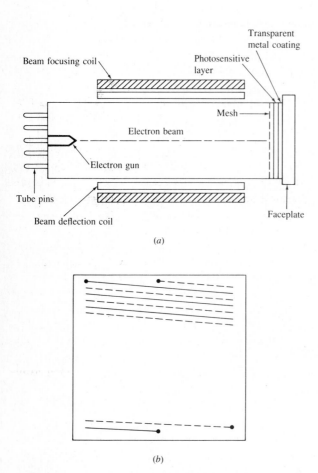

Figure 7.1 (a) Schematic of a vidicon tube. (b) Electron beam scanning pattern.

tive "target" layer is deposited onto the metal film; this layer consists of very small resistive globules whose resistance is inversely proportional to light intensity. Behind the photosensitive target there is a positively charged fine wire mesh which decelerates electrons emitted by the gun so that they reach the target surface with essentially zero velocity.

In normal operation, a positive voltage is applied to the metal coating of the faceplate. In the absence of light, the photosensitive material behaves as a dielectric, with the electron beam depositing a layer of electrons on the inner surface of the target surface to balance the positive charge on the metal coating. As the electron beam scans the surface of the target layer, the photosensitive layer thus becomes a capacitor with negative charge on the inner surface and positive charge on the other side. When light strikes the target layer, its resistance is reduced and electrons are allowed to flow and neutralize the positive charge. Since the amount of electronic charge that flows is proportional to the amount of light in any local area of the target, this effect produces an image on the target layer that is identical to the light image on the faceplate of the tube; that is, the remaining concentration of electron charge is high in dark areas and lower in light areas. As the beam again scans the target it replaces the lost charge, thus causing a current to flow in the metal layer and out one of the tube pins. This current is proportional to the number of electrons replaced and, therefore, to the light intensity at a particular location of the scanning beam. This variation in current during the electron beam scanning motion produces, after conditioning by the camera circuitry, a video signal proportional to the intensity of the input image.

The principal scanning standard used in the United States is shown in Fig. 7.1b. The electron beam scans the entire surface of the target 30 times per second, each complete scan (called a *frame*) consisting of 525 lines of which 480 contain image information. If the lines were scanned sequentially and the result shown on a TV monitor, the image would flicker perceptibly. This phenomenon is avoided by using a scan mechanism in which a frame is divided into two *interlaced fields*, each consisting of 262.5 lines and scanned 60 times each second, or twice the frame rate. The first field of each frame scans the odd lines (shown dashed in Fig. 7.1a), while the second field scans the even lines. This scanning scheme, called the RETMA (Radio-Electronics-Television Manufacturers Association) scanning convention, is the standard used for broadcast television in the United States. Other standards exist which yield higher line rates per frame, but their principle of operation is essentially the same. For example, a popular scanning approach in computer vision and digital image processing is based on 559 lines, of which 512 contain image data. Working with integer powers of 2 has a number of advantages for both hardware and software implementations.

When discussing CCD devices, it is convenient to subdivide sensors into two categories: line scan sensors and area sensors. The basic component of a line scan CCD sensor is a row of silicon imaging elements called *photosites*. Image photons pass through a transparent polycrystalline silicon gate structure and are absorbed in the silicon crystal, thus creating electron-hole pairs. The resulting photoelectrons are collected in the photosites, with the amount of charge collected at each photosite being proportional to the illumination intensity at that location. As shown in Fig. 7.2*a*, a typical line scan sensor is composed of a row of the imaging elements just discussed, two transfer gates used to clock the contents of the imaging elements into so-called transport registers, and an output gate used to clock the contents of the transport registers into an amplifier whose output is a voltage signal proportional to the contents of the row of photosites.

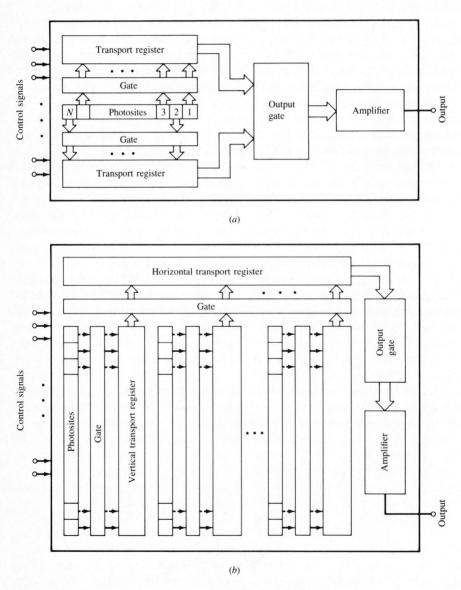

Figure 7.2 (*a*) CCD line scan sensor. (*b*) CCD area sensor.

Charge-coupled area arrays are similar to the line scan sensors, with the exception that the photosites are arranged in a matrix format and there is a gate-transport register combination between columns of photosites, as shown in Fig. 7.2*b*. The contents of odd-numbered photosites are sequentially gated into the vertical transport registers and then into the horizontal transport register. The content of this register is fed into an amplifier whose output is a line of video. Repeating this procedure for the even-numbered lines completes the second field of a TV frame. This "scanning" mechanism is repeated 30 times per second.

Line scan cameras obviously yield only one line of an input image. These devices are ideally suited for applications in which objects are moving past the sensor (as in conveyor belts). The motion of an object in the direction perpendicular to the sensor produces a two-dimensional image. Line scan sensors with resolutions ranging between 256 and 2048 elements are not uncommon. The resolutions of area sensors range between 32 × 32 at the low end to 256 × 256 elements for a medium resolution sensor. Higher-resolution devices presently in the market have a resolution on the order of 480 × 380 elements, and experimental CCD sensors are capable of achieving a resolution of 1024 × 1024 elements or higher.

Throughout this book, we will use $f(x, y)$ to denote the two-dimensional image out of a TV camera or other imaging device, where x and y denote spatial (i.e., image plane) coordinates, and the value of f at any point (x, y) is proportional to the brightness (intensity) of the image at that point. Figure 7.3 illustrates this concept, as well as the coordinate convention on which all subsequent discussions will be based. We will often use the variable z to denote intensity variations in an image when the spatial location of these variations is of no interest.

In order to be in a form suitable for computer processing, an image function $f(x, y)$ must be digitized both spatially and in amplitude (intensity). Digitization of the spatial coordinates (x, y) will be referred to as *image sampling*, while amplitude digitization will be called *intensity* or *gray-level quantization*. The latter term is applicable to monochrome images and reflects the fact that these images vary from black to white in shades of gray. The terms intensity and gray level will be used interchangeably.

Suppose that a continuous image is sampled uniformly into an array of N rows and M columns, where each sample is also quantized in intensity. This array, called a *digital image*, may be represented as

$$
f(x, y) = \begin{bmatrix}
f(0, 0) & f(0, 1) & \cdots & f(0, M-1) \\
f(1, 0) & f(1, 1) & \cdots & f(1, M-1) \\
\cdots\cdots\cdots\cdots\cdots\cdots\cdots\cdots\cdots\cdots\cdots \\
f(N-1, 0) & f(N-1, 1) & \cdots & f(N-1, M-1)
\end{bmatrix}
\qquad (7.2\text{-}1)
$$

where x and y are now discrete variables: $x = 0, 1, 2, \ldots, N-1$; $y = 0, 1, 2, \ldots, M-1$. Each element in the array is called an *image element, picture element*, or *pixel*. With reference to Fig. 7.3, it is noted that $f(0, 0)$ represents the pixel at the origin of the image, $f(0, 1)$ the pixel to its right, and so

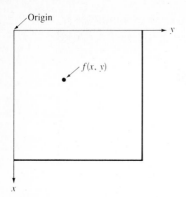

Figure 7.3 Coordinate convention for image representation. The value of any point (x, y) is given by the value (intensity) of f at that point.

on. It is common practice to let N, M, and the number of discrete intensity levels of each quantized pixel be integer powers of 2.

In order to gain insight into the effect of sampling and quantization, consider Fig. 7.4. Part (a) of this figure shows an image sampled into an array of $N \times N$ pixels with $N = 512$; the intensity of each pixel is quantized into one of 256 discrete levels. Figure 7.4b to e shows the same image, but with $N = 256$, 128, 64, and 32. In all cases the number of allowed intensity levels was kept at 256. Since the display area used for each image was the same (512×512 display points), pixels in the lower resolution images were duplicated in order to fill the entire display field. This produced a checkerboard effect that is particularly visible in the low-resolution images. It is noted that the 256×256 image is reasonably close to Fig. 7.4a, but image quality deteriorated rapidly for the other values of N.

Figure 7.5 illustrates the effect produced by reducing the number of intensity levels while keeping the spatial resolution constant at 512×512. The 256-, 128-, and 64-level images are of acceptable quality. The 32-level image, however, shows a slight degradation (particularly in areas of nearly constant intensity) as a result of using too few intensity levels to represent each pixel. This effect is considerably more visible as ridgelike structures (called *false contours*) in the image displayed with 16 levels, and increases sharply thereafter.

The number of samples and intensity levels required to produce a useful (in the machine vision sense) reproduction of an original image depends on the image itself and on the intended application. As a basis for comparison, the requirements to obtain quality comparable to that of monochrome TV pictures are on the order of 512×512 pixels with 128 intensity levels. As a rule, a minimum system for general-purpose vision work should have spatial resolution capabilities on the order of 256×256 pixels with 64 levels.

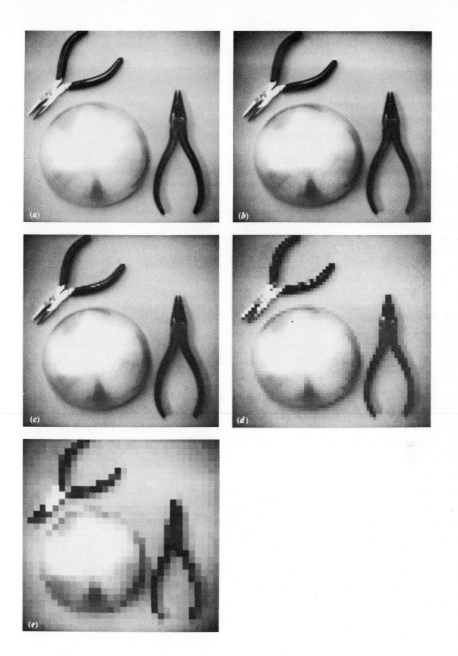

Figure 7.4 Effects of reducing sampling-grid size. (*a*) 512 × 512. (*b*) 256 × 256. (*c*) 128 × 128. (*d*) 64 × 64. (*e*) 32 × 32.

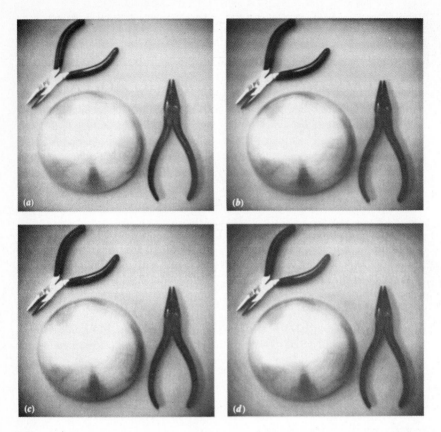

Figure 7.5 A 512 × 512 image displayed with 256, 128, 64, 32, 16, 8, 4, and 2 levels.

7.3 ILLUMINATION TECHNIQUES

Illumination of a scene is an important factor that often affects the complexity of vision algorithms. Arbitrary lighting of the environment is often not acceptable because it can result in low-contrast images, specular reflections, shadows, and extraneous details. A well-designed lighting system illuminates a scene so that the complexity of the resulting image is minimized, while the information required for object detection and extraction is enhanced.

Figure 7.6 shows four of the principal schemes used for illuminating a robot work space. The diffuse-lighting approach shown in Fig. 7.6a can be employed for objects characterized by smooth, regular surfaces. This lighting scheme is generally employed in applications where surface characteristics are important. An example is shown in Fig. 7.7. Backlighting, as shown in Fig. 7.6b, produces a black and white (binary) image. This technique is ideally suited for applications in which silhouettes of objects are sufficient for recognition or other measurements. An example is shown in Fig. 7.8.

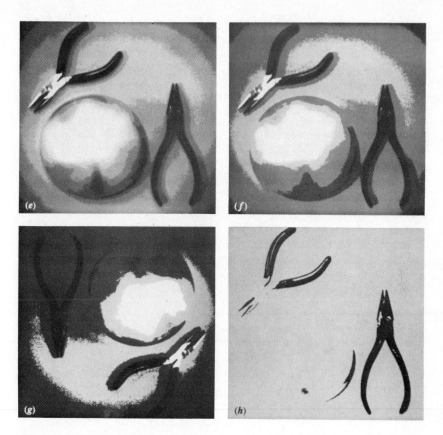

Figure 7.5 (continued)

The structured-lighting approach shown in Fig. 7.6c consists of projecting points, stripes, or grids onto the work surface. This lighting technique has two important advantages. First, it establishes a known light pattern on the work space, and disturbances of this pattern indicate the presence of an object, thus simplifying the object detection problem. Second, by analyzing the way in which the light pattern is distorted, it is possible to gain insight into the three-dimensional characteristics of the object. Two examples of the structured-lighting approach are shown in Fig. 7.9. The first shows a block illuminated by parallel light planes which become light stripes upon intersecting a flat surface. The example shown in Fig. 7.9b consists of two light planes projected from different directions, but converging on a single stripe on the surface, as shown in Fig. 7.10a. A line scan camera, located above the surface and focused on the stripe would see a continuous line of light in the absence of an object. This line would be interrupted by an object which breaks both light planes simultaneously. This particular approach is ideally suited for objects moving on a conveyor belt past the camera. As shown in Fig. 7.10b, two light sources are used to guarantee that the object will break the

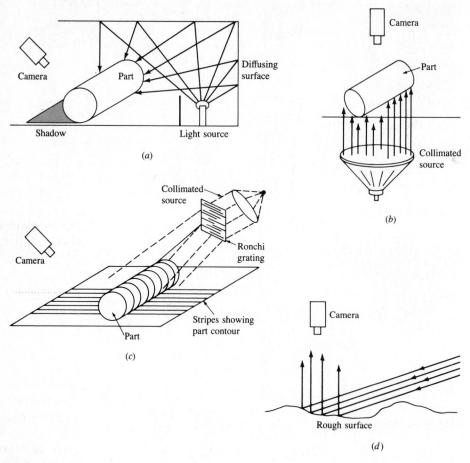

Figure 7.6 Four basic illumination schemes. (From Mundy [1977], © IEEE.)

light stripe only when it is directly below the camera. It is of interest to note that the line scan camera sees *only* the line on which the two light planes converge, but two-dimensional information can be accumulated as the object moves past the camera.

The directional-lighting approach shown in Fig. 7.6*d* is useful primarily for inspection of object surfaces. Defects on the surface, such as pits and scratches, can be detected by using a highly directed light beam (e.g., a laser beam) and measuring the amount of scatter. For flaw-free surfaces little light is scattered upward to the camera. On the other hand, the presence of a flaw generally increases the amount of light scattered to the camera, thus facilitating detection of a defect. An example is shown in Fig. 7.11.

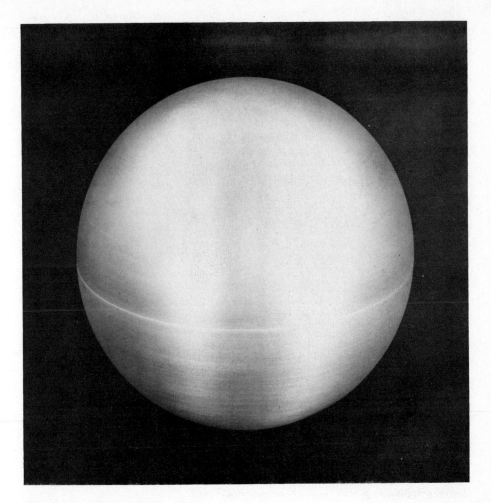

Figure 7.7 Example of diffuse lighting.

7.4 IMAGING GEOMETRY

In the following discussion we consider several important transformations used in imaging, derive a camera model, and treat the stereo imaging problem in some detail. Some of the transformations discussed in the following section were already introduced in Chap. 2 in connection with robot arm kinematics. Here, we consider a similar problem, but from the point of view of imaging.

7.4.1 Some Basic Transformations

The material in this section deals with the development of a unified representation for problems such as image rotation, scaling, and translation. All transformations

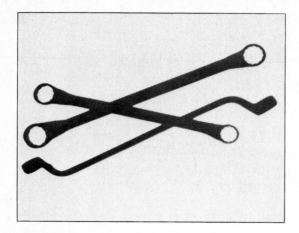

Figure 7.8 Example of backlighting.

are expressed in a three-dimensional (3D) cartesian coordinate system in which a point has coordinates denoted by (X, Y, Z). In cases involving two-dimensional images, we will adhere to our previous convention of using the lowercase representation (x, y) to denote the coordinates of a pixel. It is common terminology to refer to (X, Y, Z) as the *world coordinates* of a point.

Translation. Suppose that we wish to translate a point with coordinates (X, Y, Z) to a new location by using displacements (X_0, Y_0, Z_0). The translation is easily accomplished by using the following equations:

$$
\begin{aligned}
X^* &= X + X_0 \\
Y^* &= Y + Y_0 \\
Z^* &= Z + Z_0
\end{aligned}
\tag{7.4.1}
$$

where (X^*, Y^*, Z^*) are the coordinates of the new point. Equation (7.4-1) can be expressed in matrix form by writing:

$$
\begin{bmatrix} X^* \\ Y^* \\ Z^* \end{bmatrix} =
\begin{bmatrix} 1 & 0 & 0 & X_0 \\ 0 & 1 & 0 & Y_0 \\ 0 & 0 & 1 & Z_0 \end{bmatrix}
\begin{bmatrix} X \\ Y \\ Z \\ 1 \end{bmatrix}
\tag{7.4-2}
$$

As indicated later in this section, it is often useful to concatenate several transformations to produce a composite result, such as translation, followed by scaling, and then rotation. The notational representation of this process is

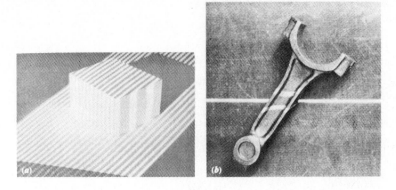

Figure 7.9 Two examples of structured lighting. (Part (*a*) is from Rocher and Keissling [1975], © Kaufmann, Inc.; part (*b*) is from Myers [1980], © IEEE.)

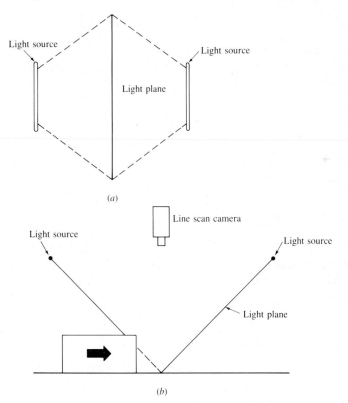

Figure 7.10 (*a*) Top view of two light planes intersecting in a line of light. (*b*) Object will be seen by the camera only when it interrupts both light planes. (Adapted from Holland [1979], © Plenum.)

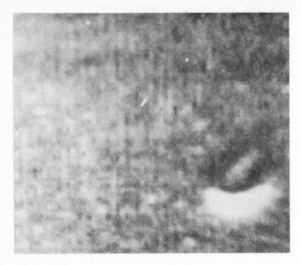

Figure 7.11 Example of directional lighting. (From Mundy [1977], © IEEE.)

simplified considerably by using square matrices. With this in mind, we write Eq. (7.4-2) in the following form:

$$
\begin{bmatrix} X^* \\ Y^* \\ Z^* \\ 1 \end{bmatrix} = \begin{bmatrix} 1 & 0 & 0 & X_0 \\ 0 & 1 & 0 & Y_0 \\ 0 & 0 & 1 & Z_0 \\ 0 & 0 & 0 & 1 \end{bmatrix} \begin{bmatrix} X \\ Y \\ Z \\ 1 \end{bmatrix} \tag{7.4-3}
$$

In terms of the values of X^*, Y^*, and Z^*, Eqs. (7.4-2) and (7.4-3) are clearly equivalent.

Throughout this section, we will use the unified matrix representation

$$
v^* = Av \tag{7.4-4}
$$

where **A** is a 4×4 transformation matrix, **v** is a column vector containing the original coordinates:

$$
v = \begin{bmatrix} X \\ Y \\ Z \\ 1 \end{bmatrix} \tag{7.4-5}
$$

and $\mathbf{v}^*$ is a column vector whose components are the transformed coordinates:

$$
\mathbf{v}^* =
\begin{bmatrix}
X^* \\
Y^* \\
Z^* \\
1
\end{bmatrix}
\tag{7.4-6}
$$

Using this notation, the matrix used for translation is given by

$$
\mathbf{T} =
\begin{bmatrix}
1 & 0 & 0 & X_0 \\
0 & 1 & 0 & Y_0 \\
0 & 0 & 1 & Z_0 \\
0 & 0 & 0 & 1
\end{bmatrix}
\tag{7.4-7}
$$

and the translation process is accomplished by using Eq. (7.4-4), so that $\mathbf{v}^* = \mathbf{Tv}$.

Scaling. Scaling by factors S_x, S_y, and S_z along the X, Y, and Z axes is given by the transformation matrix

$$
\mathbf{S} =
\begin{bmatrix}
S_x & 0 & 0 & 0 \\
0 & S_y & 0 & 0 \\
0 & 0 & S_z & 0 \\
0 & 0 & 0 & 1
\end{bmatrix}
\tag{7.4-8}
$$

Rotation. The transformations used for three-dimensional rotation are inherently more complex than the transformations discussed thus far. The simplest form of these transformations is for rotation of a point about the coordinate axes. To rotate a given point about an arbitrary point in space requires three transformations: The first translates the arbitrary point to the origin, the second performs the rotation, and the third translates the point back to its original position.

With reference to Fig. 7.12, rotation of a point about the Z coordinate axis by an angle θ is achieved by using the transformation

$$
\mathbf{R}_\theta =
\begin{bmatrix}
\cos\theta & \sin\theta & 0 & 0 \\
-\sin\theta & \cos\theta & 0 & 0 \\
0 & 0 & 1 & 0 \\
0 & 0 & 0 & 1
\end{bmatrix}
\tag{7.4-9}
$$

The rotation angle θ is measured clockwise when looking at the origin from a point on the $+Z$ axis. It is noted that this transformation affects only the values of X and Y coordinates.

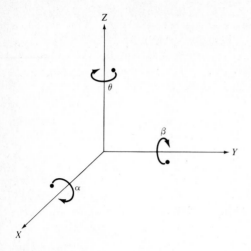

Figure 7.12 Rotation of a point about each of the coordinate axes. Angles are measured clockwise when looking along the rotation axis toward the origin.

Rotation of a point about the X axis by an angle α is performed by using the transformation

$$
\mathbf{R}_\alpha = \begin{bmatrix} 1 & 0 & 0 & 0 \\ 0 & \cos\alpha & \sin\alpha & 0 \\ 0 & -\sin\alpha & \cos\alpha & 0 \\ 0 & 0 & 0 & 1 \end{bmatrix}
\tag{7.4-10}
$$

Finally, rotation of a point about the Y axis by an angle β is achieved by using the transformation

$$
\mathbf{R}_\beta = \begin{bmatrix} \cos\beta & 0 & -\sin\beta & 0 \\ 0 & 1 & 0 & 0 \\ \sin\beta & 0 & \cos\beta & 0 \\ 0 & 0 & 0 & 1 \end{bmatrix}
\tag{7.4-11}
$$

Concatenation and Inverse Transformations. The application of several transformations can be represented by a single 4×4 transformation matrix. For example, translation, scaling, and rotation about the Z axis of a point $\mathbf{v}$ is given by

$$
\mathbf{v}^* = \mathbf{R}_\theta[\mathbf{S}(\mathbf{Tv})] = \mathbf{Av}
\tag{7.4-12}
$$

where $\mathbf{A}$ is the 4×4 matrix $\mathbf{A} = \mathbf{R}_\theta\mathbf{S}\mathbf{T}$. It is important to note that these matrices generally do not commute, and so the order of application is important.

Although our discussion thus far has been limited to transformations of a single point, the same ideas extend to transforming a set of m points simultaneously by using a single transformation. With reference to Eq. (7.4-5), let $v_1, v_2, \ldots, v_m$ represent the coordinates of m points. If we form a $4 \times m$ matrix V whose columns are these column vectors, then the simultaneous transformation of all these points by a 4×4 transformation matrix A is given by

$$V^* = AV \tag{7.4-13}$$

The resulting matrix V^* is $4 \times m$. Its ith column, v_i^*, contains the coordinates of the transformed point corresponding to v_i.

Before leaving this section, we point out that many of the transformations discussed above have inverse matrices that perform the opposite transformation and can be obtained by inspection. For example, the inverse translation matrix is given by

$$\mathbf{T}^{-1} = \begin{bmatrix} 1 & 0 & 0 & -X_0 \\ 0 & 1 & 0 & -Y_0 \\ 0 & 0 & 1 & -Z_0 \\ 0 & 0 & 0 & 1 \end{bmatrix} \tag{7.4-14}$$

Similarly, the inverse rotation matrix $\mathbf{R}_\theta^{-1}$ is given by

$$\mathbf{R}_\theta^{-1} = \begin{bmatrix} \cos(-\theta) & \sin(-\theta) & 0 & 0 \\ -\sin(-\theta) & \cos(-\theta) & 0 & 0 \\ 0 & 0 & 1 & 0 \\ 0 & 0 & 0 & 1 \end{bmatrix} \tag{7.4-15}$$

The inverse of more complex transformation matrices is usually obtained by numerical techniques.

7.4.2 Perspective Transformations

A *perspective transformation* (also called an *imaging transformation*) projects 3D points onto a plane. Perspective transformations play a central role in image processing because they provide an approximation to the manner in which an image is formed by viewing a three-dimensional world. Although perspective transformations will be expressed later in this section in a 4×4 matrix form, these transformations are fundamentally different from those discussed in the previous section because they are nonlinear in the sense that they involve division by coordinate values.

A model of the image formation process is shown in Fig. 7.13. We define the camera coordinate system (x, y, z) as having the image plane coincident with the xy plane, and optical axis (established by the center of the lens) along the z

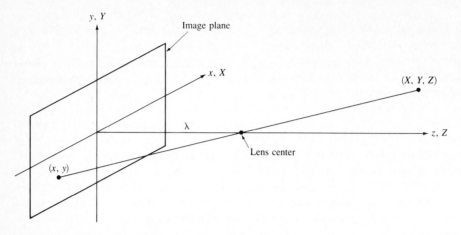

Figure 7.13 Basic model of the imaging process. The camera coordinate system (x, y, z) is aligned with the world coordinate system (X, Y, Z).

axis. Thus, the center of the image plane is at the origin, and the center of the lens is at coordinates $(0, 0, \lambda)$. If the camera is in focus for distant objects, λ is the *focal length* of the lens. In this section, it is assumed that the camera coordinate system is aligned with the world coordinate system (X, Y, Z). This restriction will be removed in the following section.

Let (X, Y, Z) be the world coordinates of any point in a 3D scene, as shown in Fig. 7.13. It will be assumed throughout the following discussion that $Z > \lambda$, that is, all points of interest lie in front of the lens. What we wish to do first is obtain a relationship that gives the coordinates (x, y) of the projection of the point (X, Y, Z) onto the image plane. This is easily accomplished by the use of similar triangles. With reference to Fig. 7.13, it follows that

$$\frac{x}{\lambda} = - \frac{X}{Z - \lambda} = \frac{X}{\lambda - Z} \tag{7.4-16}$$

and

$$\frac{y}{\lambda} = - \frac{Y}{Z - \lambda} = \frac{Y}{\lambda - Z} \tag{7.4-17}$$

where the negative signs in front of X and Y indicate that image points are actually inverted, as can be seen from the geometry of Fig. 7.13.

The image-plane coordinates of the projected 3D point follow directly from Eqs. (7.4-16) and (7.4-17):

$$x = \frac{\lambda X}{\lambda - Z} \tag{7.4-18}$$

and

$$y = \frac{\lambda Y}{\lambda - Z} \tag{7.4-19}$$

It is important to note that these equations are nonlinear because they involve division by the variable Z. Although we could use them directly as shown above, it is often convenient to express these equations in matrix form as we did in the previous section for rotation, translation, and scaling. This can be accomplished easily by using homogeneous coordinates.

The homogeneous coordinates of a point with cartesian coordinates (X, Y, Z) are defined as (kX, kY, kZ, k), where k is an arbitrary, nonzero constant. Clearly, conversion of homogeneous coordinates back to cartesian coordinates is accomplished by dividing the first three homogeneous coordinates by the fourth. A point in the cartesian world coordinate system may be expressed in vector form as

$$\mathbf{w} = \begin{bmatrix} X \\ Y \\ Z \end{bmatrix} \tag{7.4-20}$$

and its homogeneous counterpart is given by

$$\mathbf{w}_h = \begin{bmatrix} kX \\ kY \\ kZ \\ k \end{bmatrix} \tag{7.4-21}$$

If we define the *perspective transformation matrix*

$$\mathbf{P} = \begin{bmatrix} 1 & 0 & 0 & 0 \\ 0 & 1 & 0 & 0 \\ 0 & 0 & 1 & 0 \\ 0 & 0 & -\dfrac{1}{\lambda} & 1 \end{bmatrix} \tag{7.4-22}$$

Then the product $\mathbf{P}\mathbf{w}_h$ yields a vector which we shall denote by $\mathbf{c}_h$:

$$\mathbf{c}_h = \mathbf{P}\mathbf{w}_h = \begin{bmatrix} 1 & 0 & 0 & 0 \\ 0 & 1 & 0 & 0 \\ 0 & 0 & 1 & 0 \\ 0 & 0 & -\dfrac{1}{\lambda} & 1 \end{bmatrix} \begin{bmatrix} kX \\ kY \\ kZ \\ k \end{bmatrix} = \begin{bmatrix} kX \\ kY \\ kZ \\ \dfrac{-kZ}{\lambda} + k \end{bmatrix} \tag{7.4-23}$$

The elements of $\mathbf{c}_h$ are the camera coordinates in homogeneous form. As indicated above, these coordinates can be converted to cartesian form by dividing each of the first three components of c_h by the fourth. Thus, the cartesian coordinates of any point in the camera coordinate system are given in vector form by

$$
c = \begin{bmatrix} x \\ y \\ z \end{bmatrix} = \begin{bmatrix} \dfrac{\lambda X}{\lambda - Z} \\[2ex] \dfrac{\lambda Y}{\lambda - Z} \\[2ex] \dfrac{\lambda Z}{\lambda - Z} \end{bmatrix}
\tag{7.4-24}
$$

The first two components of c are the (x, y) coordinates in the image plane of a projected 3D point (X, Y, Z), as shown earlier in Eqs. (7.4-18) and (7.4-19). The third component is of no interest to us in terms of the model in Fig. 7.13. As will be seen below, this component acts as a free variable in the inverse perspective transformation.

The inverse perspective transformation maps an image point back into 3D. Thus, from Eq. (7.4-23),

$$
w_h = P^{-1} c_h
\tag{7.4-25}
$$

where P^{-1} is easily found to be

$$
P^{-1} = \begin{bmatrix} 1 & 0 & 0 & 0 \\ 0 & 1 & 0 & 0 \\ 0 & 0 & 1 & 0 \\ 0 & 0 & \dfrac{1}{\lambda} & 1 \end{bmatrix}
\tag{7.4-26}
$$

Suppose that a given image point has coordinates $(x_0, y_0, 0)$, where the 0 in the z location simply indicates the fact that the image plane is located at $z = 0$. This point can be expressed in homogeneous vector form as

$$
c_h = \begin{bmatrix} k x_0 \\ k y_0 \\ 0 \\ k \end{bmatrix}
\tag{7.4-27}
$$

Application of Eq. (7.4-25) then yields the homogeneous world coordinate vector

$$
w_h = \begin{bmatrix} k x_0 \\ k y_0 \\ 0 \\ k \end{bmatrix}
\tag{7.4-28}
$$

or, in cartesian coordinates,

$$
\mathbf{w} = \begin{bmatrix} X \\ Y \\ Z \end{bmatrix} = \begin{bmatrix} x_0 \\ y_0 \\ 0 \end{bmatrix} \tag{7.4-29}
$$

This is obviously not what one would expect since it gives $Z = 0$ for *any* 3D point. The problem here is caused by the fact that mapping a 3D scene onto the image plane is a many-to-one transformation. The image point (x_0, y_0) corresponds to the set of colinear 3D points which lie on the line that passes through $(x_0, y_0, 0)$ and $(0, 0, \lambda)$. The equations of this line in the world coordinate system are obtained from Eqs. (7.4-18) and (7.4-19); that is,

$$
X = \frac{x_0}{\lambda}(\lambda - Z) \tag{7.4-30}
$$

and

$$
Y = \frac{y_0}{\lambda}(\lambda - Z) \tag{7.4-31}
$$

These equations show that, unless we know something about the 3D point which generated a given image point (for example, its Z coordinate), we cannot completely recover the 3D point from its image. This observation, which is certainly not unexpected, can be used as a way to formulate the inverse perspective transformation simply by using the z component of $\mathbf{c}_h$ as a free variable instead of 0. Thus, letting

$$
\mathbf{c}_h = \begin{bmatrix} kx_0 \\ ky_0 \\ kz \\ k \end{bmatrix} \tag{7.4-32}
$$

we now have from Eq. (7.4-25) that

$$
\mathbf{w}_h = \begin{bmatrix} kx_0 \\ ky_0 \\ kz \\ \dfrac{kz}{\lambda} + k \end{bmatrix} \tag{7.4-33}
$$

which, upon conversion to cartesian coordinates, yields

$$\mathbf{w} = \begin{bmatrix} X \\ Y \\ Z \end{bmatrix} = \begin{bmatrix} \dfrac{\lambda x_0}{\lambda + z} \\[2mm] \dfrac{\lambda y_0}{\lambda + z} \\[2mm] \dfrac{\lambda z}{\lambda + z} \end{bmatrix} \qquad (7.4\text{-}34)$$

In other words, treating z as a free variable yields the equations

$$X = \frac{\lambda x_0}{\lambda + z}$$

$$Y = \frac{\lambda y_0}{\lambda + z} \qquad (7.4\text{-}35)$$

$$Z = \frac{\lambda z}{\lambda + z}$$

Solving for z in terms of Z in the last equation and substituting in the first two expressions yields

$$X = \frac{x_0}{\lambda}(\lambda - Z) \qquad (7.4\text{-}36)$$

$$Y = \frac{y_0}{\lambda}(\lambda - Z) \qquad (7.4\text{-}37)$$

which agrees with the above observation that recovering a 3D point from its image by means of the inverse perspective transformation requires knowledge of at least one of the world coordinates of the point. This problem will be addressed again in Sec. 7.4.5.

7.4.3 Camera Model

Equations (7.4-23) and (7.4-24) characterize the formation of an image via the projection of 3D points onto an image plane. These two equations thus constitute a basic mathematical model of an imaging camera. This model is based on the assumption that the camera and world coordinate systems are coincident. In this section we consider a more general problem in which the two coordinate systems are allowed to be separate. However, the basic objective of obtaining the image-plane coordinates of any given world point remains the same.

The situation is depicted in Fig. 7.14, which shows a world coordinate system (X, Y, Z) used to locate both the camera and 3D points (denoted by $\mathbf{w}$). This

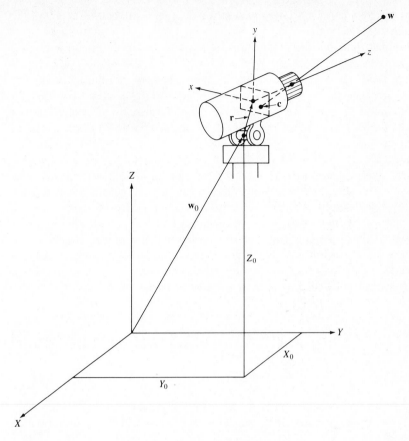

Figure 7.14 Imaging geometry with two coordinate systems.

figure also shows the camera coordinate system (x, y, z) and image points (denoted by $\mathbf{c}$). It is assumed that the camera is mounted on a gimbal which allows pan through an angle θ and tilt through an angle α. In this discussion, pan is defined as the angle between the x and X axes, and tilt as the angle between the z and Z axes. The offset of the center of the gimbal from the origin of the world coordinate system is denoted by vector $\mathbf{w}_0$, and the offset of the center of the imaging plane with respect to the gimbal center is denoted by a vector $\mathbf{r}$, with components (r_1, r_2, r_3).

The concepts developed in the last two sections provide all the necessary tools to derive a camera model based on the geometrical arrangement of Fig. 7.14. The approach is to bring the camera and world coordinate systems into alignment by applying a set of transformations. After this has been accomplished, we simply apply the perspective transformation given in Eq. (7.4-22) to obtain the image-plane coordinates of any given world point. In other words, we first reduce the problem to the geometrical arrangement shown in Fig. 7.13 before applying the perspective transformation.

Suppose that, initially, the camera was in *normal position*, in the sense that the gimbal center and origin of the image plane were at the origin of the world coordinate system, and all axes were aligned. Starting from normal position, the geometrical arrangement of Fig. 7.14 can be achieved in a number of ways. We assume the following sequence of steps: (1) displacement of the gimbal center from the origin, (2) pan of the x axis, (3) tilt of the z axis, and (4) displacement of the image plane with respect to the gimbal center.

The sequence of *mechanical* steps just discussed obviously does not affect the world points since the set of points seen by the camera after it was moved from normal position is quite different. However, we can achieve normal position again simply by applying exactly the same sequence of steps to all world points. Since a camera in normal position satisfies the arrangement of Fig. 7.13 for application of the perspective transformation, our problem is thus reduced to applying to every world point a set of transformations which correspond to the steps given above.

Translation of the origin of the world coordinate system to the location of the gimbal center is accomplished by using the following transformation matrix:

$$\mathbf{G} = \begin{bmatrix} 1 & 0 & 0 & -X_0 \\ 0 & 1 & 0 & -Y_0 \\ 0 & 0 & 1 & -Z_0 \\ 0 & 0 & 0 & 1 \end{bmatrix} \tag{7.4-38}$$

In other words, a homogeneous world point $\mathbf{w}_h$ that was at coordinates (X_0, Y_0, Z_0) is at the origin of the new coordinate system after the transformation $\mathbf{Gw}_h$.

As indicated earlier, the pan angle is measured between the x and X axes. In normal position, these two axes are aligned. In order to pan the x axis through the desired angle, we simply rotate it by θ. The rotation is with respect to the z axis and is accomplished by using the transformation matrix $\mathbf{R}_\theta$ given in Eq. (7.4-9). In other words, application of this matrix to all points (including the point $\mathbf{Gw}_h$) effectively rotates the x axis to the desired location. When using Eq. (7.4-9), it is important to keep clearly in mind the convention established in Fig. 7.12. That is, angles are considered positive when points are rotated clockwise, which implies a counterclockwise rotation of the camera about the z axis. The unrotated ($0°$) position corresponds to the case when the x and X axes are aligned.

At this point in the development the z and Z axes are still aligned. Since tilt is the angle between these two axes, we tilt the camera an angle α by rotating the z axis by α. The rotation is with respect to the x axis and is accomplished by applying the transformation matrix $\mathbf{R}_\alpha$ given in Eq. (7.4-10) to all points (including the point $\mathbf{R}_\theta \mathbf{Gw}_h$). As above, a counterclockwise rotation of the camera implies positive angles, and the $0°$ mark is where the z and Z axes are aligned.†

---

† A useful way to visualize these transformations is to construct an axis system (e.g., with pipe cleaners), label the axes x, y, and z, and perform the rotations manually, one axis at a time.

According to the discussion in Sec. 7.4.4, the two rotation matrices can be concatenated into a single matrix, $\mathbf{R} = \mathbf{R}_\alpha \mathbf{R}_\theta$. It then follows from Eqs. (7.4-9) and (7.4-10) that

$$
\mathbf{R} = \begin{bmatrix}
\cos\theta & \sin\theta & 0 & 0 \\
-\sin\theta\cos\alpha & \cos\theta\cos\alpha & \sin\alpha & 0 \\
\sin\theta\sin\alpha & -\cos\theta\sin\alpha & \cos\alpha & 0 \\
0 & 0 & 0 & 1
\end{bmatrix}
\tag{7.4-39}
$$

Finally, displacement of the origin of the image plane by vector $\mathbf{r}$ is achieved by the transformation matrix

$$
\mathbf{C} = \begin{bmatrix}
1 & 0 & 0 & -r_1 \\
0 & 1 & 0 & -r_2 \\
0 & 0 & 1 & -r_3 \\
0 & 0 & 0 & 1
\end{bmatrix}
\tag{7.4-40}
$$

Thus, by applying to $\mathbf{w}_h$ the series of transformations $\mathbf{CRGw}_h$ we have brought the world and camera coordinate systems into coincidence. The image-plane coordinates of a point $\mathbf{w}_h$ are finally obtained by using Eq. (7.4-22). In other words, a homogeneous world point which is being viewed by a camera satisfying the geometrical arrangement shown in Fig. 7.14 has the following homogeneous representation in the camera coordinate system:

$$
\mathbf{c}_h = \mathbf{PCRGw}_h
\tag{7.4-41}
$$

This equation represents a perspective transformation involving two coordinate systems.

As indicated in Sec. 7.4.2, we obtain the cartesian coordinates (x, y) of the imaged point by dividing the first and second components of $\mathbf{c}_h$ by the fourth. Expanding Eq. (7.4-41) and converting to cartesian coordinates yields

$$
x = \lambda \frac{(X - X_0)\cos\theta + (Y - Y_0)\sin\theta - r_1}{-(X - X_0)\sin\theta\sin\alpha + (Y - Y_0)\cos\theta\sin\alpha - (Z - Z_0)\cos\alpha + r_3 + \lambda}
\tag{7.4-42}
$$

and

$$
y = \lambda \frac{-(X - X_0)\sin\theta\cos\alpha + (Y - Y_0)\cos\theta\cos\alpha + (Z - Z_0)\sin\alpha - r_2}{-(X - X_0)\sin\theta\sin\alpha + (Y - Y_0)\cos\theta\sin\alpha - (Z - Z_0)\cos\alpha + r_3 + \lambda}
\tag{7.4-43}
$$

which are the image coordinates of a point $\mathbf{w}$ whose world coordinates are (X, Y, Z). It is noted that these equations reduce to Eqs. (7.4-18) and (7.4-19) when $X_0 = Y_0 = Z_0 = 0$, $r_1 = r_2 = r_3 = 0$, and $\alpha = \theta = 0°$.

Example: As an illustration of the concepts just discussed, suppose that we wish to find the image coordinates of the corner of the block shown in Fig.

7.15. The camera is offset from the origin and is viewing the scene with a pan of 135° and a tilt of 135°. We will follow the convention established above that transformation angles are positive when the camera rotates in a counterclockwise manner when viewing the origin along the axis of rotation.

Let us examine in detail the steps required to move the camera from normal position to the geometry shown in Fig. 7.15. The camera is shown in normal position in Fig. 7.16a, and displaced from the origin in Fig. 7.16b. It is important to note that, after this step, the world coordinate axes are used *only* to establish angle references. That is, after displacement of the world-coordinate origin, all rotations take place about the new (camera) axes. Figure 7.16c shows a view along the z axis of the camera to establish pan. In this case the rotation of the camera about the z axis is counterclockwise so world points are rotated about this axis in the opposite direction, which makes θ a positive angle. Figure 7.16d shows a view after pan, along the x axis of the camera to establish tilt. The rotation about this axis is counterclockwise, which makes α a positive angle. The world coordinate axes are shown dashed in the latter two figures to emphasize the fact that their only use is to establish the zero reference for the pan and tilt angles. We do not show in this figure the final step of displacing the image plane from the center of the gimbal.

The following parameter values apply to the problem:

$$X_0 = 0 \, \text{m}$$

$$Y_0 = 0 \, \text{m}$$

$$Z_0 = 1 \, \text{m}$$

$$\alpha = 135°$$

$$\theta = 135°$$

$$r_1 = 0.03 \, \text{m}$$

$$r_2 = r_3 = 0.02 \, \text{m}$$

$$\lambda = 35 \, \text{mm} = 0.035 \, \text{m}$$

The corner in question is at coordinates $(X, Y, Z) = (1, 1, 0.2)$.

To compute the image coordinates of the block corner, we simply substitute the above parameter values into Eqs. (7.4-42) and (7.4-43); that is,

$$x = \lambda \frac{-0.03}{-1.53 + \lambda}$$

and

$$y = \lambda \frac{-0.42}{-1.53 + \lambda}$$

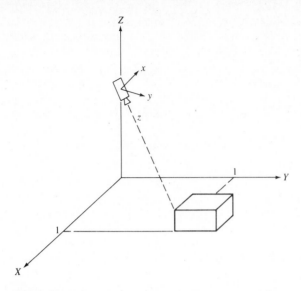

Figure 7.15 Camera viewing a 3D scene.

Substituting $\lambda = 0.035$ yields the image coordinates

$$x = 0.0007\,\text{m}$$

and
$$y = 0.009\,\text{m}$$

It is of interest to note that these coordinates are well within a 1×1 inch (0.025×0.025 m) imaging plane. If, for example, we had used a lens with a 200-mm focal length, it is easily verified from the above results that the corner of the block would have been imaged outside the boundary of a plane with these dimensions (i.e., it would have been outside the effective field of view of the camera).

Finally, we point out that all coordinates obtained via the use of Eqs. (7.4-42) and (7.4-43) are with respect to the center of the image plane. A change of coordinates would be required to use the convention established earlier, in which the origin of an image is at its top left corner. □

7.4.4 Camera Calibration

In Sec. 7.4.3 we obtained explicit equations for the image coordinates (x, y) of a world point $\mathbf{w}$. As shown in Eqs. (7.4-42) and (7.4-43), implementation of these equations requires knowledge of the focal length, camera offsets, and angles of pan and tilt. While these parameters could be measured directly, it is often more convenient (e.g., when the camera moves frequently) to determine one or more of the

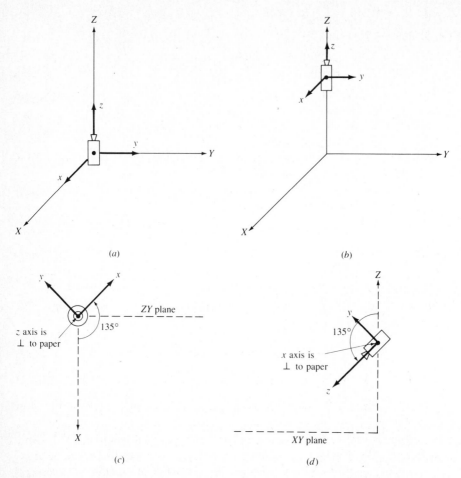

Figure 7.16 (*a*) Camera in normal position. (*b*) Gimbal center displaced from origin. (*c*) Observer view of rotation about *z* axis to determine pan angle. (*d*) Observer view of rotation about *x* axis for tilt.

parameters by using the camera itself as a measuring device. This requires a set of image points whose world coordinates are known, and the computational procedure used to obtain the camera parameters using these known points is often referred to as *camera calibration*.

With reference to Eq. (7.4-41), let $\mathbf{A} = \mathbf{PCRG}$. The elements of $\mathbf{A}$ contain all the camera parameters, and we know from Eq. (7.4-41) that $\mathbf{c}_h = \mathbf{Aw}_h$. Letting $k = 1$ in the homogeneous representation, we may write

$$
\begin{bmatrix} c_{h1} \\ c_{h2} \\ c_{h3} \\ c_{h4} \end{bmatrix} = \begin{bmatrix} a_{11} & a_{12} & a_{13} & a_{14} \\ a_{21} & a_{22} & a_{23} & a_{24} \\ a_{31} & a_{32} & a_{33} & a_{34} \\ a_{41} & a_{42} & a_{43} & a_{44} \end{bmatrix} \begin{bmatrix} X \\ Y \\ Z \\ 1 \end{bmatrix} \tag{7.4-44}
$$

From the discussion in the previous two sections we know that the camera coordinates in cartesian form are given by

$$x = \frac{c_{h1}}{c_{h4}} \qquad (7.4\text{-}45)$$

and

$$y = \frac{c_{h2}}{c_{h4}} \qquad (7.4\text{-}46)$$

Substituting $c_{h1} = xc_{h4}$ and $c_{h2} = yc_{h4}$ in Eq. (7.6-44) and expanding the matrix product yields

$$xc_{h4} = a_{11}X + a_{12}Y + a_{13}Z + a_{14}$$

$$yc_{h4} = a_{21}X + a_{22}Y + a_{23}Z + a_{24} \qquad (7.4\text{-}47)$$

$$c_{h4} = a_{41}X + a_{42}Y + a_{43}Z + a_{44}$$

where expansion of c_{h3} has been ignored because it is related to z.

Substitution of c_{h4} in the first two equations of (7.4-47) yields two equations with twelve unknown coefficients:

$$a_{11}X + a_{12}Y + a_{13}Z - a_{41}xX - a_{42}xY - a_{43}xZ - a_{44}x + a_{14} = 0 \qquad (7.4\text{-}48)$$

$$a_{21}X + a_{22}Y + a_{23}Z - a_{41}yX - a_{42}yY - a_{43}yZ - a_{44}y + a_{24} = 0 \qquad (7.4\text{-}49)$$

The calibration procedure then consists of (1) obtaining $m \geqslant 6$ world points with known coordinates (X_i, Y_i, Z_i), $i = 1, 2, \ldots, m$ (there are *two* equations involving the coordinates of these points, so at least six points are needed), (2) imaging these points with the camera in a given position to obtain the corresponding image points (x_i, y_i), $i = 1, 2, \ldots, m$, and (3) using these results in Eqs. (7.4-48) and (7.4-49) to solve for the unknown coefficients. There are many numerical techniques for finding an optimal solution to a linear system of equations such as (7.4-48) and (7.4-49) (see, for example, Noble [1969]).

7.4.5 Stereo Imaging

It was noted in Sec. 7.4.2 that mapping a 3D scene onto an image plane is a many-to-one transformation. That is, an image point does not uniquely determine the location of a corresponding world point. It is shown in this section that the missing *depth* information can be obtained by using stereoscopic (*stereo* for short) imaging techniques.

As shown in Fig. 7.17, stereo imaging involves obtaining two separate image views of an object of interest (e.g., a world point **w**). The distance between the centers of the two lenses is called the *baseline*, and the objective is to find the coordinates (X, Y, Z) of a point **w** given its image points (x_1, y_1) and (x_2, y_2). It is assumed that the cameras are identical and that the coordinate systems of both

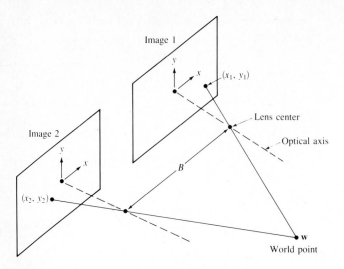

Figure 7.17 Model of the stereo imaging process.

cameras are perfectly aligned, differing only in the location of their origins, a condition usually met in practice. Recall our convention that, after the camera and world coordinate systems have been brought into coincidence, the xy plane of the image is aligned with the XY plane of the world coordinate system. Then, under the above assumption, the Z coordinate of $\mathbf{w}$ is exactly the same for both camera coordinate systems.

Suppose that we bring the first camera into coincidence with the world coordinate system, as shown in Fig. 7.18. Then, from Eq. (7.4-31), $\mathbf{w}$ lies on the line with (partial) coordinates

$$X_1 = \frac{x_1}{\lambda}(\lambda - Z_1) \tag{7.4-50}$$

where the subscripts on X and Z indicate that the first camera was moved to the origin of the world coordinate system, with the second camera and $\mathbf{w}$ following, but keeping the relative arrangement shown in Fig. 7.17. If, instead, the second camera had been brought to the origin of the world coordinate system, then we would have that $\mathbf{w}$ lies on the line with (partial) coordinates

$$X_2 = \frac{x_2}{\lambda}(\lambda - Z_2) \tag{7.4-51}$$

However, due to the separation between cameras and the fact that the Z coordinate of $\mathbf{w}$ is the same for both camera coordinate systems, it follows that

$$X_2 = X_1 + B \tag{7.4-52}$$

and

$$Z_2 = Z_1 = Z \tag{7.4-53}$$

where, as indicated above, B is the baseline distance.

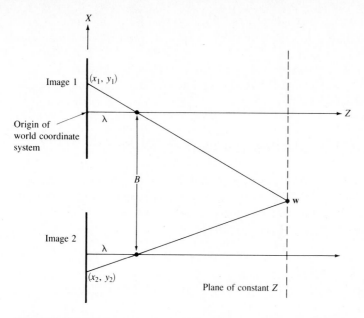

Figure 7.18 Top view of Fig. 7.17 with the first camera brought into coincidence with the world coordinate system.

Substitution of Eqs. (7.4-52) and (7.4-53) into Eqs. (7.4-50) and (7.4-51) results in the following equations:

$$X_1 + B = \frac{x_2}{\lambda} (\lambda - Z) \qquad (7.4\text{-}54)$$

and

$$X_1 = \frac{x_1}{\lambda} (\lambda - Z) \qquad (7.4\text{-}55)$$

Subtracting Eq. (7.4-55) from (7.4-54) and solving for Z yields the expression

$$Z = \lambda - \frac{\lambda B}{x_2 - x_1} \qquad (7.4\text{-}56)$$

which indicates that if the difference between the corresponding image coordinates x_2 and x_1 can be determined, and the baseline and focal length are known, calculating the Z coordinate of **w** is a simple matter. The X and Y world coordinates then follow directly from Eqs. (7.4-30) and (7.4-31) using either (x_1, y_1) or (x_2, y_2).

The most difficult task in using Eq. (7.4-56) to obtain Z is to actually find two corresponding points in different images of the same scene. Since these points are generally in the same vicinity, a frequently used approach is to select a point within a small region in one of the image views and then attempt to find the best matching region in the other view by using correlation techniques, as discussed in

Chap. 8. When the scene contains distinct features, such as prominent corners, a feature-matching approach will generally yield a faster solution for establishing correspondence.

Before leaving this discussion, we point out that the calibration procedure developed in the previous section is directly applicable to stereo imaging by simply treating the cameras independently.

7.5 SOME BASIC RELATIONSHIPS BETWEEN PIXELS

In this section we consider several primitive, but important relationships between pixels in a digital image. As in the previous sections, an image will be denoted by $f(x, y)$. When referring to a particular pixel, we will use lower-case letters, such as p and q. A subset of pixels of $f(x, y)$ will be denoted by S.

7.5.1 Neighbors of a Pixel

A pixel p at coordinates (x, y) has four *horizontal* and *vertical* neighbors whose coordinates are given by

$$(x + 1, y) \quad (x - 1, y) \quad (x, y + 1) \quad (x, y - 1)$$

This set of pixels, called the 4-*neighbors* of p, will be denoted by $N_4(p)$. It is noted that each of these pixels is a unit distance from (x, y) and also that some of the neighbors of p will be outside the digital image if (x, y) is on the border of the image.

The four *diagonal* neighbors of p have coordinates

$$(x + 1, y + 1) \quad (x + 1, y - 1) \quad (x - 1, y + 1) \quad (x - 1, y - 1)$$

and will be denoted $N_D(p)$. These points, together with the 4-neighbors defined above, are called the 8-*neighbors* of p, denoted $N_8(p)$. As before, some of the points in $N_D(p)$ and $N_8(p)$ will be outside the image if (x, y) is on the border to the image.

7.5.2 Connectivity

Let V be the set of intensity values of pixels which are allowed to be connected; for example, if only connectivity of pixels with intensities of 59, 60, and 61 is desired, then $V = \{59, 60, 61\}$. We consider three types of connectivity:

1. 4-*connectivity*. Two pixels p and q with values from V are 4-connected if q is in the set $N_4(p)$.
2. 8-*connectivity*. Two pixels p and q with values from V are 8-connected if q is in the set $N_8(p)$.
3. m-*connectivity* (mixed connectivity). Two pixels p and q with values from V are m-connected if

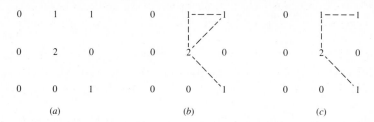

Figure 7.19 (*a*) Arrangement of pixels. (*b*) 8-neighbors of the pixel labeled "2." (*c*) m-neighbors of the same pixel.

(*a*) q is in $N_4(p)$, *or*

(*b*) q is in $N_D(p)$ *and* the set $N_4(p) \cap N_4(q)$ is empty. (This is the set of pixels that are 4-neighbors of both p and q and whose values are from V.)

Mixed connectivity is a modification of 8-connectivity and is introduced to eliminate the multiple connections which often cause difficulty when 8-connectivity is used. For example, consider the pixel arrangement shown in Fig. 7.19*a*. Assuming $V = \{1, 2\}$, the 8-neighbors of the pixel with value 2 are shown by dashed lines in Fig. 7.19*b*. It is important to note the ambiguity that results from multiple connections to this pixel. This ambiguity is removed by using m-connectivity, as shown in Fig. 7.19*c*.

A pixel p is *adjacent* to a pixel q if they are connected. We may define 4-, 8-, or m-adjacency, depending on the type of connectivity specified. Two image subsets S_1 and S_2 are adjacent if some pixel in S_1 is adjacent to some pixel in S_2.

A *path* from pixel p with coordinates (x, y) to pixel q with coordinates (s, t) is a sequence of distinct pixels with coordinates

$$(x_0, y_0), (x_1, y_1), \ldots, (x_n, y_n)$$

where $(x_0, y_0) = (x, y)$ and $(x_n, y_n) = (s, t)$, (x_i, y_i) is adjacent to (x_{i-1}, y_{i-1}), $1 \leqslant i \leqslant n$, and n is the *length* of the path. We may define 4-, 8-, or m-paths, depending on the type of adjacency used.

If p and q are pixels of an image subset S, then p is *connected* to q in S if there is a path from p to q consisting entirely of pixels in S. For any pixel p in S, the set of pixels in S that are connected to p is called a *connected component* of S. It then follows that any two pixels of a connected component are connected to each other, and that distinct connected components are disjoint.

7.5.3 Distance Measures

Given pixels p, q, and z, with coordinates (x, y), (s, t), and (u, v), respectively, we call D a *distance function* or *metric* if

1. $D(p, q) \geqslant 0 \quad [D(p, q) = 0 \text{ iff } p = q]$
2. $D(p, q) = D(q, p)$
3. $D(p, z) \leqslant D(p, q) + D(q, z)$

The *euclidean distance* between p and q is defined as

$$D_e(p, q) = [(x - s)^2 + (y - t)^2]^{1/2} \qquad (7.5\text{-}1)$$

For this distance measure, the pixels having a distance less than or equal to some value r from (x, y) are the points contained in a disk of radius r centered at (x, y).

The D_4 *distance* (also called *city-block distance*) between p and q is defined as

$$D_4(p, q) = |x - s| + |y - t| \qquad (7.5\text{-}2)$$

In this case the pixels having a D_4 distance less than or equal to some value r from (x, y) form a diamond centered at (x, y). For example, the pixels with D_4 distance $\leqslant 2$ from (x, y) (the center point) form the following contours of constant distance:

$$
\begin{array}{ccccc}
 & & 2 & & \\
 & 2 & 1 & 2 & \\
2 & 1 & 0 & 1 & 2 \\
 & 2 & 1 & 2 & \\
 & & 2 & & \\
\end{array}
$$

It is noted that the pixels with $D_4 = 1$ are the 4-neighbors of (x, y).

The D_8 *distance* (also called *chessboard distance*) between p and q is defined as

$$D_8(p, q) = \max(|x - s|, |y - t|) \qquad (7.5\text{-}3)$$

In this case the pixels with D_8 distance less than or equal to some value r form a square centered at (x, y). For example, the pixels with D_8 distance $\leqslant 2$ from (x, y) (the center point) form the following contours of constant distance:

$$
\begin{array}{ccccc}
2 & 2 & 2 & 2 & 2 \\
2 & 1 & 1 & 1 & 2 \\
2 & 1 & 0 & 1 & 2 \\
2 & 1 & 1 & 1 & 2 \\
2 & 2 & 2 & 2 & 2 \\
\end{array}
$$

The pixels with $D_8 = 1$ are the 8-neighbors of (x, y).

It is of interest to note that the D_4 distance between two points p and q is equal to the length of the shortest 4-path between these two points. Similar comments apply to the D_8 distance. In fact, we can consider both the D_4 and D_8 distances between p and q regardless of whether or not a connected path exists between them, since the definition of these distances involve only the coordinates of these points. When dealing with m-connectivity, however, the value of the distance (length of the path) between two pixels depends on the values of the pixels along the path as well as their neighbors. For instance, consider the following arrangement of pixels, where it is assumed that p, p_2, and p_4 are valued 1 and p_1 and p_3 may be valued 0 or 1:

$$p_3 \quad p_4$$
$$p_1 \quad p_2$$
$$p$$

If we only allow connectivity of pixels valued 1, and p_1 and p_3 are 0, the m-distance between p and p_4 is 2. If either p_1 or p_3 is 1, the distance is 3. If both p_1 and p_3 are 1, the distance is 4.

7.6 PREPROCESSING

In this section we discuss several preprocessing approaches used in robotic vision systems. Although the number of techniques available for preprocessing general image data is significant, only a subset of these methods satisfies the requirements of computational speed and low implementation cost, which are essential elements of an industrial vision system. The range of preprocessing approaches discussed in this section are typical of methods that satisfy these requirements.

7.6.1 Foundation

In this section we consider two basic approaches to preprocessing. The first is based on spatial-domain techniques and the second deals with frequency-domain concepts via the Fourier transform. Together, these approaches encompass most of the preprocessing algorithms used in robot vision systems.

Spatial-Domain Methods. The *spatial domain* refers to the aggregate of pixels composing an image, and spatial-domain methods are procedures that operate directly on these pixels. Preprocessing functions in the spatial domain may be expressed as

$$g(x, y) = h[f(x, y)] \qquad (7.6\text{-}1)$$

where $f(x, y)$ is the input image, $g(x, y)$ is the resulting (preprocessed) image, and h is an operator on f, defined over some neighborhood of (x, y). It is also possible to let h operate on a *set* of input images, such as performing the pixel-by-pixel sum of K images for noise reduction, as discussed in Sec. 7.6.2.

The principal approach used in defining a neighborhood about (x, y) is to use a square or rectangular subimage area centered at (x, y), as shown in Fig. 7.20. The center of the subimage is moved from pixel to pixel starting, say, at the top left corner, and applying the operator at each location (x, y) to yield $g(x, y)$. Although other neighborhood shapes, such as a circle, are sometimes used, square arrays are by far the most predominant because of their ease of implementation.

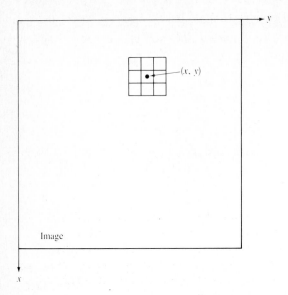

Figure 7.20 A 3 × 3 neighborhood about a point (x, y) in an image.

The simplest form of h is when the neighborhood is 1×1 and, therefore, g depends only on the value of f at (x, y). In this case h becomes an *intensity mapping* or *transformation T* of the form

$$s = T(r) \tag{7.6-2}$$

where, for simplicity, we have used s and r as variables denoting, respectively, the intensity of $f(x, y)$ and $g(x, y)$ at any point (x, y). This type of transformation is discussed in more detail in Sec. 7.6.3.

One of the spatial-domain techniques used most frequently is based on the use of so-called *convolution masks* (also referred to as *templates*, *windows*, or *filters*). Basically, a mask is a small (e.g., 3 × 3) two-dimensional array, such as the one shown in Fig. 7.20, whose coefficients are chosen to detect a given property in an image. As an introduction to this concept, suppose that we have an image of constant intensity which contains widely isolated pixels whose intensities are different from the background. These points can be detected by using the mask shown in Fig. 7.21. The procedure is as follows: The center of the mask (labeled 8) is moved around the image, as indicated above. At each pixel position in the image, we multiply every pixel that is contained within the mask area by the corresponding mask coefficient; that is, the pixel in the center of the mask is multiplied by 8, while its 8-neighbors are multiplied by -1. The results of these nine multiplications are then summed. If all the pixels within the mask area have the same value (constant background), the sum will be zero. If, on the other hand, the center of the mask is located at one of the isolated points, the sum will be different

-1	-1	-1
-1	8	-1
-1	-1	-1

Figure 7.21 A mask for detecting isolated points different from a constant background.

from zero. If the isolated point is in an off-center position, the sum will also be different from zero, but the magnitude of the response will be weaker. These weaker responses can be eliminated by comparing the sum against a threshold.

As shown in Fig. 7.22, if we let $w_1, w_2, \ldots, w_9$ represent mask coefficients and consider the 8-neighbors of (x, y), we may generalize the preceding discussion as that of performing the following operation:

$$h[f(x, y)] = w_1 f(x - 1, y - 1) + w_2 f(x - 1, y) + w_3 f(x - 1, y + 1)$$

$$+ w_4 f(x, y - 1) + w_5 f(x, y) + w_6 f(x, y + 1)$$

$$+ w_7 f(x + 1, y - 1) + w_8 f(x + 1, y)$$

$$+ w_9 f(x + 1, y + 1) \tag{7.6-3}$$

on a 3×3 neighborhood of (x, y).

w_1	w_2	w_3
$(x - 1, y - 1)$	$(x - 1, y)$	$(x - 1, y + 1)$
w_4	w_5	w_6
$(x, y - 1)$	(x, y)	$(x, y + 1)$
w_7	w_8	w_9
$(x + 1, y - 1)$	$(x + 1, y)$	$(x + 1, y + 1)$

Figure 7.22 A general 3×3 mask showing coefficients and corresponding image pixel locations.

Before leaving this section, we point out that the concept of neighborhood processing is not limited to 3×3 areas nor to the cases treated thus far. For instance, we will use neighborhood operations in subsequent discussions for noise reduction, to obtain variable image thresholds, to compute measures of texture, and to obtain the skeleton of an object.

Frequency-Domain Methods. The *frequency domain* refers to an aggregate of complex pixels resulting from taking the Fourier transform of an image. The concept of "frequency" is often used in interpreting the Fourier transform and arises from the fact that this particular transform is composed of complex sinusoids. Due to extensive processing requirements, frequency-domain methods are not nearly as widely used in robotic vision as are spatial-domain techniques. However, the Fourier transform does play an important role in areas such as the analysis of object motion and object description. In addition, many spatial techniques for enhancement and restoration are founded on concepts whose origins can be traced to a Fourier transform formulation. The material in this section will serve as an introduction to these concepts. A more extensive treatment of the Fourier transform and its properties may be found in Gonzalez and Wintz [1977].

We begin the discussion by considering discrete functions of one variable, $f(x)$, $x = 0, 1, 2, \ldots, N - 1$. The forward Fourier transform of $f(x)$ is defined as

$$F(u) = \frac{1}{N} \sum_{x=0}^{N-1} f(x)e^{-j2\pi ux/N} \tag{7.6-4}$$

for $u = 0, 1, 2, \ldots, N - 1$. In this equation $j = \sqrt{-1}$ and u is the so-called *frequency variable*. The inverse Fourier transform of $F(u)$ yields $f(x)$ back, and is defined as

$$f(x) = \sum_{u=0}^{N-1} F(u)e^{j2\pi ux/N} \tag{7.6-5}$$

for $x = 0, 1, 2, \ldots, N - 1$. The validity of these expressions, called the *Fourier transform pair*, is easily verified by substituting Eq. (7.6-4) for $F(u)$ in Eq. (7.6-5), or vice versa. In either case we would get an identity.

A direct implementation of Eq. (7.6-4) for $u = 0, 1, 2, \ldots, N - 1$ would require on the order of N^2 additions and multiplications. Use of a fast Fourier transform (FFT) algorithm significantly reduces this number to $N \log_2 N$, where N is assumed to be an integer power of 2. Similar comments apply to Eq. (7.6-5) for $x = 0, 1, 2, \ldots, N - 1$. A number of FFT algorithms are readily available in a variety of computer languages.

The two-dimensional Fourier transform pair of an $N \times N$ image is defined as

$$F(u, v) = \frac{1}{N} \sum_{x=0}^{N-1} \sum_{y=0}^{N-1} f(x, y)e^{-j2\pi(ux + vy)/N} \tag{7.6-6}$$

for $u, v = 0, 1, 2, \ldots, N - 1$, and

$$f(x, y) = \frac{1}{N} \sum_{u=0}^{N-1} \sum_{v=0}^{N-1} F(u, v) e^{j2\pi(ux + vy)/N} \qquad (7.6\text{-}7)$$

for $x, y = 0, 1, 2, \ldots, N - 1$. It is possible to show through some manipulation that each of these equations can be expressed as separate one-dimensional summations of the form shown in Eq. (7.6-4). This leads to a straightforward procedure for computing the two-dimensional Fourier transform using only a one-dimensional FFT algorithm: We first compute and save the transform of each row of $f(x, y)$, thus producing a two-dimensional array of intermediate results. These results are multiplied by N and the one-dimensional transform of each column is computed. The final result is $F(u, v)$. Similar comments apply for computing $f(x, y)$ given $F(u, v)$. The order of computation from a row-column approach can be reversed to a column-row format without affecting the final result.

The Fourier transform can be used in a number of ways by a vision system, as will be shown in Chap. 8. For example, by treating the boundary of an object as a one-dimensional array of points and computing their Fourier transform, selected values of $F(u)$ can be used as descriptors of boundary shape. The one-dimensional Fourier transform has also been used as a powerful tool for detecting object motion. Applications of the discrete two-dimensional Fourier transform in image reconstruction, enhancement, and restoration are abundant although, as mentioned earlier, the usefulness of this approach in industrial machine vision is still quite restricted due to the extensive computational requirements needed to implement this transform. We point out before leaving this section, however, that the two-dimensional, continuous Fourier transform can be computed (at the speed of light) by optical means. This approach, which requires the use of precisely aligned optical equipment, is used in industrial environments for tasks such as the inspection of finished metal surfaces. Further treatment of this topic is outside the scope of our present discussion, but the interested reader is referred to the book by Goodman [1968] for an excellent introduction to Fourier optics.

7.6.2 Smoothing

Smoothing operations are used for reducing noise and other spurious effects that may be present in an image as a result of sampling, quantization, transmission, or disturbances in the environment during image acquisition. In this section we consider several fast smoothing methods that are suitable for implementation in the vision system of a robot.

Neighborhood Averaging. Neighborhood averaging is a straightforward spatial-domain technique for image smoothing. Given an image $f(x, y)$, the procedure is to generate a smoothed image $g(x, y)$ whose intensity at every point (x, y) is obtained by averaging the intensity values of the pixels of f contained in a predefined neighborhood of (x, y). In other words, the smoothed image is obtained by using the relation

$$g(x, y) = \frac{1}{P} \sum_{(n, m) \in S} f(n, m) \qquad (7.6\text{-}8)$$

for all x and y in $f(x, y)$. S is the set of coordinates of points in the neighborhood of (x, y), including (x, y) itself, and P is the total number of points in the neighborhood. If a 3×3 neighborhood is used, we note by comparing Eqs. (7.6-8) and (7.6-3) that the former equation is a special case of the latter with $w_i = \frac{1}{9}$. Of course, we are not limited to square neighborhoods in Eq. (7.6-8) but, as mentioned in Sec. 7.6.1, these are by far the most predominant in robot vision systems.

> **Example:** Figure 7.23 illustrates the smoothing effect produced by neighborhood averaging. Figure 7.23a shows an image corrupted by noise, and Fig. 7.23b is the result of averaging every pixel with its 4-neighbors. Similarly, Figs. 7.23c through f are the results of using neighborhoods of sizes 3×3, 5×5, 7×7, and 11×11, respectively. It is noted that the degree of smoothing is strongly proportional to the size of the neighborhood used. As is true with most mask processors, the smoothed value of each pixel is determined before any of the other pixels have been changed. □

Median Filtering. One of the principal difficulties of neighborhood averaging is that it blurs edges and other sharp details. This blurring can often be reduced significantly by the use of so-called *median filters*, in which we replace the intensity of each pixel by the median of the intensities in a predefined neighborhood of that pixel, instead of by the average.

Recall that the median M of a set of values is such that half the values in the set are less than M and half the values are greater than M. In order to perform median filtering in a neighborhood of a pixel, we first sort the values of the pixel and its neighbors, determine the median, and assign this value to the pixel. For example, in a 3×3 neighborhood the median is the fifth largest value, in a 5×5 neighborhood the thirteenth largest value, and so on. When several values in a neighborhood are the same, we group all equal values as follows: Suppose that a 3×3 neighborhood has values (10, 20, 20, 20, 15, 20, 20, 25, 100). These values are sorted as (10, 15, 20, 20, 20, 20, 20, 25, 100), which results in a median of 20. A little thought will reveal that the principal function of median filtering is to force points with very distinct intensities to be more like their neighbors, thus actually eliminating intensity spikes that appear isolated in the area of the filter mask.

> **Example:** Figure 7.24a shows an original image, and Fig. 7.24b shows the same image but with approximately 20 percent of the pixels corrupted by "impulse noise." The result of neighborhood averaging over a 5×5 area is

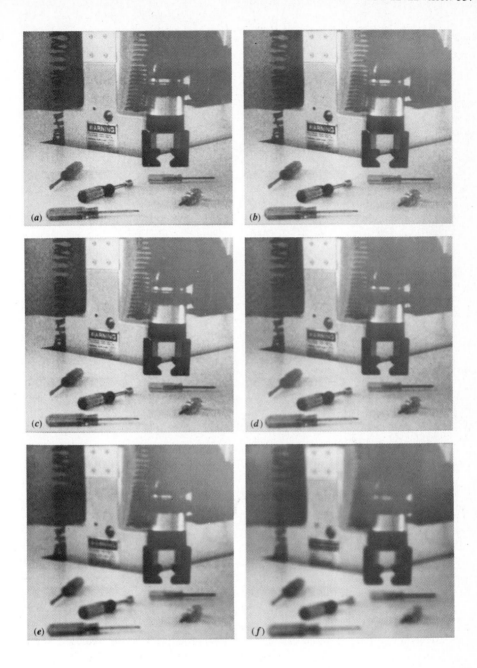

Figure 7.23 (*a*) Noisy image. (*b*) Result of averaging each pixel along with its 4-neighbors. (*c*) through (*f*) are the results of using neighborhood sizes of 3 × 3, 5 × 5, 7 × 7, and 11 × 11, respectively.

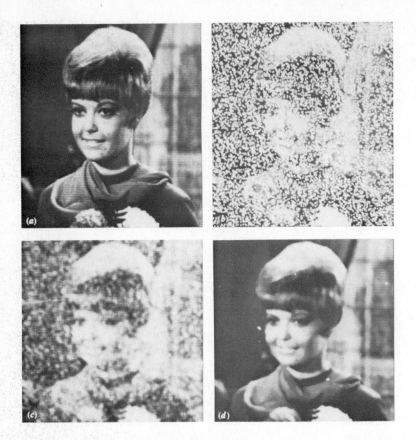

Figure 7.24 (*a*) Original image. (*b*) Image corrupted by impulse noise. (*c*) Result of 5 × 5 neighborhood averaging. (*d*) Result of 5 × 5 median filtering. (Courtesy of Martin Connor, Texas Instruments, Inc., Lewisville, Texas.)

shown in Fig. 7.24*c* and the result of a 5 × 5 median filter is shown in Fig. 7.24*d*. The superiority of the median filter over neighborhood averaging needs no explanation. The three bright dots remaining in Fig. 7.24*d* resulted from a large concentration of noise at those points, thus biasing the median calculation. Two or more passes with a median filter would eliminate those points. □

Image Averaging. Consider a noisy image $g(x, y)$ which if formed by the addition of noise $n(x, y)$ to an uncorrupted image $f(x, y)$; that is,

$$g(x, y) = f(x, y) + n(x, y) \qquad (7.6\text{-}9)$$

where it is assumed that the noise is uncorrelated and has zero average value. The objective of the following procedure is to obtain a smoothed result by adding a given set of noisy images, $g_i(x, y)$, $i = 1, 2, \ldots, K$.

If the noise satisfies the constraints just stated, it is a simple problem to show (Papoulis [1965]) that if an image $\bar{g}(x, y)$ is formed by averaging K different noisy images,

$$\bar{g}(x, y) = \frac{1}{K} \sum_{i=1}^{K} g_i(x, y) \tag{7.6-10}$$

then it follows that

$$E\{\bar{g}(x, y)\} = f(x, y) \tag{7.6-11}$$

and

$$\sigma_{\bar{g}}^2 (x, y) = \frac{1}{K} \sigma_n^2(x, y) \tag{7.6-12}$$

where $E\{\bar{g}(x, y)\}$ is the expected value of $\bar{g}$, and $\sigma_{\bar{g}}^2(x, y)$ and $\sigma_n^2(x, y)$ are the variances of $\bar{g}$ and n, all at coordinates (x, y). The standard deviation at any point in the average image is given by

$$\sigma_{\bar{g}}(x, y) = \frac{1}{\sqrt{K}} \sigma_n(x, y) \tag{7.6-13}$$

Equations (7.6-12) and (7.6-13) indicate that, as K increases, the variability of the pixel values decreases. Since $E\{\bar{g}(x, y)\} = f(x, y)$, this means that $\bar{g}(x, y)$ will approach the uncorrupted image $f(x, y)$ as the number of noisy images used in the averaging process increases.

It is important to note that the technique just discussed implicitly assumes that all noisy images are registered spatially, with only the pixel intensities varying. In terms of robotic vision, this means that all object in the work space must be at rest with respect to the camera during the averaging process. Many vision systems have the capability of performing an entire image addition in one frame time interval (i.e., one-thirtieth of a second). Thus, the addition of, say, 16 images will take on the order of ½ s. during which no motion can take place.

Example: An an illustration of the averaging method, consider the images shown in Fig. 7.25. Part (a) of this figure shows a sample noisy image and Fig. 7.25b to f show the results of averaging 4, 8, 16, 32, and 64 such images, respectively. It is of interest to note that the results are quite acceptable for $K = 32$. ☐

Smoothing Binary Images. Binary images result from using backlighting or structured lighting, as discussed in Sec. 7.3, or from processes such as edge detection or thresholding, as discussed in Secs. 7.6.4 and 7.6.5. We will use the convention of labeling dark points with a 1 and light points with a 0. Thus, since binary

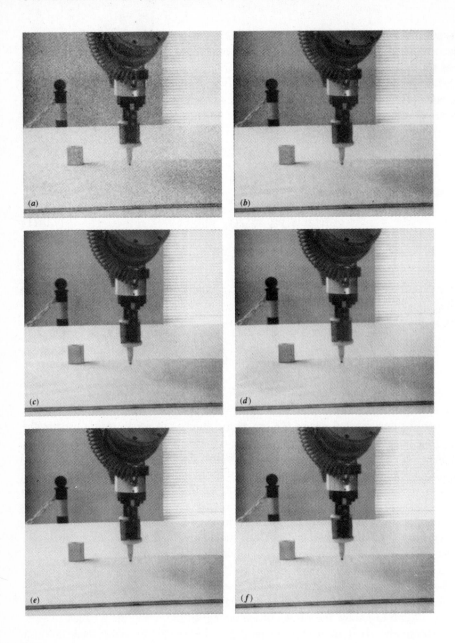

Figure 7.25 (*a*) Sample noisy image. (*b*) through (*f*) are the results of averaging 4, 8, 16, 32, and 64 such images.

images are two-valued, noise in this case produces effects such as irregular boundaries, small holes, missing corners, and isolated points.

The basic idea underlying the methods discussed in this section is to specify a boolean function evaluated on a neighborhood centered at a pixel p, and to assign to p a 1 or 0, depending on the spatial arrangement and binary values of its neighbors. Due to limitations in available processing time for industrial vision tasks, the analysis is typically limited to the 8-neighbors of p, which leads us to the 3×3 mask shown in Fig. 7.26. The smoothing approach (1) fills in small (one pixel) holes in otherwise dark areas, (2) fills in small notches in straightedge segments, (3) eliminates isolated 1's, (4) eliminates small bumps along straightedge segments, and (5) replaces missing corner points.

With reference to Fig. 7.26, the first two smoothing processes just mentioned are accomplished by using the boolean expression

$$B_1 = p + b \cdot g \cdot (d + e) + d \cdot e \cdot (b + g) \qquad (7.6\text{-}14)$$

where " $\cdot$ " and " $+$ " denote the logical AND and OR, respectively. Following the convention established above, a dark pixel contained in the mask area is assigned a logical 1 and a light pixel a logical 0. Then, if $B_1 = 1$, we assign a 1 to p, otherwise this pixel is assigned a 0. Equation (7.6.14) is applied to all pixels simultaneously, in the sense that the next value of each pixel location is determined before any of the other pixels have been changed.

Steps 3 and 4 in the smoothing process are similarly accomplished by evaluating the boolean expression

$$B_2 = p \cdot [(a + b + d) \cdot (e + g + h) + (b + c + e) \cdot (d + f + g)]$$

$$(7.6\text{-}15)$$

simultaneously for all pixels. As above, we let $p = 1$ if $B_2 = 1$ and zero otherwise.

a	b	c
d	p	e
f	g	h

Figure 7.26 Neighbors of p used for smoothing binary images. Dark pixels are denoted by 1 and light pixels by 0.

Missing top, right corner points are filled in by means of the expression

$$B_3 = \bar{p} \cdot (d \cdot f \cdot g) \cdot \overline{(a + b + c + e + h)} + p \qquad (7.6\text{-}16)$$

where overbar denotes the logical complement. Similarly, lower right, top left, and lower left missing corner points are filled in by using the expressions

$$B_4 = \bar{p} \cdot (a \cdot b \cdot d) \cdot \overline{(c + e + f + g + h)} + p \qquad (7.6\text{-}17)$$

$$B_5 = \bar{p} \cdot (e \cdot g \cdot h) \cdot \overline{(a + b + c + d + f)} + p \qquad (7.6\text{-}18)$$

and $\qquad B_6 = \bar{p} \cdot (b \cdot c \cdot e) \cdot \overline{(a + d + f + g + h)} + p \qquad (7.6\text{-}19)$

These last four expressions implement step 5 of the smoothing procedure.

Example: The concepts just discussed are illustrated in Fig. 7.27. Figure 7.27a shows a noisy binary image, and Fig. 7.27b shows the result of applying B_1. Note that the notches along the boundary and the hole in the dark area were filled in. Figure 7.27c shows the result of applying B_2 to the image in Fig. 7.27b. As expected, the bumps along the boundary of the dark area and all isolated points were removed (the image was implicitly extended with 0's for points on the image border). Finally, Fig. 7.27d shows the result of applying B_3 through B_6 to the image in Fig. 7.27c. Only B_4 had an effect in this particular case. ☐

7.6.3 Enhancement

One of the principal difficulties in many low-level vision tasks is to be able to automatically adapt to changes in illumination. The capability to compensate for effects such as shadows and "hot-spot" reflectances quite often plays a central role in determining the success of subsequent processing algorithms. In this subsection we consider several enhancement techniques which address these and similar problems. The reader is reminded that enhancement is a major area in digital image processing and scene analysis, and that our discussion of this topic is limited to sample techniques that are suitable for robot vision systems. In this context, "suitable" implies having fast computational characteristics and modest hardware requirements.

Histogram Equalization. Let the variable r represent the intensity of pixels in an image to be enhanced. It will be assumed initially that r is a normalized, continuous variable lying in the range $0 \leqslant r \leqslant 1$. The discrete case is considered later in this section.

For any r in the interval [0, 1], attention will be focused on transformations of the form

$$s = T(r) \qquad (7.6\text{-}20)$$

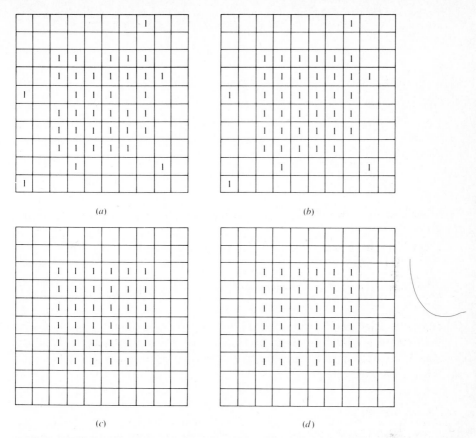

Figure 7.27 (*a*) Original image. (*b*) Result of applying B_1. (*c*) Result of applying B_2. (*d*) Final result after application of B_3 through B_6.

which produce an intensity value s for every pixel value r in the input image. It is assumed that the transformation function T satisfies the conditions:

1. $T(r)$ is single-valued and monotonically increasing in the interval $0 \leqslant T(r) \leqslant 1$.
2. $0 \leqslant T(r) \leqslant 1$ for $0 \leqslant r \leqslant 1$.

Condition 1 preserves the order from black to white in the intensity scale, and condition 2 guarantees a mapping that is consistent with the allowed 0 to 1 range of pixel values. A transformation function satisfying these conditions is illustrated in Fig. 7.28.

The inverse transformation function from s back to r is denoted by

$$r = T^{-1}(s) \qquad (7.6\text{-}21)$$

where it is assumed that $T^{-1}(s)$ satisfies the two conditions given above.

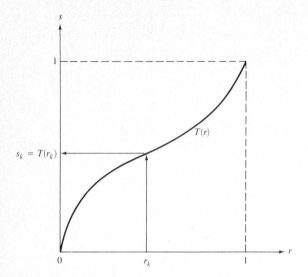

Figure 7.28 An intensity transformation function.

The intensity variables r and s are random quantities in the interval [0, 1] and, as such, can be characterized by their probability density functions (PDFs) $p_r(r)$ and $p_s(s)$. A great deal can be said about the general appearance of an image from its intensity PDF. For example, an image whose pixels have the PDF shown in Fig. 7.29a would have fairly dark characteristics since the majority of pixel values would be concentrated on the dark end of the intensity scale. On the other hand, an image whose pixels have an intensity distribution like the one shown in Fig. 7.29b would have predominant light tones.

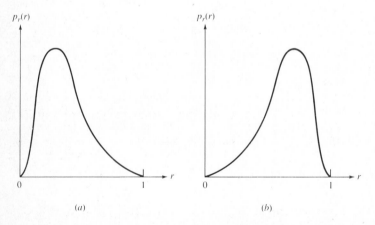

Figure 7.29 (a) Intensity PDF of a "dark" image and (b) a "light" image.

It follows from elementary probability theory that if $p_r(r)$ and $T(r)$ are known, and $T^{-1}(s)$ satisfies condition 1, then the PDF of the transformed intensities is given by

$$p_s(s) = \left[p_r(r) \frac{dr}{ds} \right]_{r=T^{-1}(s)} \tag{7.6-22}$$

Suppose that we choose a specific transformation function given by

$$s = T(r) = \int_0^r p_r(w) \, dw \qquad 0 \leqslant r \leqslant 1 \tag{7.6-23}$$

where w is a dummy variable of integration. The rightmost side of this equation is recognized as the cumulative distribution function of $p_r(r)$, which is known to satisfy the two conditions stated earlier. The derivative of s with respect to r for this particular transformation function is easily found to be

$$\frac{ds}{dr} = p_r(r) \tag{7.6-24}$$

Substitution of dr/ds into Eq. (7.6-22) yields

$$p_s(s) = \left[p_r(r) \frac{1}{p_r(r)} \right]_{r=T^{-1}(s)}$$

$$= [1]_{r=T^{-1}(s)}$$

$$= 1 \qquad 0 \leqslant s \leqslant 1 \tag{7.6-25}$$

which is a uniform density in the interval of definition of the transformed variable s. It is noted that this result is independent of the inverse transformation function. This is important because it is often quite difficult to find $T^{-1}(s)$ analytically. It is also noted that using the transformation function given in Eq. (7.6-23) yields transformed intensities that always have a flat PDF, *independent* of the shape of $p_r(r)$, a property that is ideally suited for automatic enhancement. The net effect of this transformation is to balance the distribution of intensities. As will be seen below, this process can have a rather dramatic effect on the appearance of an image.

In order to be useful for digital processing, the concepts developed above must be formulated in discrete form. For intensities that assume discrete values we deal with probabilities given by the relation

$$p_r(r_k) = \frac{n_k}{n} \qquad 0 \leqslant r_k \leqslant 1 \tag{7.6-26}$$

$$k = 0, 1, 2, \ldots, L - 1$$

where L is the number of discrete intensity levels, $p_r(r_k)$ is an estimate of the probability of intensity r_k, n_k is the number of times this intensity appears in the image, and n is the total number of pixels in the image. A plot of $p_r(r_k)$ versus r_k is usually called a *histogram*, and the technique used for obtaining a uniform histogram is known as *histogram equalization* or *histogram linearization*.

The discrete form of Eq. (7.6-23) is given by

$$s_k = T(r_k) = \sum_{j=0}^{k} \frac{n_j}{n}$$

$$= \sum_{j=0}^{k} p_r(r_j) \tag{7.6-27}$$

for $0 \leqslant r_k \leqslant 1$ and $k = 0, 1, 2, \ldots, L - 1$. It is noted from this equation that in order to obtain the mapped value s_k corresponding to r_k, we simply sum the histogram components from 0 to r_k.

The inverse discrete transformation is given by

$$r_k = T^{-1}(s_k) \qquad 0 \leqslant s_k \leqslant 1 \tag{7.6-28}$$

where both $T(r_k)$ and $T^{-1}(s_k)$ are assumed to satisfy conditions 1 and 2 stated above. Although $T^{-1}(s_k)$ is not used in histogram equalization, it plays a central role in histogram specification, as discussed below.

> **Example:** As an illustration of histogram equalization, consider the image shown in Fig. 7.30a and its histogram shown in Fig. 7.30b. The result of applying Eq. (7.6-27) to this image is shown in Fig. 7.30c and the corresponding equalized histogram is shown in Fig. 7.30d. The improvement of details is evident. It is noted that the histogram is not perfectly flat, a condition generally encountered when applying to discrete values a method derived for continuous quantities. □

Histogram Specification. Histogram equalization is ideally suited for automatic enhancement since it is based on a transformation function that is uniquely determined by the histogram of the input image. However, the method is limited in the sense that its only function is histogram linearization, a process that is not applicable when a priori information is available regarding a desired output histogram shape. Here we generalize the concept of histogram processing by developing an approach capable of generating an image with a specified intensity histogram. As will be seen below, histogram equalization is a special case of this technique.

Starting again with continuous quantities, let $p_r(r)$ and $p_z(z)$ be the original and desired intensity PDFs. Suppose that a given image is first histogram equalized by using Eq. (7.6-23); that is,

$$s = T(r) = \int_0^r p_r(w)\, dw \tag{7.6-29}$$

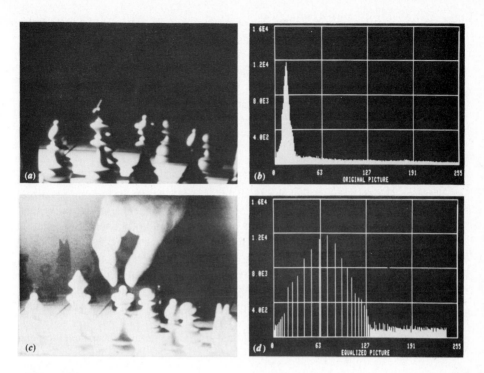

Figure 7.30 (*a*) Original image and (*b*) its histogram. (*c*) Histogram-equalized image and (*d*) its histogram. (From Woods and Gonzalez [1981], © IEEE.)

If the desired image *were* available, its levels could also be equalized by using the transformation function

$$v = G(z) = \int_0^z p_z(w)\,dw \qquad (7.6\text{-}30)$$

The inverse process, $z = G^{-1}(v)$ would then yield the desired levels back. This, of course, is a hypothetical formulation since the z levels are precisely what we are trying to obtain. It is noted, however, that $p_s(s)$ and $p_v(v)$ would be identical uniform densities since the use of Eqs. (7.6-29) and (7.6-30) guarantees a uniform density, regardless of the shape of the PDF inside the integral. Thus, if instead of using v in the inverse process, we use the inverse levels s obtained from the original image, the resulting levels $z = G^{-1}(s)$ would have the desired PDF, $p_z(z)$. Assuming that $G^{-1}(s)$ is single-valued, the procedure can be summarized as follows:

1. Equalize the levels of the original image using Eq. (7.6-29).
2. Specify the desired intensity PDF and obtain the transformation function $G(z)$ using Eq. (7.6-30).

3. Apply the inverse transformation $z = G^{-1}(s)$ to the intensity levels of the histogram-equalized image obtained in step 1.

This procedure yields an output image with the specified intensity PDF.

The two transformations required for histogram specification, $T(r)$ and $G^{-1}(s)$, can be combined into a single transformation:

$$z = G^{-1}(s) = G^{-1}[T(r)] \qquad (7.6\text{-}31)$$

which relates r to z. It is noted that, when $G^{-1}[T(r)] = T(r)$, this method reduces to histogram equalization.

Equation (7.6-31) shows that the input image need not be histogram-equalized explicitly in order to perform histogram specification. All that is required is that $T(r)$ be determined and combined with $G^{-1}(s)$ into a single transformation that is applied directly to the input image. The real problem in using the two transformations or their combined representation for continuous variables lies in obtaining the inverse function analytically. In the discrete case this problem is circumvented by the fact that the number of distinct intensity levels is usually relatively small (e.g., 256) and it becomes feasible to calculate and store a mapping for each possible integer pixel value.

The discrete formulation of the foregoing procedure parallels the development in the previous section:

$$s_k = T(r_k) = \sum_{j=0}^{k} p_r(r_j) \qquad (7.6\text{-}32)$$

$$G(z_i) = \sum_{j=0}^{i} p_z(z_j) \qquad (7.6\text{-}33)$$

and
$$z_i = G^{-1}(s_i) \qquad (7.6\text{-}34)$$

where $p_r(r_j)$ is computed from the input image, and $p_z(z_j)$ is specified.

> **Example:** An illustration of the histogram specification method is shown in Fig. 7.31. Part (a) of this figure shows the input image and Fig. 7.31b is the result of histogram equalization. Figure 7.31c shows a specified histogram and Fig. 7.31d is the result of using this histogram in the procedure discussed above. It is noted that, in this case, histogram equalization had little effect on the image. □

Local Enhancement. The histogram equalization and specification methods discussed above are global, in the sense that pixels are modified by a transformation function which is based on the intensity distribution over an entire image. While this global approach is suitable for overall enhancement, it is often necessary to enhance details over small areas. Since the number of pixels in these areas may have negligible influence on the computation of a global transformation, the use of

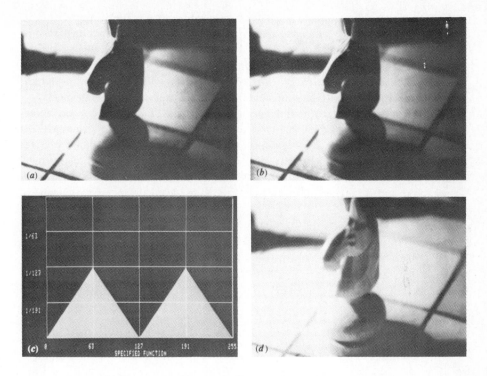

Figure 7.31 (*a*) Input image. (*b*) Result of histogram equalization. (*c*) A specified histogram. (*d*) Result of enhancement by histogram specification. (From Woods and Gonzalez [1981], © IEEE.)

global techniques seldom yields acceptable local enhancement. The solution is to devise transformation functions that are based on the intensity distribution, or other properties, in the neighborhood of every pixel in a given image.

The histogram-processing techniques developed above are easily adaptable to local enhancement. The procedure is to define an $n \times m$ neighborhood and move the center of this area from pixel to pixel. At each location, we compute the histogram of the $n \times m$ points in the neighborhood and obtain either a histogram equalization or histogram specification transformation function. This function is finally used to map the intensity of the pixel centered in the neighborhood. The center of the $n \times m$ region is then moved to an adjacent pixel location and the procedure is repeated. Since only one new row or column of the neighborhood changes during a pixel-to-pixel translation of the region, it is possible to update the histogram obtained in the previous location with the new data introduced at each motion step. This approach has obvious advantages over repeatedly computing the histogram over all $n \times m$ pixels every time the region is moved one pixel location. Another approach often used to reduce computation is to employ nonoverlapping regions, but this often produces an undesirable checkerboard effect.

Example: An illustration of local histogram equalization where the neighborhood is moved from pixel to pixel is shown in Fig. 7.32. Part (*a*) of this figure shows an image with constant background and five dark square areas. The image is slightly blurred as a result of smoothing with a 7 × 7 mask to reduce noise (see Sec. 7.6.2). Figure 7.32*b* shows the result of histogram equalization. The most striking feature in this image is the enhancement of noise, a problem that commonly occurs when using this technique on noisy images, even if they have been smoothed prior to equalization. Figure 7.32*c* shows the result of local histogram equalization using a neighborhood of size 7 × 7. Note that the dark areas have been enhanced to reveal an inner structure that was not visible in either of the previous two images. Noise was also enhanced, but its texture is much finer due to the local nature of the enhance-

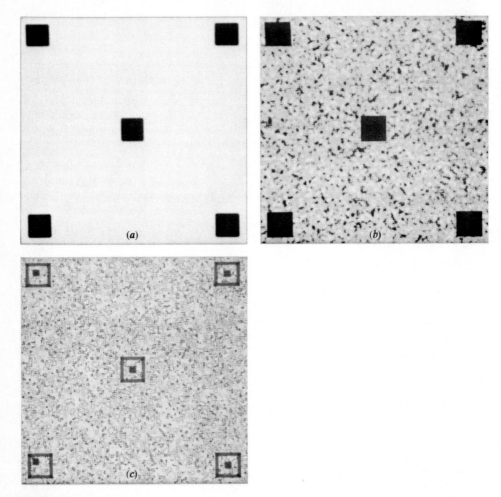

Figure 7.32 (*a*) Original image. (*b*) Result of global histogram equalization. (*c*) Result of local histogram equalization using a 7 × 7 neighborhood about each pixel.

ment approach. This example clearly demonstrates the necessity for using local enhancement when the details of interest are too small to influence significantly the overall characteristics of a global technique. □

Instead of using histograms, one could base local enhancement on other properties of the pixel intensities in a neighborhood. The intensity mean and variance (or standard deviation) are two such properties which are frequently used because of their relevance to the appearance of an image. That is, the mean is a measure of average brightness and the variance is a measure of contrast. A typical local transformation based on these concepts maps the intensity of an input image $f(x, y)$ into a new image $g(x, y)$ by performing the following transformation at each pixel location (x, y):

$$g(x, y) = A(x, y)[f(x, y) - m(x, y)] + m(x, y) \qquad (7.6\text{-}35)$$

where

$$A(x, y) = k \frac{M}{\sigma(x, y)} \quad 0 < k < 1 \qquad (7.6\text{-}36)$$

In this formulation, $m(x, y)$ and $\sigma(x, y)$ are the intensity mean and standard deviation computed in a neighborhood centered at (x, y), M is the global mean of $f(x, y)$, and k is a constant in the range indicated above.

It is important to note that A, m, and σ are variable quantities which depend on a predefined neighborhood of (x, y). Application of the local gain factor $A(x, y)$ to the difference between $f(x, y)$ and the local mean amplifies local variations. Since $A(x, y)$ is inversely proportional to the standard deviation of the intensity, areas with low contrast receive larger gain. The mean is added back in Eq. (7.6-35) to restore the average intensity level of the image in the local region. In practice, it is often desirable to add back a fraction of the local mean and to restrict the variations of $A(x, y)$ between two limits $[A_{\min}, A_{\max}]$ in order to balance out large excursions of intensity in isolated regions.

Example: The preceding enhancement approach has been implemented in hardware by Narendra and Fitch [1981], and has the capability of processing images in real time (i.e., at 30 image frames per second). An example of the capabilities of the technique using a local region of size 15×15 pixels is shown in Fig. 7.33. Note the enhancement of detail at the boundary between two regions of different overall intensities and the rendition of intensity details in each of the regions. □

7.6.4 Edge Detection

Edge detection plays a central role in machine vision, serving as the initial preprocessing step for numerous object detection algorithms. In this chapter we are

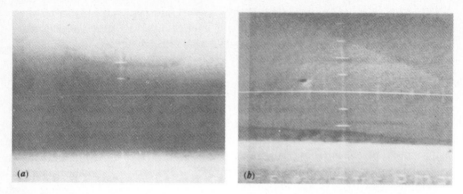

(a)   (b)

Figure 7.33 Images before and after local enhancement. (From Narendra and Fitch [1981], © IEEE.)

interested in fundamental techniques for detecting edge points. Subsequent processing of these edge points is discussed in Chap. 8.

Basic Formulation. Basically, the idea underlying most edge detection techniques is the computation of a local derivative operator. This concept can be easily illustrated with the aid of Fig. 7.34. Part (a) of this figure shows an image of a simple light object on a dark background, the intensity profile along a horizontal scan line of the image, and the first and second derivatives of the profile. It is noted from the profile that an edge (transition from dark to light) is modeled as a ramp, rather than as an abrupt change of intensity. This model is representative of the fact that edges in digital images are generally slightly blurred as a result of sampling.

The first derivative of an edge modeled in this manner is zero in all regions of constant intensity, and assumes a constant value during an intensity transition. The second derivative, on the other hand, is zero in all locations, except at the onset and termination of an intensity transition. Based on these remarks and the concepts illustrated in Fig. 7.34, it is evident that the magnitude of the first derivative can be used to detect the presence of an edge, while the sign of the second derivative can be used to determine whether an edge pixel lies on the dark (background) or light (object) side of an edge. The sign of the second derivative in Fig. 7.34a, for example, is positive for pixels lying on the dark side of both the leading and trailing edges of the object, while the sign is negative for pixels on the light side of these edges. Similar comments apply to the case of a dark object on a light background, as shown in Fig. 7.34b. It is of interest to note that identically the same interpretation regarding the sign of the second derivative is true for this case.

Although the discussion thus far has been limited to a one-dimensional horizontal profile, a similar argument applies to an edge of any orientation in an image. We simply define a profile perpendicular to the edge direction at any given point and interpret the results as in the preceding discussion. As will be shown below, the first derivative at any point in an image can be obtained by using the magnitude of the gradient at that point, while the second derivative is given by the Laplacian.

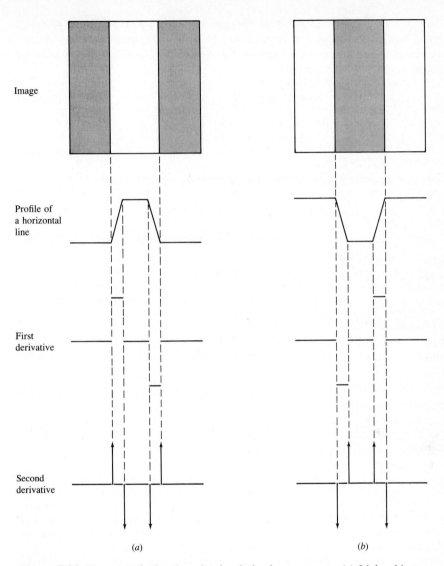

Image

Profile of
a horizontal
line

First
derivative

Second
derivative

(a)   (b)

Figure 7.34 Elements of edge detection by derivative operators. (a) Light object on a dark background. (b) Dark object on a light background.

Gradient Operators. The gradient of an image $f(x, y)$ at location (x, y) is defined as the two-dimensional vector

$$\mathbf{G}[f(x, y)] = \begin{bmatrix} G_x \\ G_y \end{bmatrix} = \begin{bmatrix} \dfrac{\partial f}{\partial x} \\[2mm] \dfrac{\partial f}{\partial y} \end{bmatrix} \qquad (7.6\text{-}37)$$

It is well known from vector analysis that the vector **G** points in the direction of maximum rate of change of f at location (x, y). For edge detection, however, we are interested in the magnitude of this vector, generally referred to as the *gradient* and denoted by $G[f(x, y)]$, where

$$G[f(x, y)] = [G_x^2 + G_y^2]^{1/2} \tag{7.6-38}$$

$$= \left[\left(\frac{\partial f}{\partial x} \right)^2 + \left(\frac{\partial f}{\partial y} \right)^2 \right]^{1/2}$$

It is common practice to approximate the gradient by absolute values:

$$G[f(x, y)] \approx |G_x| + |G_y| \tag{7.6-39}$$

This approximation is considerably easier to implement, particularly when dedicated hardware is being employed.

It is noted from Eq. (7.6-38) that computation of the gradient is based on obtaining the first-order derivatives $\partial f / \partial x$ and $\partial f / \partial y$. There are a number of ways for doing this in a digital image. One approach is to use first-order differences between adjacent pixels; that is,

$$G_x = \frac{\partial f}{\partial x} = f(x, y) - f(x - 1, y) \tag{7.6-40}$$

and

$$G_y = \frac{\partial f}{\partial y} = f(x, y) - f(x, y - 1) \tag{7.6-41}$$

A slightly more complicated definition involving pixels in a 3×3 neighborhood centered at (x, y) is given by

$$G_x = \frac{\partial f}{\partial x} = [f(x + 1, y - 1) + 2f(x + 1, y) + f(x + 1, y + 1)]$$

$$- [f(x - 1, y - 1) + 2f(x - 1, y) + f(x - 1, y + 1)]$$

$$= (g + 2h + i) - (a + 2b + c) \tag{7.6-42}$$

and

$$G_y = \frac{\partial f}{\partial y} = [f(x - 1, y + 1) + 2f(x, y + 1) + f(x + 1, y + 1)]$$

$$- [f(x - 1, y - 1) + 2f(x, y - 1) + f(x + 1, y - 1)]$$

$$= (c + 2e + i) - (a + 2d + g) \tag{7.6-43}$$

where we have used the letters a through i to represent the neighbors of point (x, y). The 3×3 neighborhood of (x, y) using this simplified notation is shown in Fig. 7.35*a*. It is noted that the pixels closest to (x, y) are weighted by 2 in these particular definitions of the digital derivative. Computing the gradient over a 3×3 area rather than using Eqs. (7.6-40) and (7.6-41) has the advantage of increased averaging, thus tending to make the gradient less sensitive to noise. It is possible to define the gradient over larger neighborhoods (Kirsch [1971]), but 3×3 operators are by far the most popular in industrial computer vision because of their computational speed and modest hardware requirements.

It follows from the discussion in Sec. 7.6.1 that G_x, as given in Eq. (7.6-42), can be computed by using the mask shown in Fig. 7.35*b*. Similarly, G_y may be obtained by using the mask shown in Fig. 7.35*c*. These two masks are commonly referred to as the *Sobel operators*. The responses of these two masks at any point (x, y) are combined using Eqs. (7.6-38) or (7.6-39) to obtain an approximation to the gradient at that point. Moving these masks throughout the image $f(x, y)$ yields the gradient at all points in the image.

There are numerous ways by which one can generate an output image, $g(x, y)$, based on gradient computations. The simplest approach is to let the value of g at coordinate (x, y) be equal to the gradient of the input image f at that point; that is,

$$g(x, y) = G[f(x, y)] \qquad (7.6\text{-}44)$$

An example of using this approach to generate a gradient image is shown in Fig. 7.36.

a	b	c
d	(x, y)	e
g	h	i

(*a*)

-1	-2	-1
0	0	0
1	2	1

(*b*)

-1	0	1
-2	0	2
-1	0	1

(*c*)

Figure 7.35 (*a*) 3×3 neighborhood of point (x, y). (*b*) Mask used to compute G_x. (*c*) Mask used to compute G_y.

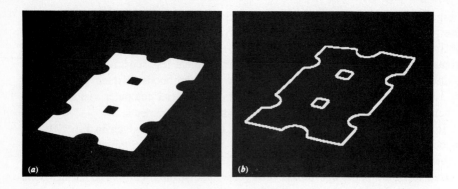

Figure 7.36 (a) Input image. (b) Result of using Eq. (7.6-44).

Another approach is to create a binary image using the following relationship:

$$g(x, y) = \begin{cases} 1 & \text{if } G[f(x, y)] > T \\ \\ 0 & \text{if } G[f(x, y)] \leqslant T \end{cases} \qquad (7.6\text{-}45)$$

where T is a nonnegative threshold. In this case, only edge pixels whose gradients exceed T are considered important. Thus, the use of Eq. (7.6-45) may be viewed as a procedure which extracts only those pixels that are characterized by significant (as determined by T) transitions in intensity. Further analysis of the resulting pixels is usually required to delete isolated points and to link pixels along proper boundaries which ultimately determine the objects segmented out of an image. The use of Eq. (7.6-45) in this context is discussed and illustrated in Sec. 8.2.1.

Laplacian Operator. The Laplacian is a second-order derivative operator defined as

$$L[f(x, y)] = \frac{\partial^2 f}{\partial x^2} + \frac{\partial^2 f}{\partial y^2} \qquad (7.6\text{-}46)$$

For digital images, the Laplacian is defined as

$$L[f(x, y)] = [f(x + 1, y) + f(x - 1, y) + f(x, y + 1) + f(x, y - 1)]$$
$$- 4 f(x, y) \qquad (7.6\text{-}47)$$

This digital formulation of the Laplacian is zero in constant areas and on the ramp section of an edge, as expected of a second-order derivative. The implementation of Eq. (7.6-47) can be based on the mask shown in Fig. 7.37.

0	1	0
1	−4	1
0	1	0

Figure 7.37 Mask used to compute the Laplacian.

Although, as indicated at the beginning of this section, the Laplacian responds to transitions in intensity, it is seldom used by itself for edge detection. The reason is that, being a second-derivative operator, the Laplacian is typically unacceptably sensitive to noise. Thus, this operator is usually delegated the secondary role of serving as a detector for establishing whether a given pixel is on the dark or light side of an edge.

7.6.5 Thresholding

Image thresholding is one of the principal techniques used by industrial vision systems for object detection, especially in applications requiring high data throughputs. In this section we are concerned with aspects of thresholding that fall in the category of low-level processing. More sophisticated uses of thresholding techniques are discussed Chap. 8.

Suppose that the intensity histogram shown in Fig. 7.38a corresponds to an image, $f(x, y)$, composed of light objects on a dark background, such that object and background pixels have intensities grouped into two dominant modes. One obvious way to extract the objects from the background is to select a threshold T which separates the intensity modes. Then, any point (x, y) for which $f(x, y) > T$ is called an object point; otherwise, the point is called a background point. A slightly more general case of this approach is shown in Fig. 7.38b. In this case the image histogram is characterized by three dominant modes (for example, two types of light objects on a dark background). Here, we can use the same basic approach and classify a point (x, y) as belonging to one object class if $T_1 < f(x, y) \leq T_2$, to the other object class if $f(x, y) > T_2$, and to the background if $f(x, y) \leq T_1$. This type of *multilevel thresholding* is generally less reliable than its single threshold counterpart because of the difficulty in establishing multiple thresholds that effectively isolate regions of interest, especially when the number of corresponding histogram modes is large. Typically, problems of this nature, if handled by thresholding, are best addressed by a single, variable threshold, as discussed in Chap. 8.

Based on the foregoing concepts, we may view thresholding as an operation that involves tests against a function T of the form

$$T = T[x, y, p(x, y), f(x, y)] \tag{7.6-48}$$

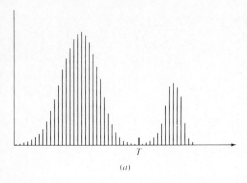

(a)

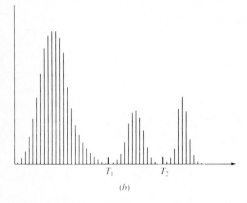

(b)

Figure 7.38 Intensity histograms that can be partioned by (a) a single threshold and (b) multiple thresholds.

where $f(x, y)$ is the intensity of point (x, y), and $p(x, y)$ denotes some local property of this point, for example, the average intensity of a neighborhood centered at (x, y). We create a thresholded image $g(x, y)$ by defining

$$g(x, y) = \begin{cases} 1 & \text{if } f(x, y) > T \\ \\ 0 & \text{if } f(x, y) \leqslant T \end{cases} \qquad (7.6\text{-}49)$$

Thus, in examining $g(x, y)$, we find that pixels labeled 1 (or any other convenient intensity level) correspond to objects, while pixels labeled 0 correspond to the background.

When T depends only on $f(x, y)$, the threshold is called *global* (Fig. 7.38a shows an example of such a threshold). If T depends on both $f(x, y)$ and $p(x, y)$, then the threshold is called *local*. If, in addition, T depends on the spatial coordinates x and y, it is called a *dynamic threshold*. We associate with low-level

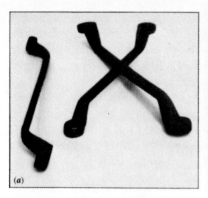

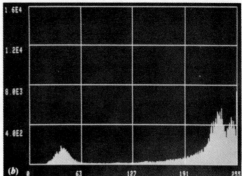

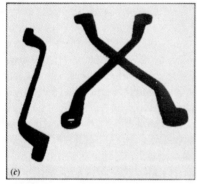

Figure 7.39 (*a*) Original image. (*b*) Histogram of intensities in the range 0 to 255. (*c*) Image obtained by using Eq. (7.6-49) with a global threshold $T = 90$.

vision those thresholding techniques which are based on a single, global value of T. Since thresholding plays a central role in object segmentation, more sophisticated formulations are associated with functions in medium-level vision. A simple example of global thresholding is shown in Fig. 7.39.

7.7 CONCLUDING REMARKS

The material presented in this chapter spans a broad range of processing functions normally associated with low-level vision. Although, as indicated in Sec. 7.1, vision is a three-dimensional problem, most machine vision algorithms, especially those used for low-level vision, are based on images of a three-dimensional scene. The range sensing methods discussed in Sec. 7.2, the structured-lighting approaches in Sec. 7.3, and the material in Sec. 7.4 are important techniques for deriving depth from image information.

Our discussion of low-level vision and other relevant topics, such as the nature of imaging devices, has been at an introductory level, and with a very directed focus toward robot vision. It is important to keep in mind that many of the areas we have discussed have a range of application much broader than this. A good example is enhancement, which for years has been an important topic in digital image processing. One of the salient features of industrial applications, however, is the ever-present (and often contradictory) requirements of low cost and high computational speeds. The selection of topics included in this chapter has been influenced by these requirements and also by the value of these topics as fundamental material which would serve as a foundation for further study in this field.

REFERENCES

Further reading on image acquisition devices may be found in Fink [1957], Herrick [1976], and Fairchild [1983]. The discussion in Sec. 7.3 is based on Mundy [1977], Holland et al. [1979], and Myers [1980]. The transformations discussed in Secs. 7.4.1 and 7.4.2 can be found in most books on computer graphics (see, for example, Newman and Sproull [1979]). Additional reading on camera modeling and calibration can be found in Duda and Hart [1973] and Yakimovsky and Cunningham [1979]. The survey article by Barnard and Fischler [1982] contains a comprehensive set of references on computational stereo.

The material in Sec. 7.5 is based on Toriwaki et al. [1979], and Rosenfeld and Kak [1982]. The discussion in Sec. 7.6.1, is adapted from Gonzalez and Wintz [1977]. For details on implementing median filters see Huang et al. [1979], Wolfe and Mannos [1979], and Chaudhuri [1983]. The concept of smoothing by image averaging is discussed by Kohler and Howell [1963]. The smoothing technique for binary images discussed in Sec. 7.6.2 is based on an early paper by Unger [1959].

The material in Sec. 7.6.3 is based on Gonzalez and Fittes [1977] and Woods and Gonzalez [1981]. For further details on local enhancement see Ketcham [1976], Harris [1977], and Narendra and Fitch [1981]. Early work on edge detection can be found in Roberts [1965]. A survey of techniques used in this area a decade later is given by Davis [1975]. More recent work in this field emphasizes computational speed, as exemplified by Lee [1983] and Chaudhuri [1983]. For an introduction to edge detection see Gonzalez and Wintz [1977]. The book by Rosenfeld and Kak [1982] contains a detailed description of threshold selection techniques. A survey paper by Weska [1978] is also of interest.

PROBLEMS

7.1 How many bits would it take to store a 512×512 image in which each pixel can have 256 possible intensity values?

7.2 Propose a technique that uses a single light sheet to determine the diameter of cylindrical objects. Assume a linear array camera with a resolution of N pixels and also that the distance between the camera and the center of the cylinders is fixed.

7.3 (*a*) Discuss the accuracy of your solution to Prob. 7.2 in terms of camera resolution (N points on a line) and maximum expected cylinder diameter, D_{max}. (*b*) What is the maximum error if $N = 2048$ pixels and $D_{max} = 1$ m?

7.4 Determine if the world point with coordinates $(1/2, 1/2, \sqrt{2}/2)$ is on the optical axis of a camera located at $(0, 0, \sqrt{2})$, panned $135°$ and tilted $135°$. Assume a 50-mm lens and let $r_1 = r_2 = r_3 = 0$.

7.5 Start with Eq. (7.4-41) and derive Eqs. (7.4-42) and (7.4-43).

7.6 Show that the D_4 distance between two points p and q is equal to the shortest 4-path between these points. Is this path unique?

7.7 Show that a Fourier transform algorithm that computes $F(u)$ can be used without modification to compute the inverse transform. (*Hint*: The answer lies on using complex conjugates).

7.8 Verify that substitution of Eq. (7.6-4) into Eq. (7.6-5) yields an identity.

7.9 Give the boolean expression equivalent to Eq. (7.6-16) for a 5×5 window.

7.10 Develop a procedure for computing the median in an $n \times n$ neighborhood.

7.11 Explain why the discrete histogram equalization technique will not, in general, yield a flat histogram.

7.12 Propose a method for updating the local histogram for use in the enhancement technique discussed in Sec. 7.6.3.

7.13 The results obtained by a single pass through an image of some two-dimensional masks can also be achieved by two passes of a one-dimensional mask. For example, the result of using a 3×3 smoothing mask with coefficients $\frac{1}{9}$ (see Sec. 7.6.2) can also be obtained by first passing through an image the mask $[1 \quad 1 \quad 1]$. The result of this pass is then followed by a pass of the mask $\begin{bmatrix} 1 \\ 1 \\ 1 \end{bmatrix}$. The final result is then scaled by $\frac{1}{9}$. Show that the Sobel masks (Fig. 7.35) can be implemented by one pass of a *differencing* mask of the form $[-1 \quad 0 \quad 1]$ (or its vertical counterpart) followed by a *smoothing* mask of the form $[1 \quad 2 \quad 1]$ (or its vertical counterpart).

7.14 Show that the digital Laplacian given in Eq. (7.6-47) is proportional (by the factor $-\frac{1}{4}$) to subtracting from $f(x, y)$ an average of the 4-neighbors of (x, y). [The process of subtracting a blurred version of $f(x, y)$ from itself is called *unsharp masking*.]

EIGHT

HIGHER-LEVEL VISION

The artist is one who gives
form to difficult visions.
Theodore Gill

8.1 INTRODUCTION

For the purpose of categorizing the various techniques and approaches used in machine vision, we introduced in Sec. 7.1 three broad subdivisions: low-, medium-, and high-level vision. Low-level vision deals with basic sensing and preprocessing, topics which were covered in some detail in Chap. 7. We may view the material in that chapter as being instrumental in providing image and other relevant information that is in a form suitable for subsequent intelligent visual processing.

Although the concept of "intelligence" is somewhat vague, particularly when one is referring to a machine, it is not difficult to conceptualize the type of behavior that we may, however grudgingly, characterize as intelligent. Several characteristics come immediately to mind: (1) the ability to extract pertinent information from a background of irrelevant details, (2) the capability to learn from examples and to generalize this knowledge so that it will apply in new and different circumstances, (3) the ability to infer facts from incomplete information, and (4) the capability to generate self-motivated goals, and to formulate plans for meeting these goals.

While it is possible to design and implement a vision system with these characteristics in a *limited* environment, we do not yet know how to endow it with a range and depth of adaptive performance that comes even close to emulating human vision. Although research in biological systems is continually uncovering new and promising concepts, the state of the art in machine vision is for the most part based on analytical formulations tailored to meet specific tasks. The time frame in which we may have machines that approach human visual and other sensory capabilities is open to speculation. It is of interest to note, however, that imitating nature is not the only solution to this problem. The reader is undoubtedly familiar with early experimental airplanes equipped with flapping wings and other birdlike features. Given that the objective is to fly between two points, our present solution is quite different from the examples provided by nature. In terms of speed and achievable altitude, this solution exceeds the capabilities of these examples by a wide margin.

As indicated in Sec. 7.1, medium-level vision deals with topics in segmentation, description, and recognition of individual objects. It will be seen in the following sections that these topics encompass a variety of approaches that are well-founded on analytical concepts. High-level vision deals with issues such as those discussed in the preceding paragraph. Our knowledge of these areas and their relationship to low- and medium-level vision is significantly more vague and speculative, leading to the formulation of constraints and idealizations intended to simplify the complexity of this task.

The material discussed in this chapter introduces the reader to a broad range of topics in state-of-the-art machine vision, with a strong orientation toward techniques that are suitable for robotic vision. The material is subdivided into four principal areas. We begin the discussion with a detailed treatment of segmentation. This is followed by a discussion of object description techniques. We then discuss the principal approaches used in the recognition stage of a vision system. We conclude the chapter with a discussion of issues on the interpretation of visual information.

8.2 SEGMENTATION

Segmentation is the process that subdivides a sensed scene into its constituent parts or objects. Segmentation is one of the most important elements of an automated vision system because it is at this stage of processing that objects are extracted from a scene for subsequent recognition and analysis. Segmentation algorithms are generally based on one of two basic principles: discontinuity and similarity. The principal approach in the first category is based on edge detection; the principal approaches in the second category are based on thresholding and region growing. These concepts are applicable to both static and dynamic (time-varying) scenes. In the latter case, however, motion can often be used as a powerful cue to improve the performance of segmentation algorithms.

8.2.1 Edge Linking and Boundary Detection

The techniques discussed in Sec. 7.6.4 detect intensity discontinuities. Ideally, these techniques should yield only pixels lying on the boundary between objects and the background. In practice, this set of pixels seldom characterizes a boundary completely because of noise, breaks in the boundary due to nonuniform illumination, and other effects that introduce spurious intensity discontinuities. Thus, edge detection algorithms are typically followed by linking and other boundary detection procedures designed to assemble edge pixels into a meaningful set of object boundaries. In the following discussion we consider several techniques suited for this purpose.

Local Analysis. One of the simplest approaches for linking edge points is to analyze the characteristics of pixels in a small neighborhood (e.g., 3×3 or 5×5)

about every point (x, y) in an image that has undergone an edge detection process. All points that are similar (as defined below) are linked, thus forming a
boundary of pixels that share some common properties.

There are two principal properties used for establishing similarity of edge pixels in this kind of analysis: (1) the strength of the response of the gradient operator
used to produce the edge pixel, and (2) the direction of the gradient. The first
property is given by the value of $G[f(x, y)]$, as defined in Eqs. (7.6-38) or (7.6-
39). Thus, we say that an edge pixel with coordinates (x', y') and in the predefined
neighborhood of (x, y) is similar in magnitude to the pixel at (x, y) if

$$|G[f(x, y)] - G[f(x', y')]| \leqslant T \tag{8.2-1}$$

where T is a threshold.

The direction of the gradient may be established from the angle of the gradient
vector given in Eq. (7.6-37). That is,

$$\theta = \tan^{-1}\left[\frac{G_y}{G_x}\right] \tag{8.2-2}$$

where θ is the angle (measured with respect to the x axis) along which the rate of
change has the greatest magnitude, as indicated in Sec. 7.6.4. Then, we say that
an edge pixel at (x', y') in the predefined neighborhood of (x, y) has an angle
similar to the pixel at (x, y) if

$$|\theta - \theta'| < A \tag{8.2-3}$$

where A is an angle threshold. It is noted that the direction of the edge at (x, y)
is, in reality, perpendicular to the direction of the gradient vector at that point.
However, for the purpose of comparing directions, Eq. (8.2-3) yields equivalent
results.

Based on the foregoing concepts, we link a point in the predefined neighborhood of (x, y) to the pixel at (x, y) if both the magnitude and direction criteria
are satisfied. This process is repeated for every location in the image, keeping a
record of linked points as the center of the neighborhood is moved from pixel to
pixel. A simple bookkeeping procedure is to assign a different gray level to each
set of linked edge pixels.

Example: As an illustration of the foregoing procedure, consider Fig. 8.1a,
which shows an image of the rear of a vehicle. The objective is to find rectangles whose sizes makes them suitable license plate candidates. The formation of these rectangles can be accomplished by detecting strong horizontal and
vertical edges. Figure 8.1b and c shows the horizontal and vertical components of the Sobel operators discussed in Sec. 7.6.4. Finally, Fig. 8.1d
shows the results of linking all points which, simultaneously, had a gradient
value greater than 25 and whose gradient directions did not differ by more

Figure 8.1 (*a*) Input image. (*b*) Horizontal component of the gradient. (*c*) Vertical component of the gradient. (*d*) Result of edge linking. (Courtesy of Perceptics Corporation.)

than 15°. The horizontal lines were formed by sequentially applying these criteria to every row of Fig. 8.1*c*, while a sequential column scan of Fig. 8.1*b* yielded the vertical lines. Further processing consisted of linking edge segments separated by small breaks and deleting isolated short segments. ☐

Global Analysis via the Hough Transform. In this section we consider the linking of boundary points by determining whether or not they lie on a curve of specified shape. Suppose initially that, given n points in the xy plane of an image, we wish to find subsets that lie on straight lines. One possible solution is to first find all lines determined by every pair of points and then find all subsets of points that are close to particular lines. The problem with this procedure is that it involves finding $n(n-1)/2 \sim n^2$ lines and then performing $n[n(n-1)]/2 \sim n^3$

comparisons of every point to all lines. This is computationally prohibitive in all but the most trivial applications.

This problem may be viewed in a different way using an approach proposed by Hough [1962] and commonly referred to as the *Hough transform*. Consider a point (x_i, y_i) and the general equation of a straight line in slope-intercept form, $y_i = ax_i + b$. There is an infinite number of lines that pass through (x_i, y_i), but they all satisfy the equation $y_i = ax_i + b$ for varying values of a and b. However, if we write this equation as $b = -x_i a + y_i$, and consider the ab plane (also called *parameter space*), then we have the equation of a *single* line for a fixed pair (x_i, y_i). Furthermore, a second point (x_j, y_j) will also have a line in parameter space associated with it, and this line will intersect the line associated with (x_i, y_i) at (a', b') where a' is the slope and b' the intercept of the line containing both (x_i, y_i) and (x_j, y_j) in the xy plane. In fact, all points contained on this line will have lines in parameter space which intercept at (a', b'). These concepts are illustrated in Fig. 8.2.

The computational attractiveness of the Hough transform arises from subdividing the parameter space into so-called *accumulator cells*, as illustrated in Fig. 8.3, where (a_{max}, a_{min}) and (b_{max}, b_{min}) are the expected ranges of slope and intercept values. Accumulator cell $A(i, j)$ corresponds to the square associated with parameter space coordinates (a_i, b_j). Initially, these cells are set to zero. Then, for every point (x_k, y_k) in the image plane, we let the parameter a equal each of the allowed subdivision values on the a axis and solve for the corresponding b using the equation $b = -x_k a + y_k$. The resulting b's are then rounded off to the nearest allowed value in the b axis. If a choice of a_p results in solution b_q, we let $A(p, q) = A(p, q) + 1$. At the end of this procedure, a value of M in cell $A(i, j)$ corresponds to M points in the xy plane lying on the line $y = a_i x + b_j$. The accuracy of the colinearity of these points is established by the number of subdivisions in the ab plane.

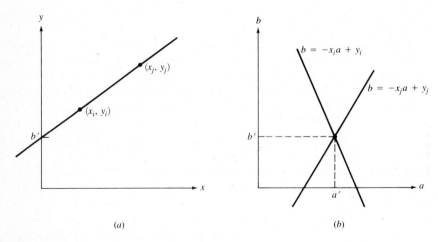

(a)   (b)

Figure 8.2 (a) xy Plane. (b) Parameter space.

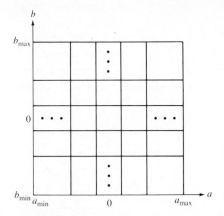

Figure 8.3 Quantization of the parameter plane into cells for use in the Hough transform.

It is noted that if we subdivide the a axis into K increments, then for every point (x_k, y_k) we obtain K values of b corresponding to the K possible values of a. Since there are n image points, this involves nK computations. Thus, the procedure just discussed is *linear* in n, and the product nK does not approach the number of computations discussed at the beginning of this section unless K approaches or exceeds n.

A problem with using the equation $y = ax + b$ to represent a line is that both the slope and intercept approach infinity as the line approaches a vertical position. One way around this difficulty is to use the normal representation of a line, given by

$$x \cos \theta + y \sin \theta = \rho \qquad (8.2\text{-}4)$$

The meaning of the parameters used in Eq. (8.2-4) is illustrated in Fig. 8.4a. The use of this representation in constructing a table of accumulators is identical to the method discussed above for the slope-intercept representation; the only difference is that, instead of straight lines, we now have sinusoidal curves in the $\theta\rho$ plane. As before, M colinear points lying on a line $x \cos \theta_i + y \sin \theta_i = \rho_j$ will yield M sinusoidal curves which intercept at (θ_i, ρ_j) in the parameter space. When we use the method of incrementing θ and solving for the corresponding ρ, the procedure will yield M entries in accumulator $A(i, j)$ associated with the cell determined by (θ_i, ρ_j). The subdivision of the parameter space is illustrated in Fig. 8.4b.

Example: An illustration of using the Hough transform based on Eq. (8.2-4) is shown in Fig. 8.5. Figure 8.5a shows an image of an industrial piece, Fig. 8.5b is the gradient image, and Fig. 8.5c shows the $\theta\rho$ plane displayed as an image in which brightness level is proportional to the number of counts in the accumulators. The abscissa in this image corresponds to θ and the ordinate to ρ, with ranges $\pm 90°$ and $\pm\rho_{\max}$, respectively. In this case, $\rho_{\max}$ was set

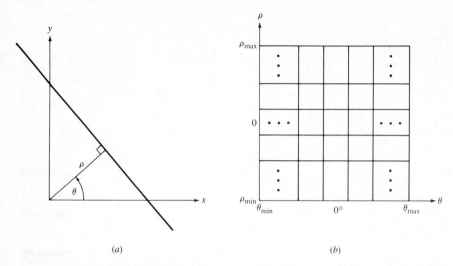

(a) (b)

Figure 8.4 (a) Normal representation of a line. (b) Quantization of the $\theta\rho$ plane into cells.

equal to the distance from corner to corner in the original image. The center of the square in Fig. 8.5c thus corresponds to $\theta = 0°$ and $\rho = 0$. It is of interest to note the bright spots (high accumulator counts) near $0°$ corresponding to the vertical lines, and near $\pm 90°$ corresponding to the horizontal lines in Fig. 8.5b. The lines detected by this method are shown in Fig. 8.5d superimposed on the original image. The discrepancy is due to the quantization error of θ and ρ in the parameter space.   □

Although our attention has been focused thus far on straight lines, the Hough transform is applicable to any function of the form $g(\mathbf{x}, \mathbf{c}) = 0$, where $\mathbf{x}$ is a vector of coordinates and $\mathbf{c}$ is a vector of coefficients. For example, the locus of points lying on the circle

$$(x - c_1)^2 + (y - c_2)^2 = c_3^2 \qquad (8.2\text{-}5)$$

can easily be detected by using the approach discussed above. The basic difference is that we now have three parameters, c_1, c_2, and c_3, which result in a three-dimensional parameter space with cubelike cells and accumulators of the form $A(i, j, k)$. The procedure is to increment c_1 and c_2, solve for the c_3 that satisfies Eq. (8.2-5), and update the accumulator corresponding to the cell associated with the triple (c_1, c_2, c_3). Clearly, the complexity of the Hough transform is strongly dependent on the number of coordinates and coefficients in a given functional representation.

Before leaving this section, we point out that further generalizations of the Hough transform to detect curves with no simple analytic representations are possible. These concepts, which are extensions of the material presented above, are treated in detail by Ballard [1981].

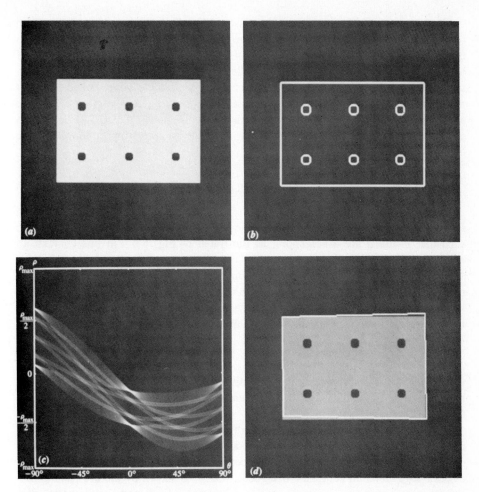

Figure 8.5 (*a*) Image of a work piece. (*b*) Gradient image. (*c*) Hough transform table. (*d*) Detected lines superimposed on the original image. (Courtesy of D. Cate, Texas Instruments, Inc.)

Global Analysis via Graph-Theoretic Techniques. The method discussed in the previous section is based on having a set of edge points obtained typically through a gradient operation. Since the gradient is a derivative, it enhances sharp variations in intensity and, therefore, is seldom suitable as a preprocessing step in situations characterized by high noise content. We now discuss a global approach based on representing edge segments in the form of a graph structure and searching the graph for low-cost paths which correspond to significant edges. This representation provides a rugged approach which performs well in the presence of noise. As might be expected, the procedure is considerably more complicated and requires more processing time than the methods discussed thus far.

We begin the development with some basic definitions. A *graph* $G = (N, A)$ is a finite, nonempty set of nodes N, together with a set A of unordered pairs of distinct elements of N. Each pair (n_i, n_j) of A is called an *arc*. A graph in which its arcs are directed is called a *directed graph*. If an arc is directed from node n_i to node n_j, then n_j is said to be a *successor* of its *parent* node n_i. The process of identifying the successors of a node is called *expansion* of the node. In each graph we will define levels, such that level 0 consists of a single node, called the *start* node, and the nodes in the last level are called *goal* nodes. A *cost* $c(n_i, n_j)$ can be associated with every arc (n_i, n_j). A sequence of nodes $n_1, n_2, \ldots, n_k$ with each node n_i being a successor of node n_{i-1} is called a *path* from n_1 to n_k, and the cost of the path is given by

$$c = \sum_{i=2}^{k} c(n_{i-1}, n_i) \tag{8.2-6}$$

Finally, we define an *edge element* as the boundary between two pixels p and q, such that p and q are 4-neighbors, as illustrated in Fig. 8.6. In this context, an *edge* is a sequence of edge elements.

In order to illustrate how the foregoing concepts apply to edge detection, consider the 3×3 image shown in Fig. 8.7, where the outer numbers are pixel coordinates and the numbers in parentheses represent intensity. With each edge element defined by pixels p and q we associate the cost

$$c(p, q) = H - [f(p) - f(q)] \tag{8.2-7}$$

where H is the highest intensity value in the image (7 in this example), $f(p)$ is the intensity value of p, and $f(q)$ is the intensity value of q. As indicated above, p and q are 4-neighbors.

The graph for this problem is shown in Fig. 8.8. Each node in this graph corresponds to an edge element, and an arc exists between two nodes if the two corresponding edge elements taken in succession can be part of an edge. The cost of each edge element, computed using Eq. (8.2-7), is shown by the arc leading into it, and goal nodes are shown in double rectangles. Each path between the start node and a goal node is a possible edge. For simplicity, it has been assumed that the edge starts in the top row and terminates in the last row, so that the first ele-

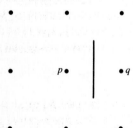

Figure 8.6 Edge element between pixels p and q.

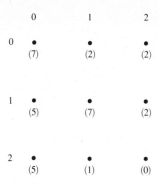

	0	1	2
0	• (7)	• (2)	• (2)
1	• (5)	• (7)	• (2)
2	• (5)	• (1)	• (0)

Figure 8.7 A 3 × 3 image.

ment of an edge can only be [(0, 0), (0, 1)] or [(0, 1), (0, 2)] and the last element [(2, 0), (2, 1)] or [(2, 1), (2, 2)]. The minimum-cost path, computed using Eq. (8.2-6), is shown dashed, and the corresponding edge is shown in Fig. 8.9.

In general, the problem of finding a minimum-cost path is not trivial from a computational point of view. Typically, the approach is to sacrifice optimality for the sake of speed, and the algorithm discussed below is representative of a class of

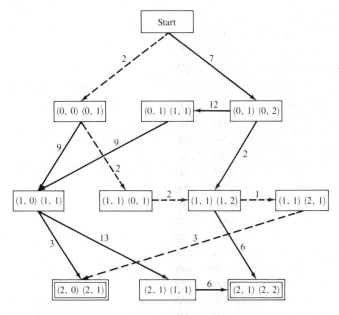

Figure 8.8 Graph used for finding an edge in the image of Fig. 8.7. The pair $(a, b)(c, d)$ in each box refers to points p and q, respectively. Note that p is assumed to be to the right of the path as the image is traversed from top to bottom. The dashed lines indicate the minimum-cost path. (Adapted from Martelli [1972], © Academic Press.)

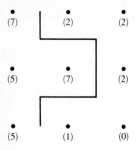

Figure 8.9 Edge corresponding to minimum-cost path in Fig. 8.8.

procedures which use heuristics in order to reduce the search effort. Let $r(n)$ be an estimate of the cost of a minimum-cost path from the start node s to a goal node, where the path is constrained to go through n. This cost can be expressed as the estimate of the cost of a minimum-cost path from s to n, plus an estimate of the cost of that path from n to a goal node; that is,

$$r(n) = g(n) + h(n) \tag{8.2-8}$$

Here, $g(n)$ can be chosen as the lowest-cost path from s to n found so far, and $h(n)$ is obtained by using any available heuristic information (e.g., expanding only certain nodes based on previous costs in getting to that node). An algorithm that uses $r(n)$ as the basis for performing a graph search is as follows:

Step 1. Mark the start node OPEN and set $g(s) = 0$.
Step 2. If no node is OPEN, exit with failure; otherwise continue.
Step 3. Mark CLOSED the OPEN node n whose estimate $r(n)$ computed from Eq. (8.2-8) is smallest. (Ties for minimum r values are resolved arbitrarily, but always in favor of a goal node.)
Step 4. If n is a goal node, exit with the solution path obtained by tracing back through the pointers; otherwise continue.
Step 5. Expand node n, generating all its successors. (If there are no successors, go to step 2.)
Step 6. If a successor n_i is not marked, set

$$r(n_i) = g(n) + c(n, n_i)$$

mark it OPEN, and direct pointers from it back to n.
Step 7. If a successor n_i is marked CLOSED or OPEN, update its value by letting

$$g'(n_i) = \min[g(n_i), g(n) + c(n, n_i)]$$

Mark OPEN those CLOSED successors whose g' values were thus lowered and redirect to n the pointers from all nodes whose g' values were lowered. Go to step 2.

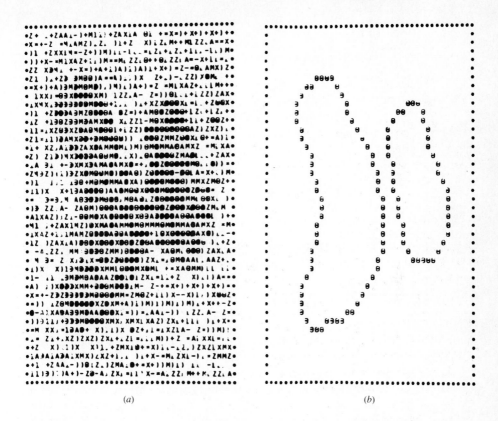

(a) (b)

Figure 8.10 (a) Noisy image. (b) Result of edge detection by using the heuristic graph search. (From Martelli [1976], © ACM.)

In general, this algorithm is not guaranteed to find a minimum-cost path; its advantage is speed via the use of heuristics. It can be shown, however, that if $h(n)$ is a lower bound on the cost of the minimal-cost path from node n to a goal node, then the procedure will indeed find an optimal path to a goal (Hart et al. [1968]). If no heuristic information is available (i.e., $h \equiv 0$) then the procedure reduces to the *uniform-cost algorithm* of Dijkstra [1959].

Example: A typical result obtainable with this procedure is shown in Fig. 8.10. Part (a) of this figure shows a noisy image and Fig. 8.10b is the result of edge segmentation by searching the corresponding graph for low-cost paths. Heuristics were brought into play by not expanding those nodes whose cost exceeded a given threshold. □

8.2.2 Thresholding

The concept of thresholding was introduced in Sec. 7.6.5 as an operation involving tests against a function of the form

$$T = T[x, \ y, \ p(x, \ y), \ f(x, \ y)] \tag{8.2-9}$$

where $f(x, \ y)$ is the intensity of point $(x, \ y)$ and $p(x, \ y)$ denotes some local property measured in a neighborhood of this point. A thresholded image, $g(x, \ y)$ is created by defining

$$g(x, \ y) = \begin{cases} 1 & \text{if } f(x, \ y) > T \\ 0 & \text{if } f(x, \ y) \leqslant T \end{cases} \tag{8.2-10}$$

so that pixels in $g(x, \ y)$ labeled 1 correspond to objects, while pixels labeled 0 correspond to the background. Equation (8.2-10) assumes that the intensity of objects is greater than the intensity of the background. The opposite condition is handled by reversing the sense of the inequalities.

Global vs. Local Thresholds. As indicated in Sec. 7.6.5, when T in Eq. (8.2-9) depends only on $f(x, \ y)$, the threshold is called *global*. If T depends on both $f(x, \ y)$ and $p(x, \ y)$, then it is called a *local threshold*. If, in addition, T depends on the spatial coordinates x and y, it is called a *dynamic threshold*.

Global thresholds have application in situations where there is clear definition between objects and background, and where illumination is relatively uniform. The backlighting and structured lighting techniques discussed in Sec. 7.3 usually yield images that can be segmented by global thresholds. For the most part, arbitrary illumination of a work space yields images that, if handled by thresholding, require some type of local analysis to compensate for effects such as nonuniformities in illumination, shadows, and reflections.

In the following discussion we consider a number of techniques for selecting segmentation thresholds. Although some of these techniques can be used for global threshold selection, they are usually employed in situations requiring local threshold analysis.

Optimum Threshold Selection. It is often possible to consider a histogram as being formed by the sum of probability density functions. In the case of a bimodal histogram the overall function approximating the histogram is given by

$$p(z) = P_1 p_1(z) + P_2 p_2(z) \tag{8.2-11}$$

where z is a random variable denoting intensity, $p_1(z)$ and $p_2(z)$ are the probability density functions, and P_1 and P_2 are called *a priori* probabilities. These last two quantities are simply the probabilities of occurrence of two types of intensity levels in an image. For example, consider an image whose histogram is shown in Fig. 8.11a. The overall histogram may be approximated by the sum of two proba-

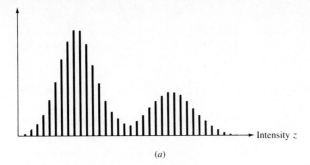

(a)

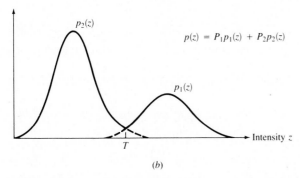

$$p(z) = P_1 p_1(z) + P_2 p_2(z)$$

(b)

Figure 8.11 (a) Intensity histogram. (b) Approximation as the sum of two probability density functions.

bility density functions, as shown in Fig. 8.11b. If it is known that light pixels represent objects and also that 20 percent of the image area is occupied by object pixels, then $P_1 = 0.2$. It is required that

$$P_1 + P_2 = 1 \tag{8.2-12}$$

which simply says that, in this case, the remaining 80 percent are background pixels.

Let us form two functions of z, as follows:

$$d_1(z) = P_1 p_1(z) \tag{8.2-13}$$

and

$$d_2(z) = P_2 p_2(z) \tag{8.2-14}$$

It is known from decision theory (Tou and Gonzalez [1974]) that the average error of misclassifying an object pixel as background, or vice versa, is minimized by using the following rule: Given a pixel with intensity value z, we substitute that value of z into Eqs. (8.2-13) and (8.2-14). Then, we classify the pixel as an object pixel if $d_1(z) > d_2(z)$ or as background pixel if $d_2(z) > d_1(z)$. The optimum threshold is then given by the value of z for which $d_1(z) = d_2(z)$. That is, setting $z = T$ in Eqs. (8.2-13) and (8.2-14), we have that the optimum threshold satisfies the equation

$$P_1 p_1(T) = P_2 p_2(T) \tag{8.2-15}$$

Thus, if the functional forms of $p_1(z)$ and $p_2(z)$ are known, we can use this equation to solve for the optimum threshold that separates objects from the background. Once this threshold is known, Eq. (8.2-10) can be used to segment a given image.

As an important illustration of the use of Eq. (8.2-15), suppose that $p_1(z)$ and $p_2(z)$ are gaussian probability density functions; that is,

$$p_1(z) = \frac{1}{\sqrt{2\pi}\sigma_1} \exp - \left[\frac{(z - m_1)^2}{2\sigma_1^2} \right] \tag{8.2-16}$$

and

$$p_2(z) = \frac{1}{\sqrt{2\pi}\sigma_2} \exp - \left[\frac{(z - m_2)^2}{2\sigma_2^2} \right] \tag{8.2-17}$$

Letting $z = T$ in these expressions, substituting into Eq. (8.2-15), and simplifying yields a quadratic equation in T:

$$AT^2 + BT + C = 0 \tag{8.2-18}$$

where

$$A = \sigma_1^2 - \sigma_2^2$$

$$B = 2(m_1 \sigma_2^2 - m_2 \sigma_1^2) \tag{8.2-19}$$

$$C = \sigma_1^2 m_2^2 - \sigma_2^2 m_1^2 + 2\sigma_1^2 \sigma_2^2 \ln \frac{\sigma_2 P_1}{\sigma_1 P_2}$$

The possibility of two solutions indicates that two threshold values may be required to obtain an optimal solution.

If the standard deviations are equal, $\sigma_1 = \sigma_2 = \sigma$, a single threshold is sufficient:

$$T = \frac{m_1 + m_2}{2} + \frac{\sigma^2}{m_1 - m_2} \ln \frac{P_2}{P_1} \tag{8.2-20}$$

If $\sigma = 0$ or $P_1 = P_2$, the optimum threshold is just the average of the means. The former condition simply means that both the object and background intensities are constant throughout the image. The latter condition means that object and background pixels are equally likely to occur, a condition met whenever the number of object pixels is equal to the number of background pixels in an image.

Example: As an illustration of the concepts just discussed, consider the segmentation of the mechanical parts shown in Fig. 8.12a, where, for the
moment, we ignore the grid superimposed on the image. Figure 8.12b shows
the result of computing a global histogram, fitting it with a bimodal gaussian
density, establishing an optimum global threshold, and finally using this threshold in Eq. (8.2-10) to segment the image. As expected, the variations in
intensity rendered this approach virtually useless. A similar approach, however, can be carried out on a local basis by subdividing the image into subimages, as defined by the grid in Fig. 8.12a.

After the image has been subdivided, a histogram is computed for each
subimage and a test of bimodality is conducted. The bimodal histograms are
fitted by a mixed gaussian density and the corresponding optimum threshold is
computed using Eqs. (8.2-18) and (8.2-19). No thresholds are computed for
subimages without bimodal histograms; instead, these regions are assigned
thresholds computed by interpolating the thresholds from neighboring subimages that are bimodal. The histograms for each subimage are shown in Fig.
8.12c, where the horizontal lines provide an indication of the relative scales of
these histograms. At the end of this procedure a second interpolation is carried out on a point-by-point manner using neighboring thresholds so that every
point is assigned a threshold value, $T(x, y)$. Note that this is a dynamic threshold since it depends on the spatial coordinates (x, y). A display of how
$T(x, y)$ varies as a function of position is shown in Fig. 8.12d.

Finally, a thresholded image is created by comparing every pixel in the
original image against its corresponding threshold. The result of using this
method in this particular case is shown in Fig. 8.12e. The improvement over
a single, global threshold is evident. It is of interest to note that this method
involves local analysis to establish the threshold for each cell, and that these
local thresholds are interpolated to create a dynamic threshold which is finally
used for segmentation. ☐

The approach developed above is applicable to the selection of multiple thresholds. Suppose that we can model a multimodal histogram as the sum of n probability density functions so that

$$p(z) = P_1 p_1(z) + \cdots + P_n p_n(z) \tag{8.2-21}$$

Then, the optimum thresholding problem may be viewed as classifying a given
pixel as belonging to one of n possible categories. The minimum-error decision
rule is now based on n functions of the form

$$d_i(z) = P_i p_i(z) \qquad i = 1, 2, \ldots, n \tag{8.2-22}$$

A given pixel with intensity z is assigned to the kth category if $d_k(z) > d_j(z)$,
$j = 1, 2, \ldots, n; j \neq k$. As before, the optimum threshold between category k

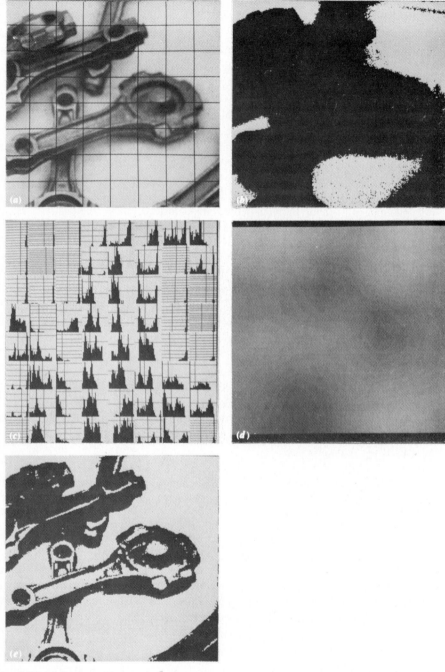

Figure 8.12 (*a*) Image of mechanical parts showing local-region grid. (*b*) Result of global thresholding. (*c*) Histograms of subimages. (*d*) Display of dynamic threshold. (*e*) Result of dynamic thresholding. (From Rosenfeld and Kak [1982], courtesy of A. Rosenfeld.)

and category j, denoted by T_{kj}, is obtained by solving the equation

$$P_k p_k(T_{kj}) = P_j p_j(T_{kj}) \qquad (8.2\text{-}23)$$

As indicated in Sec. 7.6.5, the real problem with using multiple histogram thresholds lies in establishing meaningful histogram modes.

Threshold Selection Based on Boundary Characteristics. One of the most important aspects of selecting a threshold is the capability to reliably identify the mode peaks in a given histogram. This is particularly important for automatic threshold selection in situations where image characteristics can change over a broad range of intensity distributions. Based on the discussion in the last two sections, it is intuitively evident that the chances of selecting a "good" threshold should be considerably enhanced if the histogram peaks are tall, narrow, symmetric, and separated by deep valleys.

One approach for improving the shape of histograms is to consider only those pixels that lie on or near the boundary between objects and the background. One immediate and obvious improvement is that this makes histograms less dependent on the relative size between objects and the background. For instance, the intensity histogram of an image composed of a large, nearly constant background area and one small object would be dominated by a large peak due to the concentration of background pixels. If, on the other hand, only the pixels on or near the boundary between the object and the background were used, the resulting histogram would have peaks whose heights are more balanced. In addition, the probability that a given pixel lies near the edge of an object is usually equal to the probability that it lies on the edge of the background, thus improving the symmetry of the histogram peaks. Finally, as will be seen below, using pixels that satisfy some simple measures based on gradient and Laplacian operators has a tendency to deepen the valley between histogram peaks.

The principal problem with the foregoing comments is that they implicitly assume that the boundary between objects and background is known. This information is clearly not available during segmentation since finding a division between objects and background is precisely the ultimate goal of the procedures discussed here. However, we know from the material in Sec. 7.6.4 that an indication of whether a pixel is on an edge may be obtained by computing its gradient. In addition, use of the Laplacian can yield information regarding whether a given pixel lies on the dark (e.g., background) or light (object) side of an edge. Since, as discussed in Sec. 7.6.4, the Laplacian is zero on the interior of an ideal ramp edge, we may expect in practice that the valleys of histograms formed from the pixels selected by a gradient/Laplacian criterion to be sparsely populated. This property produces the highly desirable deep valleys mentioned earlier in this section.

The gradient, $G[f(x, y)]$, at any point in an image is given by Eq. (7.6-38) or (7.6-39). Similarly, the Laplacian $L[f(x, y)]$ is given by Eq. (7.6-47). We may use these two quantities to form a three-level image, as follows:

$$
s(x, y) = \begin{cases} 0 & \text{if } G[f(x, y)] < T \\ + & \text{if } G[f(x, y)] \geq T \text{ and } L[f(x, y)] \geq 0 \quad (8.2\text{-}24) \\ - & \text{if } G[f(x, y)] \geq T \text{ and } L[f(x, y)] < 0 \end{cases}
$$

where the symbols 0, +, and − represent any three distinct gray levels, and T is a threshold. Assuming a dark object on a light background, and with reference to Fig. 7-34b, the use of Eq. (8.2-24) produces an image $s(x, y)$ in which all pixels which are not on an edge (as determined by $G[f(x, y)]$ being less than T) are labeled "0," all pixels on the dark side of an edge are labeled "+," and all pixels on the light side of an edge are labeled "−." The symbols + and − in Eq. (8.2-24) are reversed for a light object on a dark background. Figure 8.13 shows the labeling produced by Eq. (8.2-24) for an image of a dark, underlined stroke written on a light background.

The information obtained by using the procedure just discussed can be used to generate a segmented, binary image in which 1's correspond to objects of interest

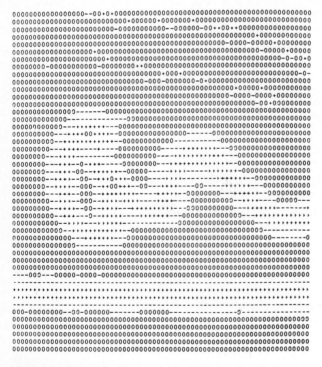

Figure 8.13 Image of a handwritten stroke coded by using Eq.(8.2-24). (From White and Rohrer [1983], ©IBM.)

and 0's correspond to the background. First we note that the transition (along a horizontal or vertical scan line) from a light background to a dark object must be characterized by the occurrence of a $-$ followed by a $+$ in $s(x, y)$. The interior of the object is composed of pixels which are labeled either 0 or $+$. Finally, the transition from the object back to the background is characterized by the occurrence of a $+$ followed by a $-$. Thus we have that a horizontal or vertical scan line containing a section of an object has the following structure:

$$(\cdots)(-, +)(0 \text{ or } +)(+, -)(\cdots)$$

where $(\cdots)$ represents any combination of $+$, $-$, or 0. The innermost parentheses contain object points and are labeled 1. All other pixels along the same scan line are labeled 0, with the exception of any sequence of (0 or $+$) bounded by $(-, +)$ and $(+, -)$.

> **Example:** As an illustration of the concepts just discussed, consider Fig. 8.14*a* which shows an image of an ordinary scenic bank check. Figure 8.15 shows the histogram as a function of gradient values for pixels with gradients greater than 5. It is noted that this histogram has the properties discussed earlier. That is, it has two dominant modes which are symmetric, nearly of the same height, and are separated by a distinct valley. Finally, Fig. 8.14*b* shows the segmented image obtained by using Eq. (8.2-24) with T near the midpoint of the valley. The result was made binary by using the sequence analysis discussed above. □

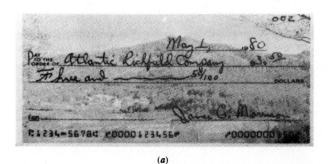

(*a*)

(*b*)

Figure 8.14 (*a*) Original image. (*b*) Segmented image. (From White and Rohrer [1983], ©IBM.)

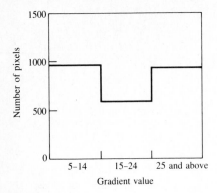

Figure 8.15 Histogram of pixels with gradients greater than 5. (From White and Rohrer [1983, ©IBM].)

Thresholds Based on Several Variables. The techniques discussed thus far deal with thresholding a single intensity variable. In some applications it is possible to use more than one variable to characterize each pixel in an image, thus enhancing the capability to differentiate not only between objects and background, but also to distinguish between objects themselves. A notable example is color sensing, where red, green, and blue (RGB) components are used to form a composite color image. In this case, each pixel is characterized by three values and it becomes possible to construct a three-dimensional histogram. The basic procedure is the same as that used for one variable. For example, given three 16-level images corresponding to the RGB components of a color sensor, we form a $16 \times 16 \times 16$ grid (cube) and insert in each cell of the cube the number of pixels whose RGB components have intensities corresponding to the coordinates defining the location of that particular cell. Each entry can then be divided by the total number of pixels in the image to form a normalized histogram.

The concept of threshold selection now becomes that of finding clusters of points in three-dimensional space, where each "tight" cluster is analogous to a dominant mode in a one-variable histogram. Suppose, for example, that we find two significant clusters of points in a given histogram, where one cluster corresponds to objects and the other to the background. Keeping in mind that each pixel now has three components and, therefore, may be viewed as a point in three-dimensional space, we can segment an image by using the following procedure: For every pixel in the image we compute the distance between that pixel and the centroid of each cluster. Then, if the pixel is closer to the centroid of the object cluster, we label it with a 1; otherwise, we label it with a 0. This concept is easily extendible to more pixel components and, certainly, to more clusters. The principal difficulty is that finding meaningful clusters generally becomes an increasingly complex task as the number of variables is increased. The reader interested in further pursuing techniques for cluster seeking can consult, for example, the book by Tou and Gonzalez [1974]. This and other related techniques for segmentation are surveyed by Fu and Mui [1981].

Example: As an illustration of the multivariable histogram approach, consider Fig. 8.16. Part (*a*) of this image is a monochrome image of a color photograph. The original color image was composed of three 16-level RGB images. For our purposes, it is sufficient to note that the scarf and one of the flowers were a vivid red, and that the hair and facial colors were light and different in spectral characteristics from the window and other background features.

Figure 8.16*b* was obtained by thresholding about a histogram cluster which was known to contain RGB components representative of flesh tones. It is important to note that the window, which in the monochrome image has a range of intensities close to those of the hair, does not appear in the segmented image because its multispectral characteristics are quite different. The fact that some small regions on top of the subject's hair appear in the segmented image indicates that their color is similar to flesh tones. Figure 8.16*c* was obtained by thresholding about a cluster close to the red axis. In this case

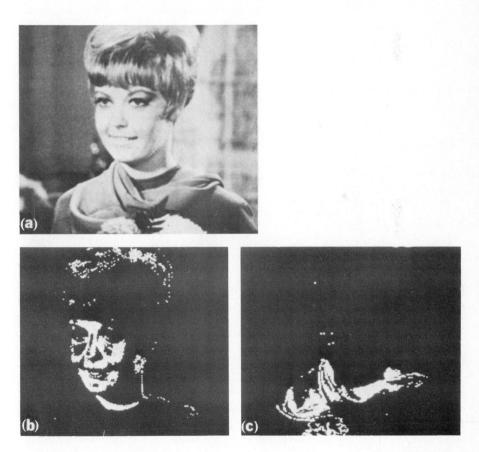

Figure 8.16 Segmentation by multivariable threshold approach. (From Gonzalez and Wintz [1977], © Addison-Wesley.)

only the scarf, the red flower, and a few isolated points appeared in the segmented image. The threshold used to obtain both results was a distance of one cell. Thus, any pixels whose components placed them within a unit distance from the centroid of the cluster under consideration were coded white. All other pixels were coded black. $\qquad\square$

8.2.3 Region-Oriented Segmentation

Basic Formulation. The objective of segmentation is to partition an image into regions. In Sec. 8.2.1 we approached this problem by finding boundaries between regions based on intensity discontinuities, while in Sec. 8.2.2 segmentation was accomplished via thresholds based on the distribution of pixel properties, such as intensity or color. In this section we discuss segmentation techniques that are based on finding the regions directly.

Let R represent the entire image region. We may view segmentation as a process that partitions R into n subregions, $R_1, R_2, \ldots, R_n$, such that

1. $\bigcup\limits_{i=1}^{n} R_i = R$
2. R_i is a connected region, $i = 1, 2, \ldots, n$
3. $R_i \cap R_j = \phi$ for all i and j, $i \neq j$
4. $P(R_i) = $ TRUE for $i = 1, 2, \ldots, n$
5. $P(R_i \cup R_j) = $ FALSE for $i \neq j$

where $P(R_i)$ is a logical predicate defined over the points in set R_i, and ϕ is the null set.

Condition 1 indicates that the segmentation must be complete; that is, every pixel must be in a region. The second condition requires that points in a region must be connected (see Sec. 7.5.2 regarding connectivity). Condition 3 indicates that the regions must be disjoint. Condition 4 deals with the properties that must be satisfied by the pixels in a segmented region. One simple example is: $P(R_i) = $ TRUE if all pixels in R_i have the same intensity. Finally, condition 5 indicates that regions R_i and R_j are different in the sense of predicate P. The use of these conditions in segmentation algorithms is discussed in the following subsections.

Region Growing by Pixel Aggregation. As implied by its name, region growing is a procedure that groups pixels or subregions into larger regions. The simplest of these approaches is *pixel aggregation*, where we start with a set of "seed" points and from these grow regions by appending to each seed point those neighboring pixels that have similar properties (e.g., intensity, texture, or color). As a simple illustration of this procedure consider Fig. 8.17a, where the numbers inside the cells represent intensity values. Let the points with coordinates (3, 2) and (3, 4) be used as seeds. Using two starting points will result in a segmentation consisting of, at most, two regions: R_1 associated with seed (3, 2) and R_2 associated

	1	2	3	4	5
1	0	0	5	6	7
2	1	1	5	8	7
3	0	1	6	7	7
4	2	0	7	6	6
5	0	1	5	6	5

(a)

a	a	b	b	b
a	a	b	b	b
a	a	b	b	b
a	a	b	b	b
a	a	b	b	b

(b)

a	a	a	a	a
a	a	a	a	a
a	a	a	a	a
a	a	a	a	a
a	a	a	a	a

(c)

Figure 8.17 Example of region growing using known starting points. (a) Original image array. (b) Segmentation result using an absolute difference of less than 3 between intensity levels. (c) Result using an absolute difference less than 8. (From Gonzalez and Wintz [1977], © Addison-Wesley.)

with seed $(3, 4)$. The property P that we will use to include a pixel in either region is that the absolute difference between the intensity of the pixel and the intensity of the seed be less than a threshold T (any pixel that satisfies this property simultaneously for both seeds is arbitrarily assigned to regions R_1). The result obtained using $T = 3$ is shown in Fig. 8.17b. In this case, the segmentation consists of two regions, where the points in R_1 are denoted by a's and the points in R_2 by b's. It is noted that any starting point in either of these two resulting regions would have yielded the same result. If, on the other hand, we had chosen $T = 8$, a single region would have resulted, as shown in Fig. 8.17c.

The preceding example, while simple in nature, points out some important problems in region growing. Two immediate problems are the selection of initial seeds that properly represent regions of interest and the selection of suitable properties for including points in the various regions during the growing process. Selecting a set of one or more starting points can often be based on the nature of the problem. For example, in military applications of infrared imaging, targets of interest are hotter (and thus appear brighter) than the background. Choosing the brightest pixels is then a natural starting point for a region-growing algorithm. When a priori information is not available, one may proceed by computing at every pixel the same set of properties that will ultimately be used to assign pixels to regions during the growing process. If the result of this computation shows clusters of values, then the pixels whose properties place them near the centroid of these clusters can be used as seeds. For instance, in the example given above, a histogram of intensities would show that points with intensity of 1 and 7 are the most predominant.

The selection of similarity criteria is dependent not only on the problem under consideration, but also on the type of image data available. For example, the analysis of land-use satellite imagery is heavily dependent on the use of color. This problem would be significantly more difficult to handle by using monochrome images alone. Unfortunately, the availability of multispectral and other complementary image data is the exception, rather than the rule, in industrial computer vision. Typically, region analysis must be carried out using a set of descriptors based on intensity and spatial properties (e.g., moments, texture) of a single image source. A discussion of descriptors useful for region characterization is given in Sec. 8.3.

It is important to note that descriptors alone can yield misleading results if connectivity or adjacency information is not used in the region growing process. An illustration of this is easily visualized by considering a random arrangement of pixels with only three distinct intensity values. Grouping pixels with the same intensity to form a "region" without paying attention to connectivity would yield a segmentation result that is meaningless in the context of this discussion.

Another important problem in region growing is the formulation of a stopping rule. Basically, we stop growing a region when no more pixels satisfy the criteria for inclusion in that region. We mentioned above criteria such as intensity, texture, and color, which are local in nature and do not take into account the "history" of region growth. Additional criteria that increase the power of a region-growing algorithm incorporate the concept of size, likeness between a candidate pixel and the pixels grown thus far (e.g., a comparison of the intensity of a candidate and the average intensity of the region), and the shape of a given region being grown. The use of these types of descriptors is based on the assumption that a model of expected results is, at least, partially available.

Region Splitting and Merging. The procedure discussed above grows regions starting from a given set of seed points. An alternative is to initially subdivide an image into a set of arbitrary, disjoint regions and then merge and/or split the regions in an attempt to satisfy the conditions stated at the beginning of this section. A split and merge algorithm which iteratively works toward satisfying these constraints may be explained as follows.

Let R represent the entire image region, and select a predicate P. Assuming a square image, one approach for segmenting R is to successively subdivide it into smaller and smaller quadrant regions such that, for any region R_i, $P(R_i) =$ TRUE. The procedure starts with the entire region R. If $P(R) =$ FALSE, we divide the image into quadrants. If P is FALSE for any quadrant, we subdivide that quadrant into subquadrants, and so on. This particular splitting technique has a convenient representation in the form of a so-called *quadtree* (i.e., a tree in which each node has exactly four descendants). A simple illustration is shown in Fig. 8.18. It is noted that the root of the tree corresponds to the entire image and that each node corresponds to a subdivision. In this case, only R_4 was subdivided further.

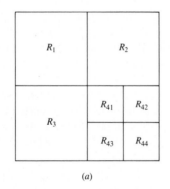

(a)

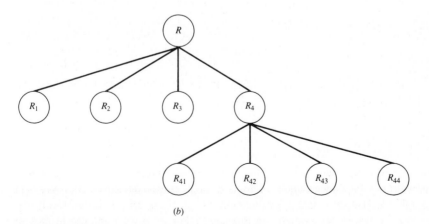

(b)

Figure 8.18 (a) Partitioned image. (b) Corresponding quadtree.

If we used only splitting, it is likely that the final partition would contain adjacent regions with identical properties. This may be remedied by allowing merging, as well as splitting. In order to satisfy the segmentation conditions stated earlier, we merge only adjacent regions whose combined pixels satisfy the predicate P; that is, we merge two adjacent regions R_i and R_k only if $P(R_i \cup R_k) =$ TRUE.

The preceding discussion may be summarized by the following procedure in which, at any step, we

1. Split into four disjoint quadrants any region R_i for which $P(R_i) =$ FALSE
2. Merge any adjacent regions R_j and R_k for which $P(R_j \cup R_k) =$ TRUE
3. Stop when no further merging or splitting is possible

A number of variations of this basic theme are possible (Horowitz and Pavlidis [1974]). For example, one possibility is to initially split the image into a set of square blocks. Further splitting is carried out as above, but merging is initially limited to groups of four blocks which are descendants in the quadtree representation and which satisfy the predicate P. When no further mergings of this type are possible, the procedure is terminated by one final merging of regions satisfying step 2 above. At this point, the regions that are merged may be of different sizes. The principal advantage of this approach is that it uses the same quadtree for splitting and merging, until the final merging step.

> **Example:** An illustration of the split and merge algorithm discussed above is shown in Fig. 8.19. The image under consideration consists of a single object and background. For simplicity, we assume that both the object and background have constant intensities and that $P(R_i) =$ TRUE if all pixels in R_i have the same intensity. Then, for the entire image region R, it follows that $P(R) =$ FALSE, so the image is split as shown in Fig. 8.19a. In the next step, only the top left region satisfies the predicate so it is not changed, while the other three quadrant regions are split into subquadrants, as shown in Fig. 8.19b. At this point several regions can be merged, with the exception of the two subquadrants that include the lower part of the object; these do not satisfy the predicate and must be split further. The results of the split and merge operation are shown in Fig. 8.19c. At this point all regions satisfy P, and merging the appropriate regions from the last split operation yields the final, segmented result shown in Fig. 8.19d. □

8.2.4 The Use of Motion

Motion is a powerful cue used by humans and other animals in extracting objects of interest from the background. In robot vision, motion arises in conveyor belt applications, by motion of a sensor mounted on a moving arm or, more rarely, by motion of the entire robot system. In this subsection we discuss the use of motion for segmentation from the point of view of image differencing.

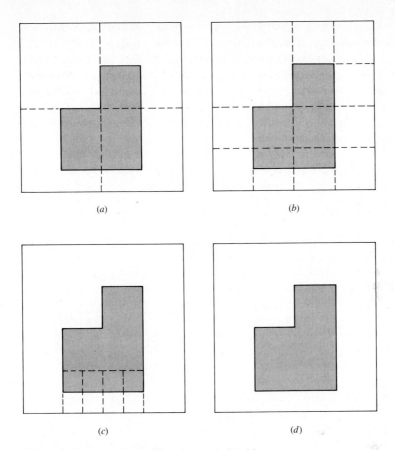

(a)   (b)

(c)   (d)

Figure 8.19 Example of split and merge algorithm.

Basic Approach. One of the simplest approaches for detecting changes between two image frames $f(x, y, t_i)$ and $f(x, y, t_j)$ taken at times t_i and t_j, respectively, is to compare the two images on a pixel-by-pixel basis. One procedure for doing this is to form a *difference image*.

Suppose that we have a reference image containing only stationary components. If we compare this image against a subsequent image having the same environment but including a moving object, the difference of the two images will cancel the stationary components, leaving only nonzero entries that correspond to the nonstationary image components.

A difference image between two images taken at times t_i and t_j may be defined as

$$d_{ij}(x, y) = \begin{cases} 1 & \text{if } |f(x, y, t_i) - f(x, y, t_j)| > \theta \\ 0 & \text{otherwise} \end{cases} \qquad (8.2\text{-}25)$$

where θ is a threshold. It is noted that $d_{ij}(x, y)$ has a 1 at spatial coordinates (x, y) only if the intensity difference between the two images is appreciably different at those coordinates, as determined by the threshold θ.

In dynamic image analysis, all pixels in $d_{ij}(x, y)$ with value 1 are considered the result of object motion. This approach is applicable only if the two images are registered and the illumination is relatively constant within the bounds established by θ. In practice, 1-valued entries in $d_{ij}(x, y)$ often arise as a result of noise. Typically, these will be isolated points in the difference image and a simple approach for their removal is to form 4- or 8-connected regions of 1's in $d_{ij}(x, y)$ and then ignore any region that has less than a predetermined number of entries. This may result in ignoring small and/or slow-moving objects, but it enhances the chances that the remaining entries in the difference image are truly due to motion.

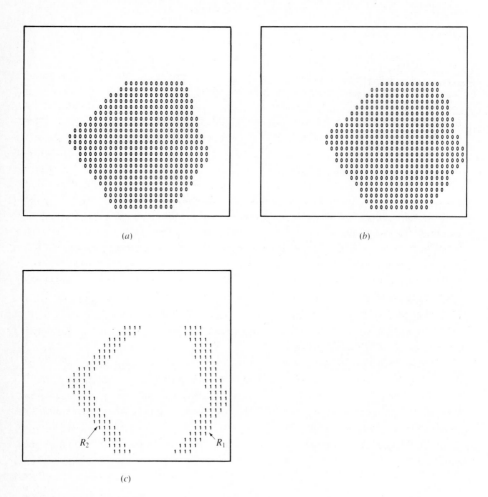

Figure 8.20 (*a*) Image taken at time t_i. (*b*) Image taken at time t_j. (*c*) Difference image. (From Jain [1981], ©IEEE.)

The foregoing concepts are illustrated in Fig. 8.20. Part (*a*) of this figure shows a reference image frame taken at time t_i and containing a single object of constant intensity that is moving with uniform velocity over a background surface, also of constant intensity. Figure 8.20*b* shows a current frame taken at time t_j, and Fig. 8.20*c* shows the difference image computed using Eq. (8.2-25) with a threshold larger than the constant background intensity. It is noted that two disjoint regions were generated by the differencing process: one region is the result of the leading edge and the other of the trailing edge of the moving object.

Accumulative Differences. As indicated above, a difference image will often contain isolated entries that are due to noise. Although the number of these entries can be reduced or completely eliminated by a thresholded connectivity analysis, this filtering process can also remove small or slow-moving objects. The approach discussed in this section addresses this problem by considering changes at a pixel location on several frames, thus introducing a "memory" into the process. The basic idea is to ignore those changes which occur only sporadically over a frame sequence and can, therefore, be attributed to random noise.

Consider a sequence of image frames $f(x, y, t_1)$, $f(x, y, t_2)$, ..., $f(x, y, t_n)$, and let $f(x, y, t_1)$ be the *reference image*. An *accumulative difference image* is formed by comparing this reference image with every subsequent image in the sequence. A counter for each pixel location in the accumulative image is incremented every time that there is a difference at that pixel location between the reference and an image in the sequence. Thus, when the *k*th frame is being compared with the reference, the entry in a given pixel of the accumulative image gives the number of times the intensity at that position was different from the corresponding pixel value in the reference image. Differences are established, for example, by use of Eq. (8.2-25).

The foregoing concepts are illustrated in Fig. 8.21. Parts (*a*) through (*e*) of this figure show a rectangular object (denoted by 0's) that is moving to the right with constant velocity of 1 pixel/frame. The images shown represent instants of time corresponding to one pixel displacement. Figure 8.21*a* is the reference image frame, Figs. 8.21*b* to *d* are frames 2 to 4 in the sequence, and Fig. 8.21*e* is the eleventh frame. Figures 8.21*f* to *i* are the corresponding accumulative images, which may be explained as follows. In Fig. 8.21*f*, the left column of 1's is due to differences between the object in Fig. 8.21*a* and the background in Fig. 8.21*b*. The right column of 1's is caused by differences between the background in the reference image and the leading edge of the moving object. By the time of the fourth frame (Fig. 8.21*d*), the first nonzero column of the accumulative difference image shows three counts, indicating three total differences between that column in the reference image and the corresponding column in the subsequent frames. Finally, Fig. 8.21*a* shows a total of 10 (represented by "A" in hexadecimal) changes at that location. The other entries in that figure are explained in a similar manner.

It is often useful to consider three types of accumulative difference images: absolute (AADI), positive (PADI), and negative (NADI). The latter two quantities

```
         9
        10   00000000
        11   00000000
        12   00000000
   (a)  13   00000000
        14   00000000
        15   00000000
        16

         9                      9
        10   00000000          10    1        1
        11   00000000          11    1        1
        12   00000000          12    1        1
   (b)  13   00000000          13    1        1          (f)
        14   00000000          14    1        1
        15   00000000          15    1        1
        16                     16

         9                      9
        10   00000000          10    21       21
        11   00000000          11    21       21
        12   00000000          12    21       21
   (c)  13   00000000          13    21       21          (g)
        14   00000000          14    21       21
        15   00000000          15    21       21
        16                     16

         9                      9
        10   00000000          10    321      321
        11   00000000          11    321      321
        12   00000000          12    321      321
   (d)  13   00000000          13    321      321          (h)
        14   00000000          14    321      321
        15   00000000          15    321      321
        16                     16

         9                              9
        10            00000000         10    A98765438887654321
        11            00000000         11    A98765438887654321
        12            00000000         12    A98765438887654321
   (e)  13            00000000         13    A98765438887654321          (i)
        14            00000000         14    A98765438887654321
        15            00000000         15    A98765438887654321
        16                             16
```

Figure 8.21 (*a*) Reference image frame. (*b*) to (*e*) Frames 2, 3, 4, and 11. (*f*) to (*i*) Accumulative difference images for frames 2, 3, 4, and 11. (From Jain [1981], ©IEEE.)

are obtained by using Eq. (8.2-25) without the absolute value and by using the reference frame instead of $f(x, y, t_i)$. Assuming that the intensities of an object are numerically greater than the background, if the difference is positive, it is compared with a positive threshold; if it is negative, the difference is compared with a negative threshold. This definition is reversed if the intensities of the object are less than the background.

Example: Figure 8.22*a* to *c* show the AADI, PADI, and NADI for a 20 × 20 pixel object whose intensity is greater than the background, and which is moving with constant velocity in a south-easterly direction. It is important to note that the spatial growth of the PADI stops when the object is displaced from its original position. In other words, when an object whose intensities are greater than the background is completely displaced from its position in the reference image, there will be no new entries generated in the positive accumulative difference image. Thus, when its growth stops, the PADI gives the initial location of the object in the reference frame. As will be seen below, this property can be used to advantage in creating a reference from a dynamic sequence of images. It is also noted in Fig. 8.22 that the AADI contains the regions of both the PADI and NADI, and that the entries in these images give an indication of the speed and direction of object movement. The images in Fig. 8.22 are shown in intensity-coded form in Fig. 8.23. □

Establishing a Reference Image. A key to the success of the techniques discussed in the previous two sections is having a reference image against which subsequent comparisons can be made. As indicated earlier, the difference between two images in a dynamic imaging problem has the tendency to cancel all stationary components, leaving only image elements that correspond to noise and to the mov-

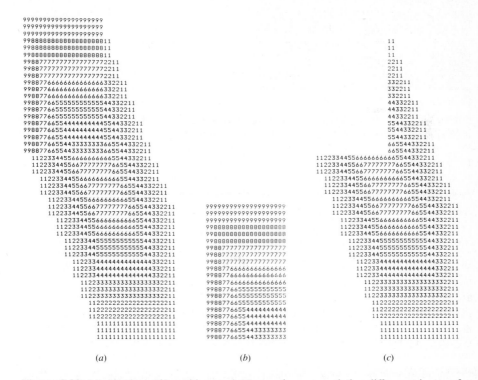

Figure 8.22 (*a*) Absolute, (*b*) positive, and (*c*) negative accumulative difference images for a 20 × 20 pixel object with intensity greater than the background and moving in a southeasterly direction. (From Jain [1983], courtesy of R. Jain.)

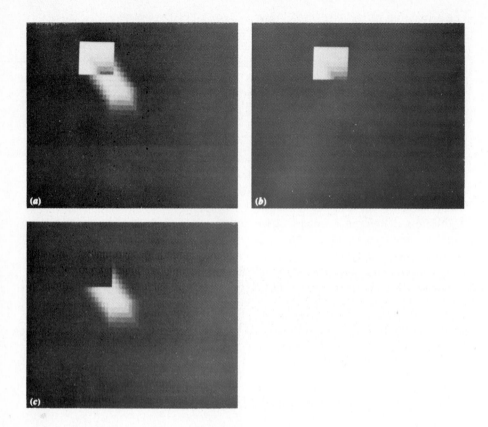

Figure 8.23 Intensity-coded accumulative difference images for Fig. 8.22. (*a*) AADI, (*b*) PADI, and (*c*) NADI. (From Jain [1983], courtesy of R. Jain.)

ing objects. The noise problem can be handled by the filtering approach discussed earlier or by forming an accumulative difference image.

In practice, it is not always possible to obtain a reference image with only stationary elements and it becomes necessary to build a reference from a set of images containing one or more moving objects. This is particularly true in situations describing busy scenes or in cases where frequent updating is required. One procedure for generating a reference image is as follows: Suppose that we consider the first image in a sequence to be the reference image. When a nonstationary component has moved completely out of its position in the reference frame, the corresponding background in the present frame can be duplicated in the location originally occupied by the object in the reference frame. When all moving objects have moved completely out of their original positions, a reference image containing only stationary components will have been created. Object displacement can be established by monitoring the growth of the PADI.

Figure 8.24 Two image frames of a traffic scene. There are two principal moving objects: a white car in the middle of the picture and a pedestrian on the lower left. (From Jain [1981], ©IEEE.)

Example: An illustration of the approach just discussed is shown in Figs. 8.24 and 8.25. Figure 8.24 shows two image frames of a traffic intersection. The first image is considered the reference, and the second depicts the same scene some time later. The principal moving features are the automobile moving from left to right and a pedestrian crossing the street in the bottom left of the picture. Removal of the moving automobile is shown in Fig. 8.25a. The pedestrian is removed in Fig. 8.25b. □

8.3 DESCRIPTION

The description problem in vision is one of extracting features from an object for the purpose of recognition. Ideally, descriptors should be independent of object size, location, and orientation and should contain enough discriminatory informa-

Figure 8.25 (a) Image with automobile removed and background restored. (b) Image with pedestrian removed and background restored. The latter image can be used as a reference. (From Jain [1981], ©IEEE.)

tion to uniquely identify one object from another. Description is a central issue in the design of vision systems in the sense that descriptors affect not only the complexity of recognition algorithms but also their performance. In Secs. 8.3.1, 8.3.2, and 8.4, respectively, we subdivide descriptors into three principal categories: boundary descriptors, regional descriptors, and descriptors suitable for representing three-dimensional structures.

8.3.1 Boundary Descriptors

Chain Codes. Chain codes are used to represent a boundary as a set of straight line segments of specified length and direction. Typically, this representation is established on a rectangular grid using 4- or 8-connectivity, as shown in Fig. 8.26. The length of each segment is established by the resolution of the grid, and the directions are given by the code chosen. It is noted that two bits are sufficient to represent all directions in the 4-code, and three bits are needed for the 8-code. Of course, it is possible to specify chain codes with more directions, but the codes shown in Fig. 8.26 are the ones most often used in practice.

To generate the chain code of a given boundary we first select a grid spacing, as shown in Fig. 8.27a. Then, if a cell is more than a specified amount (usually 50 percent) inside the boundary, we assign a 1 to that cell; otherwise, we assign it a 0. Figure 8.27b illustrates this process, where cells with value 1 are shown dark. Finally, we code the boundary between the two regions using the direction codes given in Fig. 8.26a. The result is shown in Fig. 8.27c, where the coding was started at the dot and proceeded in a clockwise direction. An alternate procedure is to subdivide the boundary into segments of equal length (i.e., each segment having the same number of pixels), connecting the endpoints of each segment with a straight line, and assigning to each line the direction closest to one of the allowed chain-code directions. An example of this approach using four directions is shown in Fig. 8.28.

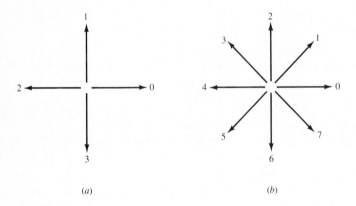

(a)   (b)

Figure 8.26 (a) 4-directional chain code. (b) 8-directional chain code.

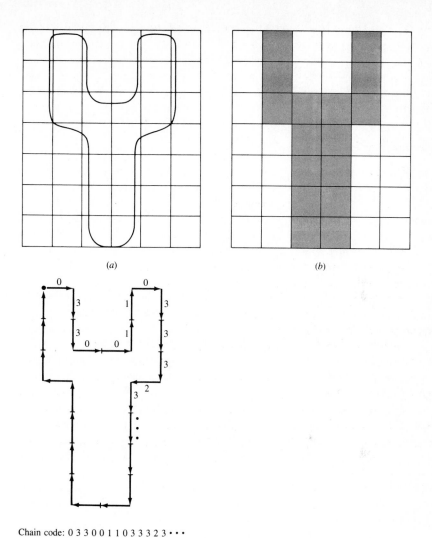

Chain code: 0 3 3 0 0 1 1 0 3 3 3 2 3 • • •

(c)

Figure 8.27 Steps in obtaining a chain code. The dot in (c) indicates the starting point.

It is important to note that the chain code of a given boundary depends upon the starting point. It is possible, however, to normalize the code by a straight-forward procedure: Given a chain code generated by starting in an arbitrary position, we treat it as a circular sequence of direction numbers and redefine the starting point so that the resulting sequence of numbers forms an integer of minimum magnitude. We can also normalize for rotation by using the first difference of the chain code, instead of the code itself. The difference is computed simply by counting (in a counterclockwise manner) the number of directions that separate two

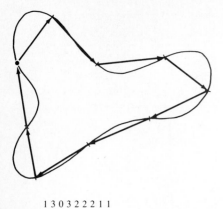

1 3 0 3 2 2 2 1 1

Figure 8.28 Generation of chain code by boundary subdivision.

adjacent elements of the code. For instance, the first difference of the 4-direction chain code 10103322 is 3133030. If we treat the code as a circular sequence, then the first element of the difference is computed using the transition between the last and first components of the chain. In this example the result is 33133030. Size normalization can be achieved by subdividing all object boundaries into the same number of equal segments and adjusting the code segment lengths to fit these subdivision, as illustrated in Fig. 8.28.

The preceding normalizations are exact only if the boundaries themselves are invariant to rotation and scale change. In practice, this is seldom the case. For instance, the same object digitized in two different orientations will in general have different boundary shapes, with the degree of dissimilarity being proportional to image resolution. This effect can be reduced by selecting chain elements which are large in proportion to the distance between pixels in the digitized image or by orienting the grid in Fig. 8.27 along the principal axes of the object to be coded. This is discussed below in the section on shape numbers.

Signatures. A signature is a one-dimensional functional representation of a boundary. There are a number of ways to generate signatures. One of the simplest is to plot the distance from the centroid to the boundary as a function of angle, as illustrated in Fig. 8.29. Signatures generated by this approach are obviously dependent on size and starting point. Size normalization can be achieved simply by normalizing the $r(\theta)$ curve to, say, unit maximum value. The starting-point problem can be solved by first obtaining the chain code of the boundary and then using the approach discussed in the previous section.

Distance vs. angle is, of course, not the only way to generate a signature. We could, for example, traverse the boundary and plot the angle between a line tangent to the boundary and a reference line as a function of position along the boundary (Ambler et al. [1975]). The resulting signature, although quite different

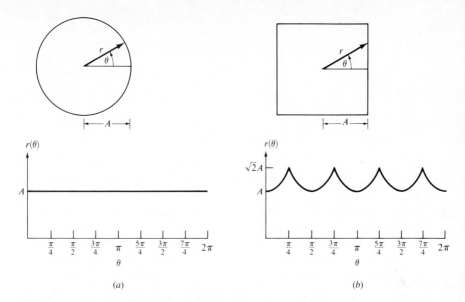

Figure 8.29 Two simple boundary shapes and their corresponding distance vs. angle signatures. In (a), $r(\theta)$ is constant, while in (b), $r(\theta) = A \sec \theta$.

from the $r(\theta)$ curve, would carry information about basic shape characteristics. For instance, horizontal segments in the curve would correspond to straight lines along the boundary since the tangent angle would be constant there. A variation of this approach is to use the so-called *slope density function* as a signature (Nahin [1974]). This function is simply a histogram of tangent angle values. Since a histogram is a measure of concentration of values, the slope density function would respond strongly to sections of the boundary with constant tangent angles (straight or nearly straight segments) and have deep valleys in sections producing rapidly varying angles (corners or other sharp inflections).

Once a signature has been obtained, we are still faced with the problem of describing it in a way that will allow us to differentiate between signatures corresponding to different boundary shapes. This problem, however, is generally easier because we are now dealing with one-dimensional functions. An approach often used to characterize a signature is to compute its *moments*. Suppose that we treat a as a discrete random variable denoting amplitude variations in a signature, and let $p(a_i)$, $i = 1, 2, \ldots, K$, denote the corresponding histogram, where K is the number of discrete amplitude increments of a. The nth moment of a about its mean is defined as

$$\mu_n(a) = \sum_{i=1}^{K} (a_i - m)^n p(a_i) \qquad (8.3\text{-}1)$$

where

$$m = \sum_{i=1}^{K} a_i p(a_i) \qquad (8.3\text{-}2)$$

The quantity m is recognized as the mean or average value of a and μ_2 as its variance. Only the first few moments are generally required to differentiate between signatures of clearly distinct shapes.

Polygonal Approximations. A digital boundary can be approximated with arbitrary accuracy by a polygon. For a closed curve, the approximation is exact when the number of segments in the polygon is equal to the number of points in the boundary so that each pair of adjacent points defines a segment in the polygon. In practice, the goal of a polygonal approximation is to capture the "essence" of the boundary shape with the fewest possible polygonal segments. Although this problem is in general not trivial and can very quickly turn into a time-consuming iterative search, there are a number of polygonal approximation techniques whose modest complexity and processing requirements makes them well-suited for robot vision applications. Several of these techniques are presented in this section.

We begin the discussion with a method proposed by Sklansky et al. [1972] for finding minimum-perimeter polygons. The procedure is best explained by means of an example. With reference to Fig. 8.30, suppose that we enclose a given boundary by a set of concatenated cells, as shown in Fig. 8.30a. We can visualize this enclosure as consisting of two walls corresponding to the outside and inside boundaries of the strip of cells, and we can think of the object boundary as a rubberband contained within the walls. If we now allow the rubberband to shrink, it will take the shape shown in Fig. 8.30b, thus producing a polygon of minimum perimeter which fits in the geometry established by the cell strip. If the cells are chosen so that each cell encompasses only one point on the boundary, then the error in each cell between the original boundary and the rubberband approximation would be at most $\sqrt{2}d$, where d is the distance between pixels. This error can be reduced in half by forcing each cell to be centered on its corresponding pixel.

Merging techniques based on error or other criteria have been applied to the problem of polygonal approximation. One approach is to merge points along a boundary until the least-squares error line fit of the points merged thus far exceeds a preset threshold. When this occurs, the parameters of the line are stored, the error is set to zero, and the procedure is repeated, merging new points along the boundary until the error again exceeds the threshold. At the end of the procedure the intersections of adjacent line segments form the vertices of a polygon. One of the principal difficulties with this method is that vertices do not generally correspond to inflections (such as corners) in the boundary because a new line is not started until the error threshold is exceeded. If, for instance, a long straight line were being tracked and it turned a corner, a number (depending on the threshold) of points past the corner would be absorbed before the threshold is exceeded.

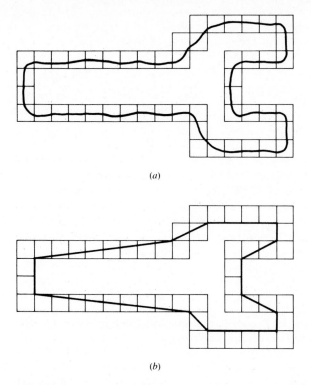

(a)

(b)

Figure 8.30 (a) Object boundary enclosed by cells. (b) Minimum-perimeter polygon.

It is possible, however, to use splitting along with merging to alleviate this difficulty.

One approach to boundary segment *splitting* is to successively subdivide a segment into two parts until a given criterion is satisfied. For instance, we might require that the maximum perpendicular distance from a boundary segment to the line joining its two endpoints not exceed a preset threshold. If it does, the furthest point becomes a vertex, thus subdividing the initial segment into two subsegments. This approach has the advantage that it "seeks" prominent inflection points. For a closed boundary, the best starting pair of points is usually the two furthest points in the boundary. An example is shown in Fig. 8.31. Part (a) of this figure shows an object boundary, and Fig. 8.31b shows a subdivision of this boundary (solid line) about its furthest points. The point marked c has the largest perpendicular distance from the top segment to line ab. Similarly, point d has the largest distance in the bottom segment. Figure 8.31c shows the result of using the splitting procedure with a threshold equal to 0.25 times the length of line ab. Since no point in the new boundary segments has a perpendicular distance (to its corresponding straight-line segment) which exceeds this threshold, the procedure terminates with the polygon shown in Fig. 8.31d.

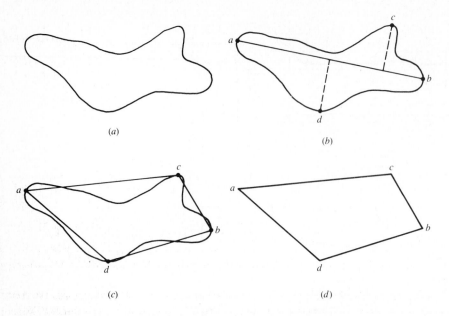

Figure 8.31 (*a*) Original boundary. (*b*) Boundary subdivided along furthest points. (*c*) Joining of vertices by straight line segments. (*d*) Resulting polygon.

We point out before leaving this section that a considerable amount of work has been done in the development of techniques which combine merging and splitting. A comprehensive discussion of these methods is given by Pavlidis [1977].

Shape Numbers. A chain-coded boundary has several first differences, depending on the starting point. The *shape number* of such a boundary, based on the 4-directional code of Fig. 8.26*a* is defined as the first difference of smallest magnitude. The *order*, n, of a shape number is defined as the number of digits in its representation. It is noted that n is even for a closed boundary, and that its value limits the number of possible different shapes. Figure 8.32 shows all the shapes of orders 4, 6, and 8, along with their chain-code representations, first differences, and corresponding shape numbers. Note that the first differences were computed by treating the chain codes as a circular sequence in the manner discussed earlier.

Although the first difference of a chain code is independent of rotation, the coded boundary in general will depend on the orientation of the coding grid shown in Fig. 8.27*a*. One way to normalize the grid orientation is as follows. The *major axis* of a boundary is the straight-line segment joining the two points furthest away from each other. The *minor axis* is perpendicular to the major axis and of length such that a box could be formed that just encloses the boundary. The ratio of the major to minor axis is called the *eccentricity* of the boundary, and the rectangle just described is called the *basic rectangle*. In most cases a unique shape number will be obtained by aligning the chain-code grid with the sides of the basic rectan-

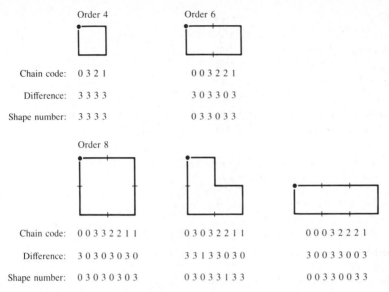

	Order 4	Order 6
Chain code:	0 3 2 1	0 0 3 2 2 1
Difference:	3 3 3 3	3 0 3 3 0 3
Shape number:	3 3 3 3	0 3 3 0 3 3

	Order 8		
Chain code:	0 0 3 3 2 2 1 1	0 3 0 3 2 2 1 1	0 0 0 3 2 2 2 1
Difference:	3 0 3 0 3 0 3 0	3 3 1 3 3 0 3 0	3 0 0 3 3 0 0 3
Shape number:	0 3 0 3 0 3 0 3	0 3 0 3 3 1 3 3	0 0 3 3 0 0 3 3

Figure 8.32 All shapes of order 4, 6, and 8. The directions are from Fig. 8.26, and the dot indicates the starting point.

gle. Freeman and Shapira [1975] give an algorithm for finding the basic rectangle of a closed, chain-coded curve.

In practice, given a desired shape order, we find the rectangle of order n whose eccentricity best approximates that of the basic rectangle, and use this new rectangle to establish the grid size. For example, if $n = 12$, all the rectangles of order 12 (i.e., those whose perimeter length is 12) are 2×4, 3×3, and 1×5. If the eccentricity of the 2×4 rectangle best matches the eccentricity of the basic rectangle for a given boundary, we establish a 2×4 grid centered on the basic rectangle and use the procedure already outlined to obtain the chain code. The shape number follows from the first difference of this code, as indicated above. Although the order of the resulting shape number will usually be equal to n because of the way the grid spacing was selected, boundaries with depressions comparable with this spacing will sometimes yield shape numbers of order greater than n. In this case, we specify a rectangle of order lower than n and repeat the procedure until the resulting shape number is of order n.

Example: Suppose that we specify $n = 18$ for the boundary shown in Fig. 8.33a. In order to obtain a shape number of this order we follow the steps discussed above. First we find the basic rectangle, as shown in Fig. 8.33b. The closest rectangle of order 18 is a 3×6 rectangle, and so we subdivide the basic rectangle as shown in Fig. 8.33c, where it is noted that the chain code directions are aligned with the resulting grid. Finally, we obtain the chain code and use its first difference to compute the shape number, as shown in Fig. 8.33d. □

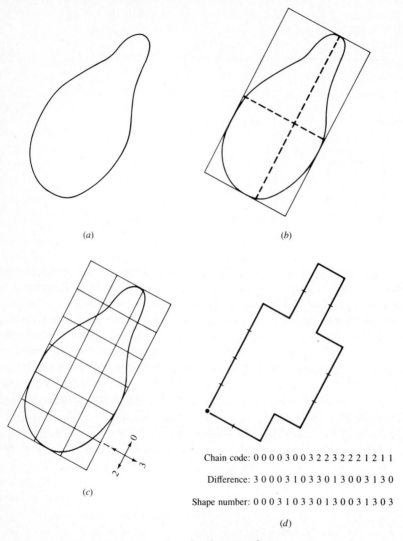

Chain code: 0 0 0 0 3 0 0 3 2 2 3 2 2 2 1 2 1 1

Difference: 3 0 0 0 3 1 0 3 3 0 1 3 0 0 3 1 3 0

Shape number: 0 0 0 3 1 0 3 3 0 1 3 0 0 3 1 3 0 3

(d)

Figure 8.33 Steps in the generation of a shape number.

Fourier Descriptors. The discrete, one-dimensional Fourier transform given in Eq. (7.6-4) can often be used to describe a two-dimensional boundary. Suppose that M points on a boundary are available. If, as shown in Fig. 8.34, we view this boundary as being in the complex plane, then each two-dimensional boundary point (x, y) is reduced to the one-dimensional complex number $x + jy$. The sequence of points along the boundary forms a function whose Fourier transform is $F(u)$, $u = 0, 1, 2, \ldots, M - 1$. If M is an integer power of 2, $F(u)$ can be computed using an FFT algorithm, as discussed in Sec. 7.6.1. The motivation for this approach is that only the first few components of $F(u)$ are generally required

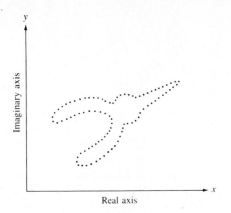

Figure 8.34 Representation of a region boundary in the frequency domain.

to distinguish between shapes that are reasonably distinct. For example, the objects shown in Fig. 8.35 can be differentiated by using less than 10 percent of the elements of the complete Fourier transform of their boundaries.

The Fourier transform is easily normalized for size, rotation, and starting point on the boundary. To change the size of a contour we simply multiply the components of $F(u)$ by a constant. Due to the linearity of the Fourier transform pair, this is equivalent to multiplying (scaling) the boundary by the same factor. Rotation by an angle θ is similarly handled by multiplying the elements of $F(u)$ by $\exp(j\theta)$. Finally, it can be shown that shifting the starting point of the contour in the spatial domain corresponds to multiplying the kth component of $F(u)$ by $\exp(jkT)$, where T is in the interval $[0, 2\pi]$. As T goes from 0 to 2π, the starting point traverses the entire contour once. This information can be used as the basis for normalization (Gonzalez and Wintz [1977]).

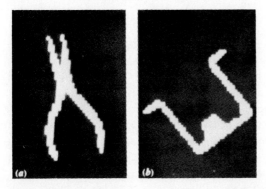

Figure 8.35 Two shapes easily distinguishable by Fourier descriptors. (From Persoon and Fu [1977], ©IEEE.)

8.3.2 Regional Descriptors

A region of interest can be described by the shape of its boundary, as discussed in Sec. 8.3.1, or by its internal characteristics, as indicated in the following discussion. It is thus important to note that the methods developed in both of these sections are applicable to region descriptions.

Some Simple Descriptors. A number of existing industrial vision systems are based on regional descriptors which are rather simple in nature and thus are attractive from a computational point of view. As might be expected, the use of these descriptors is limited to situations in which the objects of interest are so distinct that a few global descriptors are sufficient for their characterization.

The *area* of a region is defined as the number of pixels contained within its boundary. This is a useful descriptor when the viewing geometry is fixed and objects are always analyzed approximately the same distance from the camera. A typical application is the recognition of objects moving on a conveyor belt past a vision station.

The *major* and *minor* axes of a region are defined in terms of its boundary (see Sec. 8.3.1) and are useful for establishing the orientation of an object. The ratio of the lengths of these axes, called the *eccentricity* of the region, is also an important global descriptor of its shape.

The *perimeter* of a region is the length of its boundary. Although the perimeter is sometimes used as a descriptor, its most frequent application is in establishing a measure of *compactness* of a region, defined as perimeter^2/area. It is of interest to note that compactness is a dimensionless quantity (and thus is insensitive to scale changes) and that it is minimum for a disk-shaped region.

A *connected region* is a region in which all pairs of points can be connected by a curve lying entirely in the region. For a set of connected regions, some of which may have holes, it is useful to consider the *Euler number* as a descriptor. The Euler number is defined simply as the number of connected regions minus the number of holes. As an example, the Euler numbers of the letters A and B are 0 and −1, respectively. A number of other regional descriptors are discussed below.

Texture. The identification of objects or regions in an image can often be accomplished, at least partially, by the use of texture descriptors. Although no formal definition of texture exists, we intuitively view this descriptor as providing quantitative measures of properties such as smoothness, coarseness, and regularity (some examples are shown in Fig. 8.36). The two principal approaches to texture description are statistical and structural. Statistical approaches yield characterizations of textures as being smooth, coarse, grainy, and so on. Structural techniques, on the other hand, deal with the arrangement of image primitives, such as the description of texture based on regularly spaced parallel lines.

One of the simplest approaches for describing texture is to use moments of the intensity histogram of an image or region. Let z be a random variable denoting

Figure 8.36 Examples of (*a*) smooth, (*b*) coarse, and (*c*) regular texture.

discrete image intensity, and let $p(z_i)$, $i = 1, 2, \ldots, L$, be the corresponding histogram, where L is the number of distinct intensity levels. As indicated in Sec. 8.3.1, the nth moment of z about the mean is defined as

$$\mu_n(z) = \sum_{i=1}^{L} (z_i - m)^n p(z_i) \tag{8.3-3}$$

where m is the mean value of z (i.e., the average image intensity):

$$m = \sum_{i=1}^{L} z_i p(z_i) \tag{8.3-4}$$

It is noted from Eq. (8.3-3) that $\mu_0 = 1$ and $\mu_1 = 0$. The second moment [also called the *variance* and denoted by $\sigma^2(z)$] is of particular importance in texture description. It is a measure of intensity contrast which can be used to establish descriptors of relative smoothness. For example, the measure

$$R = 1 - \frac{1}{1 + \sigma^2(z)} \tag{8.3-5}$$

is 0 for areas of constant intensity [$\sigma^2(z) = 0$ if all z_i have the same value] and approaches 1 for large values of $\sigma^2(z)$. The third moment is a measure of the skewness of the histogram while the fourth moment is a measure of its relative flatness. The fifth and higher moments are not so easily related to histogram shape, but they do provide further quantitative discrimination of texture content.

Measures of texture computed using only histograms suffer from the limitation that they carry no information regarding the relative position of pixels with respect to each other. One way to bring this type of information into the texture analysis process is to consider not only the distribution of intensities but also the positions of pixels with equal or nearly equal intensity values. Let P be a position operator and let $\mathbf{A}$ be a $k \times k$ matrix whose element a_{ij} is the number of times that points with intensity z_i occur (in the position specified by P) relative to points with intensity z_j, with $1 \leqslant i, j \leqslant k$. For instance, consider an image with three intensities, $z_1 = 0$, $z_2 = 1$, and $z_3 = 2$, as follows:

$$
\begin{array}{ccccc}
0 & 0 & 0 & 1 & 2 \\
1 & 1 & 0 & 1 & 1 \\
2 & 2 & 1 & 0 & 0 \\
1 & 1 & 0 & 2 & 0 \\
0 & 0 & 1 & 0 & 1
\end{array}
$$

If we define the position operator P as "one pixel to the right and one pixel below," then we obtain the following 3×3 matrix $\mathbf{A}$:

$$
\mathbf{A} = \begin{bmatrix} 4 & 2 & 1 \\ 2 & 3 & 2 \\ 0 & 2 & 0 \end{bmatrix}
$$

where, for example, a_{11} (top left) is the number of times that a point with intensity level $z_1 = 0$ appears one pixel location below and to the right of a pixel with the same intensity, while a_{13} (top right) is the number of times that a point with level $z_1 = 0$ appears one pixel location below and to the right of a point with intensity $z_3 = 2$. It is important to note that the size of $\mathbf{A}$ is determined strictly by the number of distinct intensities in the input image. Thus, application of the concepts discussed in this section usually require that intensities be requantized into a few bands in order to keep the size of $\mathbf{A}$ manageable.

Let n be the total number of point pairs in the image which satisfy P (in the above example $n = 16$). If we define a matrix $\mathbf{C}$ formed by dividing every element of $\mathbf{A}$ by n, then c_{ij} is an estimate of the joint probability that a pair of points satisfying P will have values (z_i, z_j). The matrix $\mathbf{C}$ is called a *gray-level co-occurrence matrix*, where "gray level" is used interchangeably to denote the intensity of a monochrome pixel or image. Since $\mathbf{C}$ depends on P, it is possible to detect the presence of given texture patterns by choosing an appropriate position operator. For instance, the operator used in the above example is sensitive to bands of constant intensity running at $-45°$ (note that the highest value in $\mathbf{A}$ was $a_{11} = 4$, partially due to a streak of points with intensity 0 and running at $-45°$). In a more general situation, the problem is to analyze a given $\mathbf{C}$ matrix in order to categorize the texture of the region over which $\mathbf{C}$ was computed. A set of descriptors proposed by Haralick [1979] include

1. Maximum probability:

$$\max_{i,j} (c_{ij})$$

2. Element-difference moment of order k:

$$\sum_i \sum_j (i - j)^k c_{ij}$$

3. Inverse element-difference moment of order k:

$$\frac{\sum_i \sum_j c_{ij}}{(i - j)^k} \qquad i \neq j$$

4. Entropy:

$$-\sum_i \sum_j c_{ij} \log c_{ij}$$

5. Uniformity:

$$\sum_i \sum_j c_{ij}^2$$

The basic idea is to characterize the "content" of $\mathbf{C}$ via these descriptors. For example, the first property gives an indication of the strongest response to P (as in the above example). The second descriptor has a relatively low value when the high values of $\mathbf{C}$ are near the main diagonal since the differences $(i - j)$ are smaller there. The third descriptor has the opposite effect. The fourth descriptor is a measure of randomness, achieving its highest value when all elements of $\mathbf{C}$ are equal. Conversely, the fifth descriptor is lowest when the c_{ij} are all equal. One approach for using these descriptors is to "teach" a system representative descriptor values for a set of different textures. The texture of an unknown region is then subsequently determined by how closely its descriptors match those stored in the system memory. This approach is discussed in more detail in Sec. 8.4.

The approaches discussed above are statistical in nature. As mentioned at the beginning of this section, a second major category of texture description is based on structural concepts. Suppose that we have a rule of the form $S \rightarrow aS$ which indicates that the symbol S may be rewritten as aS (e.g., three applications of this rule would yield the string $aaaS$). If we let a represent a circle (Fig. 8.37a) and assign the meaning of "circles to the right" to a string of the form $aaa \cdots$, then the rule $S \rightarrow aS$ allows us to generate a texture pattern of the form shown in Fig. 8.37b.

Suppose next that we add some new rules to this scheme: $S \rightarrow bA$, $A \rightarrow cA$, $A \rightarrow c$, $A \rightarrow bS$, $S \rightarrow a$, such that the presence of a b means "circle down" and the presence of a c means "circle to the left." We can now generate a string of the form $aaabccbaa$ which corresponds to a three-by-three matrix of circles. Larger texture patterns, such as the one shown in Fig. 8.37c can easily be generated in the same way. (It is noted, however, that these rules can also generate structures that are not rectangular).

The basic idea in the foregoing discussion is that a simple "texture primitive" can be used to form more complex texture patterns by means of some rules which limit the number of possible arrangements of the primitive(s). These concepts lie at the heart of structural pattern generation and recognition, a topic which will be treated in considerably more detail in Sec. 8.5.

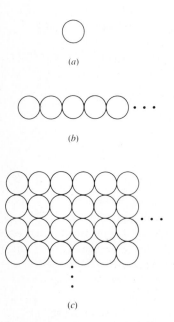

(a)

(b)

(c)

Figure 8.37 (a) Texture primitive. (b) Pattern generated by the rule $S \rightarrow aS$. (c) Two-dimensional texture pattern generated by this plus other rules.

Skeleton of a Region. An important approach for representing the structural shape of a plane region is to reduce it to a graph. This is often accomplished by obtaining the *skeleton* of the region via a thinning (also called *skeletonizing*) algorithm. Thinning procedures play a central role in a broad range of problems in computer vision, ranging from automated inspection of printed circuit boards to counting of asbestos fibers on air filters.

The skeleton of a region may be defined via the *medial axis transformation* (MAT) proposed by Blum [1967]. The MAT of a region R with border B is as follows. For each point p in R, we find its closest neighbor in B. If p has more than one such neighbor, then it is said to belong to the *medial axis* (skeleton) of R. It is important to note that the concept of "closest" depends on the definition of a distance (see Sec. 7.5.3) and, therefore, the results of a MAT operation will be influenced by the choice of a given metric. Some examples using the euclidean distance are shown in Fig. 8.38.

Although the MAT of a region yields an intuitively pleasing skeleton, a direct implementation of the above definition is typically prohibitive from a computational point of view because it potentially involves calculating the distance from every interior point to every point on the boundary of a region. A number of algorithms have been proposed for improving computational efficiency while, at the same time, attempting to produce a medial axis representation of a given region. Typically, these are thinning algorithms that iteratively delete edge points of a region subject to the constraints that the deletion of these points (1) does not remove endpoints, (2) does not break connectedness, and (3) does not cause excessive erosion of the region. Although some attempts have been made to use skeletons in gray-scale images (Dyer and Rosenfeld [1979], Salari and Siy [1984]) this type of representation is usually associated with binary data.

In the following discussion, we present an algorithm developed by Naccache and Shinghal [1984]. This procedure is fast, straightforward to implement, and, as will be seen below, yields skeletons that are in many cases superior to those

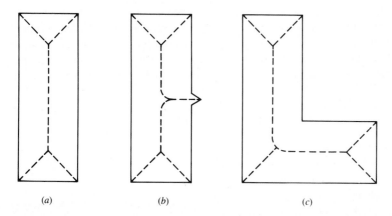

(a)   (b)   (c)

Figure 8.38 Medial axes of three simple regions.

obtained with other thinning algorithms. We begin the development with a few definitions. Assuming binary data, region points will be denoted by 1's and background points by 0's. These will be called *dark* and *light* points, respectively. An *edge point* is a dark point which has at least one light 4-neighbor. An *endpoint* is a dark point which has one and only one dark 8-neighbor. A *breakpoint* is a dark point whose deletion would break connectedness. As is true with all thinning algorithms, noise and other spurious variations along the boundary can significantly alter the resulting skeleton (Fig. 8.38b shows this effect quite clearly). Consequently, it is assumed that the boundaries of all regions have been smoothed prior to thinning by using, for example, the procedure discussed in Sec. 7.6.2.

With reference to the neighborhood arrangement shown in Fig. 8.39, the thinning algorithm identifies an edge point p as one or more of the following four types: (1) a *left edge point* having its left neighbor n_4 light; (2) a *right edge point* having n_0 light; (3) a *top edge point* having n_2 light; and (4) a *bottom edge point* having n_6 light. It is possible for p to be classified into more than one of these types. For example, a dark point p having n_0 and n_4 light will be a right edge point and a left edge point simultaneously. The following discussion initially addresses the identification (flagging) of left edge points that should be deleted. The procedure is then extended to the other types.

An edge point p is flagged if it is *not* an endpoint or breakpoint, or if its deletion would cause excessive erosion (as discussed below). The test for these conditions is carried out by comparing the 8-neighborhood of p against the windows shown in Fig. 8.40, where p and the asterisk are dark points and d and e are "don't care" points; that is, they can be either dark or light. If the neighborhood of p matches windows (a) to (c), two cases may arise: (1) If all d's are light, then p is an endpoint, or (2) if at least one of the d's is dark, then p is a breakpoint. In either case p should not be flagged.

The analysis of window (d) is slightly more complicated. If at least one d and e are dark, then p is a break point and should not be flagged. Other arrangements need to be considered, however. Suppose that all d's are light and the e's can be either dark or light. This condition yields the eight possibilities shown in Fig. 8.41. Configurations (a) through (c) make p an endpoint, and configuration (d) makes it a breakpoint. If p were deleted in configurations (e) and (f), it is easy to show by example that its deletion would cause excessive erosion in slanting regions of width 2. In configuration (g), p is what is commonly referred to as a

n_3	n_2	n_1
n_4	p	n_0
n_5	n_6	n_7

Figure 8.39 Notation for the neighbors of p used by the thinning algorithm. (From Naccache and Shinghal [1984], ©IEEE.)

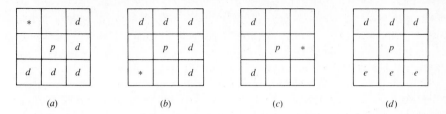

Figure 8.40 If the 8-neighborhood of a dark point p matches any of the above windows, then p is not flagged. The asterisk denotes a dark point, and d and e can be either dark or light. (From Naccache and Shinghal [1984], © IEEE.)

spur, typically due to a short tail or protrusion in a region. Since it is assumed that the boundary of the region has been smoothed initially, the appearance of a spur during thinning is considered an important description of shape and p should not be deleted. Finally, if all isolated points are removed initially, the appearance of configuration (h) during thinning indicates that a region has been reduced to a single point; its deletion would erase the last remaining portion of the region. Similar arguments apply if the roles of d and e were reversed or if the d's and e's were allowed to assume dark and light values. The essence of the preceding discussion is that any left edge point p whose 8-neighborhood matches any of the windows shown in Fig. 8.40 should not be flagged.

Testing the 8-neighborhood of p against the four windows in Fig. 8.40 has a particularly simple boolean representation given by

$$B_4 = n_0 \cdot (n_1 + n_2 + n_6 + n_7) \cdot (n_2 + \bar{n}_3) \cdot (\bar{n}_5 + n_6) \quad (8.3\text{-}6)$$

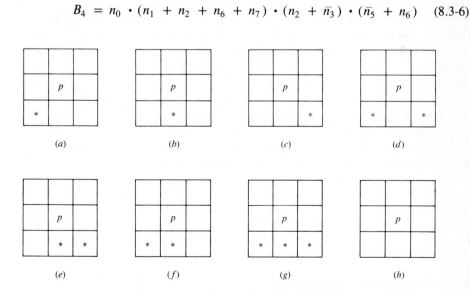

Figure 8.41 All the configurations that could exist if d is light in Fig. 8.40, and e can be dark, *, or light. (From Naccache and Shinghal [1984], © IEEE.)

where the subscript on B indicates that n_4 is light (i.e., p is a left edge point), "$\cdot$" is the logical AND, "$+$" is the logical OR, "$-$" is the logical COMPLE-MENT, and the n's are as defined in Fig. 8.39. Equation (8.3-6) is evaluated by letting dark, *previously unflagged* points be valued 1 (TRUE), and light or flagged points be valued 0 (FALSE). Then if B_4 is 1 (TRUE), we flag p. Otherwise, p is left unflagged. It is not difficult to show that these conditions on B_4 implement all four windows in Fig. 8.40 simultaneously.

Similar expressions are obtained for right edge points,

$$B_0 = n_4 \cdot (n_2 + n_3 + n_5 + n_6) \cdot (n_6 + \bar{n}_7) \cdot (\bar{n}_1 + n_2) \quad (8.3\text{-}7)$$

for top edge points,

$$B_2 = n_6 \cdot (n_0 + n_4 + n_5 + n_7) \cdot (n_0 + \bar{n}_1) \cdot (\bar{n}_3 + n_4) \quad (8.3\text{-}8)$$

and for the bottom edge points,

$$B_6 = n_2 \cdot (n_0 + n_1 + n_3 + n_4) \cdot (n_4 + \bar{n}_5) \cdot (n_0 + \bar{n}_7) \quad (8.3\text{-}9)$$

Using the above expressions, the thinning algorithm iteratively performs two scans through the data. The scanning sequence can be either along the rows or columns of the image, but the choice will generally affect the final result. In the first scan we use B_4 and B_0 to flag left and right edge points; in the second scan we use B_2 and B_6 to flag top and bottom edge points. If no new edge points were flagged during the two scans, the algorithm stops, with the unflagged points consti-tuting the skeleton; otherwise, the procedure is repeated. It is again noted that previously flagged dark points are treated as 0 in evaluating the boolean expres-sions. An alternate procedure is to set any flagged point at zero during execution of the algorithm, thus producing only skeleton and background points at the end. This approach is easier to implement, at the cost of losing all other points in the region.

Example: Figure 8.42a shows a binary region, and Fig. 8.42b shows the skeleton obtained by using the algorithm developed above. As a point of interest, Fig. 8.42c shows the skeleton obtained by applying to the same data another, well-known thinning algorithm (Pavlidis [1982]). The fidelity of the skeleton in Fig. 8.42b over that shown in Fig. 8.42c is evident. □

Moment Invariants. It was noted in Sec. 8.3.1 that Fourier descriptors which are insensitive to translation, rotation, and scale change can be used to describe the boundary of a region. When the region is given in terms of its interior points, we can describe it by a set of moments which are invariant to these effects.

Let $f(x, y)$ represent the intensity at point (x, y) in a region. The *moment* of order $(p + q)$ for the region is defined as

$$m_{pq} = \sum_x \sum_y x^p y^q f(x, y) \quad (8.3\text{-}10)$$

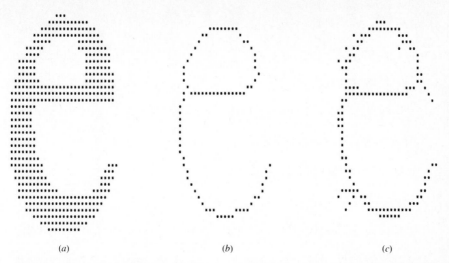

Figure 8.42 (*a*) Binary region. (*b*) Skeleton obtained using the thinning algorithm discussed in this section. (*c*) Skeleton obtained by using another algorithm. (From Naccache and Shinghal [1984], © IEEE.)

where the summation is taken over all spatial coordinates (x, y) of points in the region. The *central moment* of order $(p + q)$ is given by

$$\mu_{pq} = \sum_x \sum_y (x - \bar{x})^p (y - \bar{y})^q f(x, y) \qquad (8.3\text{-}11)$$

where

$$\bar{x} = \frac{m_{10}}{m_{00}} \qquad \bar{y} = \frac{m_{01}}{m_{00}} \qquad (8.3\text{-}12)$$

The *normalized central moments* of order $(p + q)$ are defined as

$$\eta_{pq} = \frac{\mu_{pq}}{\mu_{00}^\gamma} \qquad (8.3\text{-}13)$$

where

$$\gamma = \frac{p + q}{2} + 1 \qquad \text{for } (p + q) = 2, 3, \ldots \qquad (8.3\text{-}14)$$

The following set of *moment invariants* can be derived using only the normalized central moments of orders 2 and 3:

$$\phi_1 = \eta_{20} + \eta_{02} \qquad (8.3\text{-}15)$$

$$\phi_2 = (\eta_{20} - \eta_{02})^2 + 4\eta_{11}^2 \qquad (8.3\text{-}16)$$

$$\phi_3 = (\eta_{30} - 3\eta_{12})^2 + (3\eta_{21} - \eta_{03})^2 \tag{8.3-17}$$

$$\phi_4 = (\eta_{30} + \eta_{12})^2 + (\eta_{21} + \eta_{03})^2 \tag{8.3-18}$$

$$\phi_5 = (\eta_{30} - 3\eta_{12})(\eta_{30} + \eta_{12})[(\eta_{30} + \eta_{12})^2 - 3(\eta_{21} + \eta_{03})^2]$$
$$+ (3\eta_{21} - \eta_{03})(\eta_{21} + \eta_{03})[3(\eta_{30} + \eta_{12})^2 - (\eta_{21} + \eta_{03})^2] \tag{8.3-19}$$

$$\phi_6 = (\eta_{20} - \eta_{02})[(\eta_{30} + \eta_{12})^2 - (\eta_{21} + \eta_{03})^2]$$
$$+ 4\eta_{11}(\eta_{30} + \eta_{12})(\eta_{21} + \eta_{03}) \tag{8.3-20}$$

$$\phi_7 = (3\eta_{21} - \eta_{03})(\eta_{30} + \eta_{12})[(\eta_{30} + \eta_{12})^2 - 3(\eta_{21} + \eta_{03})^2]$$
$$+ (3\eta_{12} - \eta_{30})(\eta_{21} + \eta_{03})[3(\eta_{30} + \eta_{12})^2 - (\eta_{21} + \eta_{03})^2] \tag{8.3-21}$$

This set of moments has been shown to be invariant to translation, rotation, and scale change (Hu [1962]).

8.4 SEGMENTATION AND DESCRIPTION OF THREE-DIMENSIONAL STRUCTURES

Attention was focused in the previous two sections on techniques for segmenting and describing two-dimensional structures. In this section we consider the problem of performing these tasks on three-dimensional (3D) scene data.

As indicated in Sec. 7.1, vision is inherently a 3D problem. It is thus widely accepted that a key to the development of versatile vision systems capable of operating in unconstrained environments lies in being able to process three-dimensional scene information. Although research in this area spans more than a 10-year history, we point out that factors such as cost, speed, and complexity have inhibited the use of three-dimensional vision techniques in industrial applications.

Three-dimensional information about a scene may be obtained in three principal forms. If range sensing is used, we obtain the (x, y, z) coordinates of points on the surface of objects. The use of stereo imaging devices yields 3D coordinates, as well as intensity information about each point. In this case, we represent each point in the form $f(x, y, z)$, where the value of f at (x, y, z) gives the intensity of that point (the term *voxel* is often used to denote a 3D point and its intensity). Finally, we may *infer* 3D relationships from a single two-dimensional image of a scene. In other words, it is often possible to deduce relationships between objects such as "above," "behind," and "in front of." Since the exact 3D location of scene points generally cannot be computed from a single view, the relationships obtained from this type of analysis are sometimes referred to as 2½ D information.

8.4.1 Fitting Planar Patches to Range Data

One of the simplest approaches for segmenting and describing a three-dimensional structure given in terms of range data points (x, y, z) is to first subdivide it into small planar "patches" and then combine these patches into larger surface elements according to some criterion. This approach is particularly attractive for polyhedral objects whose surfaces are smooth with respect to the resolution of the sensed scene.

We illustrate the basic concepts underlying this approach by means of the example shown in Fig. 8.43. Part (a) of this figure shows a simple scene and Fig. 8.43b shows a set of corresponding 3D points. These points can be assembled into small surface elements by, for example, subdividing the 3D space into cells and grouping points according to the cell which contains them. Then, we fit a plane to the group of points in each cell and calculate a unit vector which is normal to the plane and passes through the centroid of the group of points in that cell. A planar

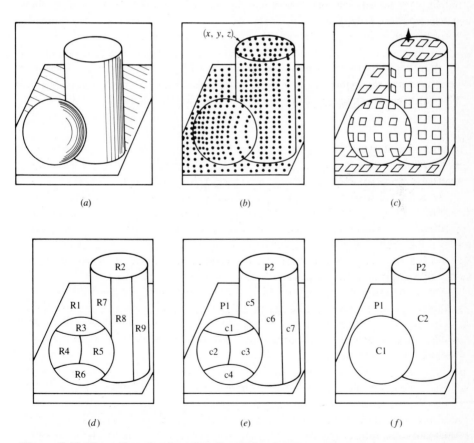

Figure 8.43 Three-dimensional surface description based on planar patches. (From Shirai [1979], © Plenum Press.)

patch is established by the intersection of the plane and the walls of the cell, with the direction of the patch being given by the unit normal, as illustrated in Fig. 8.43*c*. All patches whose directions are similar within a specified threshold are grouped into elementary regions (R), as shown in Fig. 8.43*d*. These regions are then classified as planar (P), curved (C), or undefined (U) by using the directions of the patches within each region (for example, the patches in a planar surface will all point in essentially the same direction). This type of region classification is illustrated in Fig. 8.43*e*. Finally (and this is the hardest step), the classified regions are assembled into global surfaces by grouping adjacent regions of the same classification, as shown in Fig. 8.43*f*. It is noted that, at the end of this procedure, the scene has been segmented into distinct surfaces, and that each surface has been assigned a descriptor (e.g., curved or planar).

8.4.2 Use of the Gradient

When a scene is given in terms of voxels, the 3D gradient can be used to obtain patch representations (similar to those discussed in Sec. 8.4.1) which can then be combined to form surface descriptors. As indicated in Sec. 7.6.4, the gradient vector is normal to the direction of maximum rate of change of a function, and the magnitude of this vector is proportional to the strength of that change. These concepts are just as applicable in three dimensions and they can be used to segment 3D structures in a manner analogous to that used for two-dimensional data.

Given a function $f(x, y, z)$, its gradient vector at coordinates (x, y, z) is given by

$$
\mathbf{G}[f(x, y, z)] = \begin{bmatrix} G_x \\ G_y \\ G_z \end{bmatrix} = \begin{bmatrix} \dfrac{\partial f}{\partial x} \\ \dfrac{\partial f}{\partial y} \\ \dfrac{\partial f}{\partial z} \end{bmatrix}
\tag{8.4-1}
$$

The magnitude of $\mathbf{G}$ is given by

$$
G[f(x, y, z)] = (G_x^2 + G_y^2 + G_z^2)^{1/2}
\tag{8.4-2}
$$

which, as indicated in Eq. (7.6-39), is often approximated by absolute values to simplify computation:

$$
G[f(x, y, z)] \approx |G_x| + |G_y| + |G_z|
\tag{8.4-3}
$$

The implementation of the 3D gradient can be carried out using operators analogous in form to those discussed in Sec. 7.6.4. Figure 8.44 shows a $3 \times 3 \times 3$ operator proposed by Zucker and Hummel [1981] for computing G_x. The same operator oriented along the y axis is used to compute G_y, and oriented along the z axis to compute G_z. A key property of these operators is that they yield the best (in a least-squares error sense) planar edge between two regions of different intensities in a 3D neighborhood.

The center of each operator is moved from voxel to voxel and applied in exactly the same manner as their two-dimensional counterparts, as discussed in Sec. 7.6.4. That is, the responses of these operators at any point (x, y, z) yield G_x, G_y, and G_z, which are then substituted into Eq. (8.4-1) to obtain the gradient vector at (x, y, z) and into Eq. (8.4-2) or (8.4-3) to obtain the magnitude. It is of interest to note that the operator shown in Fig. 8.44 yields a zero output in a $3 \times 3 \times 3$ region of constant intensity.

It is a straightforward procedure to utilize the gradient approach for segmenting a scene into planar patches analogous to those discussed in the previous section. It is not difficult to show that the gradient vector of a plane $ax + by + cz = 0$ has components $G_x = a$, $G_y = b$, and $G_z = c$. Since the operators discussed above yield an optimum planar fit in a $3 \times 3 \times 3$ neighborhood, it follows that the components of the vector **G** establish the direction of a planar patch in each neighborhood, while the magnitude of **G** gives an indication of abrupt changes of intensity within the patch; that is, it indicates the presence of

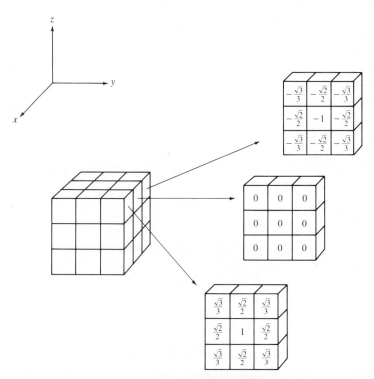

Figure 8.44 A $3 \times 3 \times 3$ operator for computing the gradient component G_x. (Adapted from Zucker and Hummel [1981], © IEEE.)

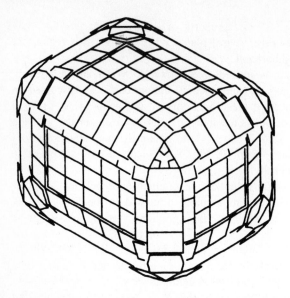

Figure 8.45 Planar patch approximation of a cube using the gradient. (From Zucker and Hummel [1981], ©IEEE.)

an intensity edge within the patch. An example of such a patch representation using the gradient operators is shown in Fig. 8.45. Since each planar patch surface passes through the center of a voxel, the borders of these patches may not always coincide. Patches that coincide are shown as larger uniform regions in Fig. 8.45.

Once patches have been obtained, they can be grouped and described in the form of global surfaces as discussed in Sec. 8.4.1. Note, however, that additional information in the form of intensity and intensity discontinuities is now available to aid the merging and description process.

8.4.3 Line and Junction Labeling

With reference to the discussion in the previous two sections, edges in a 3D scene are determined by discontinuities in range and/or intensity data. Given a set of surfaces and the edges between them, a finer description of a scene may be obtained by labeling the lines corresponding to these edges and the junctions which they form.

As illustrated in Fig. 8.46, we consider basic types of lines. A *convex line* (labeled +) is formed by the intersection of two surfaces which are part of a convex solid (e.g., the line formed by the intersection of two sides of a cube). A *concave line* (labeled −) is formed by the intersection of two surfaces belonging to two different solids (e.g., the intersection of one side of a cube with the floor). An *occluding line* (labeled with an arrow) is the edge of a surface which obscures a

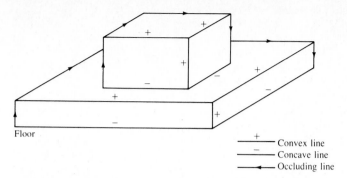

Figure 8.46 Three basic line labels.

surface. The occluding matter is to the right of the line looking in the direction of the arrow, and the occluded surface is to the left.

After the lines in a scene have been labeled, their junctions provide clues as to the nature of the 3D solids in the scene. Physical constraints allow only a few possible combinations of line labels at a junction. For example, in a polyhedral scene, no line can change its label between vertices. Violation of this rule leads to impossible physical objects, as illustrated in Fig. 8.47.

The key to using junction analysis is to form a dictionary of allowed junction types. For example, it is easily shown that the junction dictionary shown in Fig. 8.48 contains all valid labeled vertices of trihedral solids (i.e., solids in which exactly three plane surfaces come together at each vertex). Once the junctions in a scene have been classified according to their match in the dictionary, the objective is to group the various surfaces into objects. This is typically accomplished via a set of heuristic rules designed to interpret the labeled lines and sequences of neighboring junctions. The basic concept underlying this approach can be illustrated with the aid of Fig. 8.49. We note in Fig. 8.49*b* that the blob is composed entirely of an occluding boundary, with the exception of a short concave line,

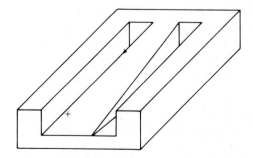

Figure 8.47 An impossible physical object. Note that one of the lines changes label from occluding to convex between two vertices.

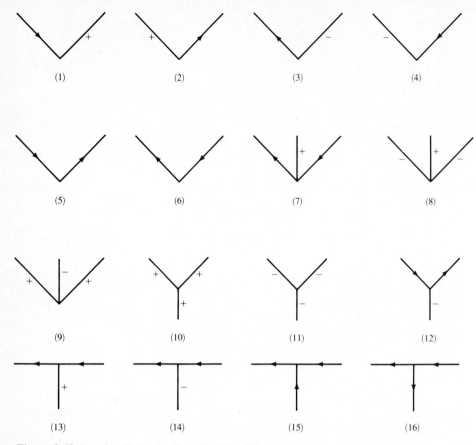

Figure 8.48 Junction dictionary for trihedral solids.

indicating where it touches the base. Thus, there is nothing in front of it and it can be extracted from the scene. We also note that there is a vertex of type (10) from the dictionary in Fig. 8.48. This is strong evidence (if we know we are dealing with trihedral objects) that the three surfaces involved in that vertex form a cube. Similar comments apply to the base after the cube surfaces are removed. Removing the base leaves the single object in the background, which completes the decomposition of the scene.

Although the preceding short explanation gives an overall view of how line and junction analysis are used to describe 3D objects in a scene, we point out that formulation of an algorithm capable of handling more complex scenes is far from a trivial task. Several comprehensive efforts in this area are referenced at the end of this chapter.

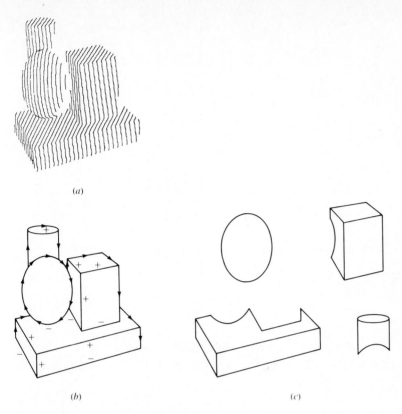

(a)

(b)   *(c)*

Figure 8.49 (*a*) Scene. (*b*) Labeled lines. (*c*) Decomposition via line and junction analysis. (Adapted from Shirai [1979], © Plenum Press.)

8.4.4 Generalized Cones

A *generalized cone* (or *cylinder*) is the volume described by a planar cross section as it is translated along an arbitrary space curve (the *spine*), held at a constant angle to the curve, and transformed according to a *sweeping rule*. In machine vision, generalized cones provide viewpoint-independent representations of three-dimensional structures which are useful for description and model-based matching purposes.

Figure 8.50 illustrates the procedure for generating generalized cones. In Fig. 8.50*a* the cross section is a ring, the spine is a straight line and the sweeping rule is to translate the cross section normal to the spine while keeping its diameter constant. The result is a hollow cylinder. In Fig. 8.50*b* we have essentially the same situation, with the exception that the sweeping rule holds the diameter of the cross section constant and then allows it to increase linearly past the midpoint of the spine.

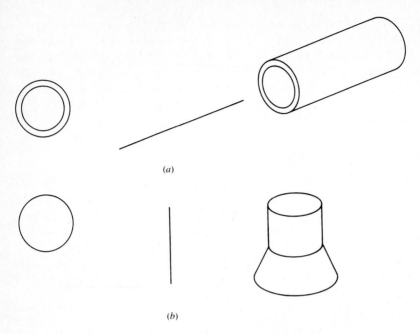

Figure 8.50 Cross sections, spines, and their corresponding generalized cones. In (a) the cross section remained constant during the sweep, while in (b) its diameter increased linearly past the midpoint in the spine.

When matching a set of 3D points against a set of known generalized cones, we first determine the center axis of the points and then find the closest set of cross sections that will fit the data as we travel along the spine. In general, considerable trial and error is required, particularly when one is dealing with incomplete data.

8.5 RECOGNITION

Recognition is a labeling process; that is, the function of recognition algorithms is to identify each segmented object in a scene and to assign a label (e.g., wrench, seal, bolt) to that object. For the most part, the recognition stages of present industrial vision systems operate on the assumption that objects in a scene have been segmented as individual units. Another common constraint is that images be acquired in a known viewing geometry (usually perpendicular to the work space). This decreases variability in shape characteristics and simplifies segmentation and description by reducing the possibility of occlusion. Variability in object orientation is handled by choosing rotation-invariant descriptors or by using the principal axis of an object to orient it in a predefined direction.

Recognition approaches in use today can be divided into two principal categories: *decision-theoretic* and *structural*. As will be seen in the following discussion, decision-theoretic methods are based on quantitative descriptions (e.g., statistical texture) while structural methods rely on symbolic descriptions and their relationships (e.g., sequences of directions in a chain-coded boundary). With a few exceptions, the procedures discussed in this section are generally used to recognize two-dimensional object representations.

8.5.1 Decision-Theoretic Methods

Decision-theoretic pattern recognition is based on the use of *decision (discriminant) functions*. Let $\mathbf{x} = (x_1, x_2, \ldots, x_n)^T$ represent a column *pattern vector* with real components, where x_i is the ith descriptor of a given object (e.g., area, average intensity, perimeter length). Given M object classes, denoted by $\omega_1, \omega_2, \ldots, \omega_M$, the basic problem in decision-theoretic pattern recognition is to identify M decision functions, $d_1(\mathbf{x}), d_2(\mathbf{x}), \ldots, d_M(\mathbf{x})$, with the property that the following relationship holds for any pattern vector $\mathbf{x}^*$ belonging to class ω_i:

$$d_i(\mathbf{x}^*) > d_j(\mathbf{x}^*) \qquad j = 1, 2, \ldots, M; \ j \neq i \qquad (8.5\text{-}1)$$

In other words, an unknown object represented by vector $\mathbf{x}^*$ is recognized as belonging to the ith object class if, upon substitution of $\mathbf{x}^*$ into all decision functions, $d_i(\mathbf{x}^*)$ yields the largest value.

The predominant use of decision functions in industrial vision systems is for *matching*. Suppose that we represent each object class by a *prototype* (or *average*) vector:

$$\mathbf{m}_i = \frac{1}{N} \sum_{k=1}^{N} \mathbf{x}_k \qquad i = 1, 2, \ldots, M \qquad (8.5\text{-}2)$$

where the $\mathbf{x}_k$ are sample vectors known to belong to class ω_i. Given an unknown $\mathbf{x}^*$, one way to determine its class membership is to assign it to the class of its closest prototype. If we use the euclidean distance to determine closeness, the problem reduces to computing the following distance measures:

$$D_j(\mathbf{x}^*) = \|\mathbf{x}^* - \mathbf{m}_j\| \qquad j = 1, 2, \ldots, M \qquad (8.5\text{-}3)$$

where $\|\mathbf{a}\| = (\mathbf{a}^T\mathbf{a})^{1/2}$ is the euclidean norm. We then assign $\mathbf{x}^*$ to class ω_i if $D_i(\mathbf{x}^*)$ is the smallest distance. It is not difficult to show that this is equivalent to evaluating the functions

$$d_j(\mathbf{x}^*) = (\mathbf{x}^*)^T\mathbf{m}_j - \tfrac{1}{2}\mathbf{m}_j^T\mathbf{m}_j \qquad j = 1, 2, \ldots, M \qquad (8.5\text{-}4)$$

and selecting the largest value. This formulation agrees with the concept of a decision function, as defined in Eq. (8.5-1).

Another application of matching is in searching for an instance of a subimage $w(x, y)$ in a larger image $f(x, y)$. At each location (x, y) of $f(x, y)$ we define the *correlation coefficient* as

$$\gamma(x, y) = \frac{\sum_s \sum_t [w(s, t) - m_w][f(s, t) - m_f]}{\left\{ \sum_s \sum_t [w(s, t) - m_w]^2 \sum_s \sum_t [f(s, t) - m_f]^2 \right\}^{1/2}} \qquad (8.5\text{-}5)$$

where it is assumed that $w(s, t)$ is centered at coordinates (x, y). The summations are taken over the image coordinates common to both regions, m_w is the average intensity of w, and m_f is the average intensity of f in the region coincident with w. It is noted that, in general, $\gamma(x, y)$ will vary from one location to the next and that its values are in the range $[-1, 1]$, with a value of 1 corresponding to a perfect match. The procedure, then, is to compute $\gamma(x, y)$ at each location

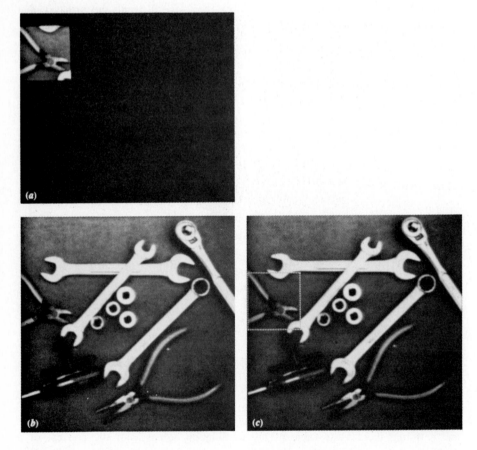

Figure 8.51 (*a*) Subimage $w(x, y)$. (*b*) Image $f(x, y)$. (*c*) Location of the best match of w in f, as determined by the largest correlation coefficient.

(x, y) and to select its largest value to determine the best match of w in f [the procedure of moving $w(x, y)$ throughout $f(x, y)$ is analogous to Fig. 7.20].

The quality of the match can be controlled by accepting a correlation coefficient only if it exceeds a preset value (for example, .9). Since this method consists of directly comparing two regions, it is clearly sensitive to variations in object size and orientation. Variations in intensity are normalized by the denominator in Eq. (8.5-5). An example of matching by correlation is shown in Fig. 8.51.

8.5.2 Structural Methods

The techniques discussed in Sec. 8.5.1 deal with patterns on a quantitative basis, ignoring any geometrical relationships which may be inherent in the shape of an object. Structural methods, on the other hand, attempt to achieve object discrimination by capitalizing on these relationships.

Central to the structural recognition approach is the decomposition of an object into *pattern primitives*. This idea is easily explained with the aid of Fig. 8.52. Part (*a*) of this figure shows a simple object boundary, and Fig. 8.52*b* shows a set of primitive elements of specified length and direction. By starting at the top left, tracking the boundary in a clockwise direction, and identifying instances of these primitives, we obtain the coded boundary shown in Fig. 8.52*c*. Basically, what we have done is represent the boundary by the string *aaabcbbbcdddcd*. The known length and direction of these primitives, together with the order in which they occur, establishes the structure of the object in terms of this particular representation. The objective of this section is to introduce the reader to techniques suitable for handling this and other types of structural pattern descriptions.

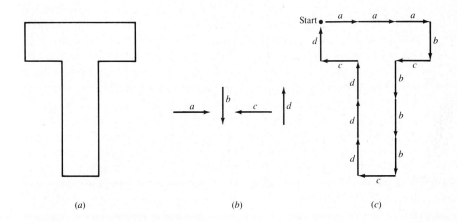

(a)   (b)   (c)

Figure 8.52 (*a*) Object boundary. (*b*) Primitives. (*c*) Boundary coded in terms of primitives, resulting in the string *aaabcbbbcdddcd*.

Matching Shape Numbers. A procedure analogous to the minimum-distance concept introduced in Sec. 8.5.1 for vector representations can be formulated for the comparison of two object boundaries that are described in terms of shape numbers. With reference to the discussion in Sec. 8.3.1, the *degree of similarity k* between two object boundaries, A and B, is defined as the largest order for which their shape numbers still coincide. That is, we have $s_4(A) = s_4(B)$, $s_6(A) = s_6(B)$, $s_8(A) = s_8(B), \ldots, \quad s_k(A) = s_k(B), \quad s_{k+2}(A) \neq s_{k+2}(B)$, $s_{k+4}(A) \neq s_{k+4}(B), \ldots$, where s indicates shape number and the subscript indicates the order. The *distance* between two shapes A and B is defined as the inverse of their degree of similarity:

$$D(A, B) = \frac{1}{k} \tag{8.5-6}$$

This distance satisfies the following properties:

(a) $D(A, B) \geqslant 0$

(b) $D(A, B) = 0 \quad$ iff $A = B$ \hfill (8.5-7)

(c) $D(A, C) \leqslant \max\,[D(A, B), D(B, C)]$

In order to compare two shapes, we can use either k or D. If the degree of similarity is used, then we know from the above discussion that the larger k is, the more similar the shapes are (note that k is infinite for identical shapes). The reverse is true when the distance measure is used.

Example: As an illustration of the preceding concepts, suppose that we wish to find which of the five shapes (A, B, D, E, F) in Fig. 8.53a best matches shape C. This is analogous to having five prototype shapes whose identities are known and trying to determine which of these constitutes the best match to an unknown shape. The search may be visualized with the aid of the *similarity tree* shown in Fig. 8.53b. The root of the tree corresponds to the lowest degree of similarity considered, which in this example is 4. As shown in the tree, all shapes are identical up to degree 8, with the exception of shape A. That is, the degree of similarity of this shape with respect to all the others is 6. Proceeding down the tree we find that shape D has degree 8 with respect to the remaining shapes, and so on. In this particular case, shape F turned out to be a unique match for C and, furthermore, their degree of similarity is higher than any of the other shapes. If E had been the unknown, a unique match would have also been found, but with a lower degree of similarity. If A had been the unknown, all we could have said using this method is that it is similar to the other five figures with degree 6. The same information can be summarized in the form of a *similarity matrix*, as shown in Fig. 8.53c. $\qquad\square$

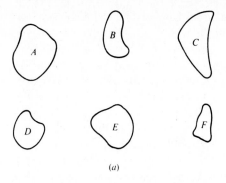

(a)

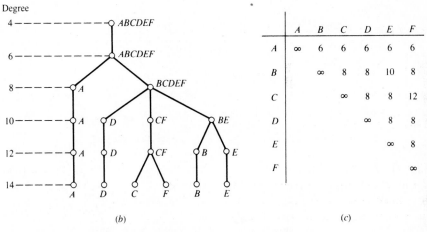

(b)

	A	B	C	D	E	F
A	∞	6	6	6	6	6
B		∞	8	8	10	8
C			∞	8	8	12
D				∞	8	8
E					∞	8
F						∞

(c)

Figure 8.53 (a) Shapes. (b) Similarity tree. (c) Similarity matrix. (From Bribiesca and Guzman [1980], © Pergamon Press.)

String Matching. Suppose that two object contours C_1 and C_2 are coded into strings $a_1 a_2 \cdots a_n$ and $b_1 b_2 \cdots b_m$, respectively. Let A represent the number of matches between the two strings, where we say that a match has occurred in the jth position if $a_j = b_j$. The number of symbols that do not match up is given by

$$B = \max(|C_1|, |C_2|) - A \qquad (8.5\text{-}8)$$

where $|C|$ is the length (number of symbols) of string C. It can be shown that $B = 0$ if and only if C_1 and C_2 are identical.

A simple measure of similarity between strings C_1 and C_2 is defined as the ratio

$$R = \frac{A}{B} = \frac{A}{\max(|C_1|, |C_2|) - A} \qquad (8.5\text{-}9)$$

Based on the above comment regarding B, R is infinite for a perfect match and zero when none of the symbols in C_1 and C_2 match (i.e., $A = 0$ in this case). Since the matching is done on a symbol-by-symbol basis, the starting point on each boundary when creating the string representation is important. Alternatively, we can start at arbitrary points on each boundary, shift one string (with wraparound), and compute Eq. (8.5-9) for each shift. The number of shifts required to perform all necessary comparisons is max ($|C_1|$, $|C_2|$).

Example: Figure 8.54a and b shows a sample boundary from each of two classes of objects. The boundaries were approximated by a polygonal fit (Fig. 8.54c and d) and then strings were formed by computing the interior angle

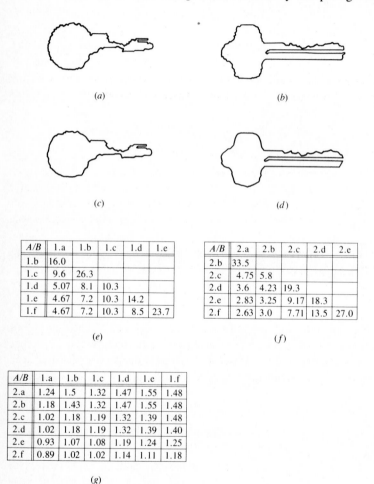

(a)

(b)

(c)

(d)

A/B	1.a	1.b	1.c	1.d	1.e
1.b	16.0				
1.c	9.6	26.3			
1.d	5.07	8.1	10.3		
1.e	4.67	7.2	10.3	14.2	
1.f	4.67	7.2	10.3	8.5	23.7

(e)

A/B	2.a	2.b	2.c	2.d	2.e
2.b	33.5				
2.c	4.75	5.8			
2.d	3.6	4.23	19.3		
2.e	2.83	3.25	9.17	18.3	
2.f	2.63	3.0	7.71	13.5	27.0

(f)

A/B	1.a	1.b	1.c	1.d	1.e	1.f
2.a	1.24	1.5	1.32	1.47	1.55	1.48
2.b	1.18	1.43	1.32	1.47	1.55	1.48
2.c	1.02	1.18	1.19	1.32	1.39	1.48
2.d	1.02	1.18	1.19	1.32	1.39	1.40
2.e	0.93	1.07	1.08	1.19	1.24	1.25
2.f	0.89	1.02	1.02	1.14	1.11	1.18

(g)

Figure 8.54 (a), (b) Sample boundaries of two different object classes. (c), (d) Their corresponding polygonal approximations. (e)–(g) Tabulations of $R = A/B$. (Adapted from Sze and Yang [1981], © IEEE.)

between the polygon segments as the polygon was traversed in a clockwise direction. Angles were coded into one of eight possible symbols which correspond to $45°$ increments, $s_1: 0° < \theta \leqslant 45°$, $s_2: 45° < \theta \leqslant 90°$, ..., $s_8: 315° < \theta \leqslant 360°$.

The results of computing the measure R for five samples of object 1 against themselves are shown in Fig. 8.54e, where the entries correspond to values of $R = A/B$ and, for example, the notation 1.c refers to the third string for object class 1. Figure 8.54f shows the results for the strings of the second object class. Finally, Fig. 8.54g is a tabulation of R values obtained by comparing strings of one class against the other. The important thing to note is that all values of R in this last table are considerably smaller than any entry in the preceding two tables, indicating that the R measure achieved a high degree of discrimination between the two classes of objects. For instance, if string 1.a had been an unknown, the smallest value in comparing it with the other strings of class 1 would have been 4.67. By contrast, the largest value in a comparison against class 2 would have been 1.24. Thus, classification of this string into class 1 based on the maximum value of R would have been a simple, unambiguous matter. $\qquad \square$

Syntactic Methods. Syntactic techniques are by far the most prevalent concepts used for handling structural recognition problems. Basically, the idea behind syntactic pattern recognition is the specification of structural pattern primitives and a set of rules (in the form of a *grammar*) which govern their interconnection. We consider first string grammars and then extend these ideas to higher-dimensional grammars.

String Grammars. Suppose that we have two classes of objects, ω_1 and ω_2, which are represented as strings of primitives, as outlined at the beginning of Sec. 8.5.2. We may interpret each primitive as being a symbol permissible in some grammar, where a *grammar* is a set of rules of syntax (hence the name syntactic pattern recognition) for the generation of *sentences* formed from the given symbols. In the context of the present discussion, these sentences are strings of symbols which in turn represent patterns. It is further possible to envision two grammars, G_1 and G_2, whose rules are such that G_1 only allows the generation of sentences which correspond to objects of class ω_1 while G_2 only allows generation of sentences corresponding to objects of class ω_2. The set of sentences generated by a grammar G is called its *language*, and denoted by $L(G)$.

Once the two grammars G_1 and G_2 have been established, the syntactic pattern recognition process is, in principle, straightforward. Given a sentence representing an unknown pattern, the problem is one of deciding in which language the pattern represents a valid sentence. If the sentence belongs to $L(G_1)$, we say that the pattern belongs to object class ω_1. Similarly, we say that the object comes from class ω_2 if the sentence is in $L(G_2)$. A unique decision cannot be made if the sentence belongs to both $L(G_1)$ and $L(G_2)$. If the sentence is found to be invalid over both languages it is rejected.

When there are more than two pattern classes, the syntactic classification approach is the same as described above, except that more grammars (at least one per class) are involved in the process. In this case the pattern is assigned to class ω_i if it is a sentence of *only* $L(G_i)$. A unique decision cannot be made if the sentence belongs to more than one language, and (as above) a pattern is rejected if it does not belong to any of the languages under consideration.

When dealing with strings, we define a grammar as the four-tuple

$$G = (N, \Sigma, P, S) \qquad (8.5\text{-}10)$$

where

$$N = \text{finite set of } nonterminals \text{ or variables}$$
$$\Sigma = \text{finite set of } terminals \text{ or constants}$$
$$P = \text{finite set of } productions \text{ or rewriting rules}$$
$$S \text{ in } N = \text{the } starting\ symbol$$

It is required that N and Σ be disjoint sets. In the following discussion nonterminals will be denoted by capital letters: $A, B, \ldots, S, \ldots$. Lowercase letters at the beginning of the alphabet will be used for terminals: $a, b, c, \ldots$. Strings of terminals will be denoted by lowercase letters toward the end of the alphabet: v, w, x, y, z. Strings of mixed terminals and nonterminals will be denoted by lowercase Greek letters: $\alpha, \beta, \theta, \ldots$. The *empty sentence* (the sentence with no symbols) will be denoted by λ. Finally, given a set V of symbols, we will use the notation V^* to denote the set of all sentences composed of elements from V.

String grammars are characterized primarily by the form of their productions. Of particular interest in syntactic pattern recognition are *regular grammars*, whose productions are always of the form $A \rightarrow aB$ or $A \rightarrow a$ with A and B in N, and a in Σ, and *context-free* grammars, with productions of the form $A \rightarrow \alpha$ with A in N, and α in the set $(N \cup \Sigma)^* - \lambda$; that is, α can be any string composed of terminals and nonterminals, except the empty string.

Example: The preceding concepts are best clarified by an example. Suppose that the object shown in Fig. 8.55a is represented by its skeleton, and that we define the primitives shown in Fig. 8.55b to describe the structure of this and similar skeletons. Consider the grammar $G = (N, \Sigma, P, S)$ with $N = \{A, B, S\}$, $\Sigma = \{a, b, c\}$, and production rules

1. $S \rightarrow aA$
2. $A \rightarrow bA$
3. $A \rightarrow bB$
4. $B \rightarrow c$

where the terminals a, b, and c are as shown in Fig. 8.55b. As indicated earlier, S is the starting symbol from which we generate all strings in $L(G)$.

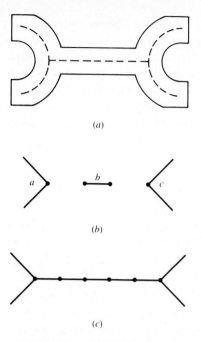

(a)

(b)

(c)

Figure 8.55 (a) Object represented by its skeleton. (b) Primitives. (c) Structure generated using a regular string grammar.

If, for instance, we apply production 1 followed by two applications of production 2, we obtain: $S \Rightarrow aA \Rightarrow abA \Rightarrow abbA$, where " $\Rightarrow$ " indicates a string derivation starting from S and using production rules from P. It is noted that we interpret the production $S \rightarrow aA$ and $A \rightarrow bA$ as "S can be rewritten as aA" and "A can be rewritten as bA." Since we have a nonterminal in the string $abbA$ and a rule which allows us to rewrite it, we can continue the derivation. For example, if we apply production 2 two more times, followed by production 3 and then production 4, we obtain the string $abbbbbc$ which corresponds to the structure shown in Fig. 8.55c. It is important to note that no nonterminals are left after application of production 4 so the derivation terminates after this production is used. A little thought will reveal that the grammar given above has the language $L(G) = \{ab^nc \,|\, n \geq 1\}$, where b^n indicates n repetitions of the symbol b. In other words, G is capable of generating the skeletons of wrenchlike structures with bodies of arbitrary length within the resolution established by the length of primitive b. $\square$

Use of Semantics. In the above example we have implicitly assumed that the interconnection between primitives takes place only at the dots shown in Fig. 8.55b. In more complicated situations the rules of connectivity, as well as other information regarding factors such as primitive length and direction, and the number of times a production can be applied, must be made explicit. This is usually accomplished via the use of *semantics*. Basically, syntax establishes the structure of an object or expression, while semantics deal with its meaning. For example, the FORTRAN statement $A = B/C$ is syntactically correct, but it is semantically correct only if $C \neq 0$.

In order to fix these ideas, suppose that we attach semantic information to the wrench grammar just discussed. This information may be attached to the productions as follows:

Production	Semantic information
$S \rightarrow aA$	Connections to a are made only at the dot. The direction of a, denoted by θ, is given by the direction of the perpendicular bisector of the line joining the endpoints of the two undotted segments. These line segments are 3 cm each.
$A \rightarrow bA$	Connections to b are made only at the dots. No multiple connections are allowed. The direction of b must be the same as that of a and the length of b is 0.25 cm. This production cannot be applied more than 10 times.
$A \rightarrow bB$	The direction of a and b must be the same. Connections must be simple and made only at the dots.
$B \rightarrow c$	The direction of c and a must be the same. Connections must be simple and made only at the dots.

It is noted that, by using semantic information, we are able to use a few rules of syntax to describe a broad (although limited as desired) class of patterns. For instance, by specifying the direction θ, we avoid having to specify primitives for each possible object orientation. Similarly, by requiring that all primitives be oriented in the same direction, we eliminate from consideration nonsensical wrenchlike structures.

Recognition. Thus far, we have seen that grammars are *generators* of patterns. In the following discussion we consider the problem of recognizing if a given pattern string belongs to the language $L(G)$ generated by a grammar G. The basic concepts underlying syntactic recognition can be illustrated by the development of mathematical models of computing machines, called *automata*. Given an input pattern string, these automata have the capability of recognizing whether or not the pattern belongs to a specified language or class. We will focus attention only on *finite automata*, which are the recognizers of languages generated by regular grammars.

A *finite automaton* is defined as a five-tuple

$$A = (Q, \Sigma, \delta, q_0, F) \tag{8.5-11}$$

where Q is a finite, nonempty set of *states*, Σ is a finite input *alphabet*, δ is a mapping from $Q \times \Sigma$ (the set of ordered pairs formed from elements of Q and Σ) into the collection of all subsets of Q, q_0 is the *starting state*, and F (a subset of Q) is a set of *final* or *accepting states*. The terminology and notation associated with Eq. (8.5-11) are best illustrated by a simple example.

Example: Consider an automaton given by Eq. (8.5-11) where $Q = \{q_0, q_1, q_2\}$, $\Sigma = \{a, b\}$, $F = \{q_0\}$, and the mappings are given by $\delta(q_0, a) = \{q_2\}$, $\delta(q_0, b) = \{q_1\}$, $\delta(q_1, a) = \{q_2\}$, $\delta(q_1, b) = \{q_0\}$, $\delta(q_2, a) = \{q_0\}$, $\delta(q_2, b) = \{q_1\}$. If, for example, the automaton is in state q_0 and an a is input, its state changes to q_2. Similarly, if a b is input next, the automaton moves to state q_1, and so forth. It is noted that, in this case, the initial and final states are the same. ☐

A *state diagram* for the automaton just discussed is shown in Fig. 8.56. The state diagram consists of a node for each state, and directed arcs showing the possible transitions between states. The final state is shown as a double circle and each arc is labeled with the symbol that causes that transition. A string w of terminal symbols is said to be *accepted* or *recognized* by the automaton if, starting in state q_0, the sequence of symbols in w causes the automaton to be in a final state after the last symbol in w has been input. For example, the automaton in Fig. 8.56 recognizes the string $w = abbabb$, but rejects the string $w = aabab$.

There is a one-to-one correspondence between regular grammars and finite automata. That is, a language is recognized by a finite automaton if and only if it is generated by a regular grammar. The procedure for obtaining the automaton corresponding to a given regular grammar is straightforward. Let the grammar be denoted by $G = (N, \Sigma, P, X_0)$, where $X_0 \equiv S$, and suppose that N is composed of X_0 plus n additional nonterminals $X_1, X_2, \ldots, X_n$. The state set Q for the automaton is formed by introducing $n + 2$ states, $\{q_0, q_1, \ldots, q_n, q_{n+1}\}$ such that q_i corresponds to X_i for $0 \leqslant i \leqslant n$ and q_{n+1} is the final state. The set of

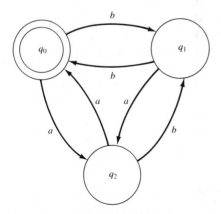

Figure 8.56 A finite automaton.

input symbols is identical to the set of terminals in G. The mappings in δ are defined by two rules based on the production of G; namely, for each i and j, $0 \leqslant i \leqslant n$, $0 \leqslant j \leqslant n$:

1. If $X_i \rightarrow aX_j$ is in P, then δ (q_i, a) contains q_j.
2. If $S_i \rightarrow a$ is in P, then δ (q_i, a) contains q_{n+1}.

On the other hand, given a finite automaton $A = (Q, \Sigma, \delta, q_0, F)$, we obtain the corresponding regular grammar $G = (N, \Sigma, P, X_0)$ by letting N be identified with the state set Q, with the starting symbol X_0 corresponding to q_0, and the productions of G obtained as follows:

1. If q_j is in $\delta(q_i, a)$, there is a production $X_i \rightarrow aX_j$ in P.
2. If a state in F is in $\delta(q_i, a)$, there is a production $X_i \rightarrow a$ in P.

Example: The finite automaton corresponding to the wrench grammar given earlier is obtained by writing the productions as $X_0 \rightarrow aX_1$, $X_1 \rightarrow bX_1$, $X_1 \rightarrow bX_2$, $X_2 \rightarrow c$. Then, from the above discussion, we have $A = (Q, \Sigma, \delta, q_0, F)$ with $Q = \{q_0, q_1, q_2, q_3\}$, $\Sigma = \{a, b, c\}$, $F = \{q_3\}$, and mappings $\delta(q_0, a) = \{q_1\}$, $\delta(q_1, b) = \{q_1, q_2\}$, $\delta(q_2, c) = \{q_3\}$. For completeness, we can write $\delta(q_0, b) = \delta(q_0, c) = \delta(q_1, a) = \delta(q_1, c) = \delta(q_2, a) = \delta(q_2, b) = \phi$, where ϕ is the null set, indicating that these transitions are not defined for this automaton. □

Higher-Dimensional Grammars. The grammars discussed above are best suited for applications where the connectivity of primitives can be conveniently expressed in a stringlike manner. In the following discussion we consider two examples of grammars capable of handling more general interconnections between primitives and subpatterns.

A *tree grammar* is defined as the five-tuple

$$G = (N, \Sigma, P, r, S) \tag{8.5-12}$$

where N and Σ are, as before, sets of nonterminals and terminals, respectively; S is the start symbol which can, in general, be a tree; P is a set of productions of the form $T_i \rightarrow T_j$, where T_i and T_j are trees; and r is a *ranking function* which denotes the number of direct descendants of a node whose label is a terminal in the grammar. An *expansive* tree grammar has productions of the form

where $A, A_1, \ldots, A_n$ are nonterminals, and a is a terminal.

Example: The skeleton of the structure shown in Fig. 8.57a can be generated by means of a tree grammar with productions

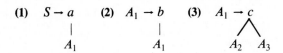

(1) $S \rightarrow a$
   |
   A_1

(2) $A_1 \rightarrow b$
   |
   A_1

(3) $A_1 \rightarrow c$
   $\bigwedge$
   A_2 A_3

(4) $A_2 \rightarrow d$
   |
   A_2

(5) $A_2 \rightarrow e$

(6) $A_3 \rightarrow e$
   |
   A_3

(7) $A_3 \rightarrow a$

where connectivity between linear primitives is head to tail, and connections to the circle primitive can be made anywhere on its circumference. The ranking functions in this case are $r(a) = \{0, 1\}$, $r(b) = r(d) = r(e) = \{1\}$, $r(c) = \{2\}$. It is noted that restricting the use of productions 2, 4, and 6 to

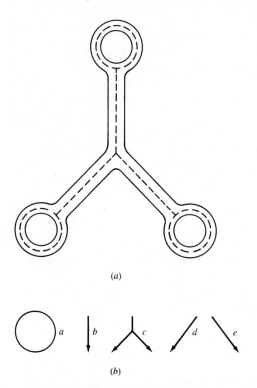

(a)

a $\quad | b$ $\quad \bigwedge c$ $\quad / d$ $\quad \diagdown e$

(b)

Figure 8.57 (a) An object and (b) primitives used for representing the skeleton by means of a tree grammar.

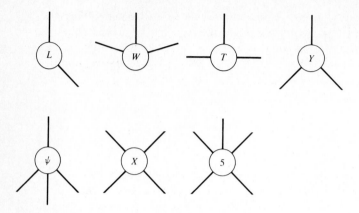

Figure 8.58 Vertex primitives. (From Gips [1974], © Pergamon Press.)

be applied the same number of times generates a structure in which all three legs are of the same length. Similarly, requiring that productions 4 and 6 be applied the same number of times produces a structure that is symmetrical about the vertical axis in Fig. 8.57a. ☐

We conclude this section with a brief discussion of a grammar proposed by Gips [1974] for generating three-dimensional objects consisting of cube structures. As in the previous discussion, the key to object generation by syntactic techniques is the specification of a set of primitives and their interconnections. In this case, the primitives are the vertices shown in Fig. 8.58. Vertices of type T are further classified as either T_1 or T_3, using local information. If a T vertex is contained in a parallelogram of vertices it is classified as type T_3. If a T vertex is not contained in a parallelogram of vertices, it is classified as type T_1. Figure 8.59 shows this classification.

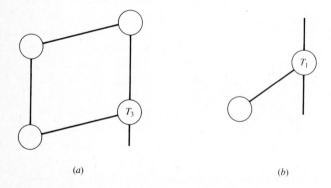

(a)   (b)

Figure 8.59 Further classification of T vertices.

The rules of the grammar consist of specifying valid interconnections between structures, as detailed in Fig. 8.60. The vertices denoted by double circles denote central vertices of the end cube of an object where further connections can be made. This is illustrated in Fig. 8.61 which shows a typical derivation using the rules of Fig. 8.60. Figure 8.62 illustrates the range of structures that can be generated with these rules.

8.6 INTERPRETATION

In this discussion, we view interpretation as the process which endows a vision system with a higher level of cognition about its environment than that offered by any of the concepts discussed thus far. When viewed in this way, interpretation clearly encompasses all these methods as an integral part of understanding a visual scene. Although this is one of the most active research topics in machine vision, the reader is reminded of the comments made in Secs. 7.1 and 8.1 regarding the fact that our understanding of this area is really in its infancy. In this section we touch briefly upon a number of topics which are representative of current efforts toward advancing the state of the art in machine vision.

The power of a machine vision system is determined by its ability to extract meaningful information from a scene under a broad range of viewing conditions and using minimal knowledge about the objects being viewed. There are a number of factors which make this type of processing a difficult task, including variations in illumination, occluding bodies, and viewing geometry. In Sec. 7.3 we spent considerable time discussing techniques designed to reduce variability in illumination and thus provide a relatively constant input to a vision system. The back- and structured-lighting approaches discussed in that section are indicative of the extreme levels of specialization employed by current industrial systems to reduce the difficulties associated with arbitrary lighting of the work space. Among these difficulties we find shadowing effects which complicate edge finding, and the introduction of nonuniformities on smooth surfaces which often results in their being detected as distinct bodies. Clearly, many of these problems result from the fact that relatively little is known about modeling the illumination-reflectance properties of 3D scenes. The line and junction labeling techniques discussed in Sec. 8.4 represent an attempt in this direction, but they fall short of explaining the interaction of illumination and reflectivity in quantitative terms. A more promising approach is based on mathematical models which attempt to infer intrinsic relationships between illumination, reflectance, and surface characteristics such as orientation (Horn [1977]; Marr [1979]; Katsushi and Horn [1981]).

Occlusion problems come into play when we are dealing with a multiplicity of objects in an unconstrained working environment. Consider, for example, the scene shown in Fig. 8.63. A human observer would have little difficulty, say, in determining the presence of two wrenches behind the sockets. For a machine, however, interpretation of this scene is a totally different story. Even if the system were able to perform a perfect segmentation of object clusters from the back-

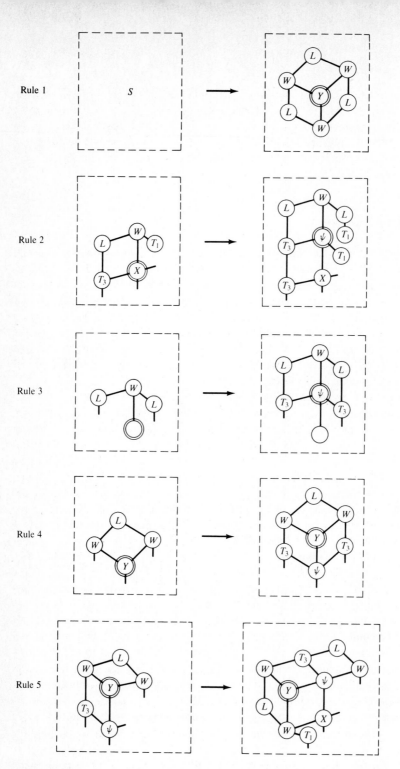

Figure 8.60 Rules used to generate three-dimensional structures. The blank circles indicate that more than one vertex type is allowed. (Adapted from Gips [1974], © Pergamon Press.)

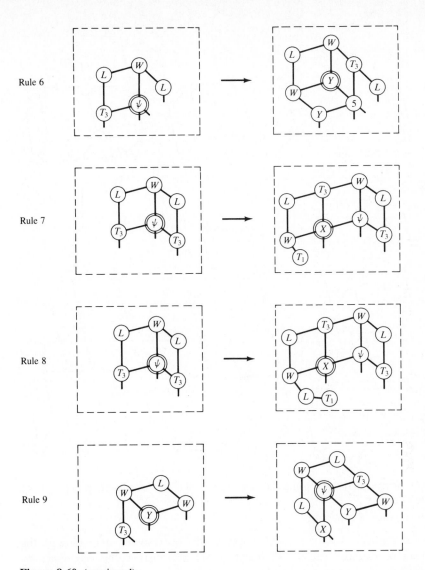

Figure 8.60 (continued)

ground, all the two-dimensional procedures discussed thus far for description and recognition would perform poorly on most of the occluded objects. The three-dimensional descriptors discussed in Sec. 8.4 would have a better chance, but even they would yield incomplete information. For instance, several of the sockets would appear as partial cylindrical surfaces, and the middle wrench would appear as two separate objects.

Processing scenes such as the one shown in Fig. 8.63 requires the capability to obtain descriptions which inherently carry shape and volumetric information, and procedures for establishing relationships between these descriptions, even when

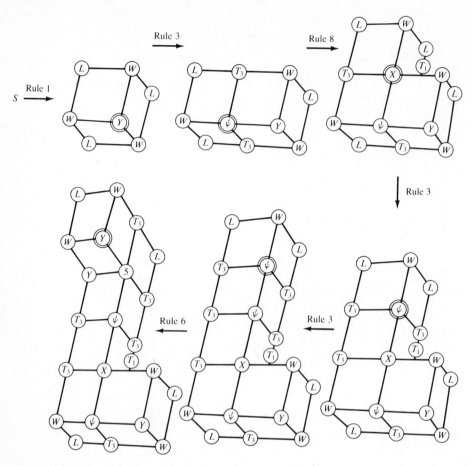

Figure 8.61 Sample derivation using the rules in Fig. 8.60. (Adapted from Gips [1974], © Pergamon Press.)

they are incomplete. Ultimately, these issues will be resolved only through the development of methods capable of handling 3D information obtained either by means of direct measurements or via geometric reasoning techniques capable of inferring (but not necessarily quantifying) 3D relationships from intensity imagery.

As an example of this type of reasoning, the reader would have little difficulty in arriving at a detailed interpretation of the objects in Fig. 8.63 with the exception of the object occluded by the screwdriver. The capability to know when interpretation of a scene or part of a scene is not an achievable task is just as important as correctly analyzing the scene. The decision to look at the scene from a different viewpoint (Fig. 8.64) to resolve the issue would be a natural reaction in an intelligent observer.

One of the most promising approaches in this direction is research in model-driven vision (Brooks [1981]). The basic idea behind this approach is to base the

Figure 8.62 Sample three-dimensional structures generated by the rules given in Fig. 8.60. (From Gips [1974], © Pergamon Press.)

interpretation of a scene on discovering instances of matches between image data and 3D models of volumetric primitives or entire objects of interest. Vision based on 3D models has another important advantage: It provides an approach for handling variances in viewing geometry. Variability in the appearance of an object when viewed from different positions is one of the most serious problems in machine vision. Even in two-dimensional situations where the viewing geometry is fixed, object orientation can strongly influence recognition performance if not handled properly (the reader will recall numerous comments made about this in Sec.

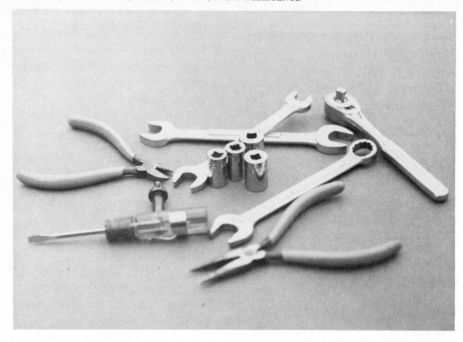

Figure 8.63 Oblique view of a three-dimensional scene.

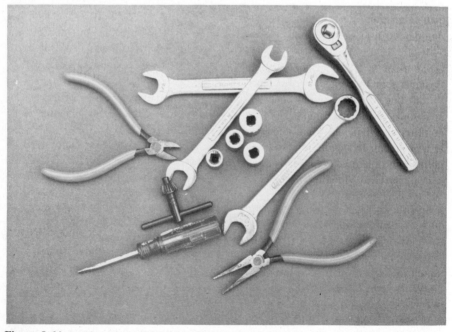

Figure 8.64 Another view of the scene shown in Fig. 8.63.

8.3). One of the advantages of a model-driven approach is that, depending on a known viewing geometry, it is possible to project the 3D model onto an imaging plane (see Sec. 7.4) in that orientation and thus simplify the match between an unknown object and what the system would expect to see from a given viewpoint.

8.7 CONCLUDING REMARKS

The focus of the discussion in this chapter is on concepts and techniques of machine vision with a strong bias toward industrial applications. As indicated in Sec. 8.2, segmentation is one of the most important processes in the early stages of a machine vision system. Consequently, a significant portion of this chapter is dedicated to this topic. Following segmentation, the next task of a vision system is to form a set of descriptors which will uniquely identify the objects of a particular class. As indicated in Sec. 8.3, the key in selecting descriptors is to minimize their dependence on object size, location, and orientation.

Although vision is inherently a three-dimensional problem, most present industrial systems operate on image data which are often idealized via the use of specialized illumination techniques and a fixed viewing geometry. The problems encountered when these constraints are relaxed are addressed briefly in Secs. 8.4 and 8.6.

Our treatment of recognition techniques has been at an introductory level. This is a broad area in which dozens of books and thousands of articles have been written. The references at the end of this chapter provide a pointer for further reading on both the decision-theoretic and structural aspects of pattern recognition and related topics.

REFERENCES

Further reading on the local analysis concepts discussed in Sec. 8.2.1 may be found in the book by Rosenfeld and Kak [1982]. The Hough transform was first proposed by P. V. C. Hough [1962] in a U.S. patent and later popularized by Duda and Hart [1972]. A generalization of the Hough transform for detecting arbitrary shape has been proposed by Ballard [1981]. The material on graph-theoretic techniques is based on two papers by Martelli [1972, 1976]. Another interesting approach based on a minimum-cost search is given in Ramer [1975]. Additional reading on graph searching techniques may be found in Nilsson [1971, 1980]. Edge following may also be approached from a dynamic programming point of view. For further details on this topic see Ballard and Brown [1982].

The optimum thresholding approach discussed in Sec. 8.2.2 was first utilized by Chow and Kaneko [1972] for detecting boundaries in cineagiograms (x-ray pictures of a heart which has been injected with a dye). Further reading on optimum discrimination may be found in Tou and Gonzalez [1974]. The book by Rosenfeld and Kak [1982] contains a number of approaches for threshold selection and evaluation. Our use of boundary characteristics for thresholding is based on a paper by White and Rohrer [1983]. The discussion on using several variables for thresholding is from Gonzalez and Wintz [1977].

An overview of region-oriented segmentation (Sec. 8.2.3) is given in a paper by Zucker [1976]. Additional reading on this topic may be found in Barrow and Tenenbaum [1977], Brice and Fennema [1970], Horowitz and Pavlidis [1974], and Ohlander et al. [1979]. The concept of a quad tree was originally called *regular decomposition* (Klinger [1972, 1976]). The material in Sec. 8.2.4 is based on two papers by Jain [1981, 1983]. Other approaches to dynamic scene analysis may be found in Thompson and Barnard [1981], Nagel [1981], Rajala et al. [1983], Webb and Aggarwal [1981], and Aggarwal and Badler [1980].

The chain code representation discussed in Sec. 8.3.1 was first proposed by Freeman [1961, 1974]. Further reading on signatures may be found in Ambler et al. [1975], Nahim [1974], and Ballard and Brown [1982]. The book by Pavlidis [1977] contains a comprehensive discussion on techniques for polygonal approximations. The discussion on shape numbers is based on the work of Bribiesca and Guzman [1980] and Bribiesca [1981]. Further reading on Fourier descriptors may be found in Zahn and Roskies [1972], Persoon and Fu [1977], and Gonzalez and Wintz [1977]. For a discussion of 3D Fourier descriptors see Wallace and Mitchell [1980].

Further reading for the material in Sec. 8.3.2 may be found in Gonzalez and Wintz [1977]. Texture descriptors have received a great deal of attention during the past few years. For further reading on the statistical aspects of texture see Haralick et al. [1973], Bajcsy and Lieberman [1976], Haralick [1978], and Cross and Jain [1983]. On structural texture, see Lu and Fu [1978] and Timita et al. [1982]. The material on skeletons is based on a paper by Naccache and Shinghal [1984], which also contains an extensive set of references to other work on skeletons. Davies and Plummer [1981] address some fundamental issues on thinning which complement our discussion of this topic. The extraction of a skeleton using Fourier descriptors is discussed by Persoon and Fu [1977]. The moment-invariant approach is due to Hu [1962]. This technique has been extended to three dimensions by Sadjadi and Hall [1980].

The approach discussed in Sec. 8.4.1 has been used by Shirai [1979] for segmenting range data. The gradient operator discussed in Sec. 8.4.2 was developed by Zucker and Hummel [1981]. Early work on line and junction labeling for scene analysis (Sec. 8.4.3) may be found in Roberts [1965] and Guzman [1969]. A more comprehensive utilization of these ideas may be found in Waltz [1972, 1976]. For a more recent survey of work in this areas see Barrow and Tenenbaum [1981]. For further details on generalized cones (Sec. 8.4.4) see Agin [1972], Nevatia and Binford [1977], Marr [1979], and Shani [1980].

For further reading on the decision-theoretic approach discussed in Sec. 8.5.1 see the book by Tou and Gonzalez [1974]. The material in Sec. 8.5.2 dealing with matching shape numbers is based on a paper by Bribiesca and Guzman [1980]. The string matching results are from Sze and Yang [1981]. For further reading on structural pattern recognition see the books by Pavlidis [1977], Gonzalez and Thomason [1978], and Fu [1982].

Further reading for the material in Sec. 8.6 may be found in Dodd and Rossol [1979] and in Ballard and Brown [1982]. A set of survey papers on the topics discussed in that section has been compiled by Brady [1981].

PROBLEMS

8.1 (a) Develop a general procedure for obtaining the normal representation of a line given its slope-intercept equation $y = ax + b$. (b) Find the normal representation of the line $y = -2x + 1$.

8.2 (a) Superimpose on Fig. 8.7 all the possible edges given by the graph in Fig. 8.8. (b) Compute the cost of the minimum-cost path.

8.3 Find the edge corresponding to the minimum-cost path in the subimage shown below, where the numbers in parentheses indicate intensity. Assume that the edge starts on the first column and ends in the last column.

$$
\begin{array}{c c c c}
 & 0 & 1 & 2 \\
0 & \cdot & \cdot & \cdot \\
 & (2) & (1) & (0) \\
1 & \cdot & \cdot & \cdot \\
 & (1) & (1) & (7) \\
2 & \cdot & \cdot & \cdot \\
 & (6) & (8) & (2)
\end{array}
$$

8.4 Suppose that an image has the following intensity distributions, where $p_1(z)$ corresponds to the intensity of objects and $p_2(z)$ corresponds to the intensity of the background. Assuming that $P_1 = P_2$, find the optimum threshold between object and background pixels.

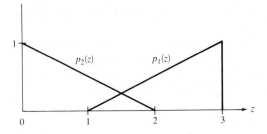

8.5 Segment the image on page 448 using the split and merge procedure discussed in Sec. 8.2.3. Let $P(R_i) =$ TRUE if all pixels in R_i have the same intensity. Show the quadtree corresponding to your segmentation.

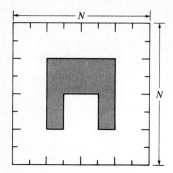

8.6 (*a*) Show that redefining the starting point of a chain code so that the resulting sequence of numbers forms an integer of minimum magnitude makes the code independent of where we initially start on the boundary. (*b*) What would be the normalized starting point of the chain code 11076765543322?

8.7 (*a*) Show that using the first difference of a chain code normalizes it to rotation, as explained in Section 8.3.1. (*b*) Compute the first difference of the code 0101030303323232212111.

8.8 (*a*) Plot the signature of a square boundary using the tangent angle method discussed in Sec. 8.3.1. (*b*) Repeat for the slope density function. Assume that the square is aligned with the x and y axes and let the x axis be the reference line. Start at the corner closest to the origin.

8.9 Give the fewest number of moment descriptors that would be needed to differentiate between the shapes shown in Fig. 8.29.

8.10 (*a*) Show that the rubberband polygonal approximation approach discussed in Sec. 8.3.1 yields a polygon with minimum perimeter. (*b*) Show that if each cell corresponds to a pixel on the boundary, then the maximum possible error in that cell is $\sqrt{2}d$, where d is the grid distance between pixels.

8.11 (*a*) What would be the effect on the resulting polygon if the error threshold were set to zero in the merging method discussed in Sec. 8.3.1? (*b*) What would be the effect on the splitting method?

8.12 (*a*) What is the order of the shape number in each of the following figures? (*b*) Obtain the shape number for the fourth figure.

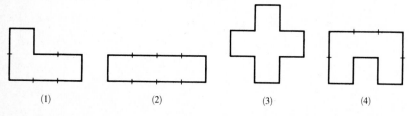

(1)   (2)   (3)   (4)

8.13 Compute the mean and variance of a four-level image with histogram $p(z_1) = 0.1$, $p(z_2) = 0.4$, $p(z_3) = 0.3$, $p(z_4) = 0.2$. Assume that $z_1 = 0$, $z_2 = 1$, $z_3 = 2$, and $z_4 = 3$.

8.14 Obtain the gray-level co-occurrence matrix of a 5×5 image composed of a checkerboard of alternating 1's and 0's if (*a*) P is defined as "one pixel to the right," and (*b*) "two pixels to the right." Assume that the top, left pixel has value 0.

8.15 Consider a checkerboard image composed of alternating black and white squares, each of size $m \times m$. Give a position operator that would yield a diagonal co-occurrence matrix.

8.16 (*a*) Show that the medial axis of a circular region is a single point at its center. (*b*) Sketch the medial axis of a rectangle, the region between two concentric circles, and an equilateral triangle.

8.17 (*a*) Show that the boolean expression given in Eq. (8.3-6) implements the conditions given by the four windows in Fig. 8.40. (*b*) Draw the windows corresponding to B_0 in Eq. (8.3-7).

8.18 Draw a trihedral object which has a junction of the form

8.19 Show that using Eq. (8.5-4) to classify an unknown pattern vector $\mathbf{x}^*$ is equivalent to using Eq. (8.5-3).

8.20 Show that $D(A, B) = 1/k$ satisfies the three conditions given in Eq. (8.5-7).

8.21 Show that $B = \max(|C_1|, |C_2|) - A$ in Eq. (8.5-8) is zero if and only if C_1 and C_2 are identical strings.

ROBOT PROGRAMMING LANGUAGES

Observers are not led by the same physical
evidence to the same picture of the
universe unless their linguistic backgrounds
are similar or can in some way be calibrated.
Benjamin Lee Whorf

9.1 INTRODUCTION

The discussion in the previous chapters focused on kinematics, dynamics, control, trajectory planning, sensing, and vision for computer-based manipulators. The algorithms used to accomplish these functions are usually embedded in the controlling software modules. A major obstacle in using manipulators as general-purpose assembly machines is the lack of suitable and efficient communication between the user and the robotic system so that the user can direct the manipulator to accomplish a given task. There are several ways to communicate with a robot, and three major approaches to achieve it are *discrete word recognition, teach and playback,* and *high-level programming languages.*

Current state-of-the-art speech recognition systems are quite primitive and generally speaker-dependent. These systems can recognize a set of discrete words from a limited vocabulary and usually require the user to pause between words. Although it is now possible to recognize discrete words in real time due to faster computer components and efficient processing algorithms, the usefulness of discrete word recognition to describe a robot task is quite limited in scope. Moreover, speech recognition generally requires a large memory or secondary storage to store speech data, and it usually requires a training period to build up speech templates for recognition.

Teach and playback, also known as *guiding,* is the most commonly used method in present-day industrial robots. The method involves teaching the robot by leading it through the motions the user wishes the robot to perform. Teach and playback is typically accomplished by the following steps: (1) leading the robot in slow motion using manual control through the entire assembly task and recording the joint angles of the robot at appropriate locations in order to replay the motion; (2) editing and playing back the taught motion; and (3) if the taught motion is correct, then the robot is run at an appropriate speed in a repetitive mode.

Leading the robot in slow motion usually can be achieved in several ways: using a joystick, a set of pushbuttons (one for each joint), or a master-slave manipulator system. Presently, the most commonly used system is a manual box with pushbuttons. With this method, the user moves the robot manually through the space, and presses a button to record any desired angular position of the manipulator. The set of angular positions that are recorded form the set-points of the trajectory that the manipulator has traversed. These position set-points are then interpolated by numerical methods, and the robot is "played back" along the smoothed trajectory. In the edit-playback mode, the user can edit the recorded angular positions and make sure that the robot will not collide with obstacles while completing the task. In the run mode, the robot will run repeatedly according to the edited and smoothed trajectory. If the task is changed, then the above three steps are repeated. The advantages of this method are that it requires only a relatively small memory space to record angular positions and it is simple to learn. The main disadvantage is that it is difficult to utilize this method for integrating sensory feedback information into the control system.

High-level programming languages provide a more general approach to solving the human-robot communication problem. In the past decade, robots have been successfully used in such areas as arc welding and spray painting using guiding (Engelberger [1980]). These tasks require no interaction between the robot and the environment and can be easily programmed by guiding. However, the use of robots to perform assembly tasks requires high-level programming techniques because robot assembly usually relies on sensory feedback, and this type of unstructured interaction can only be handled by conditionally programmed methods.

Robot programming is substantially different from traditional programming. We can identify several considerations which must be handled by any robot programming method: The objects to be manipulated by a robot are three-dimensional objects which have a variety of physical properties; robots operate in a spatially complex environment; the description and representation of three-dimensional objects in a computer are imprecise; and sensory information has to be monitored, manipulated, and properly utilized. Current approaches to programming can be classified into two major categories: *robot-oriented programming* and *object-oriented*, or *task-level programming*.

In robot-oriented programming, an assembly task is explicitly described as a sequence of robot motions. The robot is guided and controlled by the program throughout the entire task with each statement of the program roughly corresponding to one action of the robot. On the other hand, task-level programming describes the assembly task as a sequence of positional goals of the objects rather than the motion of the robot needed to achieve these goals, and hence no explicit robot motion is specified. These approaches are discussed in detail in the following two sections.

9.2 CHARACTERISTICS OF ROBOT-LEVEL LANGUAGES

The most common approach taken in designing robot-level language is to extend an existing high-level language to meet the requirements of robot programming.

To a certain extent, this approach is ad hoc and there are no guidelines on how to implement the extension.

We can easily recognize several key characteristics that are common to all robot-oriented languages by examining the steps involved in developing a robot program. Consider the task of inserting a bolt into a hole (Fig. 9.1). This requires moving the robot to the feeder, picking up the bolt, moving it to the beam and inserting the bolt into one of the holes. Typically, the steps taken to develop the program are:

1. The workspace is set up and the parts are fixed by the use of fixtures and feeders.
2. The location (orientation and position) of the parts (*feeder*, *beam*, etc.) and their features (*beam_bore*, *bolt_grasp*, etc.) are defined using the data structures provided by the language.†
3. The assembly task is partitioned into a sequence of actions such as moving the robot, grasping objects, and performing an insertion.
4. Sensory commands are added to detect abnormal situations (such as inability to locate the bolt while grasping) and monitor the progress of the assembly task.

† The reader will recall that the use of the underscore symbol is a common practice in programming languages to provide an effective identity in a variable name and thus improve legibility.

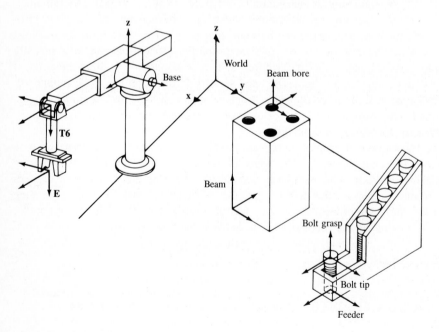

Figure 9.1 A simple robotic insertion task.

5. The program is debugged and refined by repeating steps 2 to 4.

The important characteristics we recognized are position specification (step 2), motion specification (step 3), and sensing (step 4). These characteristics are discussed in detail in this section.

We will use the languages AL (Mujtaba et al. [1982]) and AML (Taylor et al. [1983]) as examples. The choice of using these two languages is not arbitrary. AL has influenced the design of many robot-oriented languages and is still actively being developed. It provides a large set of commands to handle the requirements of robot programming and it also supports high-level programming features. AML is currently available as a commercial product for the control of IBM's robots and its approach is different from AL. Its design philosophy is to provide a system environment where different robot programming interfaces may be built. Thus, it has a rich set of primitives for robot operations and allows the users to design high-level commands according to their particular needs. These two languages represent the state of the art in robot-oriented programming languages. A brief description of the two languages is shown in Table 9.1.

Table 9.1 A brief summary of the AL and AML robot programming languages

AL was developed by Stanford University. Currently AL can be executed on a VAX computer and real-time control of the arms are performed on a stand alone PDP-11. Its characteristics are:

> High-level language with features of ALGOL and Pascal
> Supports both robot-level and task-level specification
> Compiled into low-level language and interpreted on a real time control machine
> Has real-time programming language constructs like synchronization, concurrent execution, and on-conditions
> ALGOL like data and control structure
> Support for world modeling

AML was developed by IBM. It is the control language for the IBM RS-1 robot. It runs on a Series-1 computer (or IBM personal computer) which also controls the robot. The RS-1 robot is a cartesian manipulator with 6 degrees of freedom. Its first three joints are prismatic and the last three joints are rotary. Its characteristics are:

> Provides an environment where different user-interface can be built
> Supports features of LISP-like and APL-like constructs
> Supports data aggregation
> Supports joint-space trajectory planning subject to position and velocity constraints
> Provides absolute and relative motions
> Provides sensor monitoring that can interrupt motion

Table 9.2 AL and AML definitions for base frames

AL:

> *base* ← FRAME(nilrot, VECTOR(20, 0, 15)*inches);
> *beam* ← FRAME(ROT(Z, 90*deg), VECTOR(20, 15, 0)*inches);
> *feeder* ← FRAME(nilrot, VECTOR(25, 20, 0)*inches);

Notes: nilrot is a predefined frame which has value ROT(Z, 0*deg).
> The "←" is the assignment operator in AL.
> A semicolon terminates a statement.
> The "*" is a type-dependent multiplication operator. Here, it is used to append units
> to the elements of the vector.

AML:

> *base* = <<20, 0, 15>, EULERROT(<0, 0, 0>)>;
> *beam* = <<20, 15, 0>, EULERROT(<0, 0, 90>)>;
> *feeder* = <<25, 20, 0>, EULERROT(<0, 0, 0>)>;

Note: EULERROT is a subroutine which forms the rotation matrix given the angles.

9.2.1 Position Specification

In robot assembly, the robot and the parts are generally confined to a well-defined workspace. The parts are usually restricted by fixtures and feeders to minimize positional uncertainities. Assembly from a set of randomly placed parts requires vision and is not yet a common practice in industry.

The most common approach used to describe the orientation and the position of the objects in the workspace is by coordinate frames. They are usually represented as 4×4 homogeneous transformation matrices. A frame consists of a 3×3 submatrix (specifying the orientation) and a vector (specifying the position) which are defined with respect to some base frame. Table 9.2 shows AL and AML definitions for the three frames *base*, *beam*, and *feeder* shown in Fig. 9.1. The approach taken by AL is to provide predefined data structures for frames (FRAME), rotational matrices (ROT), and vectors (VECTOR), all of them in cartesian coordinates. On the other hand, AML provides a general structure called an *aggregate* which allows the user to design his or her own data structures. The AML frames defined in Table 9.2 are in cartesian coordinates and the format is <vector, matrix>, where vector is an aggregate of three scalars representing position and matrix is an aggregate of three vectors representing orientation.

In order to further explain the notation used in Table 9.2, the first statement in AL means the establishment of the coordinate frame *base,* whose principal axes are parallel (nilrot implies no rotation) to the principal axes of the reference frame and whose origin is at location (20, 0, 15) inches from the origin of the reference frame. The second statement in AL establishes the coordinate frame *beam,* whose principal axes are rotated 90° about the Z axis of the reference frame, and whose origin is at location (20, 15, 0) inches from the origin of the reference frame. The third statement has the same meaning as the first, except for location. The meaning of the three statements in AML is exactly the same as for those in AL.

A convenient way of referring to the features of an object is to define a frame (with respect to the object's base frame) for it. An advantage of using a homogeneous transformation matrix is that defining frames relative to a base frame can be simply done by postmultiplying a transformation matrix to the base frame. Table 9.3 shows the AL and AML statements used to define the features *T6*, *E*, *bolt_tip*, *bolt_grasp*, and *beam_bore* with respect to their base frames, as indicated in Fig. 9.1. AL provides a matrix multiplication operator (∗) and a data structure TRANS (a transformation which consists of a rotation and a translation operation) to represent transformation matrices. AML has no built-in matrix multiplication operator, but a system subroutine, DOT, is provided.

In order to illustrate the meaning of the statements in Table 9.3, the first AL statement means the establishment of the coordinate frame *T6*, whose principal axes are rotated 180° about the *X* axis of the *base* coordinate frame, and whose origin is at location (15, 0, 0) inches from the origin of the *base* coordinate frame. The second statement establishes the coordinate frame *E*, whose principal axes are parallel (nilrot implies no rotation) to the principal axes of the *T6* coordinate frame, and whose origin is at location (0, 0, 5) inches from the origin of the *T6* coordinate frame. Similar comments apply to the other three AL statements. The meaning of the AML statements is the same as those for AL.

Figure 9.2*a* shows the relationships between the frames we have defined in Tables 9.2 and 9.3. Note that the frames defined for the arm are not needed for AL because AL uses an implicit frame to represent the position of the end-effector and does not allow access to intermediate frames (*T6*, *E*). As parts are moved or

Table 9.3 AL and AML definitions for feature frames

AL:

 T6 ← *base* ∗ TRANS(ROT(X, 180∗deg), VECTOR(15, 0, 0)∗inches);

 E ← *T6* ∗ TRANS(nilrot, VECTOR(0, 0, 5)∗inches);

 bolt_tip ← *feeder* ∗ TRANS(nilrot, nilvect)

 bolt_grasp ← *bolt_tip* ∗ TRANS(nilrot, VECTOR(0, 0, 1)∗inches);

 beam_bore ← *beam* ∗ TRANS(nilrot, VECTOR(0, 2, 3)∗inches);

Note: nilvect is a predefined vector which has value VECTOR(0, 0, 0)∗inches.

AML:

 T6 = DOT(*base*, < <15, 0, 0>, EULERROT(<180, 0, 0>)>);

 E = DOT(*T6*, < <0, 0, 5>, EULERROT(<0, 0, 0>)>);

 bolt_tip = DOT(*feeder*, < <0, 0, 0>, EULERROT(<0, 0, 0>)>);

 bolt_grasp = DOT(*bolt_tip*, < <0, 0, 1>, EULERROT(<0, 0, 0>)>);

 beam_bore = DOT(*beam*, < <0, 2, 3>, EULERROT(<0, 0, 0>)>);

Note: DOT is a subroutine that multiplies two matrices.

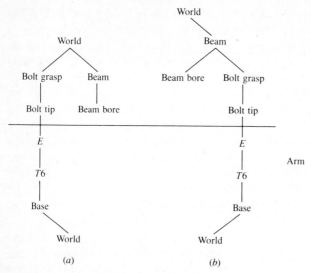

Figure 9.2 Relationships between the frames.

are attached to other objects, the frames are adjusted to reflect the current state of the world (see Fig. 9.2*b*).

Another way of acquiring the position and orientation of an object is by using the robot as a pointing device to gather the information interactively. POINTY (Grossman and Taylor [1978]), a system designed for AL, allows the user to lead the robot through the workspace (by hand or by a pendant) and, by pointing the hand (equipped with a special tool) to objects, it generates AL declarations similar to those shown in Tables 9.2 and 9.3. This eliminates the need to measure the distances and angles between frames, which can be quite tedious.

Although coordinate frames are quite popular for representing robot configurations, they do have some limitations. The natural way to represent robot configurations is in the joint-variable space rather than the cartesian space. Since the inverse kinematics problem gives nonunique solutions, the robot's configuration is not uniquely determined given a point in the cartesian space. As the number of features and objects increases, the relationships between coordinate frames become complicated and difficult to manage. Furthermore, the number of computations required also increases significantly.

9.2.2 Motion Specification

The most common operation in robot assembly is the pick-and-place operation. It consists of moving the robot from an initial configuration to a grasping configuration, picking up an object, and moving to a final configuration. The motion is usually specified as a sequence of positional goals for the robot to attain. However, only specifying the initial and final configurations is not sufficient. The path is planned by the system without considering the objects in the workspace and obstacles may be present on the planned path. In order for the system to generate a collision-free path, the programmer must specify enough intermediate or *via* points on the path. For example, in Fig. 9.3, if a straight line motion were used from point *A* to point *C*, the robot would collide with the beam. Thus, intermediate point *B* must be used to provide a safe path.

The positional goals can be specified either in the joint-variable space or in the cartesian space, depending on the language. In AL, the motion is specified by using the MOVE command to indicate the destination frame the arm should move to. Via points can be specified by using the keyword "VIA" followed by the frame of the via point (see Table 9.4). AML allows the user to specify motion in the joint-variable space and the user can write his or her own routines to specify motions in the cartesian space. Joints are specified by joint numbers (1 through 6) and the motion can be either relative or absolute (see Table 9.4).

One disadvantage of this type of specification is that the programmer must preplan the entire motion in order to select the intermediate points. The resulting path may produce awkward and inefficient motions. Furthermore, describing a complex path as a sequence of points produces an unnecessarily long program.

As the robot's hand departs from its starting configuration or approaches its final configuration, physical constraints, such as an insertion, which require the hand to travel along an axis, and environmental constraints, such as moving in a crowded area, may prohibit certain movement of the robot. The programmer must have control over various details of the motion such as speed, acceleration, deceleration, approach and departure directions to produce a safe motion. Instead

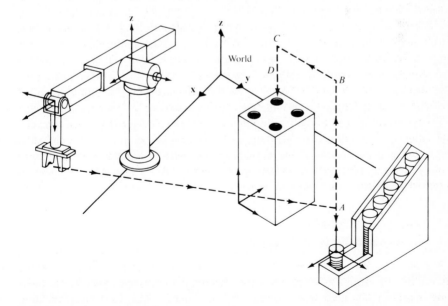

Figure 9.3 Trajectory of the robot.

Table 9.4 Examples of AL and AML motion statements

AL:

    { Move arm from rest to frame A and then to *bolt_grasp* }
    MOVE *barm* TO A;
    MOVE *barm* TO *bolt_grasp*;

    { Another way of specifying the above movement }
    MOVE *barm* TO *bolt_grasp* VIA A;

    { Move along the current Z axis by 1 inch, i.e., move relative }
    MOVE *barm* TO $\otimes$ $-$ 1*Z*inches;

Notes: *barm* is the name of the robot arm.
       $\otimes$ indicates the current location of the arm which is equivalent to *base* $*$ *T6* $*$ *E*.
       Statements inside brackets { $\cdots$ } are comments.

AML:

    -- Move joint 1 and 4 to 10 inches and 20 degrees, respectively (absolute move)
    MOVE($<1, 4>$, $<10, 20>$);

    -- Move joints 1, 3 and 6 by 1 inch, 2 inches, and 5 degrees, respectively (relative move)
    DMOVE($<1, 3, 6>$, $<1, 2, 5>$);

Notes: Statements preceeded by "--" are comments.

---

of separate commands, the usual approach is to treat them as constraints to be satisfied by the move command. AL provides a keyword "WITH" to attach constraint clauses to the move command. The constraints can be an approach vector, departure vector, or a time limit. Table 9.5 shows the AL statements for moving the robot from *bolt_grasp* to A with departure direction along $+Z$ of *feeder* and time duration of 5 seconds (i.e., move slowly). In AML, aggregates of the form $<$speed, acceleration, deceleration$>$ can be added to the MOVE statement to specify speed, acceleration, and deceleration of the robot.

    In general, gripper motions have to be tailored according to the environment and the task. Most languages provide simple commands on gripper motion so that sophisticated motions can be built using them. For a two-fingered gripper, one can either move the fingers apart (open) or move them together (close). Both AML and AL use a predefined variable to indicate the gripper (*bhand* corresponds to *barm* for AL and GRIPPER for AML). Using the OPEN (for AL) and MOVE (for AML) primitives, the gripper can be programmed to move to a certain opening (see Table 9.5).

9.2.3 Sensing and Flow of Control

The location and the dimension of the objects in the workspace can be identified only to a certain degree of accuracy. For the robot to perform tasks in the presence of these uncertainties, sensing must be performed. The sensory information gathered also acts as a feedback from the environment, enabling the robot to exam-

Table 9.5 Examples of AL and AML motion statements

AL:

> { Move arm from *bolt_grasp* to *A* }
> MOVE *barm* TO *A*
> WITH DEPARTURE = Z WRT *feeder*
> WITH DURATION = 5*seconds;
>
> { Open the hand to 2.5 inches }
> OPEN *bhand* TO 2.5*inches;

Note: WRT (means with respect to) generates a vector in the specified frame.

AML:

> -- Move joint 1 and 4 to 10 inches and 20 degrees, respectively, with speed
> 1 inch/second,
> -- Acceleration and deceleration 1 inch/second^2
> MOVE(<1, 4>, <10, 20>, <1, 1, 1>);
>
> -- Open the hand to 2.5 inches
> MOVE(GRIPPER, 2.5);

---

ine and verify the state of the assembly. Sensing in robot programming can be classified into three types:

1. *Position sensing* is used to identify the current position of the robot. This is usually done by encoders that measure the joint angles and compute the corresponding hand position in the workspace.
2. *Force* and *tactile sensing* can be used to detect the presence of objects in the workspace. Force sensing is used in compliant motion to provide feedback for force-controlled motions. Tactile sensing can be used to detect slippage while grasping an object.
3. *Vision* is used to identify objects and provide a rough estimate of their position.

There is no general consensus on how to implement sensing commands, and each language has its own syntax. AL provides primitives like FORCE(axis) and TORQUE(axis) for force sensing. They can be specified as conditions like FORCE(Z) > 3*ounces in the control commands. AML provides a primitive called MONITOR which can be specified in the motion commands to detect asynchronous events. The programmer can specify the sensors to monitor and, when the sensors are triggered, the motion is halted (see Table 9.6). It also has position-sensing primitives like QPOSITION (joint numbers†) which returns the current position of the joints. Most languages do not explicitly support vision, and the user has to provide modules to handle vision information.

---

† Specified as an aggregate like <1, 5> which specifies joints 1 and 5.

Table 9.6 Force sensing and compliant motion

AL:

    { Test for presence of hole with force sensing }
    MOVE *barm* TO $\otimes$ $-1*Z*$inches ON FORCE(Z) $>$ 10$*$ounces
       DO ABORT("No Hole");

    { Insert bolt, exert downward force while complying side forces }
    MOVE *barm* TO beam_bore
    WITH FORCE(Z) = $-$10$*$ounces   WITH FORCE(X) = 0$*$ounces
    WITH FORCE(Y) = 0$*$ounces   WITH DURATION = 3$*$seconds;

AML:

    -- Define a monitor for the force sensors SLP and SLR; Monitor triggers if the sensor
    -- values exceed the range 0 and F
    fmons = MONITOR($<$SLP, SRP$>$, 1, 0, F);

    -- Move joint 3 by 1 inch and stop if *fmons* is triggered
    DMOVE($<$3$>$, $<$1$>$, *fmons*);

Note: The syntax for monitor is MONITOR(sensors, test type, limit1, limit2).

One of the primary uses of sensory information is to initiate or terminate an action. For example, a part arriving on a conveyor belt may trip an optical sensor and activate the robot to pick up the part, or an action may be terminated if an abnormal condition has occurred. Table 9.6 illustrates the use of force sensing information to detect whether the hand has positioned correctly above the hole. The robot arm is moved downward slightly and, as it descends, the force exerted on the hand along the Z axis of the hand coordinate frame is returned by FORCE(Z). If the force exceeds 10 ounces, then this indicates that the hand missed the hole and the task is aborted.

The flow of a robot program is usually governed by the sensory information acquired. Most languages provide the usual decision-making constructs like "if_ then_ else_", "case_", "do_ until_", and "while_ do_" to control the flow of the program under different conditions.

Certain tasks require the robot to comply with external constraints. For example, insertion requires the hand to move along one direction only. Any sideward forces may generate unwanted friction which would impede the motion. In order to perform this compliant motion, force sensing is needed. Table 9.6 illustrates the use of AL's force sensing commands to perform the insertion task with compliance. The compliant motion is indicated by quantifying the motion statement with the amount of force allowed in each direction of the hand coordinate frame. In this case, forces are applied only along the Z axis of this frame.

9.2.4 Programming Support

A language without programming support (editor, debugger, etc.) is useless to the user. A sophisticated language must provide a programming environment that allows the user to support it. Complex robot programs are difficult to develop and can be difficult to debug. Moreover, robot programming imposes additional requirements on the development and debugging facilitates:

1. *On-line modification and immediate restart.* Since robot tasks requires complex motions and long execution time, it is not always feasible to restart the program upon failure. The robot programming system must have the ability to allow programs to be modified on-line and restart at any time.
2. *Sensor outputs and program traces.* Real-time interactions between the robot and the environment are not always repeatable; the debugger should be able to record sensor values along with program traces.
3. *Simulation.* This feature allows testing of programs without actually setting up robot and workspace. Hence, different programs can be tested more efficiently.

The reader should realize by now that programming in a robot-oriented language is tedious and cumbersome. This is further illustrated by the following example.

Example: Table 9.7 shows a complete AL program for performing the insertion task shown diagramatically in Fig. 9.1. The notation and meaning of the statements have already been explained in the preceding discussion. Keep in mind that a statement is not considered terminated until a semicolon is encountered. □

Table 9.7 An AL program for performing an insertion task

BEGIN insertion

{ set the variables }
bolt_diameter ← 0.5∗inches;
bolt_height ← 1∗inches;
tries ← 0;
grasped ← false;
{ Define base frames }
beam ← FRAME(ROT(Z, 90∗deg), VECTOR(20, 15, 0)∗inches);
feeder ← FRAME(nilrot, VECTOR(25, 20, 0)∗inches);

{ Define feature frames }
bolt_grasp ← *feeder* ∗ TRANS(nilrot, nilvect);
bolt_tip ← *bolt_grasp* ∗ TRANS(nilrot, VECTOR(0, 0, 0.5)∗inches);
beam_bore ← *beam* ∗ TRANS(nilrot, VECTOR(0, 0, 1)∗inches);

{ Define via points frames }
A ← *feeder* ∗ TRANS(nilrot, VECTOR(0, 0, 5)∗inches);
B ← *feeder* ∗ TRANS(nilrot, VECTOR(0, 0, 8)∗inches);
C ← *beam_bore* ∗ TRANS(nilrot, VECTOR(0, 0, 5)∗inches);
D ← *beam_bore* ∗ TRANS(nilrot, *bolt_height*∗Z);

Table 9.7 (continued)

{ Open the hand }
OPEN *bhand* TO *bolt_diameter* + 1*inches;

{ Position the hand just above the bolt }
MOVE *barm* TO *bolt_grasp* VIA *A*
    WITH APPROACH = −Z WRT *feeder;*

{ Attempt to grasp the bolt }
DO
    CLOSE *bhand* TO 0.9*bolt_diameter;*
    IF *bhand* < *bolt_diameter* THEN BEGIN{ failed to grasp the bolt, try again }
    OPEN *bhand* TO bolt_diameter + 1*inches;
    MOVE *barm* TO ⊗ − 1*Z*inches;
    END ELSE *grasped* ← true;
    tries ← *tries* + 1;
UNTIL *grasped* OR (*tries* > 3);

{ Abort the operation if the bolt is not grasped in three tries. }
IF NOT *grasped* THEN ABORT("failed to grasp bolt");

{ Move the arm to *B* }
MOVE *barm* TO *B*
    VIA *A*
    WITH DEPARTURE = Z WRT *feeder*;

{ Move the arm to *D* }
MOVE *barm* TO *D* VIA *C*
    WITH APPROACH = −Z WRT *beam_bore*;

{ Check whether the hole is there }
MOVE *barm* TO ⊗ − 0.1*Z*inches ON FORCE(Z) > 10*ounces
    DO ABORT("No hole");

{ Do insertion with compliance }
MOVE *barm* TO *beam_bore* DIRECTLY
    WITH FORCE(Z) = −10*ounces
    WITH FORCE(X) = 0*ounces
    WITH FORCE(Y) = 0*ounces
    WITH DURATION = 5*seconds;

END insertion.

9.3 CHARACTERISTICS OF TASK-LEVEL LANGUAGES

A completely different approach in robot programming is by task-level programming. The natural way to describe an assembly task is in terms of the objects being manipulated rather than by the robot motions. Task-level languages make use of this fact and simplify the programming task.

A task-level programming system allows the user to describe the task in a high-level language (task specification); a task planner will then consult a database (world models) and transform the task specification into a robot-level program (robot program synthesis) that will accomplish the task. Based on this description, we can conceptually divide task planning into three phases: world modeling, task specification, and program synthesis. It should be noted that these three phases are not completely independent, in fact, they are computationally related.

Figure 9.4 shows one possible architecture for the task planner. The task specification is decomposed into a sequence of subtasks by the task decomposer and information such as initial state, final state, grasping position, operand, specifications, and attachment relations are extracted. The subtasks then pass through the subtask planner which generates the required robot program.

The concept of task planning is quite similar to the idea of automatic program generation in artificial intelligence. The user supplies the input-output requirements of a desired program, and the program generator then generates a program that will produce the desired input-output behavior (Barr et al. [1981, 1982]).

Task-level programming, like automatic program generation, is, in the research stage with many problems still unsolved. In the remaining sections we will discuss the problems encountered in task planning and some of the solutions that have been proposed to solve them.

9.3.1 World Modeling

World modeling is required to describe the geometric and physical properties of the objects (including the robot) and to represent the state of the assembly of objects in the workspace.

Geometric and Physical Models. For the task planner to generate a robot program that performs a given task, it must have information about the objects and the robot itself. These include the geometric and physical properties of the objects which can be represented by models.

A geometric model provides the spatial information (dimension, volume, shape) of the objects in the workspace. As discussed in Chap. 8, numerous techniques exist for modeling three-dimensional objects (Baer et al. [1979], Requicha [1980]). The most common approach is constructive solid geometry (CSG), where objects are defined as constructions or combinations, using regularized set operations (such as union, intersection), of primitive objects (such as cube, cylinder). The primitives can be represented in various ways:

1. A set of edges and points
2. A set of surfaces
3. Generalized cylinders
4. Cell decomposition

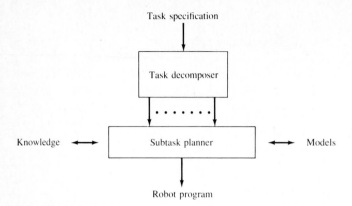

Figure 9.4 Task planner.

In the AUTOPASS system (Lieberman and Wesley [1977]), objects are modeled by utilizing a modeling system called GDP (geometric design processor) (Wesley et al. [1980]) which uses a procedural representation to describe objects. The basic idea is that each object is represented by a procedure name and a set of parameters. Within this procedure, the shape of the object is defined by calls to other procedures representing other objects or set operations.

GDP provides a set of primitive objects (all of them are polyhedra) which can be cuboid, cylinder, wedge, cone, hemisphere, laminum, and revolute. These primitives are internally represented as a list of surfaces, edges, and points which are defined by the parameters in the corresponding procedure. For example,

CALL SOLID(CUBOID, "Block", xlen, ylen, zlen);

will invoke the procedure SOLID to define a rectangular box called Block with dimensions xlen, ylen, and zlen. More complicated objects can then be defined by calling other procedures and applying the MERGE subroutine to them. Table 9.8 shows a description of the bolt used in the insertion task discussed in Sec. 9.2.

Physical properties such as inertia, mass, and coefficient of friction may limit the type of motion that the robot can perform. Instead of storing each of the properties explicitly, they can be derived from the object model. However, no model can be 100 percent accurate and identical parts may have slight differences in their physical properties. To deal with this, tolerances must be introduced into the model (Requicha [1983]).

Representing World States. The task planner must be able to stimulate the assembly steps in order to generate the robot program. Each assembly step can be succinctly represented by the current state of the world. One way of representing these states is to use the configurations of all the objects in the workspace.

AL provides an attachment relation called AFFIX that allows frames to be attached to other frames. This is equivalent to physically attaching a part to

Table 9.8 GDP description of a bolt

Bolt: PROCEDURE(shaft_height, shaft_radius, shaft_nfacets, head_height, head_radius, head_nfacets);

/* define parameters */
DECLARE
      shaft_height,   /* height of shaft */
      head_height,   /* height of head */
      shaft_radius,   /* radius of shaft */
      head_radius,   /* radius of head */
      shaft_nfacets,   /* number of shaft faces */
      head_nfacets,   /* number of head faces */

/* specify floating point for above variables */
FLOAT;

/* define shape of the shaft */
CALL SOLID(CYLIND, "Shaft", shaft_height, shaft_radius, shaft_nfacets);

/* define shape of head */
CALL SOLID(CYLIND, "Head", head_height, head_radius, head_nfacets);

/* perform set union to get bolt */
CALL MERGE("Shaft", "Head", union);

END Bolt.

Note: The notation /* · · · */ indicates a comment.

---

another part and if one of the parts moves, the other parts attached will also move. AL automatically updates the locations of the frames by multiplying the appropriate transformations. For example,

    AFFIX *beam_bore* TO *beam* RIGIDLY;
    beam_bore = FRAME(nilrot, VECTOR(1,0,0)*inches);

describes that the frame *beam_bore* is attached to the frame *beam*.

    AUTOPASS uses a graph to represent the world state. The nodes of the graph represents objects and the edges represent relationships. The relations can be one of:

1. *Attachment*. An object can be rigidly, nonrigidly, or conditionally attached to another object. The first two of these have a function similar to the AFFIX statement in AL. Conditionally attachment means that the object is supported by the gravity (but not strictly attached).
2. *Constraints*. Constraint relationships represent physical constraints between objects which can be translational or rotational.
3. *Assembly component*. This is used to indicate that the subgraph linked by this edge is an assembly part and can be referenced as an object.

As the assembly proceeds, the graph is updated to reflect the current state of the assembly.

9.3.2 Task Specification

Task specification is done with a high-level language. At the highest level one would like to have natural languages as the input, without having to give the assembly steps. An entire task like building a water pump could then be specified by the command "build water pump." However, this level of input is still quite far away. Not even omitting the assembly sequence is possible. The current approach is to use an input language with a well-defined syntax and semantics, where the assembly sequence is given.

An assembly task can be described as a sequence of states of the world model. The states can be given by the configurations of all the objects in the workspace, and one way of specifying configurations is to use the spatial relationships between the objects. For example, consider the block world shown in Fig. 9.5. We define a spatial relation AGAINST to indicate that two surfaces are touching each other. Then the statements in Table 9.9 can be used to describe the two situations depicted in Fig. 9.5. If we assume that state A is the initial state and state B is the goal state, then they can be used to represent the task of picking up *Block3* and placing it on top of *Block2*. If state A is the goal state and state B is the initial state, then they would represent the task of removing *Block3* from the stack of blocks and placing it on the table. The advantage of using this type of representation is that they are easy to interpret by a human, and therefore, easy to specify and modify. However, a serious limitation of this method is that it does not specify all the necessary information needed to describe an operation. For example, the torque required to tighten a bolt cannot be incorporated into the state description.

An alternate approach is to describe the task as a sequence of symbolic operations on the objects. Typically, a set of spatial constraints on the objects are also given to eliminate any ambiguity. This form of description is quite similar to those used in an industrial assembly sheet. Most robot-oriented languages have adopted this type of specification.

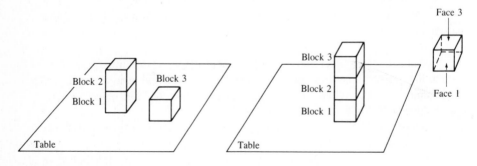

Figure 9.5 Block world.

Table 9.9 State description of block world

State A:	State B:
(*Block1_face1* AGAINST *table*)	(*Block1_face1* AGAINST *Table*)
(*Block1_face3* AGAINST *Block2_face1*)	(*Block1_face3* AGAINST *Block2_face1*)
(*Block3_face1* AGAINST *Table*)	(*Block2_face3* AGAINST *Block3_face1*)

AL provides a limited way of describing a task using this method. With the AFFIX statements, an object frame can be attached to *barm* to indicate that the hand is holding the object. Then moving the object to another point can be described by moving the object frame instead of the arm. For example, the inserting process in Fig. 9.1 can be specified as

AFFIX *bolt_tip* TO *barm;*
MOVE *bolt_tip* TO *beam_bore;*

Popplestone et al. [1978] have proposed a language called RAPT which uses contact relations AGAINST, FIT, and COPLANAR to specify the relationship between object features. Object features, which can be planar or spherical faces, cylindrical shafts and holes, edges, and vertices, are defined by coordinate frames similar to those used in AL. For example, the two operations in the block world example can be described as:

PLACE *Block3* SO THAT (*Block2_face3* AGAINST *Block3_face1*)
PLACE *Block3* SO THAT (*Block3_face1* AGAINST *Table*)

The spatial relationships are then extracted and solved for the configuration constraints on the objects required to perform the task.

AUTOPASS also uses this type of specification but it has a more elaborate syntax. It divides its assembly related statements into three groups:

1. *State change statement*: Describes an assembly operation such as placement and adjustment of parts.
2. *Tools statement*: Describes the type of tools to use.
3. *Fastener statement*: Describes a fastening operation.

The syntax of these statements is complicated (see Table 9.10). For example,

PLACE *bolt* ON *beam* SUCH THAT *bolt_tip* IS ALIGNED WITH *beam_bore*;
DRIVE IN *bolt* AT *bolt_grasp* SUCH THAT TORQUE IS EQ 12.0 IN-LBS
USING *air_driver*;

would be used to describe the operation of inserting a bolt and tightening it.

Table 9.10 The syntax of the state change and tool statements in AUTOPASS

State change statement

PLACE <object> <preposition> <object> <grasping> <final-condition>
     <constraint> <then-hold>

where

<object>	Is a symbolic name for the object.
<preposition>	Is either IN or ON; it is used to determine the type of operation.
<grasping>	Specifies how the object should be grasped.
<constraint>	Specifies the constraints to be met during the execution of the command.
<then-hold>	Indicates that the hand is to remain in position on completion of the command.

Tool statement

OPERATE <tool> <load-list> <at-position> <attachment> <final-condition>
     <tool-parameters> <then-hold>

where

<tool>	Specifies the tool to be used.
<load-list>	Specifies the list of accessories.
<at-position>	Specifies where the tool is to be operated.
<attachment>	Specifies new attachment.
<final-condition>	Specifies the final condition to be satisfied at the completion of the command.
<tool-parameters>	Specifies tool operation parameters such as direction of rotation and speed.
<then-hold>	Indicates that the hand is to remain in position on completion of the command.

9.3.3 Robot Program Synthesis

The synthesis of a robot program from a task specification is one of the most important and most difficult phases of task planning. The major steps in this phase are grasping planning, motion planning, and plan checking. Before the task planner can perform the planning, it must first convert the symbolic task specification into a usable form. One approach is to obtain configuration constraints from the symbolic relationships. The RAPT interpreter extracts the symbolic relationships and forms a set of matrix equations with the constraint parameters of the objects as unknowns. These equations are then solved symbolically by using a set of rewrite rules to simplify them. The result obtained is a set of constraints on the configurations of each object that must be satisfied to perform the operation.

Grasping planning is probably the most important problem in task planning because the way the object is grasped affects all subsequent operations. The way the robot can grasp an object is constrained by the geometry of the object being grasped and the presence of other objects in the workspace. A usable grasping configuration is one that is reachable and stable. The robot must be able to reach the object without colliding with other objects in the workspace and, once grasped, the object must be stable during subsequent motions of the robot.

Typically, the method used to choose a grasp configuration is a variation of the following procedure:

1. A set of candidate grasping configurations are chosen based on:

    Object geometry (e.g., for a parallel-jaw gripper, a good place to grasp is on either side of parallel surfaces).

    Stability (one heuristic is to have the center of mass of the object lie within the fingers).

    Uncertainty reduction.

2. The set is then pruned according to whether they:

    Are reachable by the robot.

    Would lead to collisions with other objects.

3. The final configuration is selected among the remaining configurations (if any) such that:

    It would lead to the most stable grasp.

    It would be the most unlikely to have a collision in the presence of position errors.

Most of the current methods for grasp planning focus only on finding reachable grasping positions, and only a subset of constraints are considered. Grasping in the presence of uncertainties is more difficult and often involves the use of sensing.

After the object is grasped, the robot must move the object to its destination and accomplish the operation. This motion can be divided into four phases:

1. A guarded departure from the current configuration
2. A free motion to the desired configuration without collision
3. A guarded approach to the destination
4. A compliant motion to achieve the goal configuration

One of the important problems here is planning the collision-free motion. Several algorithms have been proposed for planning collision-free path and they can be grouped into three classes:

1. *Hypothesis and test.* In this method, a candidate path is chosen and the path is tested for collision at a set of selected configurations. If a collision occurs, a correction is made to avoid the collision (Lewis and Bejczy [1973]). The main advantage of this method is its simplicity and most of the tools needed are already available in the geometric modeling system. However, generating the correction is difficult, particularly when the workspace is clustered with obstacles.

2. *Penalty functions.* This method involves defining penalty functions whose values depend on the proximity of the obstacles. These functions have the characteristic that, as the robot gets closer to the obstacles, their values increase. A total penalty function is computed by adding all the individual penalty functions and possibly a penalty term relating to minimum path. Then the derivatives of the total penalty function with respect to the configuration parameters are estimated and the collision-free path is obtained by following the local minima of the total penalty function. This method has the advantage that adding obstacles and constraints is easy. However, the penalty functions generally are difficult to specify.

3. *Explicit free space.* Several algorithms have been proposed in this class. Lozano-Perez [1982] proposed to represent the free space (space free of obstacles) in terms of the robot's configuration (configuration space). Conceptually, the idea is equivalent to transforming the robot's hand holding the object into a point, and expanding the obstacles in the workspace appropriately. Then, finding a collision-free path amounts to finding a path that does not intersect any of the expanded obstacles. This algorithm performs reasonably well when only translation is considered. However, with rotation, approximations must be made to generate the configuration space and the computations required increase significantly. Brooks [1983a, 1983b] proposed another method by representing the free space as overlapping generalized cones and the volume swept by the moving object as a function of its orientation. Then, finding the collision-free path reduces to comparing the swept volume of the object with the swept volume of the free space.

Generating the compliant motion is another difficult and important problem. Current work has been based on using the task kinematics to constrain the legal robot configurations to lie on a C-surface† (Mason [1981]) in the robot's configuration space. Then generating compliant motions is equivalent to finding a hybrid position/force control strategy that guarantees the path of the robot to stay on the C surface.

9.4 CONCLUDING REMARKS

We have discussed the characteristics of robot-oriented languages and task-level programming languages. In robot-oriented languages, an assembly task is explicitly described as a sequence of robot motions. The robot is guided and controlled by the program throughout the entire task with each statement of the program roughly corresponding to one action of the robot. On the other hand, task-level

---

† A C-surface is defined on a C-frame. It is a task configuration which allows only partial freedom in position. Along its tangent is the positional freedom and along its normal is the force freedom. A C-frame is an orthogonal coordinate system in the cartesian space. The frame is so chosen that the task freedoms are defined to be translation along and rotation about each of the three principal axes.

Table 9.11a Comparison of various existing robot control languages

Language	AL	AML	AUTOPASS	HELP	JARS	MAPLE
Institute	Stanford	IBM	IBM	GE	JPL	IBM
Robot controlled	PUMA Stanford Arm	IBM Arm	IBM	Allegro	PUMA Stanford Arm	IBM
Robot-or object-level	Mix	Robot	Object	Robot	Robot	Robot
Language basis	Concurrent Pascal	Lisp, APL, Pascal	PL/I	Pascal	Pascal	PL/I
Compiler or interpreter	Both	Interpreter	Both	Interpreter	Compiler	Interpreter
Geometric data type	Frame	Aggregate	Model	None	Frame	None
Motion specified by	Frame	Joints	Implicit	Joints	Joints, frame	Translation, rotation
Control structure	Pascal	Pascal	PL/I	Pascal	Pascal	PL/I
Sensing command	Position, force	Position	Force, tactile	Force, vision	Proximity, vision	Force, proximity
Parallel processing	COBEGIN, semaphores	None	IN PARALLEL	Semaphores	None	IN PARALLEL
Multiple robot	Yes	No	No	Yes	No	Yes
References	1	2	3	4	5	6

1. Mujtaba et al. [1982].
2. Taylor et al. [1983].
3. Lieberman and Wesley [1977].
4. *Automation Systems A12 Assembly Robot Operator's Manual*, P50VE025, General Electric Co., Bridgeport, Conn., February 1982.
5. Craig [1980].
6. Darringer and Blasgen [1975].

languages describe the assembly task as a sequence of positional goals of the objects rather than the motion of the robot needed to achieve these goals, and hence no explicit robot motion is specified. Two existing robot programming languages, AL and AML, were used to illustrate the characteristics of robot-oriented languages. We conclude that a robot-oriented language is difficult to use because it requires the user to program each detailed robot motion in completing a task. Task-level languages are much easier to use. However, many problems in task-level languages, such as task planning, object modeling, obstacle avoidance,

Table 9.11*b*

Language	MCL	PAL	RAIL	RPL	VAL
Institute	McDonnell Douglas	Purdue	Automatix	SRI	Unimate
Robot controlled	Cincinnati Milacron T-3	Stanford Arm	Custom-designed Cartesian arm	PUMA	PUMA
Robot-or object-level	Robot	Robot	Robot	Robot	Robot
Language basis	APT	Transform base	Pascal	Fortran, Lisp	Basic
Complier or interpreter	Compiler	Interpreter	Interpreter	Both	Interpreter
Geometric data type	Frame	Frame	Frame	None	Frame
Motion specified by	Translation, rotation	Frame	Joints, frame	Joints	Joints, frame
Control structure	If-then-else while-do	If-then-else	Pascal	Fortran	If-then
Sensing command	Position	Force	Force, vision	Position, vision	Position, force
Parallel processing	INPAR	None	None	None	Semaphores
Multiple robot	Yes	No	No	No	No
References	1	2	3	4	5

1. Oldroyd [1981].
2. Takase et al. [1981].
3. Franklin and Vanderbrug [1982].
4. Park [1981].
5. *User's Guide to VAL,* version 11, second edition, Unimation, Inc., Danbury, Conn., 1979.

trajectory planning, sensory information utilization, and grasping configurations, must be solved before they can be used effectively. We conclude this chapter with a comparison of various languages, as shown in Table 9.11*a* and *b*.

REFERENCES

Further reading in robot-level programming can be found in Bonner and Shin [1982], Geschke [1983], Gruver et al. [1984], Lozano-Perez [1983a], Oldroyd [1981], Park [1981], Paul [1976, 1981], Popplestone et al. [1978, 1980], Shimano [1979], Synder [1985], Takase et al. [1981], and Taylor et al. [1983]. Further reading in task-level programming can be found in Binford [1979], Darringer and Blasgen [1975], Finkel et al. [1975], Lieberman and Wesley [1977], and Mujtaba et al. [1982]. Languages for describing objects can be found in Barr et al. [1981, 1982], Grossman and Taylor [1978], Lieberman and Wesley [1977], and Wesley et al. [1980]. Takase et al. [1981] presented a homogeneous transformation matrix equation in describing a task sequence to a manipulator.

Various obstacle avoidance algorithms embedded in the programming languages can be found in Brooks [1983*a*, 1983*b*], Brooks and Lozano-Perez [1983], Lewis and Bejczy [1973], Lozano-Perez [1983*a*], Lozano-Perez and Wesley [1979]. In task planning, Lozano-Perez [1982, 1983*b*] presented a configuration space approach for moving an object through a crowded workspace.

Future robot programming languages will incorporate techniques in artificial intelligence (Barr et al. [1981, 1982]) and utilize "knowledge" to perform reasoning (Brooks [1981]) and planning for robotic assembly and manufacturing.

PROBLEMS

9.1 Write an AL statement for defining a coordinate frame *grasp* which can be obtained by rotating the coordinate frame *block* through an angle of 65° about the Y axis and then translating it 4 and 6 inches in the X and Y axes, respectively.

9.2 Repeat Prob. 9.1 with an AML statement.

9.3 Write an AL program to palletize nine parts from a feeder to a tray consisting of a 3 × 3 array of bins. Assume that the locations of the feeder and tray are known. The program has to index the location for each pallet and signal the user when the tray is full.

9.4 Repeat Prob. 9.3 with an AML program.

9.5 Repeat Prob. 9.3 with a VAL program.

9.6 Repeat Prob. 9.3 with an AUTOPASS program.

9.7 *Tower of Hanoi problem.* Three pegs, *A, B,* and *C,* whose coordinate frames are, respectively, $(\mathbf{x}_A, \mathbf{y}_A, \mathbf{z}_A)$, $(\mathbf{x}_B, \mathbf{y}_B, \mathbf{z}_B)$, and $(\mathbf{x}_C, \mathbf{y}_C, \mathbf{z}_C)$, are at a known location from the reference coordinate frame $(\mathbf{x}_0, \mathbf{y}_0, \mathbf{z}_0)$, as shown in the figure below. Initially, peg A has two disks of different sizes, with disks having smaller diameters always on the top of disks with larger diameters. You are asked to write an AL program to control a robot equipped with a special suction gripper (to pick up the disks) to move the two disks from peg A to peg C so that at any instant of time disks of smaller diameters are always on the top of disks with larger diameters. Each disk has an equal thickness of 1 inch.

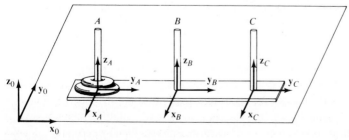

9.8 Repeat Prob. 9.7 with an AML program.

ROBOT INTELLIGENCE AND TASK PLANNING

*That which is apprehended by intelligence and
reason is always in the same state; but
that which is conceived by opinion with the
help of sensation and without reason, is
always is a process of becoming and
perishing and never really is.*
Timaeus, in the "Dialogues of Plato"

10.1 INTRODUCTION

A basic problem in robotics is *planning* motions to solve some prespecified task, and then *controlling* the robot as it executes the commands necessary to achieve those actions. Here, planning means deciding on a course of action before acting. This action synthesis part of the robot problem can be solved by a problem-solving system that will achieve some stated goal, given some initial situation. A plan is, thus, a representation of a course of action for achieving the goal.

Research on robot problem solving has led to many ideas about problem-solving systems in artificial intelligence. In a typical formulation of a robot problem we have a robot that is equipped with sensors and a set of primitive actions that it can perform in some easy-to-understand world. Robot actions change one state, or configuration, of the world into another. In the "blocks world," for example, we imagine a world of several labeled blocks resting on a table or on each other and a robot consisting of a TV camera and a moveable arm and hand that is able to pick up and move blocks. In some problems the robot is a mobile vehicle with a TV camera that performs tasks such as pushing objects from place to place through an environment containing other objects.

In this chapter, we briefly introduce several basic methods in problem solving and their applications to robot planning.

10.2 STATE SPACE SEARCH

One method for finding a solution to a problem is to try out various possible approaches until we happen to produce the desired solution. Such an attempt involves essentially a trial-and-error search. To discuss solution methods of this sort, it is helpful to introduce the notion of problem states and operators. *A prob-*

lem state, or simply *state,* is a particular problem situation or configuration. The set of all possible configurations is the space of problem states, or the *state space.* An *operator,* when applied to a state, transforms the state into another state. A *solution* to a problem is a sequence of operators that transforms an initial state into a goal state.

It is useful to imagine the space of states reachable from the initial state as a graph containing nodes corresponding to the states. The nodes of the graph are linked together by arcs that correspond to the operators. A solution to a problem could be obtained by a search process that first applies operators to the initial state to produce new states, then applies operators to these, and so on until the goal state is produced. Methods of organizing such a search for the goal state are most conveniently described in terms of a graph representation.

10.2.1 Introductory Examples

Before proceeding with a discussion of graph search techniques, we consider briefly some basic examples as a means of introducing the reader to the concepts discussed in this chapter.

Blocks World. Consider that a robot's world consists of a table T and three blocks, A, B, and C. The initial state of the world is that blocks A and B are on the table, and block C is on top of block A (see Fig. 10.1). The robot is asked to change the initial state to a goal state in which the three blocks are stacked with block A on top, block B in the middle, and block C on the bottom. The only operator that the robot can use is MOVE X from Y to Z, which moves object X from the top of object Y onto object Z. In order to apply the operator, it is required that (1) X, the object to be moved, be a block with nothing on top of it, and (2) if Z is a block, there must be nothing on it.

We can simply use a graphical description like the one in Fig. 10.1 as the state representation. The operator MOVE X from Y to Z is represented by MOVE(X,Y,Z). A graph representation of the state space search is illustrated in Fig. 10.2. If we remove the dotted lines in the graph (that is, the operator is not to be used to generate the same operation more than once), we obtain a state space

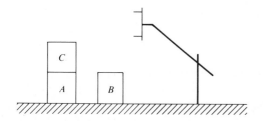

Figure 10.1 A configuration of robot and blocks.

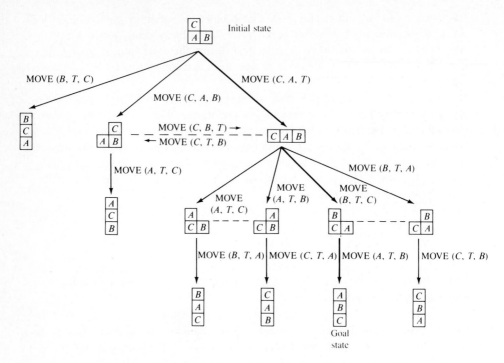

Figure 10.2 State space search graph.

search tree. It is easily seen from Fig. 10.2 that a solution that the robot can obtain consists of the following operator sequence:

$$MOVE(C,A,T), \ MOVE(B,T,C), \ MOVE(A,T,B)$$

Path Selection. Suppose that we wish to move a long thin object A through a crowded two-dimensional environment as shown in Fig. 10.3. To map motions of the object once it is grasped by a robot arm, we may choose the state space representation (x, y, α) where

x = horizontal coordinate of the object $\quad 1 \leqslant x \leqslant 5$

y = vertical coordinate of the object $\quad 1 \leqslant y \leqslant 3$

α = orientation of the object

$$\alpha = \begin{cases} 0 & \text{if object } A \text{ is parallel to } x \text{ axis} \\ 1 & \text{if object } A \text{ is parallel to } y \text{ axis} \end{cases}$$

Both position and orientation of the object are quantized. The operators or robot

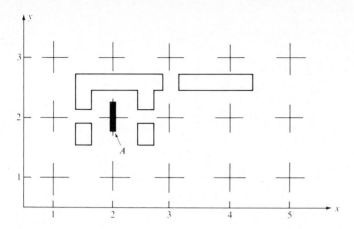

Figure 10.3 Physical space for Example 2.

commands are:

MOVE $\pm$ x direction one unit

MOVE $\pm$ y direction one unit

ROTATE 90°

The state space appears in Fig. 10.4. We assume for illustration that each "move" is of length 2, and each "rotate" of length 3. Let the object A be initially at location (2,2), oriented parallel to the y axis, and the goal is to move A to (3,3) and oriented parallel to the x axis. Thus the initial state is (2,2,1) and the goal state is (3,3,0).

There are two equal-length solution paths, shown in Fig. 10.5 and visualized on a sketch of the task site in Fig. 10.6. These paths may not look like the most direct route. Closer examination, however, reveals that these paths, by initially moving the object away from the goal state, are able to save two rotations by utilizing a little more distance.

Monkey-and-Bananas Problem. A monkey† is in a room containing a box and a bunch of bananas (Fig. 10.7). The bananas are hanging from the ceiling out of reach of the monkey. How can the monkey get the bananas?

The four-element list (W,x,Y,z) can be selected as the state representation, where

W = horizontal position of the monkey

x = 1 or 0, depending on whether the monkey is on top of the box or not, respectively

---

† It is noted that the monkey could be a mobile robot.

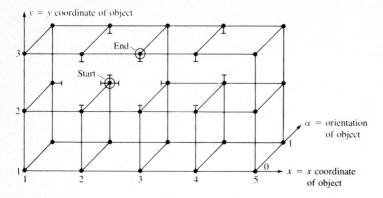

Figure 10.4 Graph of the problem in Fig. 10.3.

Y = horizontal position of the box

z = 1 or 0, depending on whether the monkey has grasped the bananas or not, respectively

The operators in this problem are:

1. goto(U). Monkey goes to horizontal position U, or in the form of a *production rule,*

$$(W,0,Y,z) \xrightarrow{\text{goto}(U)} (U,0,Y,z)$$

That is, state $(W,0,Y,z)$ can be transformed into $(U,0,Y,z)$ by the applying operator goto(U).

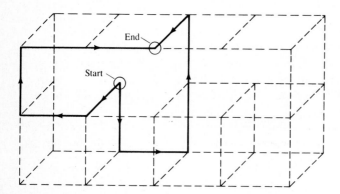

Figure 10.5 Solution to the graph of Fig. 10.4.

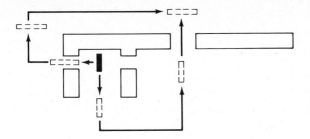

Figure 10.6 Visualization of solution path for Fig. 10.5.

2. pushbox(V). Monkey pushes the box to horizontal position V, or

$$(W,0,W,z) \xrightarrow{\text{pushbox}(V)} (V,0,V,z)$$

It should be noted from the left side of the production rule that, in order to apply the operator pushbox(V), the monkey should be at the same position W as the box, but not on top of it. Such a condition imposed on the applicability of an operator is called the *precondition* of the production rule.

3. climbbox. Monkey climbs on top of the box, or

$$(W,0,W,z) \xrightarrow{\text{climbbox}} (W,1,W,z)$$

It should be noted that, in order to apply the operator climbbox, the monkey must be at the same position W as the box, but not on top of it.

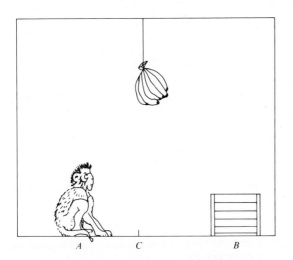

Figure 10.7 Monkey-and-bananas problem.

4. grasp. Monkey grasps the bananas, or

$$(C,1,C,0) \overset{\text{grasp}}{\longrightarrow} (C,1,C,1)$$

where C is the location on the floor directly under the bananas. It should be noted that in order to apply the operator grasp, the monkey and the box should both be at position C and the monkey should already be on the top of the box.

It is noted that both the applicability and the effects of the operators are expressed by the production rules. For example, in rule 2, the operator pushbox(V) is only applicable when its precondition is satisfied. The effect of the operator is that the monkey has pushed the box to position V. In this formulation, the set of goal states is described by any list whose last element is 1.

Let the initial state be $(A,0,B,0)$. The only operator that is applicable is goto (U), resulting in the next state $(U,0,B,0)$. Now three operators are applicable; they are goto(U), pushbox(V) and climbbox (if $U = B$). Continuing to apply all operators applicable at every state, we produce the state space in terms of the graph representation shown in Fig. 10.8. It can be easily seen that the sequence of operators that transforms the initial state into a goal state consists of goto(B), pushbox(C), climbbox, and grasp.

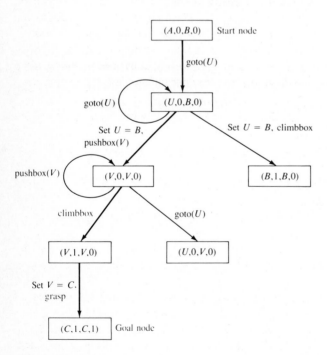

Figure 10.8 Graph representation for the monkey-and-bananas problem.

10.2.2 Graph-Search Techniques

For small graphs, such as the one shown in Fig. 10.8, a solution path from the initial state to a goal state can be easily obtained by inspection. For a more complicated graph a formal search process is needed to move through the state (problem) space until a path from an initial state to a goal state is found. One way to describe the search process is to use *production systems*. A production system consists of:

1. A *database* that contains the information relevant to the particular task. Depending on the application, this database may be as simple as a small matrix of numbers or as complex as a large, relational indexed file structure.
2. A set of *rules* operating on the database. Each rule consists of a left side that determines the applicability of the rule or precondition, and a right side that describes the action to be performed if the rule is applied. Application of the rule changes the database.
3. A *control strategy* that specifies which rules should be applied and ceases computation when a termination condition on the database is satisfied.

In terms of production system terminology, a graph such as the one shown in Fig. 10.8 is generated by the control strategy. The various databases produced by rule applications are actually represented as nodes in the graph. Thus, a graph-search control strategy can be considered as a means of finding a path in a graph from a (start) node representing the initial database to one (goal node) representing a database that satisfies the termination (or goal) condition of the production system.

A general graph-search procedure can be described as follows.

Step 1. Create a search graph G consisting solely of the start node s. Put s on a list called OPEN.

Step 2. Create a list called CLOSED that is initially empty.

Step 3. LOOP: if OPEN is empty, exit with failure.

Step 4. Select the first node on OPEN, remove it from OPEN, and put it on CLOSED. Call this mode n.

Step 5. If n is a goal node, exit successfully with the solution obtained by tracing a path along the pointers from n to s in G (pointers are established in step 7).

Step 6. Expand node n, generating the set M of its successors that are not ancestors of n. Install these members of M as successors of n in G.

Step 7. Establish a pointer to n from those members of M that were not already in OPEN or CLOSED. Add these members of M to OPEN. For each member of M that was already on OPEN or CLOSED, decide whether or not to redirect its pointer to *n*. For each member of *M* already on CLOSED, decide for each of its descendants in *G* whether or not to redirect its pointer†.

Step 8. Reorder the list OPEN, either according to some arbitrary criterion or according to heuristic merit.

Step 9. Go to LOOP.

If no heuristic information from the problem domain is used in ordering the nodes on OPEN, some arbitrary criterion must be used in step 8. The resulting search procedure is called *uninformed* or *blind*. The first type of blind search procedure orders the nodes on OPEN in increasing order of their depth in the search tree.‡ The search that results from such an ordering is called breadth-first search. It has been shown that breadth-first search is guaranteed to find a shortest-length path to a goal node, providing that a path exists. The second type of blind search orders the nodes on OPEN in descending order of their depth in the search tree. The deepest nodes are put first in the list. Nodes of equal depth are ordered arbitrarily. The search that results from such an ordering is called depth-first search. To prevent the search process from running away along some fruitless path forever, a depth bound is set. No node whose depth in the search tree exceeds this bound is ever generated.

The blind search methods described above are exhaustive search techniques for finding paths from the start node to a goal node. For many tasks it is possible to use task-dependent information to help reduce the search. This class of search procedures is called *heuristic* or *best-first* search, and the task-dependent information used is called heuristic information. In step 8 of the graph search procedure, heuristic information can be used to order the nodes on OPEN so that the search expands along those sectors of the graph thought to be the most promising. One important method uses a real-valued evaluation function to compute the "promise" of the nodes. Nodes on OPEN are ordered in increasing order of their values of the evaluation function. Ties among the nodes are resolved arbitrarily, but always in favor of goal nodes. The choice of evaluation function critically determines search results. A useful best-first search algorithm is the so-called A* algorithm described below.

Let the evaluation function *f* at any node *n* be

$$f(n) = g(n) + h(n)$$

where $g(n)$ is a measure of the cost of getting from the start node to node *n*, and

---

† If the graph being searched is a tree, then none of the successors generated in step 6 has been generated previously. Thus, the members of *M* are not already on either OPEN or CLOSED. In this case, each member of *M* is added to OPEN and is installed in the search tree as successors of *n*. If the graph being searched is not a tree, it is possible that some of the members of *M* have already been generated, that is, they may already be on OPEN or CLOSED.

‡ To promote earlier termination, goal nodes should be put at the very beginning of OPEN.

$h(n)$ is an estimate of the additional cost from node n to a goal node. That is, $f(n)$ represents an estimate of the cost of getting from the start node to a goal node along the path constrained to go through node n.

The A* Algorithm

Step 1. Start with OPEN containing only the start node. Set that node's g value to 0, its h value to whatever it is, and its f value to $h + 0$, or h. Set CLOSED to the empty list.

Step 2. Until a goal node is found, repeat the following procedure: If there are no nodes on OPEN, report failure. Otherwise, pick the node on OPEN with the lowest f value. Call it BESTNODE. Remove it from OPEN. Place it on CLOSED. See if BESTNODE is a goal node. If so, exit and report a solution (either BESTNODE if all we want is the node, or the path that has been created between the start node and BESTNODE if we are interested in the path). Otherwise, generate the successors of BEST-NODE, but do not set BESTNODE to point to them yet. (First we need to see if any of them have already been generated.) For each such SUC-CESSOR, do the following:

   a. Set SUCCESSOR to point back to BESTNODE. These back links will make it possible to recover the path once a solution is found.

   b. Compute $g(\text{SUCCESSOR}) = g(\text{BESTNODE}) +$ cost of getting from BESTNODE to SUCCESSOR.

   c. See if SUCCESSOR is the same as any node on OPEN (i.e., it has already been generated but not processed). If so, call that node OLD. Since this node already exists in the graph, we can throw SUCCES-SOR away, and add OLD to the list of BESTNODE's successors. Now we must decide whether OLD's parent link should be reset to point to BESTNODE. It should be if the path we have just found to SUCCESSOR is cheaper than the current best path to OLD (since SUCCESSOR and OLD are really the same node). So see whether it is cheaper to get to OLD via its current parent or to SUCCESSOR via BESTNODE, by comparing their g values. If OLD is cheaper (or just as cheap), then we need do nothing. If SUCCESSOR is cheaper, then reset OLD's parent link to point to BESTNODE, record the new cheaper path in $g(\text{OLD})$, and update $f(\text{OLD})$.

   d. If SUCCESSOR was not on OPEN, see if it is on CLOSED. If so, call the node on CLOSED OLD, and add OLD to the list of BESTNODE's successors. Check to see if the new path or the old path is better just as in step 2*c*, and set the parent link and g and f values appropriately. If we have just found a better path to OLD, we must propagate the improvement to OLD's successors. This is a bit tricky. OLD points to its successors. Each successor in turn points to its suc-cessors, and so forth, until each branch terminates with a node that either is still on OPEN or has no successors. So to propagate the new cost downward, do a depth-first traversal of the tree starting at OLD, changing each node's g value (and thus also its f value), terminating each branch when you reach either a node with no successors or a node to which an equivalent or better path has already been found. This condition is easy to check for. Each node's parent link points back to its best known parent. As we propagate down to a node, see if its parent points to the node we are coming from. If so, continue the propagation. If not, then its g value already reflects the better path of which it is part. So the propagation may stop here. But it is possible that with the new value of g being propagated downward, the path we are following may become better than the path through the current parent. So compare the two. If the path through the current parent is still better, stop the propagation. If the path we are propagating through is now better, reset the parent and continue propagation.

e. If SUCCESSOR was not already on either OPEN or CLOSED, then put it on OPEN, and add it to the list of BESTNODE's successors. Compute $f(\text{SUCCESSOR}) = g(\text{SUCCESSOR}) + h(\text{SUCCESSOR})$.

It is easy to see that the A* algorithm is essentially the graph search algorithm using $f(n)$ as the evaluation function for ordering nodes. Note that because $g(n)$ and $h(n)$ must be added, it is important that $h(n)$ be a measure of the cost of getting from node n to a goal node.

The objective of a search procedure is to discover a path through a problem space from an initial state to a goal state. There are two directions in which such a search could proceed: (1) forward, from the initial states, and (2) backward, from the goal states. The rules in the production system model can be used to reason forward from the initial states and to reason backward from the goal states. To reason forward, the left sides or the preconditions are matched against the current state and the right side (the results) are used to generate new nodes until the goal is reached. To reason backward, the right sides are matched against the current state and the left sides are used to generate new nodes representing new goal states to be achieved. This continues until one of these goal states is matched by a initial state.

By describing a search process as the application of a set of rules, it is easy to describe specific search algorithms without reference to the direction of the search. Of course, another possibility is to work both forward from the initial state and backward from the goal state simultaneously until two paths meet somewhere in between. This strategy is called *bidirectional search*.

10.3 PROBLEM REDUCTION

Another approach to problem solving is *problem reduction*. The main idea of this approach is to reason backward from the problem to be solved, establishing sub-

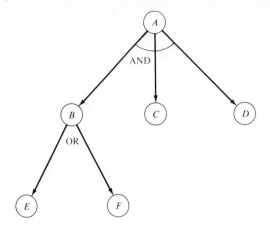

Figure 10.9 An AND/OR graph.

problems and sub-subproblems until, finally, the original problem is reduced to a set of trivial primitive problems whose solutions are obvious. A problem-reduction operator transforms a problem description into a set of reduced or successor problem descriptions. For a given problem description there may be many reduction operators that are applicable. Each of these produces an alternative set of subproblems. Some of the subproblems may not be solvable, however, so we may have to try several operators in order to produce a set whose members are all solvable. Thus it again requires a search process.

The reduction of problem to alternative sets of successor problems can be conveniently expressed by a graphlike structure. Suppose problem A can be solved by solving all of its three subproblems B, C, and D; an AND arc will be marked on the incoming arcs of the nodes B, C, and D. The nodes B, C, and D are called AND nodes. On the other hand, if problem B can be solved by solving any one of the subproblems E and F, an OR arc will be used. These relationships can be shown by the AND/OR graph shown in Fig. 10.9. It is easily seen that the search methods discussed in Sec. 10.2 are for OR graphs through which we want to find a single path from the start node to a goal node.

Example: An AND/OR graph for the monkey-and-bananas problem is shown in Fig. 10.10. Here, the problem configuration is represented by a triple (S,F,G), where S is the set of starting states, F is the set of operators, and G the set of goal states. Since the operator set F does not change in this problem and the initial state is $(A,0,B,0)$, we can suppress the symbol F and denote the problem simply by $(\{(A,0,B,0)\},G)$. One way of selecting problem-reduction operators is through the use of a *difference*. Loosely speaking, the difference for (S,F,G) is a partial list of reasons why the goal test defining the set G is failed by the member of S. (If some member of S is in G, the problem is solved and there is no difference.)

From the example in Sec. 10.2.1, F = $\{f_1, f_2, f_3, f_4\}$ = $\{goto(U),$ pushbox(V), climbbox, grasp$\}$. First, we calculate the difference for the initial problem. The reason that the list $(A,0,B,0)$ fails to satisfy the goal test is that the last element is not 1. The operator relevant to reduce this difference is f_4 = grasp. Using f_4 to reduce the initial problem, we obtain the following pair of subproblems: $(\{(A,0,B,0)\}, G_{f_4})$ and $(\{f_4(s_1)\}G)$, where G_{f_4} is the set of state descriptions to which the operator f_4 is applicable and s_1 is that state in G_{f_4} obtained as a consequence of solving $(\{(A,0,B,0)\}, G_{f_4})$.

To solve the problem $(\{(A,0,B,0)\}, G_{f_4})$, we first calculate its difference. The state described by $(A,0,B,0)$ is not in G_{f_4} because (1) the box is not at C, (2) the monkey is not at C, and (3) the monkey is not on the box. The operators relevant to reduce these differences are, respectively, f_2 = pushbox(C), f_1 = goto(C), and f_3 = climbbox. Applying operator f_2 results in the subproblems $(\{(A,0,B,0)\}, G_{f_2})$ and $(f_2(s_{11}), G_{f_2})$, where $s_{11} \in G_{f_2}$ is obtained as a consequence of solving the first subproblem.

Since $(\{(A,0,B,0)\}, G_{f_2})$ must be solved first, we calculate its difference. The difference is that the monkey is not at B, and the relevant operator is f_1 = goto(B). This operator is then used to reduce the problem to a pair of subproblems $(\{(A,0,B,0)\}, G_{f_1})$ and $(f_1(s_{111}), G_{f_2})$. Now the first of these problems is primitive; its difference is zero since $(A,0,B,0)$ is in the domain of f_1 and f_1 is applicable to solve this problem. Note that $f_1(s_{111})$ = $(B,0,B,0)$ so the second problem becomes $(\{(B,0,B,0)\}, G_{f_2})$. This problem is also primitive since $(B,0,B,0)$ is in the domain of f_2, and f_2 is applicable to solve this problem. This process of completing the solution of problems generated earlier is continued until the initial problem is solved. □

In an AND/OR graph, one of the nodes, called the *state node*, corresponds to the original problem description. Those nodes in the graph corresponding to primitive problem descriptions are called *terminal nodes*. The objective of the search process carried out on an AND/OR graph is to show that the start node is solved. The definition of a solved node can be given recursively as follows:

1. The terminal nodes are solved nodes since they are associated with primitive problems.
2. If a nonterminal node has OR successors, then it is a solved node if and only if at least one of its successors is solved.
3. If a nonterminal node has AND successors, then it is a solved node if and only if all of its successors are solved.

A *solution graph* is the subgraph of solved nodes that demonstrates that the start node is solved. The task of the production system or the search process is to find a solution graph from the start node to the terminal nodes. Roughly speaking, a solution graph from node n to a set of nodes N of an AND/OR graph is analo-

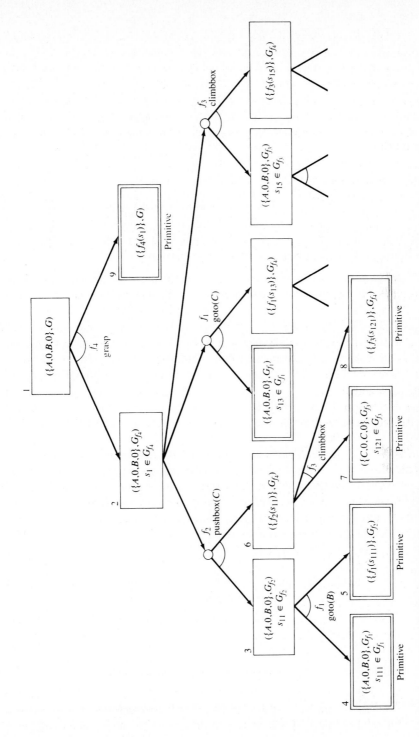

Figure 10.10 AND/OR graph for monkey-and-bananas problem.

gous to a path in an ordinary graph. It can be obtained by starting with node *n* and selecting exactly one outgoing arc. From each successor node to which this arc is directed, we continue to select one outgoing arc, and so on until eventually every successor thus produced is an element of *N*.

In order to find solutions in an AND/OR graph, we need an algorithm similar to A*, but with the ability to handle the AND arc appropriately. Such an algorithm for performing heuristic search of an AND/OR graph is the so-called AO* algorithm.

The AO* Algorithm

Step 1. Let *G* consist only of the node representing the initial state. (Call this node INIT.) Compute *h*(INIT).

Step 2. Until INIT is labeled SOLVED or until INIT's *h* value becomes greater than FUTILITY, repeat the following procedure:

    a. Trace the marked arcs from INIT and select for expansion one of the as yet unexpanded nodes that occurs on this path. Call the selected node NODE.

    b. Generate the successors of NODE. If there are none, then assign FUTILITY as the *h* value of NODE. This is equivalent to saying that NODE is not solvable. If there are successors, then for each one (called SUCCESSOR) that is not also an ancestor of NODE do the following:

        (1) Add SUCCESSOR to *G*.

        (2) If SUCCESSOR is a terminal node, label it SOLVED and assign it an *h* value of 0.

        (3) If SUCCESSOR is not a terminal node, compute its *h* value.

    c. Propagate the newly discovered information up the graph by doing the following: Let *S* be a set of nodes that have been marked SOLVED or whose *h* values have been changed and so need to have values propagated back to their parents. Initialize *S* to NODE. Until *S* is empty, repeat the following procedure:

        (1) Select from *S* a node none of whose descendants in *G* occurs in *S*. (In other words, make sure that for every node we are going to process, we process it before processing any of its ancestors.) Call this node CURRENT, and remove it from *S*.

        (2) Compute the cost of each of the arcs emerging from CURRENT. The cost of each arc is equal to the sum of the *h* values of each of the nodes at the end of the arc plus whatever the cost of the arc itself is. Assign as CURRENT's new *h* value the minimum of the costs just computed for the arcs emerging from it.

        (3) Mark the best path out of CURRENT by marking the arc that had the minimum cost as computed in the previous step.

        (4) Mark CURRENT SOLVED if all of the nodes connected to it through the new marked arc have been labeled SOLVED.

(5) If CURRENT has been marked SOLVED or if the cost of CURRENT was just changed, then its new status must be propagated back up the graph. So add to S all the ancestors of CURRENT.

It is noted that rather than the two lists, OPEN and CLOSED, that were used in the A* algorithm, the AO* algorithm uses a single structure G, representing the portion of the search graph that has been explicitly generated so far. Each node in the graph points both down to its immediate successors and up to its immediate predecessors. Each node in the graph is associated with an h value, an estimate of the cost of a path from itself to a set of solution nodes. The g value (the cost of getting from the start node to the current node) is not stored as in the A* algorithm, and h serves as the estimate of goodness of a node. A quantity FUTILITY is needed. If the estimated cost of a solution becomes greater than the value of FUTILITY, then the search is abandoned. FUTILITY should be selected to correspond to a threshold such that any solution with a cost above it is too expensive to be practical, even if it could ever be found.

A breadth-first algorithm can be obtained from the AO* algorithm by assigning $h \equiv 0$.

10.4 USE OF PREDICATE LOGIC

Robot problem solving requires the capability for representing, retrieving, and manipulating sets of statements. The language of logic or, more specifically, the first-order predicate calculus, can be used to express a wide variety of statements. The logical formalism is appealing because it immediately suggests a powerful way of deriving new knowledge from old (i.e., mathematical deduction). In this formalism, we can conclude that a new statement is true by proving that it follows from the statements that are already known to be true. Thus the idea of a proof, as developed in mathematics as a rigorous way of demonstrating the truth of an already believed proposition, can be extended to include deduction as a way of deriving answers to questions and solutions to problems.†

Let us first explore the use of propositional logic as a way of representing knowledge. Propositional logic is appealing because it is simple to deal with and there exists a decision procedure for it. We can easily represent real-world facts as logical *propositions* written as *well-formed formulas* (wffs) in propositional logic, as shown in the following:

It is raining.
RAINING

---

† At this point, readers who are unfamiliar with propositional and predicate logic may want to consult a good introductory logic text before reading the rest of this chapter. Readers who want a more complete and formal presentation of the material in this section should consult the book by Chang and Lee [1973].

It is sunny.

SUNNY

It is foggy.

FOGGY

If it is raining then it is not sunny.

RAINING $\rightarrow$ ~SUNNY

Using these propositions, we could, for example, deduce that it is not sunny if it is raining. But very quickly we run up against the limitations of propositional logic. Suppose we want to represent the obvious fact stated by the sentence

John is a man

We could write

JOHNMAN

But if we also wanted to represent

Paul is a man

we would have to write something such as

PAULMAN

which would be a totally separate assertion, and we would not be able to draw any conclusions about similarities between John and Paul. It would be much better to represent these facts as

MAN(JOHN)

MAN(PAUL)

since now the structure of the representation reflects the structure of the knowledge itself. We are in even more difficulty if we try to represent the sentence

All men are mortal

because now we really need quantification unless we are willing to write separate statements about the mortality of every known man.

So we appear to be forced to move to predicate logic as a way of representing knowledge because it permits representations of things that cannot reasonably be represented in propositional logic. In predicate logic, we can represent real-world facts as *statements* written as wffs. But a major motivation for choosing to use logic at all was that if we used logical statements as a way of representing knowledge, then we had available a good way of reasoning with that knowledge. In this section, we briefly introduce the language and methods of predicate logic.

The elementary components of predicate logic are *predicate symbols, variable symbols, function symbols,* and *constant symbols.* A predicate symbol is used to represent a relation in a domain of discourse. For example, to represent the fact "Robot is in Room r_1," we might use the simple *atomic formula*:

INROOM(ROBOT,r_1)

In this atomic formula, ROBOT and r_1 are constant symbols. In general, atomic formulas are composed of predicate symbols and terms. A constant symbol is the simplest kind of term and is used to represent objects or entities in a domain of discourse. Variable symbols are terms also, and they permit us to be indefinite about which entity is being referred to, for example, INROOM(x, y). Function symbols denote functions in the domain of discourse. For example, the function symbol *mother* can be used to denote the mapping between an individual and his or her female parent. We might use the following atomic formula to represent the sentence "John's mother is married to John's father."

$$\text{MARRIED[father(JOHN), mother(JOHN)]}$$

An atomic formula has value T (true) just when the corresponding statements about the domain are true and it has the value F (false) just when the corresponding statement is false. Thus INROOM(ROBOT,r_1) has value T, and INROOM(ROBOT,r_2) has value F. Atomic formulas are the elementary building blocks of predicate logic. We can combine atomic formulas to form more complex wffs by using connectives such as $\wedge$ (and), $\vee$ (or), and $\Longrightarrow$ (implies). Formulas built by connecting other formulas by $\wedge$'s are called *conjunctions*. Formulas built by connecting other formulas by $\vee$'s are called *disjunctions*. The connective " $\Longrightarrow$ " is used for representing "if-then" statements, e.g., as in the sentence "If the monkey is on the box, then the monkey will grasp the bananas":

$$\text{ON(MONKEY, BOX)} \Longrightarrow \text{GRASP(MONKEY, BANANAS)}$$

The symbol " $\sim$ " (not) is used to negate the truth value of a formula; that is, it changes the value of a wff from T to F and vice versa. The (true) sentence "Robot is not in Room r_2" might be represented as

$$\sim \text{INROOM(ROBOT,}r_2)$$

Sometimes an atomic formula, $P(x)$, has value T for all possible values of x. This property is represented by adding the universal quantifier ($\forall x$) in front of $P(x)$. If $P(x)$ has value T for at least one value of x, this property is represented by adding the existential quantifier ($\exists x$) in front of $P(x)$. For example, the sentence "All robots are gray" might be represented by

$$(\forall x)[\text{ROBOT}(x) \Longrightarrow \text{COLOR}(x,\text{GRAY})]$$

The sentence "There is an object in Room r_1" might be represented by

$$(\exists x)\text{INROOM}(x,r_1)$$

If P and Q are two wffs, the truth values of composite expressions made up of these wffs are given by the following table:

P	Q	$P \vee Q$	$P \wedge Q$	$P \Longrightarrow Q$	$\sim P$
T	T	T	T	T	F
F	T	T	F	T	T
T	F	T	F	F	F
F	F	F	F	T	T

If the truth values of two wffs are the same regardless of their interpretation, these two wffs are said to be equivalent. Using the truth table, we can establish the following equivalences:

$\sim(\sim P)$	is equivalent to	P
$P \vee Q$	is equivalent to	$\sim P \Rightarrow Q$

deMorgan's laws:

$\sim(P \vee Q)$	is equivalent to	$\sim P \wedge \sim Q$
$\sim(P \wedge Q)$	is equivalent to	$\sim P \vee \sim Q$

Distributive laws:

$P \wedge (Q \vee R)$	is equivalent to	$(P \wedge Q) \vee (P \wedge R)$
$P \vee (Q \wedge R)$	is equivalent to	$(P \vee Q) \wedge (P \vee R)$

Commutative laws:

$P \wedge Q$	is equivalent to	$Q \wedge P$
$P \vee Q$	is equivalent to	$Q \vee P$

Associative laws:

$(P \wedge Q) \vee R$	is equivalent to	$P \wedge (Q \wedge R)$
$(P \vee Q) \vee R$	is equivalent to	$P \vee (Q \vee R)$

Contrapositive law:

$P \Rightarrow Q$	is equivalent to	$\sim Q \Rightarrow \sim P$

In addition, we have

$\sim (\exists x)P(x)$	is equivalent to	$(\forall x)[\sim P(x)]$
$\sim (\forall x)P(x)$	is equivalent to	$(\exists x)[\sim P(x)]$

In predicate logic, there are rules of inference that can be applied to certain wffs and sets of wffs to produce new wffs. One important inference rule is *modus ponens*, that is, the operation to produce the wff W_2 from wffs of the form W_1 and $W_1 \Rightarrow W_2$. Another rule of inference, *universal specialization*, produces the wff $W(A)$ from the wff $(\forall x)W(x)$, where A is any constant symbol. Using modus ponens and universal specialization together, for example, produces the wff $W_2(A)$ from the wffs $(\forall x)[W_1(x) \Rightarrow W_2(x)]$ and $W_1(A)$.

Inference rules are applied to produce derived wffs from given ones. In the predicate logic, such derived wffs are called *theorems*, and the sequence of inference rule applications used in the derivation constitutes a *proof* of the theorem. In artificial intelligence, some problem-solving tasks can be regarded as the task of finding a proof for a theorem. The sequence of inferences used in the proofs gives a solution to the problem.

Example: The state space representation of the monkey-and-bananas problem can be modified so that the states are described by wffs. We assume that, in this example, there are three operators—grasp, climbbox, and pushbox.

Let the initial state s_0 be described by the following set of wffs:

$$\sim \text{ONBOX}$$

$$AT(BOX,B)$$
$$AT(BANANAS,C)$$
$$\sim HB$$

The predicate ONBOX has value T only when the monkey is on top of the box, and the predicate HB has value T only when the monkey has the bananas.

The effects of the three operators can be described by the following wffs:

1. grasp

$$(\forall s)\{ONBOX(s) \wedge AT(BOX,C,s) \implies HB(grasp(s))\}$$

meaning "For all s, if the monkey is on the box and the box is at C in state s, then the monkey will have the bananas in the state attained by applying the operator grasp to state s." It is noted that the value of grasp(s) is the new state resulting when the operator is applied to state s.

2. climbbox

$$(\forall s)\{ONBOX(climbbox(s))\}$$

meaning "For all s, the monkey will be on the box in the state attained by applying the operator climbbox to state s."

3. pushbox

$$(\forall x \forall s)\{\sim ONBOX(s) \implies AT(BOX,x,pushbox(x,s))\}$$

meaning "For all x and s, if the monkey is not on the box in state s, then the box will be at position x in the state attained by applying the operator pushbox(x) to state s."
The goal wff is

$$(\exists s)HB(s)$$

This problem can now be solved by a theorem-proving process to show that the monkey can have the bananas (Nilsson [1971]). □

10.5 MEANS-ENDS ANALYSIS

So far, we have discussed several search methods that reason either forward or backward but, for a given problem, one direction or the other must be chosen. Often, however, a mixture of the two directions is appropriate. Such a mixed strategy would make it possible to solve the main parts of a problem first and then go back and solve the small problems that arise in connecting the big pieces together.

A technique known as *means-ends analysis* allows us to do that. The technique centers around the detection of the difference between the current state and the goal state. Once such a difference is determined, an operator that can reduce the difference must be found. It is possible that the operator may not be applicable to the current state. So a subproblem of getting to a state in which it can be applied is generated. It is also possible that the operator does not produce exactly the goal state. Then we have a second subproblem of getting from the state it does produce to the goal state. If the difference was determined correctly, and if the operator is really effective at reducing the difference, then the two subproblems should be easier to solve than the original problem. The means-ends analysis is applied recursively to the subproblems. From this point of view, the means-ends analysis could be considered as a problem-reduction technique.

In order to focus the system's attention on the big problems first, the differences can be assigned priority levels. Differences of higher priority can then be considered before lower priority ones. The most important data structure used in the means-ends analysis is the "goal." The goal is an encoding of the current problem situation, the desired situation, and a history of the attempts so far to change the current situation into the desired one. Three main types of goals are provided:

Type 1. *Transform* object A into object B.
Type 2. *Reduce* a difference between object A and object B by modifying object A.

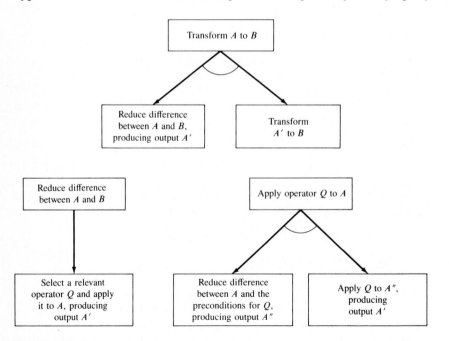

Figure 10.11 Methods for means-ends analysis.

Type 3. *Apply* operator Q to object A.

Associated with the goal types are methods or procedures for achieving them. These methods, shown in a simplified form in Fig. 10.11, can be interpreted as problem-reduction operators that give rise either to AND nodes, in the case of *transform* or *apply*, or to OR nodes, in the case of a *reduce* goal.

The first program to exploit means-ends analysis was the general problem solver (GPS). Its design was motivated by the observation that people often use this technique when they solve problems. For GPS, the initial task is represented as a *transform* goal, in which A is the initial object or state and B the desired object or the goal state. The recursion stops if, for a *transform* goal, there is no difference between A and B, or for an *apply* goal the operator Q is immediately applicable. For a *reduce* goal, the recursion may stop, with failure, when all relevant operators have been tried and have failed.

In trying to transform object A into object B, the *transform* method uses a matching process to discover the differences between the two objects. The difference with the highest priority is the one chosen for reduction. A difference-operator table lists the operators relevant to reducing each difference.

Consider a simple robot problem in which the available operators are listed as follows:

	Preconditions	Operator	Results
1.	AT(ROBOT,OBJ) $\wedge$ LARGE(OBJ) $\wedge$ CLEAR(OBJ) $\wedge$ HANDEMPTY	PUSH(OBJ,LOC) $\longrightarrow$	AT(OBJ,LOC) $\wedge$ AT(ROBOT,LOC)
2.	AT(ROBOT,OBJ) $\wedge$ SMALL(OBJ)	CARRY(OBJ,LOC) $\longrightarrow$	AT(OBJ,LOC) $\wedge$ AT(ROBOT,LOC)
3.	None	WALK(LOC) $\longrightarrow$	AT(ROBOT,LOC)
4.	AT(ROBOT,OBJ)	PICKUP(OBJ) $\longrightarrow$	HOLDING(OBJ)
5.	HOLDING(OBJ)	PUTDOWN(OBJ) $\longrightarrow$	~ HOLDING(OBJ)
6.	AT(ROBOT,OBJ2) $\wedge$ HOLDING(OBJ1)	PLACE(OBJ1,OBJ2) $\longrightarrow$	ON(OBJ1,OBJ2)

Fig. 10.12 shows the difference-operator table that describes when each of the operators is appropriate. Notice that sometimes there may be more than one operator that can reduce a given difference, and a given operator may be able to reduce more than one difference.

Difference \ Operator	PUSH	CARRY	WALK	PICKUP	PUTDOWN	PLACE
Move object	√	√				
Move robot			√			
Clear object				√		
Get object on object						√
Get hand empty					√	√
Be holding object				√		

Figure 10.12 A difference-operator table.

Suppose that the robot were given the problem of moving a desk with two objects on it from one room to another. The objects on top must also be moved. The main difference between the initial state and the goal state would be the location of the desk. To reduce the difference, either PUSH or CARRY could be chosen. If CARRY is chosen first, its preconditions must be met. This results in two more differences that must be reduced: the location of the robot and the size of the desk. The location of the robot can be handled by applying operator WALK, but there are no operators that can change the size of an object. So the path leads to a dead end. Following the other possibility, operator PUSH will be attempted.

PUSH has three preconditions, two of which produce differences between the initial state and the goal state. Since the desk is already large, one precondition creates no difference. The robot can be brought to the correct location by using the operator WALK, and the surface of the desk can be cleared by applying operator PICKUP twice. But after one PICKUP, an attempt to apply the second time results in another difference—the hand must be empty. The operator PUTDOWN can be applied to reduce that difference.

Once PUSH is performed, the problem is close to the goal state, but not quite. The objects must be placed back on the desk. The operator PLACE will put them there. But it cannot be applied immediately. Another difference must be eliminated, since the robot must be holding the objects. The operator PICKUP can be applied. In order to apply PICKUP, the robot must be at the location of the objects. This difference can be reduced by applying WALK. Once the robot is at the location of the two objects, it can use PICKUP and CARRY to move the objects to the other room.

The order in which differences are considered can be critical. It is important that significant differences be reduced before less critical ones. Section 10.6 describes a robot problem-solving system, STRIPS, which uses the means-ends analysis.

10.6 PROBLEM SOLVING

The simplest type of robot problem-solving system is a production system that uses the state description as the database. State descriptions and goals for robot problems can be constructed from logical statements. As an example, consider the robot hand and configurations of blocks shown in Fig. 10.1. This situation can be represented by the conjunction of the following statements:

CLEAR(B)   Block B has a clear top
CLEAR(C)   Block C has a clear top
ON(C,A)   Block C is on block A
ONTABLE(A)   Block A is on the table
ONTABLE(B)   Block B is on the table
HANDEMPTY   The robot hand is empty

The goal is to construct a stack of blocks in which block B is on block C and block A is on block B. In terms of logical statements, we may describe the goal as ON(B,C) $\wedge$ ON(A,B).

Robot actions change one state, or configuration, of the world into another. One simple and useful technique for representing robot action is employed by a robot problem-solving system called STRIPS (Fikes and Nilsson [1971]). A set of rules is used to represent robot actions. Rules in STRIPS consist of three components. The first is the *precondition* that must be true before the rule can be applied. It is usually expressed by the left side of the rule. The second component is a list of predicates called the *delete list*. When a rule is applied to a state description, or database, delete from the database the assertions in the delete list. The third component is called the *add list*. When a rule is applied, add the assertions in the add list to the database. The MOVE action for the block-stacking example is given below:

MOVE(X,Y,Z)   Move object X from Y to Z
Precondition: CLEAR(X), CLEAR(Z), ON(X,Y)
Delete list: ON(X,Y), CLEAR(Z)
Add list: ON(X,Z), CLEAR(Y)

If MOVE is the only operator or robot action available, the search graph (or tree) shown in Fig. 10.2 is generated.

Consider a more concrete example with the initial database shown in Fig. 10.1 and the following four robot actions or operations in STRIPS-form:

1. PICKUP(X)
   Precondition and delete list: ONTABLE(X), CLEAR(X), HANDEMPTY
   Add list: HOLDING(X)
2. PUTDOWN(X)
   Precondition and delete list: HOLDING(X)
   Add list: ONTABLE(X), CLEAR(X), HANDEMPTY

3. STACK(X,Y)
   Precondition and delete list: HOLDING(X), CLEAR(Y)
   Add list: HANDEMPTY, ON(X,Y), CLEAR(X)
4. UNSTACK(X,Y)
   Precondition and delete list: HANDEMPTY, CLEAR(X), ON(X,Y)
   Add list: HOLDING(X), CLEAR(Y)

Suppose that our goal is ON(B,C) $\wedge$ ON(A,B). Working forward from the initial state description shown in Fig. 10.1, we obtain the complete state space for this problem as shown in Fig. 10.13, with a solution path between the initial state and the goal state indicated by dark lines. The solution sequence of actions consists of: {UNSTACK(C,A), PUTDOWN(C), PICKUP(B), STACK(B,C), PICKUP(A), STACK(A,B)}. It is called a "plan" for achieving the goal.

If a problem-solving system knows how each operator changes the state of the world or the database and knows the preconditions for an operator to be executed, it can apply means-ends analysis to solve problems. Briefly, this technique involves looking for a difference between the current state and a goal state and trying to find an operator that will reduce the difference. A *relevant operator* is one whose add list contains formulas that would remove some part of the difference. This continues recursively until the goal state has been reached. STRIPS and most other planners use means-ends analysis.

We have just seen how STRIPS computes a specific plan to solve a particular robot problem. The next step is to generalize the specific plan by replacing constants by new parameters. In other words, we wish to elevate the particular plan to a *plan schema*. The need for a plan generalization is apparent in a learning system. For the purpose of saving plans so that portions of them can be used in a later planning process, the preconditions and effects of any portion of the plan need to be known. To accomplish this, plans are stored in a *triangle table* with rows and columns corresponding to the operators of the plan. The triangle table reveals the structure of a plan in a fashion that allows parts of the plan to be extracted later in solving related problems.

An example of a triangle table is shown in Fig. 10.14. Let the leftmost column be called the zeroth column; then the jth column is headed by the jth operator in the sequence. Let the top row be called the first row. If there are N operators in the plan sequence, then the last row is the $(N + 1)$th row. The entries in cell (i, j) of the table, for $j > 0$ and $i < N + 1$, are those statements added to the state description by the jth operator that survive as preconditions of the ith operator. The entries in cell $(i, 0)$ for $i < N + 1$ are those statements in the initial state description that survive as preconditions of the ith operator. The entries in the $(N + 1)$th row of the table are then those statements in the original state description, and those added by the various operators, that are components of the goal.

Triangle tables can easily be constructed from the initial state description, the operators in the sequence, and the goal description. These tables are concise and convenient representations for robot plans. The entries in the row to the left of the

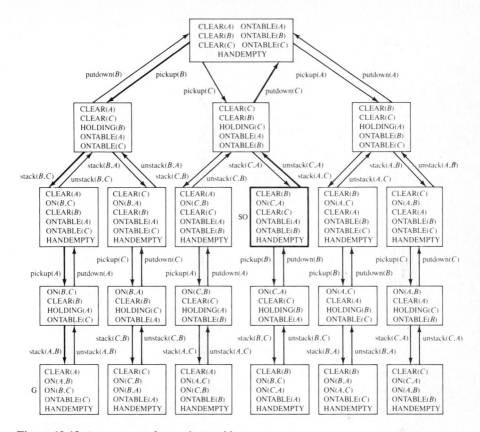

Figure 10.13 A state space for a robot problem.

ith operator are precisely the preconditions of the operator. The entries in the column below the ith operator are precisely the add formula statements of that operator that are needed by subsequent operators or that are components of the goal.

Let us define the ith *kernel* as the intersection of all rows below, and including, the ith row with all columns to the left of the ith column. The fourth kernel is outlined by double lines in Fig. 10.14. The entries in the ith kernel are then precisely the conditions that must be matched by a state description in order that the sequence composed of the ith and subsequent operators be applicable and achieve the goal. Thus, the first kernel (i.e., the zeroth column), contains those conditions of the initial state needed by subsequent operators and by the goal; the $(N + 1)$th kernel [i.e., the $(N + 1)$th row] contains the goal conditions themselves. These properties of triangle tables are very useful for monitoring the actual execution of robot plans.

Since robot plans must ultimately be executed in the real world by a mechanical device, the execution system must acknowledge the possibility that the actions

	0	1	2	3	4	5	6
1	HANDEMPTY CLEAR(C) ON(C,A)	1 unstack(C,A)					
2		HOLDING(C)	2 putdown(C)				
3	ONTABLE(B) CLEAR(B)		HANDEMPTY	3 pickup(B)			
4			CLEAR(C)	HOLDING(B)	4 stack(B,C)		
5	ONTABLE(A)	CLEAR(A)			HANDEMPTY	5 pickup(A)	
6					CLEAR(B)	HOLDING(A)	6 stack(A,B)
7					ON(B,C)		ON(A,B)

Figure 10.14 A triangle table.

in the plan may not accomplish their intended tasks and that mechanical tolerances may introduce errors as the plan is executed. As actions are executed, unplanned effects might either place us unexpectedly close to the goal or throw us off the track. These problems could be dealt with by generating a new plan (based on an updated state description) after each execution step, but obviously, such a strategy would be too costly, so we instead seek a scheme that can intelligently monitor progress as a given plan is being executed.

The kernels of triangle tables contain just the information needed to realize such a plan execution system. At the beginning of a plan execution, we know that the entire plan is applicable and appropriate for achieving the goal because the statements in the first kernel are matched by the initial state description, which was used when the plan was created. (Here we assume that the world is static; that is, no changes occur in the world except those initiated by the robot itself.) Now suppose the system has just executed the first $i - 1$ actions of a plan sequence. Then, in order for the remaining part of the plan (consisting of the ith and subsequent actions) to be both applicable and appropriate for achieving the goal, the statements in the ith kernel must be matched by the new current state description. (We assume that a sensory perception system continuously updates the state description as the plan is executed so that this description accurately models the current state of the world.) Actually, we can do better than merely check to see if the expected kernel matches the state description after an action; we can look for the highest numbered matching kernel. Then, if an unanticipated effect places us closer to the goal, we need only execute the appropriate remaining actions; and if an execution error destroys the results of previous actions, the appropriate actions can be reexecuted.

To find the appropriate matching kernel, we check each one in turn starting with the highest numbered one (which is the last row of the table) and work backward. If the goal kernel (the last row of the table) is matched, execution halts; otherwise, supposing the highest numbered matching kernel is the ith one, then we know that the ith operator is applicable to the current state description. In this case, the system executes the action corresponding to this ith operator and checks the outcome, as before, by searching again for the highest numbered matching kernel. In an ideal world, this procedure merely executes in order each action in the plan. In a real-world situation, on the other hand, the procedure has the flexibility to omit execution of unnecessary actions or to overcome certain kinds of failures by repeating the execution of appropriate actions. Replanning is initiated when there are no matching kernels.

As an example of how this process might work, let us return to our block-stacking problem and the plan represented by the triangle table in Fig. 10.14. Suppose that the system executes actions corresponding to the first four operators and that the results of these actions are as planned. Now suppose that the system attempts to execute the pickup block A action, but the execution routine (this time) mistakes block B for block A and picks up block B instead. [Assume again that the perception system accurately updates the state description by adding HOLDING(B) and deleting ON(B,C); in particular, it does not add HOLDING(A).] If there were no execution error, the sixth kernel would now be matched; the result of the error is that the highest numbered matching kernel is now kernel 4. The action corresponding to STACK(B,C) is thus reexecuted, putting the system back on track.

The fact that the kernels of triangle tables overlap can be used to advantage to scan the table efficiently for the highest numbered matching kernel. Starting in the bottom row, we scan the table from left to right, looking for the first cell that contains a statement that does not match the current state description. If we scan the whole row without finding such a cell, the goal kernel is matched; otherwise, if we find such a cell in column i, the number of the highest numbered matching kernel cannot be greater than i. In this case, we set a *boundary* at column i and move up to the next-to-bottom row and begin scanning this row from left to right, but not past column i. If we find a cell containing an unmatched statement, we reset the column boundary and move up another row to begin scanning that row, etc. With the column boundary set to k, the process terminates by finding that the kth kernel is the highest numbered matching kernel when it completes a scan of the kth row (from the bottom) up to the column boundary.

Example: Consider the simple task of fetching a box from an adjacent room by a robot vehicle. Let the initial state of the robot's world model be as shown in Fig. 10.15. Assume that there are two operators, GOTHRU and PUSHTHRU.

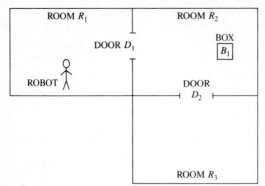

Initial data base M_0:
  INROOM (ROBOT,R_1)
  CONNECTS (D_1,R_1,R_2)
  CONNECTS (D_2,R_2,R_3)
  BOX (B_1)
  INROOM (B_1,R_2)

$(\forall x \ \forall y \ \forall z)$ [CONNECTS $(x)(y)(z)$ → CONNECTS $(x)(y)(z)$]

Goal G_0:
  $(\exists x)$ [BOX(x) ∧ INROOM (x,R_1)]

Figure 10.15 Initial world model.

GOTHRU(d,r_1,r_2) Robot goes through door d from room r_1 into room r_2
Precondition: INROOM(ROBOT,r_1) ∧ CONNECTS(d,r_1,r_2) the robot is in
  room r_1 and door d connects room r_1 to room r_2
Delete list: INROOM(ROBOT,S) for any value of S
Add list: INROOM(ROBOT,r_2)

PUSHTHRU(b,d,r_1,r_2) Robot pushes object b through door d from room r_1
  into room r_2
Precondition: INROOM(b,r_1) ∧ INROOM(ROBOT,r_1)
  ∧ CONNECTS(d,r_1,r_2)
Delete list: INROOM(ROBOT,S), INROOM(b,S)
Add list: INROOM(ROBOT,r_2), INROOM(b,r_2)

The difference-operator table is shown in Fig. 10.16. ☐

When STRIPS is given the problem, it first attempts to achieve the goal G_0
from the initial state M_0. This problem cannot be solved immediately. However,
if the initial database contains a statement INROOM(B_1,R_1), the problem-solving
process could continue. STRIPS finds the operator PUSHTHRU(B_1,d,r_1,R_1)
whose effect can provide the desired statement. The precondition G_1 for
PUSHTHRU is

G_1: INROOM(B_1,r_1) ∧ INROOM(ROBOT,r_1) ∧ CONNECTS(d,r_1,R_1)

Operator Difference	GOTHRU	PUSHTHRU
Location of box		√
Location of robot	√	
Location of box and robot		√

Figure 10.16 Difference-operator table.

From the means-ends analysis, this precondition is set up as a subgoal and STRIPS tries to accomplish it from M_0.

Although no immediate solution can be found to solve this problem, STRIPS finds that if $r_1 = R_2$, $d = D_1$ and the current database contains INROOM(ROBOT, R_2), the process could continue. Again STRIPS finds the operator GOTHRU(d,r_1,R_2) whose effect can produce the desired statement. Its precondition is the next subgoal, namely:

G_2: INROOM(ROBOT,r_1) $\wedge$ CONNECTS(d,r_1,R_2)

Using the substitutions $r_1 = R_1$ and $d = D_1$, STRIPS is able to accomplish G_2. It therefore applies GOTHRU(D_1,R_1,R_2) to M_0 to yield

M_1: INROOM(ROBOT,R_2), CONNECTS(D_1,R_1,R_2)

CONNECTS(D_2,R_2,R_3), BOX(B_1)

INROOM(B_1,R_2), . . .

$(\forall x)(\forall y)(\forall z)$ [CONNECTS(x,y,z) $\Rightarrow$ CONNECTS(x,z,y)]

Now STRIPS attempts to achieve the subgoal G_1 from the new database M_1. It finds the operator PUSHTHRU(B_1,D_1,R_2,R_1) with the substitutions $r_1 = R_2$ and $d = D_1$. Application of this operator to M_1 yields

M_2: INROOM(ROBOT,R_1), CONNECTS(D_1,R_1,R_2)

CONNECTS(D_1,R_2,R_3), BOX(B_1)

INROOM(B_1,R_1), . . .

$(\forall x \forall y \forall z)[$CONNECTS($x,y,z$) $\Rightarrow$ CONNECTS(x,z,y)]

1	INROOM(ROBOT,R_1) CONNECTS(D_1,R_1,R_2)	GOTHRU(D_1,R_1,R_2)	
2	INROOM(B_1,R_2) CONNECTS(D_1,R_1,R_2) CONNECTS(x,y,z) → CONNECTS(x,y,z)	INROOM(ROBOT,R_2)	PUSHTHRU(B_1,D_1,R_2,R_1)
3			INROOM(ROBOT,R_1) INROOM(B_1,R_1)

Figure 10.17 Triangle table.

Next, STRIPS attempts to accomplish the original goal G_0 from M_2. This attempt is successful and the final operator sequence is

$$\text{GOTHRU}(D_1,R_1,R_2), \ \text{PUSHTHRU}(B_1,D_1,R_2,R_1)$$

We would like to generalize the above plan so that it could be free from the specific constants D_1, R_1, R_2, and B_1 and used in situations involving arbitrary doors, rooms, and boxes. The triangle table for the plan is given in Fig. 10.17, and the triangle table for the generalized plan is shown in Fig. 10.18. Hence the plan could be generalized as follows:

$$\text{GOTHRU}(d_1,r_1,r_2)$$

$$\text{PUSHTHRU}(b,d_2,r_2,r_3)$$

and could be used to go from one room to an adjacent second room and push a box to an adjacent third room.

10.7 ROBOT LEARNING

We have discussed the use of triangle tables for generalized plans to control the execution of robot plans. Triangle tables for generalized plans can also be used by STRIPS to extract a relevant operator sequence during a subsequent planning process. Conceptually, we can think of a single triangle table as representing a family of generalized operators. Upon the selection by STRIPS of a relevant add list, we must extract from this family an economical parameterized operator achieving the add list. Recall that the $(i + 1)$th row of a triangle table (excluding the first cell) represents the add list, $A_{1,\ldots,i}$, of the ith head of the plan, i.e., of the sequence

$OP_1, \ldots, OP_i$. An n-step plan presents STRIPS with n alternative add lists, any one of which can be used to reduce a difference encountered during the normal planning process. STRIPS tests the relevance of each of a generalized plan's add lists in the usual fashion, and the add lists that provide the greatest reduction in the difference are selected. Often a given set of relevant statements will appear in more than one row of the table. In that case only the lowest-numbered row is selected, since this choice results in the shortest operator sequence capable of producing the desired statements.

Suppose that STRIPS selects the ith add list $A_1, \ldots, _i$, $i < n$. Since this add list is achieved by applying in sequence $OP_1, \ldots, OP_i$, we will obviously not be interested in the application of $OP_{i+1}, \ldots, OP_n$, and will therefore not be interested in establishing any of the preconditions for these operators. In general, some steps of a plan are needed only to establish preconditions for subsequent steps. If we lost interest in a tail of a plan, then the relevant instance of the generalized plan need not contain those operators whose sole purpose is to establish preconditions for the tail. Also, STRIPS will, in general, have used only some subset of $A_1, \ldots, _i$ in establishing the relevance of the ith head of the plan. Any of the first i operators that does not add some statement in this subset, or help establish the preconditions for some operator that adds a statement in the subset, is not needed in the relevant instance of the generalized plan.

In order to obtain a robot planning system that can not only speed up the planning process but can also improve its problem-solving capability to handle more complex tasks, one could design the system with a learning capability. STRIPS uses a generalization scheme for machine learning. Another form of learning would be the updating of the information in the difference-operator table from the system's experience.

Learning by analogy has been considered as a powerful approach and has been applied to robot planning. A robot planning system with learning, called PULP-I, has been proposed (Tangwongsan and Fu [1979]). The system uses an analogy between a current unplanned task and any known similar tasks to reduce the search

1	INROOM(ROBOT.p_2) CONNECTS(p_3.p_2.p_5)	GOTHRU(p_3.p_2.p_5)	
2	INROOM(p_6.p_5) CONNECTS(p_8.p_9.p_5) CONNECTS(x.y.z) → CONNECTS(x.y.z)	INROOM(ROBOT.p_5)	PUSHTHRU(p_6.p_8.p_5.p_9)
3			INROOM(ROBOT.p_9) INROOM(p_6.p_9)

Figure 10.18 Triangle table for generalized plan.

for a solution. A semantic network, instead of predicate logic, is used as the internal representation of tasks. Initially a set of basic task examples is stored in the system as knowledge based on past experience. The analogy of two task statements is used to express the similarity between them and is determined by a semantic matching procedure. The matching algorithm measures the semantic "closeness"; the smaller the value, the closer the meaning. Based on the semantic matching measure, past experience in terms of stored information is retrieved and a candidate plan is formed. Each candidate plan is then checked by its operators' preconditions to ensure its applicability to the current world state. If the plan is not applicable, it is simply dropped out of the candidacy. After the applicability check, several candidate plans might be found. These candidate plans are listed in ascending order according to their evaluation values of semantic matching. The one with the smallest value of semantic matching has the top priority and must be at the beginning of the candidate list. Of course, if no candidate is found, the system terminates with failure.

Computer simulation of PULP-I has shown a significant improvement of planning performance. This improvement is not merely in the planning speed but also in the capability of forming complex plans from the learned basic task examples.

10.8 ROBOT TASK PLANNING

The robot planners discussed in the previous section require only a description of the initial and final states of a given task. These planning systems typically do not specify the detailed robot motions necessary to achieve an operation. These systems issue robot commands such as: PICKUP(A) and STACK(X,Y) without specifying the robot path. In the foreseeable future, however, robot task planners will need more detailed information about intermediate states than these systems provide. But they can be expected to produce a much more detailed robot program. In other words, a task planner would transform the task-level specifications into manipulator-level specifications. To carry out this transformation, the task planner must have a description of the objects being manipulated, the task environment, the robot carrying out the task, the initial state of the environment, and the desired final (goal) state. The output of the task planner would be a robot program to achieve the desired final state when executed in the specified initial state.

There are three phases in task planning: modeling, task specification, and manipulator program synthesis. The world model for a task must contain the following information: (1) geometric description of all objects and robots in the task environment; (2) physical description of all objects; (3) kinematic description of all linkages; and (4) descriptions of robot and sensor characteristics. Models of task states also must include the configurations of all objects and linkages in the world model.

10.8.1 Modeling

The geometric description of objects is the principal component of the world model. The major sources of geometric models are computer-aided design (CAD) systems and computer vision. There are three major types of three-dimensional object representation schemes (Requicha and Voelcker [1982]):

1. Boundary representation
2. Sweep representation
3. Volumetric representation

There are three types of volumetric representations: (1) spatial occupancy, (2) cell decomposition, and (3) constructive solid geometry (CSG). A system based on constructive solid geometry has been suggested for task planning. In CSG, the basic idea is that complicated solids are constructed by performing set operations on a few types of primitive solids. The object in Fig. 10.19a can be described by the structure given in Fig. 10.19b.

The legal motions of an object are constrained by the presence of other objects in the environment, and the form of the constraints depends on the shapes of the objects. This is the reason why a task planner needs geometric descriptions of objects. There are additional constraints on motion imposed by the kinematic structure of the robot itself. The kinematic models provide the task planner with the information required to plan manipulator motions that are consistent with external constraints.

Many of the physical characteristics of objects play important roles in planning robot operations. The mass and inertia of parts, for example, determine how fast they can be moved or how much force can be applied to them before they fall over. Another important aspect of a robot system is its sensing capabilities. For task planning, vision enables the robot to obtain the configuration of an object to some specified accuracy at execution time; force sensing allows the use of compliant motions; touch information could serve in both capacities. In addition to sensing, there are many individual characteristics of manipulators that must be described; for example, velocity and acceleration bounds, and positioning accuracy of each of the joints.

10.8.2 Task Specification

A model state is given by the configurations of all the objects in the environment; tasks are actually defined by sequences of states of the world model. There are three methods for specifying configurations: (1) using a CAD system to position models of the objects at the desired configurations, (2) using the robot itself to specify robot configurations and to locate features of the objects, and (3) using symbolic spatial relationships among object features to constrain the configurations of objects. Methods 1 and 2 produce numerical configurations which are difficult

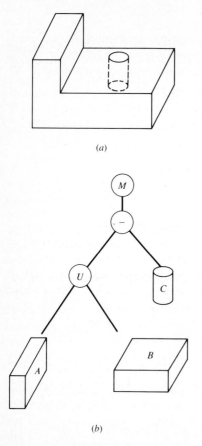

Figure 10.19 Constructive solid geometry (CSG). Attributes of *A* and *B*: length, width, height; attributes of *C*: radius, height. Set relational operators: $\cup$, union, $\cap$, intersection, $-$, difference.

to interpret and modify. In the third method, a configuration is described by a set of symbolic spatial relationships that are required to hold between objects in that configuration.

Since model states are simply sets of configurations and task specifications are sequences of model states, given symbolic spatial relationships for specifying configurations, we should be able to specify tasks. Assume that the model includes names for objects and object features. The first step in the task planning process is transforming the symbolic spatial relationships among object features to equations on the configuration parameters of objects in the model. These equations must then be simplified as much as possible to determine the legal ranges of configurations of all objects. The symbolic form of the relationships is also used during program synthesis.

10.8.3 Manipulator Program Synthesis

The synthesis of a manipulator program from a task specification is the crucial phase of task planning. The major steps involved in this phase are grasp planning, motion planning, and error detection. The output of the synthesis phase is a program composed of grasp commands, several kinds of motion specifications, and error tests. This program is generally in a manipulator-level language for a particular manipulator and is suitable for repeated execution without replanning.

10.9 BASIC PROBLEMS IN TASK PLANNING

10.9.1 Symbolic Spatial Relationships

The basic steps in obtaining configuration constraints from symbolic spatial relationships are: (1) defining a coordinate system for objects and object features, (2) defining equations of object configuration parameters for each of the spatial relationships among features, (3) combining the equations for each object, and (4) solving the equations for the configuration parameters of each object. Consider the following specification, given in the state depicted in Fig. 10.20:

$$\text{PLACE Block1}(f_1 \text{ against } f_3) \text{ and } (f_2 \text{ against } f_4)$$

The purpose is to obtain a set of equations that constrain the configuration of Block1 relative to the known configuration of Block2. That is, the face f_1 of Block2 must be against the face f_3 of Block1 and the face f_2 of Block2 must be against the face f_4 of Block1.

Each object and feature has a set of axes embedded in it, as shown in Fig. 10.21. Configurations of entities are the 4×4 transformation matrices:

$$
f_1 = \begin{bmatrix} 1 & 0 & 0 & 0 \\ 0 & 1 & 0 & 0 \\ 0 & 0 & 1 & 0 \\ 0 & 1 & 1 & 1 \end{bmatrix}
\qquad
f_2 = \begin{bmatrix} 0 & 1 & 0 & 0 \\ -1 & 0 & 0 & 0 \\ 0 & 0 & 1 & 0 \\ 1 & 0 & 1 & 1 \end{bmatrix}
$$

$$
f_3 = \begin{bmatrix} 1 & 0 & 0 & 0 \\ 0 & 1 & 0 & 0 \\ 0 & 0 & 1 & 0 \\ 1 & 1 & 1 & 1 \end{bmatrix}
\qquad
f_4 = \begin{bmatrix} 0 & -1 & 0 & 0 \\ 1 & 0 & 0 & 0 \\ 0 & 0 & 1 & 0 \\ 1 & 0 & 1 & 1 \end{bmatrix}
$$

Let $\text{twix}(\theta)$ be the transformation matrix for a rotation of angle θ around the x axis, $\text{trans}(x,y,z)$ the matrix for a translation x, y, and z, and let M be the matrix for the rotation around the y axis that rotates the positive x axis into the negative x axis, with $M = M^{-1}$. Each *against* relationship between two faces, say, face f on

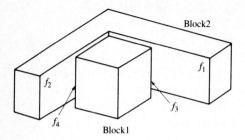

Figure 10.20 Illustration for spatial relationships among objects.

object A and face g on object B, generates the following constraint on the configuration of the two objects:

$$A = f^{-1}M \text{ twix}(\theta) \ (0, \ y, \ z)gB \tag{10.9-1}$$

The two *against* relations in the example of Fig. 10.15 generate the following equations:

$$\text{Block1} = f_3^{-1}M \text{ twix}(\theta_1) \text{ trans}(0, y_1, z_1)f_1 \text{Block2}$$

$$\text{Block1} = f_4^{-1}M \text{ twix}(\theta_2) \text{ trans}(0, y_2, z_2)f_2 \text{Block2} \tag{10.9-2}$$

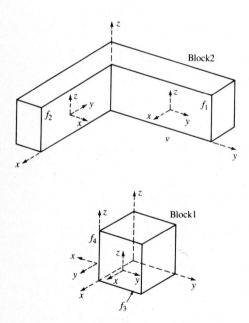

Figure 10.21 Axes embedded in objects and features from Fig. 10.20.

Equation (10.9-2) consists of two independent constraints on the configuration of Block1 that must be satisfied simultaneously. Setting the two expressions equal to each other and removing the common term, Block2, we get

$$f_3^{-1}M \text{ twix}(\theta_1) \text{ trans}(0,y_1,z_1) f_1 = f_4^{-1}M \text{ twix}(\theta_2) \text{ trans}(0,y_2,z_2) f_2 \qquad (10.9\text{-}3)$$

Applying the rewrite rules, (10.9-3) is transformed to

$$f_3^{-1}M \text{ twix}(\theta_1) \text{ trans}(0,y_1 + 1,z_1 + 1)(f_2')^{-1}$$

$$\times \text{ trans}(0,-y_1,-z_2) \text{ twix}(-\theta_2)M^{-1} f_4 = I \qquad (10.9\text{-}4)$$

where the primed matrix denotes the rotational component of the transformation, obtained by setting the last row of the matrix to [0,0,0,1].

It can be shown that the rotational and translational components of this type of equation can be solved independently. The rotational equation can be obtained by replacing each of the trans matrices by the identity and only using the rotational components of other matrices. The rotational equation for Eq. (10.9-4) is

$$(f_3')^{-1}M \text{ twix}(\theta_1)(f_2')^{-1} \text{ twix}(-\theta_2)M(f_4') = I \qquad (10.9\text{-}5)$$

Since $f_3' = I$, (10.9-5) can be rewritten as

$$\text{twix}(\theta_1)(f_2')^{-1} \text{ twix}(-\theta_2) = M(f_4')^{-1}M \qquad (10.9\text{-}6)$$

Also, since

$$(f_2') = M(f_4')^{-1}M = \begin{bmatrix} 0 & -1 & 0 & 0 \\ 1 & 0 & 0 & 0 \\ 0 & 0 & 1 & 0 \\ 0 & 0 & 0 & 1 \end{bmatrix}$$

Eq. (10.9-6) is satisfiable and we can choose $\theta_2 = 0$. Letting $\theta_2 = 0$ in Eq. (10.9-6), we obtain

$$\text{twix}(\theta_1) = M(f_4')^{-1}M(f_2') = I \qquad (10.9\text{-}7)$$

From Eq. (10.9-7) we conclude that $\theta_1 = 0$. Thus, Eq. (10.9-2) becomes

$$\text{Block1} = \begin{bmatrix} -1 & 0 & 0 & 0 \\ 0 & 1 & 0 & 0 \\ 0 & 0 & -1 & 0 \\ 1 & y_1 & 2+z_1 & 1 \end{bmatrix} = \begin{bmatrix} -1 & 0 & 0 & 0 \\ 0 & 1 & 0 & 0 \\ 0 & 0 & -1 & 0 \\ 2-y_2 & 0 & 2+z_2 & 1 \end{bmatrix} \qquad (10.9\text{-}8)$$

Equating the corresponding matrix terms, we obtain

$$2 - y_2 = 1$$

$$y_1 = 0$$

$$2 + z_1 = 2 + z_2$$

Hence, $y_2 = 1$, $y_1 = 0$, and $z_1 = z_2$; that is, the position of Block1 has 1 degree of freedom corresponding to translations along the z axis.

The method used in the above example was proposed by Ambler and Popplestone [1975]. The contact relationships treated there include *against*, *fits*, and *coplanar* among features that can be planar or spherical faces, cylindrical shafts and holes, edges and vertices. Taylor [1976] extended this approach to noncontact relationships such as for a peg in a hole of diameter greater than its own, an object in a box, or a feature in contact with a region of another feature. These relationships give rise to inequality constraints on the configuration parameters. They can be used to model, for example, the relationship of the position of the tip of a screwdriver in the robot's gripper to the position errors in the robot joints and the slippage of the screwdriver in the gripper. After simplifying the equalities and inequalities, a set of linear constraints are derived by using differential approximations for the rotations around a nominal configuration. The values of the configuration parameters satisfying the constraints can be bounded by applying linear programming techniques to the linearized constraint equations.

10.9.2 Obstacle Avoidance

The most common robot motions are transfer movements for which the only constraint is that the robot and whatever it is carrying should not collide with objects in the environment. Therefore, an ability to plan motions that avoid obstacles is essential to a task planner. Several obstacle avoidance algorithms have been proposed in different domains. In this section, we briefly review those algorithms that deal with robot obstacle avoidance in three dimensions. The algorithms for robot obstacle avoidance can be grouped into the following classes: (1) hypothesize and test, (2) penalty function, and (3) explicit free space.

The hypothesize and test method was the earliest proposal for robot obstacle avoidance. The basic method consists of three steps: first, hypothesize a candidate path between the initial and final configuration of the robot manipulator; second, test a selected set of configurations along the path for possible collisions; third, if a possible collision is found, propose an avoidance motion by examining the obstacle(s) that would cause the collision. The entire process is repeated for the modified motion.

The main advantage of the hypothesize and test technique is its simplicity. The method's basic computational operations are detecting potential collisions and modifying proposed paths to avoid collisions. The first operation, detecting potential collisions, amounts to the ability to detect nonnull geometric intersections between the manipulator and obstacle models. This capability is part of the repertoire of most geometric modeling systems. We have pointed out in Sec. 10.8 that the second operation, modifying a proposed path, can be very difficult. Typical proposals for path modification rely on approximations of the obstacles, such as enclosing spheres. These methods work fairly well when the obstacles are sparsely located so that they can be dealt with one at a time. When the space is cluttered, however, attempts to avoid a collision with one obstacle will typically lead to another collision with a different obstacle. Under such conditions, a more accurate detection of potential collisions could be accomplished by using the information from vision and/or proximity sensors.

The second class of algorithms for obstacle avoidance is based on defining a penalty function on manipulator configurations that encodes the presence of objects. In general, the penalty is infinite for configurations that cause collisions and drops off sharply with distance from obstacles. The total penalty function is computed by adding the penalties from individual obstacles and, possibly, adding a penalty term for deviations from the shortest path. At any configuration, we can compute the value of the penalty function and estimate its partial derivatives with respect to the configuration parameters. On the basis of this local information, the path search function must decide which sequence of configurations to follow. The decision can be made so as to follow local minima in the penalty function. These minima represent a compromise between increasing path length and approaching too close to obstacles. The penalty function methods are attractive because they provide a relatively simple way of combining the constraints from multiple objects. This simplicity, however, is achieved only by assuming a circular or spherical robot; only in this case will the penalty function be a simple transformation of the obstacle shape. For more realistic robots, such as two-link manipulator, the penalty function for an obstacle would have to be defined as a transformation of the configuration space obstacle. Otherwise, motions of the robot that reduce the value of the penalty function will not necessarily be safe. The distinction between these two types of penalty functions is illustrated in Fig. 10.22. It is noted that in Fig. 10.22a moving along decreasing values of the penalty function is safe, whereas in Fig. 10.22b moving the tip of the manipulator in the same way leads to a collision.

An approach proposed by Khatib [1980] is intermediate between these two extremes. The method uses a penalty function which satisfies the definition of a potential field; the gradient of this field at a point on the robot is interpreted as a repelling force acting on that point. In addition, an attractive force from the destination is added. The motion of the robot results from the interaction of these two forces, subject to kinematic constraints. By using many points of the robot, rather than a single one, it is possible to avoid many situations such as those depicted in Fig. 10.22.

The key drawback of using penalty functions to plan safe paths is the strictly local information that they provide for path searching. Pursuing the local minima of the penalty function can lead to situations where no further progress can be

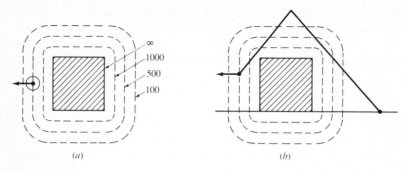

Figure 10.22 Illustration of penalty function for (*a*) simple circular robot and (*b*) the two-link manipulator. (Numbers in the figure indicate values of the penalty function.)

made. In these cases, the algorithm must choose a previous configuration where the search is to be resumed, but in a different direction from the previous time. These backup points are difficult to identify from local information. This suggests that the penalty function method might be combined profitably with a more global method of hypothesizing paths. Penalty functions are more suitable for applications that require only small modifications to a known path.

The third class of obstacle avoidance algorithms builds explicit representations of subsets of robot configurations that are free of collisions, the *free space*. Obstacle avoidance is then the problem of finding a path, within these subsets, that connects the initial and final configurations. The algorithms differ primarily on the basis of the particular subsets of free-space which they represent and in the representation of these subsets. The advantage of free space methods is that their use of an explicit characterization of free space allows them to define search methods that are guaranteed to find paths if one exists within the known subset of free space. Moreover, it is feasible to search for short paths, rather than simply finding the first path that is safe. The disadvantage is that the computation of the free space may be expensive. In particular, other methods may be more efficient for uncluttered spaces. However, in relatively cluttered spaces other methods will either fail or expend an undue amount of effort in path searching.

10.9.3 Grasp Planning

A typical robot operation begins with the robot grasping an object; the rest of the operation is influenced by choices made during grasping. Several proposals for choosing collision-free grasp configurations on objects exist, but other aspects of the general problem of planning grasp motions have received little attention. In this section, *target object* refers to the object to be grasped. The surfaces on the robot used for grasping, such as the inside of the fingers, are *gripping surfaces*. The manipulator configuration which has it grasping the target object at that object's initial configuration is the *initial-grasp configuration*. The manipulator configuration that places the target object at its destination is the *final-grasp configuration.*

There are three principal considerations in choosing a grasp configuration for objects whose configuration is known. The first is *safety*: the robot must be safe at the initial and final grasp configurations. The second is *reachability*: the robot must be able to reach the initial grasp configuration and, with the object in the hand, to find a collision-free path to the final grasp configuration. The third is *stability*: the grasp should be stable in the presence of forces exerted on the grasped object during transfer motions and parts mating operations. If the initial configuration of the target object is subject to substantial uncertainty, an additional consideration in grasping is *certainty*: the grasp motion should reduce the uncertainty in the target object's configuration.

Choosing grasp configurations that are safe and reachable is related to obstacle avoidance; there are significant differences, however. The first difference is that the goal of grasp planning is to identify a single configuration, not a path. The second difference is that grasp planning must consider the detailed interaction of the manipulator's shape and that of the target object. Note that candidate grasp configurations are those having the gripping surfaces in contact with the target object while avoiding collisions between the manipulator and other objects. The third difference is that grasp planning must deal with the interaction of the choice of grasp configuration and the constraints imposed by subsequent operations involving the grasped object. Because of these differences, most existing methods for grasp planning treat it independently from obstacle avoidance.

Most approaches to choosing safe grasps can be viewed as instances of the following method:

1. Choose a set of candidate grasp configurations. The choice can be based on considerations of object geometry, stability, or uncertainty reduction. For parallel-jaw grippers, a common choice is grasp configurations that place the grippers in contact with a pair of parallel surfaces of the target object. An additional consideration in choosing the surfaces is to minimize the torques about the axis between the grippers.
2. The set of candidate grasp configurations is then pruned by removing those that are not reachable by the robot or lead to collisions. Existing approaches to grasp planning differ primarily on the collision-avoidance constraints, for example:
   a. Potential collisions of gripper and neighboring objects at initial-grasp configuration.
   b. *p* convexity (indicates that all the matter near a geometric entity lies to one side of a specified plane).
   c. Existence of collision-free path to initial-grasp configuration.
   d. Potential collisions of whole manipulator and neighboring objects at initial-grasp configuration, potential collisions of gripper and neighboring objects at final-grasp configuration, potential collisions of whole manipulator and neighboring objects at final-grasp configuration, and existence of collision-free paths from initial to final-grasp configuration.

3. After pruning, a choice is made among the remaining configurations, if any. One possibility is choosing the configuration that leads to the most stable grasp; another is choosing the one least likely to cause a collision in the presence of position error or uncertainty.

It is not difficult to see that sensory information (vision, proximity, torque, or force) should be very useful in determining a stable and collision-free grasp configuration.

10.10 EXPERT SYSTEMS AND KNOWLEDGE ENGINEERING

Most techniques in the area of artificial intelligence fall far short of the competence of humans or even animals. Computer systems designed to see images, hear sounds, and understand speech can only claim limited success. However, in one area of artificial intelligence—that of reasoning from knowledge in a limited domain—computer programs can not only approach human performance but in some cases they can exceed it.

These programs use a collection of facts, rules of thumb, and other knowledge about a given field, coupled with methods of applying those rules, to make inferences. They solve problems in such specialized fields as medical diagnosis, mineral exploration, and oil-well log interpretation. They differ substantially from conventional computer programs because their tasks have no algorithmic solutions and because often they must make conclusions based on incomplete or uncertain information. In building such expert systems, researchers have found that amassing a large amount of knowledge, rather than sophisticated reasoning techniques, is responsible for most of the power of the system. Such high-performance expert systems, previously limited to academic research projects, are beginning to enter the commercial marketplace.

10.10.1 Construction of an Expert System

Not all fields of knowledge are suitable at present for building expert systems. For a task to qualify for "knowledge engineering," the following prerequisites must be met:

1. There must be at least one human expert who is acknowledged to perform the task well.
2. The primary sources of the expert's abilities must be special knowledge, judgment, and experience.
3. The expert must be able to articulate that special knowledge, judgement, and experience and also explain the methods used to apply it to a particular task.
4. The task must have a well-bounded domain of application.

Sometimes an expert system can be built that does not exactly match these prerequisites; for example, the abilities of several human experts, rather than one, might be brought to bear on a problem.

The structure of an expert system is modular. Facts and other knowledge about a particular domain can be separated from the inference procedure—or control structure—for applying those facts, while another part of the system—the global database—is the model of the "world" associated with a specific problem, its status, and its history. It is desirable, though not yet common, to have a natural-language interface to facilitate the use of the system both during development and in the field. In some sophisticated systems, an explanation module is also included, allowing the user to challenge the system's conclusions and to examine the underlying reasoning process that led to them.

An expert system differs from more conventional computer programs in several important respects. In a conventional computer program, knowledge pertinent to the problem and methods for using this knowledge are intertwined, so it is difficult to change the program. In an expert system there is usually a clear separation of general knowledge about the problem (the knowledge base) from information about the current problem (the input data) and methods (the inference machine) for applying the general knowledge to the problem. With this separation the program can be changed by simple modifications of the knowledge base. This is particularly true of rule-based systems, where the system can be changed by the simple addition or subtraction of rules in the knowledge base.

10.10.2 Rule-Based Systems

The most popular approach to representing the domain knowledge (both facts and heuristics) needed for an expert system is by production rules (also referred to as SITUATION-ACTION rules or IF-THEN rules). A simple example of a production rule is: IF the power supply on the space shuttle fails, AND a backup power supply is available, AND the reason for the first failure no longer exists, THEN switch to the backup power supply. Rule-based systems work by applying rules, noting the results, and applying new rules based on the changed situation. They can also work by directed logical inference, either starting with the initial evidence in a situation and working toward a solution, or starting with hypotheses about possible solutions and working backward to find existing evidence—or a deduction from existing evidence—that supports particular hypothesis.

One of the earliest and most often applied expert systems is Dendral (Barr et al. [1981, 1982]). It was devised in the late 1960s by Edward A. Feigenbaum and Joshua Lederberg at Stanford University to generate plausible structural representations of organic molecules from mass spectrogram data. The approach called for:

1. Deriving constraints from the data
2. Generating candidate structures
3. Predicting mass spectrographs for candidates
4. Comparing the results with data

This rule-based system, chaining forward from the data, illustrates the very common AI problem-solving approach of "generation and test." Dendral has been used as a consultant by organic chemists for more than 15 years. It is currently recognized as an expert in mass-spectral analysis.

One of the best-known expert systems is MYCIN (Barr et al. [1981, 1982]), design by Edward Shortliffe at Stanford University in the mid-1970s. It is an interactive system that diagnoses bacterial infections and recommends antibiotic therapy. MYCIN represents expert judgmental reasoning as condition-conclusions rules, linking patient data to infection hypotheses, and at the same time it provides the expert's "certainty" estimate for each rule. It chains backward from hypothesized diagnoses, using rules to estimate the certainty factors of conclusions based on the certainty factors of their antecedents, to see if the evidence supports a diagnosis. If there is not enough information to narrow the hypotheses, it asks the physician for additional data, exhaustively evaluating all hypotheses. When it has finished, MYCIN matches treatments to all diagnoses that have high certainty values..

Another rule-based system, R1, has been very successful in configuring VAX computer systems from a customer's order of various standard and optional components. The initial version of R1 was developed by John McDermott in 1979 at Carnegie-Mellon University, for Digital Equipment Corp. Because the configuration problem can be solved without backtracking and without undoing previous steps, the system's approach is to break the problem up into the following subtasks and do each of them in order:

1. Correct mistakes in order.
2. Put components into CPU cabinets.
3. Put boxes in Unibus cabinets and put components in boxes.
4. Put panels in Unibus cabinets.
5. Lay out system floor plan.
6. Do the cabling.

At each point in the configuration development, several rules for what to do next are usually applicable. Of the applicable rules, R1 selects the rule having the most IF clauses for its applicability, on the assumption that that rule is more specialized for the current situation. (R1 is written in OPS 5, a special language for executing production rules.) The system now has about 1200 rules for VAXs, together with information about some 1000 VAX components. The total system has about 3000 rules and knowledge about PDP-11 as well as VAX components.

10.10.3 Remarks

The application areas of expert systems include medical diagnosis and prescription, medical-knowledge automation, chemical-data interpretation, chemical and biological synthesis, mineral and oil exploration, planning and scheduling, signal interpretation, military threat assessment, tactical targeting, space defense, air-traffic control, circuit analysis, VLSI design, structure damage assessment, equipment fault diagnosis, computer-configuration selection, speech understanding, computer-aided instruction, knowledge-base access and management, manufacturing process planning and scheduling, and expert-system construction.

There appear to be few constraints on the ultimate use of expert systems. However, the nature of their design and construction is changing. The limitations of rule-based systems are becoming apparent: not all knowledge can be structured as empirical associations. Such associations tend to hide causal relationships, and they are also inappropriate for highlighting structure and function. The newer expert systems contain knowledge about causality and structure. These systems promise to be considerably more robust than current systems and may yield correct answers often enough to be considered for use in autonomous systems, not just as intelligent assistants.

Another change is the increasing trend toward non-rule-based systems. Such systems, using semantic networks, frames, and other knowledge-representation structures, are often better suited for causal modeling. By providing knowledge representations more appropriate to the specific problem, they also tend to simplify the reasoning required. Some expert systems, using the "blackboard" approach, combine rule-based and non-rule-based portions which cooperate to build solutions in an incremental fashion, with each segment of the program contributing its own particular expertise.

10.11 CONCLUDING REMARKS

The discussion in this chapter emphasizes the problem-solving or planning aspect of a robot. A robot planner attempts to find a path from our initial robot world to a final robot world. The path consists of a sequence of operations that are considered primitive to the system. A solution to a problem could be the basis of a corresponding sequence of physical actions in the physical world. Planning should certainly be regarded as an intelligent function of a robot.

In late 1971 and early 1972, two main approaches to robot planning were proposed. One approach, typified by the STRIPS system, is to have a fairly general robot planner which can solve robot problems in a great variety of worlds. The second approach is to select a specific robot world and, for that world, to write a computer program to solve problems. The first approach, like any other general problem-solving process in artificial intelligence, usually requires extensive computing power for searching and inference in order to solve a reasonably complex real-world problem, and, hence, has been regarded computationally infeasible. On the other hand, the second approach lacks generality, in that a new set of computer programs must be written for each operating environment and, hence, significantly limits the robot's flexibility in real-world applications.

In contrast to high-level robot task planning usually requires more detailed and numerical information describing the robot world. Existing methods for task planning are considered computationally infeasible for real-time practical applications.

Powerful and efficient task planning algorithms are certainly in demand. Again, special-purpose computers can be used to speed up the computations in order to meet the real-time requirements.

Robot planning, which provides the intelligence and problem-solving capability to a robot system, is still a very active area of research. For real-time robot applications, we still need powerful and efficient planning algorithms that will be executed by high-speed special-purpose computer systems.

REFERENCES

Further general reading for the material in this chapter can be found in Barr et al. [1981, 1982], Nilsson [1971, 1980], and Rich [1983]. The discussion in Secs. 10.2 and 10.3 is based on the material in Whitney [1969], Nilsson [1971], and Winston [1984]. Further basic reading for Sec. 10.4 may be found in Chang and Lee [1973]. Complementary reading for the material in Secs. 10.5 and 10.6 may be found in Fikes and Nilsson [1971] and Rich [1983]. Additional reading and references for the material in Sec. 10.7 can be found in Tangwongsan and Fu [1979].

Early representative references on robot task planning (Secs. 10.8 and 10.9) are Doran [1970], Fikes et al. [1972], Siklossy and Dreussi [1973], Ambler and Popplestone [1975], and Taylor [1976]. More recent work may be found in Khatib [1980], Requicha and Voelcher [1982], and Davis and Comacho [1984]. Additional reading for the material in Sec. 10.10 may be found in Nau [1983], Hayer-Roth et al. [1983], and Weiss and Allanheld [1984].

PROBLEMS

10.1 Suppose that three missionaries and three cannibals seek to cross a river from the right bank to the left bank by boat. The maximum capacity of the boat is two persons. If the missionaries are outnumbered at any time by the cannibals, the cannibals will eat the missionaries. Propose a computer program to find a solution for the safe crossing of all six persons. *Hint*: Using the state-space representation and search methods described in Sec. 10.2, one can represent the state description by (N_m, N_c), where N_m, N_c are the number of missionaries and cannibals in the left bank, respectively. The initial state is (0,0), i.e., no missionary and cannibal are on the left bank, the goal state is (3,3) and the possible intermediate states are (0,1), (0,2), (0,3), (1,1), (2,2), (3,0), (3,1), (3,2).

10.2 Imagine that you are a high school geometry student and find a proof for the theorem: "The diagonals of a parallelogram bisect each other." Use an AND/OR graph to chart the steps in your search for a proof. Indicate the solution subgraph that constitutes a proof of the theorem.

10.3 Represent the following sentences by predicate logic wffs. (*a*) A formula whose main connective is a $\Rightarrow$ is equivalent to some formula whose main connective is a $\lor$. (*b*) A robot is intelligent if it can perform a task which, if performed by a human, requires intelligence. (*c*) If a block is on the table, then it is not also on another block.

10.4 Show how the monkey-and-bananas problem can be represented so that STRIPS would generate a plan consisting of the following actions: go to the box, push the box under the bananas, climb the box, grasp the bananas.

10.5 Show, step by step, how means-ends analysis could be used to solve the robot planning problem described in the example at the end of Sec. 10.4.

10.6 Show how the monkey-and-bananas problem can be represented so that STRIPS would generate a plan consisting of the following actions: go to the box, push the box under the bananas, climb the box, grab the bananas. Use means-ends analysis as the control strategy.

VECTORS AND MATRICES

This appendix contains a review of basic vector and matrix algebra. In the following discussion, vectors are represented by lowercase bold letters, while matrices are in uppercase bold type.

A.1 SCALARS AND VECTORS

The quantities of physics can be divided into two classes, namely, those having magnitude only and those having magnitude *and* direction. A quantity characterized by magnitude only is called a *scalar*. Time, mass, density, length, and coordinates are scalars. A scalar is usually represented by a real number with some unit of measurement. Scalars can be compared only if they have the same units.

A quantity which is characterized by direction as well as magnitude is called a *vector*. Force, moment, velocity, and acceleration are examples of vectors. Usually, a vector is represented graphically by a directed line segment whose length and direction correspond to the magnitude and direction of the quantity under consideration. Vectors can be compared only if they have the same physical meaning and dimensions.

Two vectors **a** and **b** are equal if they have the same length and direction. The notation $-\mathbf{a}$ is used to represent a vector having the same magnitude as **a** but in the opposite direction. Associated with vector **a** is a positive scalar equal to its magnitude. This is represented as $|\mathbf{a}|$. If a is the magnitude or length of the vector **a**, then

$$a = |\mathbf{a}| \tag{A.1}$$

A unit vector **a** has unit length in the assigned direction,

$$|\mathbf{a}| = 1 \tag{A.2}$$

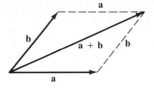 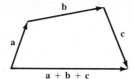

Figure A.1 Vector addition.

Any vector **a** in three-dimensional space can be normalized to a unit vector as

$$\frac{\mathbf{a}}{\|\mathbf{a}\|} = \frac{\mathbf{a}}{\sqrt{a_x^2 + a_y^2 + a_z^2}} \tag{A.3}$$

where a_x, a_y, and a_z are components of **a** along the principal axes.

A.2 ADDITION AND SUBTRACTION OF VECTORS

Addition of two vectors **a** and **b** is *commutative*, that is,

$$\mathbf{a} + \mathbf{b} = \mathbf{b} + \mathbf{a} \tag{A.4}$$

This can be verified easily by drawing a parallelogram having **a** and **b** as consecutive sides, as in Fig. A.1.

Addition of three or more vectors is *associative*,

$$(\mathbf{a} + \mathbf{b}) + \mathbf{c} = \mathbf{a} + (\mathbf{b} + \mathbf{c}) = \mathbf{a} + \mathbf{b} + \mathbf{c} \tag{A.5}$$

This can be seen by constructing a polygon having those vectors as consecutive sides and drawing a vector from an initial point of the first to the terminal point of the last (see Fig. A.1).

The difference between two vectors **a** and **b**, denoted by **a** − **b**, is defined as the vector extending from the end of **b** to the end of **a**, with the vectors **a** and **b** drawn from the same origin point (see Fig. A.2).

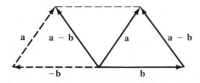

Figure A.2 Vector subtraction.

A.3 MULTIPLICATION BY SCALARS

Multiplication of a vector **a** by a scalar m means lengthening the magnitude of vector **a** by $|m|$ times with the same direction as **a** if $m > 0$ and the opposite direction if $m < 0$. Thus,

$$\mathbf{b} = m\mathbf{a} \tag{A.6}$$

and
$$|\mathbf{b}| = |m|\,|\mathbf{a}| \tag{A.7}$$

The following rules are applicable to the multiplication of vectors by scalars:

(1) $m(n\mathbf{a}) = mn\mathbf{a}$

(2) $m(\mathbf{a} + \mathbf{b}) = m\mathbf{a} + m\mathbf{b}$ (A.8)

(3) $(m + n)\mathbf{a} = m\mathbf{a} + n\mathbf{a}$

where m and n are scalars.

A.4 LINEAR VECTOR SPACE

A linear vector space V is a nonempty set of vectors defined over a real number field F, which satisfies the following conditions of vector addition and multiplication by scalars:

1. For any two vector elements of V, the sum is also a vector element belonging to V.
2. For any two vector elements of V, vector addition is commutative.
3. For any three vector elements of V, vector addition is associative.
4. There is a unique element called the zero vector in V (denoted by **0**) such that for every element $\mathbf{a} \in V$

$$\mathbf{0} + \mathbf{a} = \mathbf{a} + \mathbf{0} = \mathbf{a}$$

5. For every vector element $\mathbf{a} \in V$, there is a unique vector $(-\mathbf{a}) \in V$ such that

$$\mathbf{a} + (-\mathbf{a}) = \mathbf{0}$$

6. For every vector element $\mathbf{a} \in V$ and for any scalar $m \in F$, the product of m and **a** is another vector element in V. If $m = 1$, then

$$m\mathbf{a} = 1\,\mathbf{a} = \mathbf{a}1 = \mathbf{a}$$

7. For any scalars m and n in F, and any vectors **a** and **b** in V, multiplication by scalars is distributive.

$$m(\mathbf{a} + \mathbf{b}) = m\mathbf{a} + m\mathbf{b}$$

$$(m + n)\mathbf{a} = m\mathbf{a} + n\mathbf{a}$$

8. For any scalars m and n in F, and any vector $\mathbf{a}$ in V,

$$m(n\mathbf{a}) = (mn)\mathbf{a} = mn\mathbf{a}$$

Examples of linear vector space are the sets of all real one-, two- or three-dimensional vectors.

A.5 LINEAR DEPENDENCE AND INDEPENDENCE

A finite set of vectors $\{\mathbf{x}_1, \mathbf{x}_2, \ldots, \mathbf{x}_n\}$ in V is *linearly dependent* if and only if there exist n scalars $\{c_1, c_2, c_3, \ldots, c_n\}$ in F (not all equal to zero) such that

$$c_1\mathbf{x}_1 + c_2\mathbf{x}_2 + c_3\mathbf{x}_3 + \cdots + c_n\mathbf{x}_n = \mathbf{0} \tag{A.9}$$

If the only way to satisfy this equation is for each scalar c_i to be identically equal to zero, then the set of vectors $\{\mathbf{x}_i\}$ are said to be *linearly independent*.

Two linearly dependent vectors in a three-dimensional space are *collinear*. That is, they lie in the same line. Three linearly dependent vectors in three-dimensional space are *coplanar*, that is, they lie in the same plane.

> **Example:** Let $\mathbf{a} = (1, 2, 0)^T$, $\mathbf{b} = (0, 3, 2)^T$, and $\mathbf{c} = (3, 0, -4)^T$ be vectors in a three-dimensional vector space. Then, the three vectors $\mathbf{a}$, $\mathbf{b}$, $\mathbf{c}$ constitute a linearly dependent set in the three-dimensional vector space because
>
> $$3\mathbf{a} - 2\mathbf{b} - \mathbf{c} = \mathbf{0}$$
>
> These vectors also are coplanar. □

A.6 LINEAR COMBINATIONS, BASIS VECTORS, AND DIMENSIONALITY

If there exists a subset of vectors $\{\mathbf{e}_1, \mathbf{e}_2, \ldots, \mathbf{e}_n\}$ in V and a set of scalars $\{c_1, c_2, \ldots, c_n\}$ in F such that every vector $\mathbf{x}$ in V, can be expressed as

$$\mathbf{x} = c_1\mathbf{e}_1 + c_2\mathbf{e}_2 + \cdots + c_n\mathbf{e}_n = \sum_{i=1}^{n} c_i\mathbf{e}_i \tag{A.10}$$

then we say that $\mathbf{x}$ is a *linear combination* of the vectors $\{\mathbf{e}_i\}$. The set of vectors $\{\mathbf{e}_i\}$ is said to *span* the vector space V.

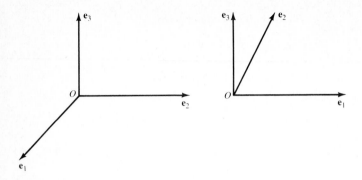

Figure A.3 Coordinate systems.

The basis vectors for a vector space V are a set of linearly independent vectors that span the vector space V. In other words, the basis vectors are the minimum number of vectors that span the vector space V. One can choose different sets of basis vectors to span a vector space V. However, once a set of basis vectors are chosen to span a vector space V, every vector $\mathbf{x} \in V$ can be expressed uniquely as a linear combination of the basis vectors.

The dimension of a vector space V is equal to the number of basis vectors that span the vector space V. Thus, an n-dimensional linear vector space has n basis vectors. We shall use the notation V_n to represent a vector space of dimension n.

A.7 CARTESIAN COORDINATE SYSTEMS

Given a set of n basis vectors $\{\mathbf{e}_1, \mathbf{e}_2, \ldots, \mathbf{e}_n\}$ for a vector space V_n, it follows from Sec. A.6 that any vector $\mathbf{r} \in V_n$ can be expressed uniquely as a linear combination of the basis vectors,

$$\mathbf{r} = r_1\mathbf{e}_1 + r_2\mathbf{e}_2 + \cdots + r_n\mathbf{e}_n \tag{A.11}$$

In particular, if $n = 3$, then any triple of noncoplanar vectors can serve as basis vectors.

In a three-dimensional vector space, if a set of basis vectors $\{\mathbf{e}_1, \mathbf{e}_2, \mathbf{e}_3\}$ are all drawn from a common origin O, then these vectors form an *oblique* coordinate system with axes $OX, OY,$ and OZ drawn along the basis vectors (see Fig. A.3). By properly choosing the direction of the basis vectors, one can form various coordinate frames commonly used in engineering work.

If the basis vectors $\{\mathbf{e}_1, \mathbf{e}_2, \mathbf{e}_3\}$ are orthogonal to each other, that is, if they intersect at right angles at the origin O, then they form a *rectangular* or *cartesian* coordinate system. Furthermore, if each of the basis vectors is of unit length, the coordinate system is called *orthonormal*. In this case, we usually use $\{\mathbf{i}, \mathbf{j}, \mathbf{k}\}$ to denote the basis vectors instead of $\{\mathbf{e}_1, \mathbf{e}_2, \mathbf{e}_3\}$ (see Fig. A.4).

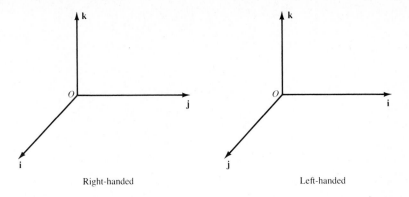

Figure A.4 Cartesian coordinate system.

If the basis vectors $\{i, j, k\}$ of an orthonormal coordinate system are chosen in the directions along the principal axes and a right-handed rotation of $90°$ about OZ carries OX into OY, then the coordinate system is called a *right-handed coordinate system*. Similarly, if the basis vectors are chosen in the directions along the principal axes and a left-handed rotation of $90°$ about OZ carries OX into OY, then the coordinate system is called a *left-handed coordinate system*. Throughout this book, we use only right-handed coordinate systems.

A.8 PRODUCT OF TWO VECTORS

In addition to the product of a scalar and a vector, two other types of vector product are of importance. The first is the *inner* or *dot* or *scalar* product. The other is the *vector* or *cross* product.

A.8.1 Inner Product (Dot Product or Scalar Product)

The inner product of two vectors **a** and **b** results in a scalar and is defined as

$$\mathbf{a} \cdot \mathbf{b} = |\mathbf{a}| \, |\mathbf{b}| \cos \theta \tag{A.12}$$

where θ is the angle between the two vectors (see Fig. A.5). The scalar quantity

$$b = |\mathbf{b}| \cos \theta \tag{A.13}$$

is the component of **b** along **a**; that is, b is numerically equal to the projection of **b** on **a**. It is positive if the projection is in the same direction as **a**, and negative if it is in the opposite direction.

The scalar product

$$\mathbf{a} \cdot \mathbf{b} = |\mathbf{a}| \, |\mathbf{b}| \cos \theta = |\mathbf{a}| b = |\mathbf{b}| \, |\mathbf{a}| \cos \theta = |\mathbf{b}| a \tag{A.14}$$

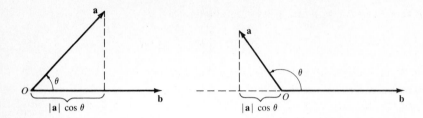

Figure A.5 Scalar product.

is equal to the product of the magnitude of **a** and the component of **b** along **a**. It is also equal to the product of $|\mathbf{b}|$ and the component of **a** along **b**. Hence, the scalar product is *commutative*:

$$\mathbf{a} \cdot \mathbf{b} = \mathbf{b} \cdot \mathbf{a} \tag{A.15}$$

If the scalar product of **a** and **b** is zero, then either (or both) of the vectors is zero or they are perpendicular to each other because $\cos(\pm 90°) = 0$. Thus, two nonzero vectors **a** and **b** are orthogonal if and only if their scalar product is zero.

Since the inner product may be zero when neither vector is zero, it follows that *division by a vector* is prohibited. Thus, if

$$\mathbf{a} \cdot \mathbf{b} = \mathbf{a} \cdot \mathbf{c} \tag{A.16}$$

one cannot conclude that $\mathbf{b} = \mathbf{c}$ but merely that $\mathbf{a} \cdot (\mathbf{b} - \mathbf{c}) = 0$, and so $\mathbf{b} - \mathbf{c}$ is a zero vector or orthogonal to **a**.

If **a** and **b** have the same direction and $\theta = 0°$, then $\cos\theta = 1$ and $\mathbf{a} \cdot \mathbf{b}$ is equal to the product of the lengths of the two vectors. In particular if $\mathbf{a} = \mathbf{b}$, then

$$\mathbf{a} \cdot \mathbf{b} = \mathbf{a} \cdot \mathbf{a} = |\mathbf{a}| \, |\mathbf{a}| = |\mathbf{a}|^2 \tag{A.17}$$

is the square of the length of **a**.

The dot product of vectors is *distributive over addition*, that is,

$$\mathbf{a} \cdot (\mathbf{b} + \mathbf{c}) = \mathbf{a} \cdot \mathbf{b} + \mathbf{a} \cdot \mathbf{c} \tag{A.18}$$

and
$$(\mathbf{b} + \mathbf{c}) \cdot \mathbf{a} = \mathbf{b} \cdot \mathbf{a} + \mathbf{c} \cdot \mathbf{a} \tag{A.19}$$

A.8.2 Vector Product (Cross Product)

The vector or cross product of two vectors **a** and **b** is defined as the vector **c**,

$$\mathbf{c} = \mathbf{a} \times \mathbf{b} \tag{A.20}$$

which is orthogonal to both **a** and **b** and has magnitude

$$|\mathbf{c}| = |\mathbf{a}|\,|\mathbf{b}|\sin\theta \tag{A.21}$$

The vector **c** is so directed that a right-handed rotation about **c** through an angle θ of less than $180°$ carries **a** into **b**, where θ is the angle between **a** and **b** (see Fig. A.6). In Fig. A.6, since $h = |\mathbf{b}|\sin\theta$, the cross product $\mathbf{a} \times \mathbf{b}$ has a magnitude equal to the area of the parallelogram formed with sides **a** and **b**.

The cross product $\mathbf{a} \times \mathbf{b}$ can be considered as the result obtained by projecting **b** on the plane $WXYZ$ perpendicular to the plane of **a** and **b**, rotating the projection $90°$ in the positive direction about **a** and then multiplying the resulting vector by $|\mathbf{a}|$.

The cross product of $\mathbf{b} \times \mathbf{a}$ has the same magnitude as $\mathbf{a} \times \mathbf{b}$ but in the opposite direction of rotation as $\mathbf{a} \times \mathbf{b}$. Thus,

$$\mathbf{b} \times \mathbf{a} = -(\mathbf{a} \times \mathbf{b}) \tag{A.22}$$

and the cross product are *not* commutative. If vectors **a** and **b** are parallel, then θ is $0°$ or $180°$ and

$$|\mathbf{a} \times \mathbf{b}| = |\mathbf{a}|\,|\mathbf{b}|\sin\theta = 0 \tag{A.23}$$

Conversely, if the cross product is zero, then one (or both) of the vectors is zero or else they are parallel.

Also, we note that the cross product is *distributed over addition*, that is,

$$\mathbf{a} \times (\mathbf{b} + \mathbf{c}) = \mathbf{a} \times \mathbf{b} + \mathbf{a} \times \mathbf{c} \tag{A.24}$$

and $$(\mathbf{b} + \mathbf{c}) \times \mathbf{a} = \mathbf{b} \times \mathbf{a} + \mathbf{c} \times \mathbf{a} \tag{A.25}$$

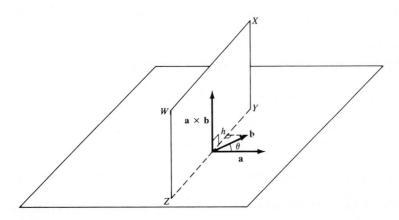

Figure A.6 Cross product.

Applying the scalar and cross product to the unit vectors **i**, **j**, **k** along the principal axes of a right-handed cartesian coordinate system, we have

$$\mathbf{i} \cdot \mathbf{i} = \mathbf{j} \cdot \mathbf{j} = \mathbf{k} \cdot \mathbf{k} = 1$$

$$\mathbf{i} \cdot \mathbf{j} = \mathbf{j} \cdot \mathbf{k} = \mathbf{k} \cdot \mathbf{i} = 0$$

$$\mathbf{i} \times \mathbf{i} = \mathbf{j} \times \mathbf{j} = \mathbf{k} \times \mathbf{k} = \mathbf{0} \qquad (A.26)$$

$$\mathbf{i} \times \mathbf{j} = -\mathbf{j} \times \mathbf{i} = \mathbf{k}$$

$$\mathbf{j} \times \mathbf{k} = -\mathbf{k} \times \mathbf{j} = \mathbf{i}$$

$$\mathbf{k} \times \mathbf{i} = -\mathbf{i} \times \mathbf{k} = \mathbf{j}$$

Using the definition of components and Eq. (A.26), the scalar product of **a** and **b** can be written as

$$\mathbf{a} \cdot \mathbf{b} = (a_1\mathbf{i} + a_2\mathbf{j} + a_3\mathbf{k}) \cdot (b_1\mathbf{i} + b_2\mathbf{j} + b_3\mathbf{k})$$

$$= a_1 b_1 + a_2 b_2 + a_3 b_3$$

$$= \mathbf{a}^T\mathbf{b} \qquad (A.27)$$

where $\mathbf{a}^T$ indicates the transpose of **a** (see Sec. A.12). The cross product of **a** and **b** can be written as a determinant operation (see Sec. A.15),

$$\mathbf{a} \times \mathbf{b} = \begin{vmatrix} \mathbf{i} & \mathbf{j} & \mathbf{k} \\ a_1 & a_2 & a_3 \\ b_1 & b_2 & b_3 \end{vmatrix}$$

$$= (a_2 b_3 - a_3 b_2)\mathbf{i} + (a_3 b_1 - a_1 b_3)\mathbf{j} + (a_1 b_2 - a_2 b_1)\mathbf{k} \qquad (A.28)$$

A.9 PRODUCTS OF THREE OR MORE VECTORS

For scalar or vector product of three or more vectors, we usually encounter the following types:

$$(\mathbf{a} \cdot \mathbf{b})\mathbf{c} \qquad \mathbf{a} \cdot (\mathbf{b} \times \mathbf{c}) \qquad \mathbf{a} \times (\mathbf{b} \times \mathbf{c}) \qquad (A.29)$$

The product $(\mathbf{a} \cdot \mathbf{b})\mathbf{c}$ is simply the product of a scalar $(\mathbf{a} \cdot \mathbf{b})$ and the vector **c**. The resultant vector has a magnitude of $|\mathbf{a} \cdot \mathbf{b}| \, |\mathbf{c}|$ and a direction which is the same as the vector **c** or opposite to it, according to whether $(\mathbf{a} \cdot \mathbf{b})$ is positive or negative.

The *scalar triple product,* $\mathbf{a} \cdot (\mathbf{b} \times \mathbf{c})$, is a scalar whose magnitude equals the volume of a parallelepiped with the vectors $\mathbf{a}$, $\mathbf{b}$, and $\mathbf{c}$ as coterminous edges (see Fig. A.7); that is,

$$\mathbf{a} \cdot (\mathbf{b} \times \mathbf{c}) = |\mathbf{a}|\,|\mathbf{b}|\,|\mathbf{c}|\sin\theta\cos\alpha \tag{A.30}$$

$$= hA = \text{volume of parallelepiped}$$

where h and A are, respectively, the height and area of the parallelepiped. Expressing the vectors in terms of their components in a three-dimensional vector space yields

$$\mathbf{a} \cdot (\mathbf{b} \times \mathbf{c}) = (a_x\mathbf{i} + a_y\mathbf{j} + a_z\mathbf{k}) \cdot \begin{vmatrix} \mathbf{i} & \mathbf{j} & \mathbf{k} \\ b_x & b_y & b_z \\ c_x & c_y & c_z \end{vmatrix}$$

$$= a_x(b_yc_z - b_zc_y) + a_y(b_zc_x - b_xc_z) + a_z(b_xc_y - b_yc_x)$$

$$= \begin{vmatrix} a_x & a_y & a_z \\ b_x & b_y & b_z \\ c_x & c_y & c_z \end{vmatrix} \tag{A.31}$$

Note that the parentheses around the vector product $\mathbf{b} \times \mathbf{c}$ can be taken out without confusion as we cannot interpret $\mathbf{a} \cdot \mathbf{b} \times \mathbf{c}$ as $(\mathbf{a} \cdot \mathbf{b}) \times \mathbf{c}$, which is meaningless. We also observe that the volume of a parallelepiped is independent of the face chosen as its base (see Fig. A.7). Thus,

$$\mathbf{a} \cdot \mathbf{b} \times \mathbf{c} = \mathbf{b} \cdot \mathbf{c} \times \mathbf{a} = \mathbf{c} \cdot \mathbf{a} \times \mathbf{b} \tag{A.32}$$

and $\quad \mathbf{a} \cdot \mathbf{b} \times \mathbf{c} = -\mathbf{a} \cdot \mathbf{c} \times \mathbf{b} = -\mathbf{c} \cdot \mathbf{b} \times \mathbf{a} = -\mathbf{b} \cdot \mathbf{a} \times \mathbf{c} \tag{A.33}$

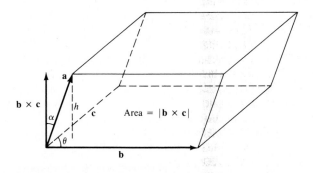

Figure A.7 Scalar triple product.

These results can be readily shown from the properties of determinants (see Sec. A.15). Also, $\mathbf{a} \cdot \mathbf{b} \times \mathbf{c} \equiv \mathbf{a} \times \mathbf{b} \cdot \mathbf{c}$, which is usually written as $(\mathbf{abc})$ because there is no confusion on the position of dot and cross operators. Equation (A.32) indicates a cyclic permutation on these vectors, while Eq. (A.33) indicates an anticyclic permutation. An illustration of cyclic permutation is shown in Fig. A.8. By following a clockwise traversal along the circle, we obtain Eq. (A.32). Similarly, reversing the direction of the arrows, we obtain Eq. (A.33). Finally, the scalar triple vector can be used to prove linear dependence of three coplanar vectors. If three vectors $\mathbf{a}$, $\mathbf{b}$, and $\mathbf{c}$ are coplanar, then $(\mathbf{abc}) = 0$. Hence, if two of the three vectors are equal, the scalar triple product vanishes. It follows that if $\mathbf{e}_1$, $\mathbf{e}_2$, and $\mathbf{e}_3$ are basis vectors for a vector space V_3, then $(\mathbf{e}_1\mathbf{e}_2\mathbf{e}_3) \neq 0$ and they form a right-handed coordinate system if $(\mathbf{e}_1\mathbf{e}_2\mathbf{e}_3) > 0$ and a left-handed coordinate system if $(\mathbf{e}_1\mathbf{e}_2\mathbf{e}_3) < 0$.

The *vector triple product*, $\mathbf{a} \times (\mathbf{b} \times \mathbf{c})$, is a vector perpendicular to $(\mathbf{b} \times \mathbf{c})$ and lying in the plane of $\mathbf{b}$ and $\mathbf{c}$ (see Fig. A.9). Suppose that the vectors $\mathbf{a}$, $\mathbf{b}$, and $\mathbf{c}$ are noncollinear. Then, the vector $\mathbf{a} \times (\mathbf{b} \times \mathbf{c})$ lying in the plane of $\mathbf{b}$ and $\mathbf{c}$ can be expressed as a linear combination of $\mathbf{b}$ and $\mathbf{c}$; that is,

$$\mathbf{a} \times (\mathbf{b} \times \mathbf{c}) = m\mathbf{b} + n\mathbf{c} \tag{A.34}$$

Since the vector $\mathbf{a} \times (\mathbf{b} \times \mathbf{c})$ is also perpendicular to the vector $\mathbf{a}$, then dotting both sides of Eq. (A.34) with the vector $\mathbf{a}$ will yield zero:

$$\mathbf{a} \cdot [\mathbf{a} \times (\mathbf{b} \times \mathbf{c})] = m(\mathbf{a} \cdot \mathbf{b}) + n(\mathbf{a} \cdot \mathbf{c}) = 0 \tag{A.35}$$

Thus,

$$\frac{m}{\mathbf{a} \cdot \mathbf{c}} = \frac{-n}{\mathbf{a} \cdot \mathbf{b}} = \lambda \tag{A.36}$$

where m, n, and λ are scalars and Eq. (A.34) becomes

$$\mathbf{a} \times (\mathbf{b} \times \mathbf{c}) = \lambda[(\mathbf{a} \cdot \mathbf{c})\mathbf{b} - (\mathbf{a} \cdot \mathbf{b})\mathbf{c}] \tag{A.37}$$

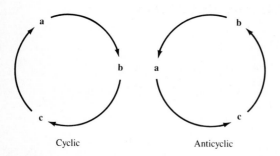

Cyclic   Anticyclic

Figure A.8 Cyclic permutation.

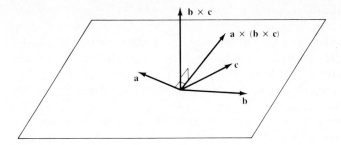

Figure A.9 Vector triple product.

It can be shown that $\lambda \equiv 1$, so the vector triple product becomes

$$\mathbf{a} \times (\mathbf{b} \times \mathbf{c}) = (\mathbf{a} \cdot \mathbf{c})\mathbf{b} - (\mathbf{a} \cdot \mathbf{b})\mathbf{c} \qquad (A.38)$$

Also, using Eq. (A.22) it follows that

$$(\mathbf{a} \times \mathbf{b}) \times \mathbf{c} = -\mathbf{c} \times (\mathbf{a} \times \mathbf{b}) = -(\mathbf{c} \cdot \mathbf{b})\mathbf{a} + (\mathbf{c} \cdot \mathbf{a})\mathbf{b} \qquad (A.39)$$

More complicated cases involving four or more products can be simplified by the use of triple products. For example,

$$(\mathbf{a} \times \mathbf{b}) \times (\mathbf{c} \times \mathbf{d}) = (\mathbf{a} \times \mathbf{b} \cdot \mathbf{d})\mathbf{c} - (\mathbf{a} \times \mathbf{b} \cdot \mathbf{c})\mathbf{d}$$

$$= (\mathbf{abd})\mathbf{c} - (\mathbf{abc})\mathbf{d} \qquad (A.40)$$

and

$$(\mathbf{a} \times \mathbf{b}) \cdot (\mathbf{c} \times \mathbf{d}) = \mathbf{a} \cdot \mathbf{b} \times (\mathbf{c} \times \mathbf{d})$$

$$= \mathbf{a} \cdot [(\mathbf{b} \cdot \mathbf{d})\mathbf{c} - (\mathbf{b} \cdot \mathbf{c})\mathbf{d}]$$

$$= (\mathbf{b} \cdot \mathbf{d})(\mathbf{a} \cdot \mathbf{c}) - (\mathbf{b} \cdot \mathbf{c})(\mathbf{a} \cdot \mathbf{d}) \qquad (A.41)$$

A.10 DERIVATIVES OF VECTOR FUNCTIONS

The derivative of a vector $\mathbf{r}(t)$ means

$$\frac{d\mathbf{r}}{dt} \triangleq \lim_{\Delta t \to 0} \frac{\mathbf{r}(t + \Delta t) - \mathbf{r}(t)}{\Delta t} = \lim_{\Delta t \to 0} \frac{\Delta \mathbf{r}}{\Delta t} \qquad (A.42)$$

It follows from Eqs. (A.11) and (A.42) that

$$\frac{d\mathbf{r}}{dt} = \left[\frac{dr_x}{dt} \right] \mathbf{i} + \left[\frac{dr_y}{dt} \right] \mathbf{j} + \left[\frac{dr_z}{dt} \right] \mathbf{k} \qquad (A.43)$$

and the *n*th derivative of $\mathbf{r}(t)$ is

$$\frac{d^n \mathbf{r}}{dt^n} = \left[\frac{d^n r_x}{dt^n}\right] \mathbf{i} + \left[\frac{d^n r_y}{dt^n}\right] \mathbf{j} + \left[\frac{d^n r_z}{dt^n}\right] \mathbf{k} \qquad (A.44)$$

Using Eq. (A.42), the following rules for differentiating vector functions can be obtained:

(1) $\quad \dfrac{d}{dt}(\mathbf{a} \pm \mathbf{b}) = \dfrac{d\mathbf{a}}{dt} \pm \dfrac{d\mathbf{b}}{dt}$ $\qquad (A.45)$

(2) $\quad \dfrac{d}{dt}(m\mathbf{a}) = m\dfrac{d\mathbf{a}}{dt} \qquad$ where m is a scalar

(3) $\quad \dfrac{d}{dt}(\mathbf{a} \cdot \mathbf{b}) = \left[\dfrac{d\mathbf{a}}{dt}\right] \cdot \mathbf{b} + \mathbf{a} \cdot \left[\dfrac{d\mathbf{b}}{dt}\right]$

(4) $\quad \dfrac{d}{dt}(\mathbf{a} \times \mathbf{b}) = \left[\dfrac{d\mathbf{a}}{dt}\right] \times \mathbf{b} + \mathbf{a} \times \left[\dfrac{d\mathbf{b}}{dt}\right]$

(5) $\quad \dfrac{d}{dt}(\mathbf{abc}) = \left[\left[\dfrac{d\mathbf{a}}{dt}\right]\mathbf{bc}\right] + \mathbf{a}\left[\left[\dfrac{d\mathbf{b}}{dt}\right]\mathbf{c}\right] + \left[\mathbf{ab}\left[\dfrac{d\mathbf{c}}{dt}\right]\right]$

(6) $\quad \dfrac{d}{dt}[\mathbf{a} \times (\mathbf{b} \times \mathbf{c})] = \left[\dfrac{d\mathbf{a}}{dt} \times (\mathbf{b} \times \mathbf{c})\right]$

$$+ \left[\mathbf{a} \times \left[\dfrac{d\mathbf{b}}{dt} \times \mathbf{c}\right]\right] + \left[\mathbf{a} \times \left[\mathbf{b} \times \dfrac{d\mathbf{c}}{dt}\right]\right]$$

A.11 INTEGRATION OF VECTOR FUNCTIONS

If $d\mathbf{a}/dt = \mathbf{b}(t)$, the integral of the vector $\mathbf{b}(t)$ means

$$\mathbf{a}(t) = \int \mathbf{b}(\tau)d\tau + \mathbf{c} \qquad (A.46)$$

where $\mathbf{c}$ is a constant vector.

If $\mathbf{b}(t)$ can be expressed in a rectangular coordinate system, then

$$a_x(t) = \int b_x(\tau)\, d\tau + c_x$$

$$a_y(t) = \int b_y(\tau)\, d\tau + c_y \qquad (A.47)$$

$$a_z(t) = \int b_z(\tau)\, d\tau + c_z$$

A.12 MATRIX ALGEBRA

In the remainder of this appendix, we shall discuss another important mathematical tool, matrices, which is essential for the analysis of robotic mechanism.

A *matrix* $\mathbf{A}$ (or $\mathbf{A}_{m \times n}$) of order m by n is a rectangular array of real or complex numbers (called elements) arranged in m rows and n columns.

$$
\mathbf{A} = [a_{ij}] = \begin{bmatrix} a_{11} & a_{12} & \cdots & a_{1n} \\ a_{21} & a_{22} & \cdots & a_{2n} \\ \cdot & \cdot & & \cdot \\ \cdot & \cdot & & \cdot \\ \cdot & \cdot & & \cdot \\ a_{m1} & a_{m2} & \cdots & a_{mn} \end{bmatrix} \quad \begin{aligned} i &= 1, 2, \ldots, m \\ j &= 1, 2, \ldots, n \end{aligned}
$$

(A.48)

Unless it is otherwise noted, we will assume that $\mathbf{A}$ is a real matrix. A matrix consisting of a single column (row) is called a column (row) matrix. Both column and row matrices are often referred to as vectors.

The *transpose* of a matrix $\mathbf{A}$, denoted by $\mathbf{A}^T$, is defined to be the matrix whose row number is identical with the column number of $\mathbf{A}$. In other words, if

$$
\mathbf{A} = \begin{bmatrix} a_{11} & a_{12} & \cdots & a_{1n} \\ a_{21} & a_{22} & \cdots & a_{2n} \\ \cdot & \cdot & & \cdot \\ \cdot & \cdot & & \cdot \\ \cdot & \cdot & & \cdot \\ a_{m1} & a_{m2} & \cdots & a_{mn} \end{bmatrix} \quad \begin{aligned} i &= 1, 2, \ldots, m \\ j &= 1, 2, \ldots, n \end{aligned}
$$

(A.49)

then

$$
\mathbf{A}^T = \begin{bmatrix} a_{11} & a_{21} & \cdots & a_{m1} \\ a_{12} & a_{22} & \cdots & a_{m2} \\ \cdot & \cdot & & \cdot \\ \cdot & \cdot & & \cdot \\ \cdot & \cdot & & \cdot \\ a_{1n} & a_{2n} & \cdots & a_{mn} \end{bmatrix} \quad \begin{aligned} i &= 1, 2, \ldots, n \\ j &= 1, 2, \ldots, m \end{aligned}
$$

(A.50)

In particular, the transpose of a column matrix is a row matrix and vice versa.

A *square* matrix of order n has an equal number of rows and columns (i.e., $m = n$). A *diagonal* matrix is a square matrix of order n whose off-diagonal elements are zero. That is, the elements

$$a_{ij} = 0 \quad \text{if } i \neq j \quad \text{for } i, j = 1, 2, \ldots, n \qquad \text{(A.51)}$$

A *unit* matrix of order n is a diagonal matrix whose diagonal elements are all unity. That is, $a_{ij} = 1$ if $i = j$ and $a_{ij} = 0$ if $i \neq j$. This matrix is called the *identity* matrix and denoted by $\mathbf{I}_n$ or $\mathbf{I}_{n \times n}$.

A *symmetric* matrix is a square matrix of order n whose transpose is identical to itself. That is, $\mathbf{A} \equiv \mathbf{A}^T$ or $a_{ij} \equiv a_{ji}$ for all i and j. If the elements of a square matrix are such that

$$a_{ij} = -a_{ji} \quad \text{and} \quad a_{ii} = 0 \quad i, j = 1, 2, \ldots, n \qquad \text{(A.52)}$$

then the matrix is called a *skew* matrix. It is noted that if $\mathbf{A}$ is skew, then $\mathbf{A} = -\mathbf{A}^T$.

Any nonsymmetric square matrix $\mathbf{A}$ can be made into a symmetric matrix $\mathbf{C}$ by letting

$$\mathbf{C} = \frac{\mathbf{A} + \mathbf{A}^T}{2} \qquad \text{(A.53)}$$

A *null* matrix is a matrix whose elements are all identically equal to zero. Two matrices of the same order are equal if their respective elements are equal. That is, if $a_{ij} = b_{ij}$ for all i and j, then $\mathbf{A} = \mathbf{B}$.

A.13 SUMMATION OF MATRICES

Two matrices $\mathbf{A}$ and $\mathbf{B}$ of the same order can be added (subtracted) forming a resultant matrix $\mathbf{C}$ of the same order by adding (subtracting) corresponding elements. Thus,

$$\mathbf{A} + \mathbf{B} = \mathbf{C} \quad \text{or} \quad a_{ij} + b_{ij} = c_{ij} \quad \text{for all } i, j \qquad \text{(A.54)}$$

and $\quad \mathbf{A} - \mathbf{B} = \mathbf{C} \quad \text{or} \quad a_{ij} - b_{ij} = c_{ij} \quad \text{for all } i, j \qquad \text{(A.55)}$

Matrix addition has similar properties as real number addition:

(1) $\mathbf{A} + \mathbf{B} = \mathbf{B} + \mathbf{A}$

(2) $(\mathbf{A} + \mathbf{B}) + \mathbf{C} = \mathbf{A} + (\mathbf{B} + \mathbf{C})$

(3) $\mathbf{A} + \mathbf{0} = \mathbf{A} \quad$ ($\mathbf{0}$ is the zero or null matrix) $\qquad \text{(A.56)}$

(4) $\mathbf{A} + (-\mathbf{A}) = \mathbf{0}$

A.14 MATRIX MULTIPLICATION

The product of a scalar and a matrix is formed by multiplying every element of **A** by the scalar. Thus,

$$kA = Ak = [ka_{ij}] = [a_{ij}k] \qquad i = 1, 2, \ldots, m$$

$$j = 1, 2, \ldots, n$$

The following rules hold for the product of any $(m \times n)$ matrices and any scalars:

(1) $a(A + B) = aA + aB$

(2) $(a + b)A = aA + bA$

(3) $a(bA) = (ab)A$ (A.57)

(4) $1A = A$

where a and b are scalars.

Two matrices can be multiplied together only if they are *conformable*. That is, if $AB = C$, then the number of column of **A** must be equal to the number of row of **B** and the resultant matrix **C** has the row and column number equal to those of **A** and **B**, respectively. Thus,

$$(A_{m \times n})(B_{n \times p}) = C_{m \times p} \qquad \text{or} \qquad c_{ij} = \sum_{k=1}^{n} a_{ik}b_{kj} \qquad (A.58)$$

In Eq. (A.58), we can either say **B** is *premultiplied* by **A** or **A** is *postmultiplied* by **B** to obtain **C**. In order to obtain the element at the ith row and jth column of **C**, we sum the product terms of the corresponding elements in the ith row of **A** and the jth column of **B**, as in Eq. (A.58). In other words, we postmultiply the ith row of **A** by the jth column of **B**. In general, matrix multiplication is *not commutative* even if the matrices are conformable. That is, if **A** and **B** are square matrices of order n, then

$$AB \neq BA$$

If $AB = BA$, then we say the matrices are *commutative*. The unit matrix commutes with any square matrix.

$$IA = AI = A \qquad (A.59)$$

Matrix multiplication is associative and distributive with respect to matrix addition; that is,

(1) $(kA)B = k(AB) = A(kB)$

(2) $\mathbf{A}(\mathbf{BC}) = (\mathbf{AB})\mathbf{C}$

(3) $(\mathbf{A} + \mathbf{B})\mathbf{C} = \mathbf{AC} + \mathbf{BC}$ (A.60)

(4) $\mathbf{C}(\mathbf{A} + \mathbf{B}) = \mathbf{CA} + \mathbf{CB}$

assuming that the matrix multiplications are defined. From rule (2), we see that for the product of three matrices, we can either postmultiply $\mathbf{B}$ by $\mathbf{C}$ or premultiply $\mathbf{B}$ by $\mathbf{A}$ first and then multiply the result by the remaining matrix. In general, $\mathbf{AB} = \mathbf{0}$ does not imply that $\mathbf{A} = \mathbf{0}$ or $\mathbf{B} = \mathbf{0}$. It is worthwhile to note the following rules for the product of matrices:

(1) $(\text{matrix})_{m \times n} (\text{matrix})_{n \times p} = (\text{matrix})_{m \times p}$

(2) $(\text{matrix})_{m \times n} (\text{column matrix})_{n \times 1} = (\text{column matrix})_{m \times 1}$

(3) $(\text{row matrix})_{1 \times n} (\text{column matrix})_{n \times 1} = \text{scalar}$

(4) $(\text{row matrix})_{1 \times m} (\text{matrix})_{m \times n} = (\text{row matrix})_{1 \times n}$

(5) $(\text{column matrix})_{m \times 1} (\text{row matrix})_{1 \times n} = (\text{matrix})_{m \times n}$

Sometimes in matrix addition or multiplication, it is more convenient to *partition* the matrices into submatrices which are manipulated according to the above rules of matrix algebra.

A.15 DETERMINANTS

The *determinant* of an $n \times n$ matrix $\mathbf{A}$ is denoted by

$$|\mathbf{A}| = \begin{vmatrix} a_{11} & a_{12} & \cdots & a_{1n} \\ a_{21} & a_{22} & \cdots & a_{2n} \\ \cdot & \cdot & & \cdot \\ \cdot & \cdot & & \cdot \\ \cdot & \cdot & & \cdot \\ a_{n1} & a_{n2} & \cdots & a_{nn} \end{vmatrix}$$ (A.61)

and is equal to the sum of the products of the elements of any row or column and their respective cofactors, that is,

$$|\mathbf{A}| = \sum_{j=1}^{n} a_{ij} A_{ij} = \sum_{i=1}^{n} a_{ij} A_{ij}$$ (A.62)

Here, A_{ij} is the cofactor of a_{ij}, which can be obtained as

$$A_{ij} = (-1)^{i+j} M_{ij} \qquad \text{(A.63)}$$

where M_{ij} is the complementary minor, obtained by deleting the elements in the ith row and the jth column of $|\mathbf{A}|$. In other words, if

$$|\mathbf{A}| = \begin{vmatrix} a_{11} & a_{12} & \cdots & a_{1j} & \cdots & a_{1n} \\ a_{21} & a_{22} & \cdots & a_{2j} & \cdots & a_{2n} \\ \cdot & \cdot & & & & \cdot \\ \cdot & \cdot & & & & \cdot \\ \cdot & \cdot & & & & \cdot \\ a_{i1} & a_{i2} & \cdots & a_{ij} & \cdots & a_{in} \\ \cdot & & & & & \cdot \\ \cdot & & & & & \cdot \\ \cdot & & & & & \cdot \\ a_{n1} & a_{n2} & \cdots & a_{nj} & \cdots & a_{nn} \end{vmatrix}$$

and we delete the elements in the ith row and jth column, then

$$|M_{ij}| = \begin{vmatrix} a_{11} & a_{12} & \cdots & a_{1,j-1} & a_{1,j+1} & \cdots & a_{1n} \\ \cdot & & & & & & \cdot \\ \cdot & & & & & & \cdot \\ \cdot & & & & & & \cdot \\ a_{i-1,1} & a_{i-1,2} & \cdots & a_{i-1,j-1} & a_{i-1,j+1} & \cdots & a_{i-1,n} \\ a_{i+1,1} & a_{i+1,2} & \cdots & a_{i+1,j-1} & a_{i+1,j+1} & \cdots & a_{i+1,n} \\ \cdot & & & & & & \cdot \\ \cdot & & & & & & \cdot \\ \cdot & & & & & & \cdot \\ a_{n1} & a_{n2} & \cdots & a_{n,j-1} & a_{n,j+1} & \cdots & a_{nn} \end{vmatrix}$$

From the above definition, a determinant of order n depends upon n determinants of order $n-1$, each of which in turn depends upon $n-1$ determinants of order $n-2$, and so on, until reaching a determinant of order 1, which is a scalar.

A simple diagonal method can be used to evaluate the determinants of order 2 and 3. For $n = 2$ we have

$$|\mathbf{A}| = \begin{vmatrix} a_{11} & a_{12} \\ a_{21} & a_{22} \end{vmatrix} = a_{11}a_{22} - a_{21}a_{12} \tag{A.64}$$

For a third order determinant, it can be evaluated as

$$|\mathbf{A}| = \begin{vmatrix} a_{11} & a_{12} & a_{13} \\ a_{21} & a_{22} & a_{23} \\ a_{31} & a_{32} & a_{33} \end{vmatrix}$$

$$= a_{11}a_{22}a_{33} + a_{12}a_{23}a_{31} + a_{13}a_{32}a_{21} - a_{31}a_{22}a_{13}$$

$$- a_{12}a_{21}a_{33} - a_{11}a_{32}a_{23} \tag{A.65}$$

The following properties are useful for simplifying the evaluation of determinants:

1. If all the elements of any row (or column) of $\mathbf{A}$ are zero, then $|\mathbf{A}| = 0$.
2. $|\mathbf{A}| = |\mathbf{A}^T|$.
3. If any two rows (or columns) of $\mathbf{A}$ are interchanged, then the sign of its determinant is changed.
4. If $\mathbf{A}$ and $\mathbf{B}$ are of order n, then $|\mathbf{AB}| = |\mathbf{A}| |\mathbf{B}|$.
5. If all the elements of any row (or column) of $\mathbf{A}$ are multiplied by a scalar k, then the determinant is scaled by k.
6. If the rank (see Sec. A.16) of a matrix $\mathbf{A}$ of order n is less than n, its determinant is zero.
7. If a multiple of any row (or column) is added to other row (or column), then the determinant remains unchanged.

Example: Let

$$\mathbf{A} = \begin{bmatrix} 1 & a & a^2 \\ 1 & b & b^2 \\ 1 & c & c^2 \end{bmatrix}$$

Then,

$$
|\mathbf{A}| = \begin{vmatrix} 1 & a & a^2 \\ 1 & b & b^2 \\ 1 & c & c^2 \end{vmatrix} = \begin{vmatrix} 1 & a & a^2 \\ 0 & b-a & b^2-a^2 \\ 0 & c-a & c^2-a^2 \end{vmatrix}
$$

$$
= (b-a)(c^2-a^2) - (c-a)(b^2-a^2)
$$

$$
= (a-b)(b-c)(c-a)
$$

This is the Vandermonde determinant of order 3. □

A.16 RANK OF A MATRIX

In general, if the rows of a square matrix $\mathbf{A}$ of order n are linearly independent, then the determinant of that matrix is nonzero, and the matrix is said to be *nonsingular*. If the determinant of a square matrix of order n is zero, then the matrix is *singular* and the rows of the matrix are *not* linearly independent. Thus, the determinant can be used as a test for matrix singularity.

The rank of a matrix $\mathbf{A}$ of order $m \times n$ is equal to the order of the largest submatrix of $\mathbf{A}$ with nonzero determinant. Thus, a matrix of order $m \times n$ can have a rank equal to the smaller value of m and n, or less. The rank of a matrix indicates the number of linearly independent rows (or columns) in the matrix.

A.17 ADJOINT AND INVERSE MATRICES

If $\mathbf{A}$ is a square matrix and A_{ij} is the cofactor of a_{ij} in $|\mathbf{A}|$, then the transpose of the matrix formed from the cofactors $\mathbf{A}_{ij}$ is called the *adjoint* of $\mathbf{A}$, and

$$
[A_{ij}]^T = [A_{ji}] \qquad i, j = 1, 2, \ldots, n \tag{A.66}
$$

Sometimes, the adjoint of $\mathbf{A}$ is denoted by adj $\mathbf{A}$.

The inverse of a nonsingular square matrix $\mathbf{A}$, $\mathbf{A}^{-1}$, is the adjoint of $\mathbf{A}$ divided by the determinant of $\mathbf{A}$, that is,

$$
\mathbf{A}^{-1} = \frac{[A_{ij}]^T}{|\mathbf{A}|} = \frac{\text{adj } \mathbf{A}}{|\mathbf{A}|} \tag{A.67}
$$

The product (in either order) of a nonsingular $n \times n$ matrix $\mathbf{A}$ and its inverse is the identity matrix $\mathbf{I}_n$; that is,

$$
\mathbf{A}\mathbf{A}^{-1} = \mathbf{A}^{-1}\mathbf{A} = \mathbf{I}_n \tag{A.68}
$$

Thus, from Eqs. (A.67) and (A.68),

$$(\text{adj } \mathbf{A})\mathbf{A} = \mathbf{A}(\text{adj } \mathbf{A}) = |\mathbf{A}|\mathbf{I}_n \qquad (A.69)$$

and

$$|\mathbf{A}^{-1}| = \frac{1}{|\mathbf{A}|} \qquad (A.70)$$

If $\mathbf{A}_1, \mathbf{A}_2, \ldots, \mathbf{A}_n$ are square matrices of order n, then the inverse of their product is the product of the inverse of each matrix in reverse order:

$$(\mathbf{A}_1 \mathbf{A}_2 \ \cdots \ \mathbf{A}_n)^{-1} = \mathbf{A}_n^{-1} \mathbf{A}_{n-1}^{-1} \ \cdots \ \mathbf{A}_2^{-1} \mathbf{A}_1^{-1} \qquad (A.71)$$

Similarly, if the matrix product of $\mathbf{A}_1 \mathbf{A}_2 \ \cdots \ \mathbf{A}_n$ is conformable, then the transpose of their product is the product of the transpose of each matrix in reverse order:

$$(\mathbf{A}_1 \mathbf{A}_2 \mathbf{A}_3 \ \cdots \ \mathbf{A}_n)^T = (\mathbf{A}_n)^T (\mathbf{A}_{n-1})^T \ \cdots \ (\mathbf{A}_2)^T (\mathbf{A}_1)^T \qquad (A.72)$$

In general, a 2×2 matrix

$$\mathbf{A} = \begin{bmatrix} a & c \\ b & d \end{bmatrix}$$

has the inverse

$$\mathbf{A}^{-1} = \frac{1}{ad - bc} \begin{bmatrix} d & -c \\ -b & a \end{bmatrix}$$

Similarly, a 3×3 matrix

$$\mathbf{A} = \begin{bmatrix} a & b & c \\ d & e & f \\ g & h & i \end{bmatrix}$$

has the inverse

$$\mathbf{A}^{-1} = \frac{1}{aei + dhc + gfb - afh - dbi - gec}$$

$$\times \begin{bmatrix} (ei - fh) & -(bi - ch) & (bf - ce) \\ -(di - fg) & (ai - cg) & -(af - cd) \\ (dh - ge) & -(ah - bg) & (ae - bd) \end{bmatrix}$$

An important result called *the matrix inversion lemma*, may be stated as follows:

$$[\mathbf{A}^{-1} + \mathbf{B}^T \mathbf{C} \mathbf{B}]^{-1} = \mathbf{A} - \mathbf{A} \mathbf{B}^T [\mathbf{B} \mathbf{A} \mathbf{B}^T + \mathbf{C}^{-1}]^{-1} \mathbf{B} \mathbf{A} \qquad (A.73)$$

The proof of this result is left as an exercise.

A.18 TRACE OF A MATRIX

The trace of a square matrix $\mathbf{A}$ of order n is the sum of its principal diagonal elements,

$$\text{Trace } \mathbf{A} \equiv \text{Tr}(\mathbf{A}) = \sum_{i=1}^{n} a_{ii} \qquad (A.74)$$

Some useful properties of the trace operator on matrices are:

$$\text{Tr}(\mathbf{A}) = \text{Tr}(\mathbf{A}^T) \qquad (A.75)$$

$$\text{Tr}(\mathbf{A} + \mathbf{B}) = \text{Tr}(\mathbf{A}) + \text{Tr}(\mathbf{B}) \qquad (A.76)$$

$$\text{Tr}(\mathbf{AB}) = \text{Tr}(\mathbf{BA}) \qquad (A.77)$$

$$\text{Tr}(\mathbf{ABC}^T) = \text{Tr}(\mathbf{CB}^T\mathbf{A}^T) \qquad (A.78)$$

REFERENCES

Further reading for the material in this appendix may be found in Frazer et al. [1960], Bellman [1970], Pipes [1963], Thrall and Tornheim [1963], and Noble [1969].

MANIPULATOR JACOBIAN

In resolved motion control, (Chap. 5) one needs to determine how each infinitesimal joint motion affects the infinitesimal motion of the manipulator hand. One advantage of resolved motion is that there exists a linear mapping between the infinitesimal joint motion space and the infinitesimal hand motion space. This mapping is defined by a jacobian. This appendix reviews three methods for obtaining the jacobian for a six-link manipulator with rotary or sliding joints.

B.1 VECTOR CROSS PRODUCT METHOD

Let us define the position, linear velocity, and angular velocity vectors of the manipulator hand with respect to the base coordinate frame ($\mathbf{x}_0$, $\mathbf{y}_0$, $\mathbf{z}_0$), respectively:

$$\mathbf{p}(t) \triangleq [p_x(t), p_y(t), p_z(t)]^T$$

$$\mathbf{v}(t) \triangleq [v_x(t), v_y(t), v_z(t)]^T \qquad \text{(B.1)}$$

$$\mathbf{\Omega}(t) \triangleq [\omega_x(t), \omega_y(t), \omega_z(t)]^T$$

where, as before, the superscript T denotes the transpose operation. Based on the moving coordinate frame concept (Whitney [1972]), the linear and angular velocities of the hand can be obtained from the velocities of the lower joints:

$$\begin{bmatrix} \mathbf{v}(t) \\ \mathbf{\Omega}(t) \end{bmatrix} = \mathbf{J}(\mathbf{q})\dot{\mathbf{q}}(t) = [\mathbf{J}_1(\mathbf{q}), \mathbf{J}_2(\mathbf{q}), \ldots, \mathbf{J}_6(\mathbf{q})]\dot{\mathbf{q}}(t) \qquad \text{(B.2)}$$

where $\mathbf{J}(\mathbf{q})$ is a 6×6 matrix whose ith column vector $\mathbf{J}_i(\mathbf{q})$ is given by (Whitney [1972]):

$$
\mathbf{J}_i(\mathbf{q}) = \begin{cases} \begin{bmatrix} \mathbf{z}_{i-1} \times {}^{i-1}\mathbf{p}_6 \\ \mathbf{z}_{i-1} \end{bmatrix} & \text{if joint } i \text{ is rotational} \\[2em] \begin{bmatrix} \mathbf{z}_{i-1} \\ 0 \end{bmatrix} & \text{if joint } i \text{ is translational} \end{cases}
$$

(B.3)

and $\dot{\mathbf{q}}(t) = [\dot{q}_1(t), \ldots, \dot{q}_6(t)]^T$ is the joint velocity vector of the manipulator, $\times$ indicates cross product, ${}^{i-1}\mathbf{p}_6$ is the position of the origin of the hand coordinate frame from the $(i-1)$th coordinate frame expressed in the base coordinate frame, and $\mathbf{z}_{i-1}$ is the unit vector along the axis of motion of joint i expressed in the base coordinate frame.

For a six-link manipulator with rotary joints, the jacobian can be found to be:

$$
\mathbf{J}(\theta) = \begin{bmatrix} \mathbf{z}_0 \times {}^0\mathbf{p}_6 & \mathbf{z}_1 \times {}^1\mathbf{p}_6 & \cdots & \mathbf{z}_5 \times {}^5\mathbf{p}_6 \\ \mathbf{z}_0 & \mathbf{z}_1 & \cdots & \mathbf{z}_5 \end{bmatrix}
$$

(B.4)

For the PUMA robot manipulator shown in Fig. 2.11 and its link coordinate transformation matrices in Fig. 2.13, the elements of the jacobian are found to be:

$$
\mathbf{J}_1(\theta) = \begin{bmatrix} -S_1[d_6(C_{23}C_4S_5 + S_{23}C_5) + S_{23}d_4 + a_3C_{23} + a_2C_2] - C_1(d_6S_4S_5 + d_2) \\ C_1[d_6(C_{23}C_4S_5 + S_{23}C_5) + S_{23}d_4 + a_3C_{23} + a_2C_2] - S_1(d_6S_4S_5 + d_2) \\ 0 \\ 0 \\ 0 \\ 1 \end{bmatrix}
$$

$$
\mathbf{J}_2(\theta) = \begin{bmatrix} d_6C_1S_4S_5 + d_2C_1 \\ d_6S_1S_4S_5 + d_2S_1 \\ J_{2z} \\ -S_1 \\ C_1 \\ 0 \end{bmatrix}
$$

where

$$J_{2z} = -S_1 [d_6 S_{23} C_4 S_5 - d_6 C_{23} C_5 - d_4 C_{23} + a_3 S_{23} + a_2 S_2]$$
$$- C_1 [d_6 C_{23} C_4 S_5 + d_6 S_{23} C_5 + d_4 S_{23} + a_3 C_{23} + a_2 C_2]$$

$$\mathbf{J}_3(\boldsymbol{\theta}) = \begin{bmatrix} d_6 C_1 S_4 S_5 \\ d_6 S_1 S_4 S_5 \\ J_{3z} \\ -S_1 \\ C_1 \\ 0 \end{bmatrix}$$

where

$$J_{3z} = -S_1 [d_6 S_3 C_4 S_5 - d_6 C_3 C_5 - d_4 C_3 + a_3 S_3]$$
$$- C_1 [d_6 C_3 C_4 S_5 + d_6 S_3 C_5 + d_4 S_3 + a_3 C_3]$$

$$\mathbf{J}_4(\boldsymbol{\theta}) = \begin{bmatrix} S_1 S_{23} (d_6 C_5 + d_4) - d_6 C_{23} S_4 S_5 \\ d_6 C_{23} C_4 S_5 - C_1 S_{23} (d_6 C_5 + d_4) \\ d_6 C_1 S_{23} S_4 S_5 - d_6 S_1 S_{23} C_4 S_5 \\ C_1 S_{23} \\ S_1 S_{23} \\ C_{23} \end{bmatrix}$$

$$\mathbf{J}_5(\boldsymbol{\theta}) = \begin{bmatrix} d_6 S_{23} S_4 C_5 \\ d_6 S_{23} S_4 S_5 \\ d_6 C_1 C_{23} S_4 C_5 + d_6 S_1 C_4 C_5 + d_6 S_1 C_{23} S_4 S_5 - d_6 C_1 C_4 S_5 \\ -C_1 C_{23} S_4 - S_1 C_4 \\ -S_1 C_{23} S_4 + C_1 C_4 \\ S_{23} S_4 \end{bmatrix}$$

$$\mathbf{J}_6(\boldsymbol{\theta}) = \begin{bmatrix} d_6 (S_1 C_{23} C_4 + C_1 S_4) S_5 + d_6 S_1 S_{23} C_5 \\ -d_6 (C_1 C_{23} C_4 - S_1 S_4) S_5 - d_6 C_1 S_{23} C_5 \\ 0 \\ (C_1 C_{23} C_4 - S_1 S_4) S_5 + C_1 S_{23} C_5 \\ (S_1 C_{23} C_4 + C_1 S_4) S_5 + S_1 S_{23} C_5 \\ -S_{23} C_4 S_5 + C_{23} C_5 \end{bmatrix}$$

where $S_i = \sin \theta_i$, $C_i = \cos \theta_i$, $S_{ij} = \sin (\theta_i + \theta_j)$, and $C_{ij} = \cos (\theta_i + \theta_j)$.

If it is desired to control the manipulator hand along or about the hand coordinate axes, then one needs to express the linear and angular velocities in hand coordinates. This can be accomplished by premultiplying the $\mathbf{v}(t)$ and $\boldsymbol{\Omega}(t)$ by the 3×3 rotation matrix $[^0\mathbf{R}_6]^T$, where $^0\mathbf{R}_6$ is the hand rotation matrix which relates the orientation of the hand coordinate frame to the base coordinate frame. Thus,

$$
\begin{bmatrix} ^6\mathbf{v}_0(t) \\ ^6\boldsymbol{\Omega}_0(t) \end{bmatrix} = \begin{bmatrix} [^0\mathbf{R}_6]^T & \mathbf{0} \\ \mathbf{0} & [^0\mathbf{R}_6]^T \end{bmatrix} \begin{bmatrix} \mathbf{v}(t) \\ \boldsymbol{\Omega}(t) \end{bmatrix} = \begin{bmatrix} ^6\mathbf{R}_0 & \mathbf{0} \\ \mathbf{0} & ^6\mathbf{R}_0 \end{bmatrix} [\mathbf{J}(\mathbf{q})]\dot{\mathbf{q}}(t) \quad \text{(B.5)}
$$

where $\mathbf{0}$ is a 3×3 zero matrix.

B.2 DIFFERENTIAL TRANSLATION AND ROTATION METHOD

Paul [1981] utilizes 4×4 homogeneous transformation matrices to obtain differential translation and rotation about a coordinate frame from which the jacobian of the manipulator is derived. Given a link coordinate frame $\mathbf{T}$, a differential change in $\mathbf{T}$ corresponds to a differential translation and rotation along and about the base coordinates; that is,

$$
\mathbf{T} + d\mathbf{T} = \begin{bmatrix} 1 & 0 & 0 & d_x \\ 0 & 1 & 0 & d_y \\ 0 & 0 & 1 & d_z \\ 0 & 0 & 0 & 1 \end{bmatrix} \begin{bmatrix} 1 & -\delta_z & \delta_y & 0 \\ \delta_z & 1 & -\delta_x & 0 \\ -\delta_y & \delta_x & 1 & 0 \\ 0 & 0 & 0 & 1 \end{bmatrix} \mathbf{T} \quad \text{(B.6)}
$$

or

$$
d\mathbf{T} = \left[\begin{bmatrix} 1 & 0 & 0 & d_x \\ 0 & 1 & 0 & d_y \\ 0 & 0 & 1 & d_z \\ 0 & 0 & 0 & 1 \end{bmatrix} \begin{bmatrix} 1 & -\delta_z & \delta_y & 0 \\ \delta_z & 1 & -\delta_x & 0 \\ -\delta_y & \delta_x & 1 & 0 \\ 0 & 0 & 0 & 1 \end{bmatrix} - \begin{bmatrix} 1 & 0 & 0 & 0 \\ 0 & 1 & 0 & 0 \\ 0 & 0 & 1 & 0 \\ 0 & 0 & 0 & 1 \end{bmatrix} \right] \mathbf{T}
$$

$$
= \Delta \mathbf{T} \quad \text{(B.7)}
$$

where

$$
\Delta \triangleq \begin{bmatrix} 1 & 0 & 0 & d_x \\ 0 & 1 & 0 & d_y \\ 0 & 0 & 1 & d_z \\ 0 & 0 & 0 & 1 \end{bmatrix} \begin{bmatrix} 1 & -\delta_z & \delta_y & 0 \\ \delta_z & 1 & -\delta_x & 0 \\ -\delta_y & \delta_x & 1 & 0 \\ 0 & 0 & 0 & 1 \end{bmatrix} - \begin{bmatrix} 1 & 0 & 0 & 0 \\ 0 & 1 & 0 & 0 \\ 0 & 0 & 1 & 0 \\ 0 & 0 & 0 & 1 \end{bmatrix} \quad \text{(B.8)}
$$

$\boldsymbol{\delta} = (\delta_x, \delta_y, \delta_z)^T$ is the differential rotation about the principal axes of the base coordinate frame and $\mathbf{d} = (d_x, d_y, d_z)^T$ is the differential translation along the principal axes of the base coordinate frame. Similarly, a differential change in **T** can be expressed to correspond with a differential translation and rotation along and about the coordinate frame **T**:

$$
\mathbf{T} + d\mathbf{T} = \mathbf{T}
\begin{bmatrix}
1 & 0 & 0 & d_x \\
0 & 1 & 0 & d_y \\
0 & 0 & 1 & d_z \\
0 & 0 & 0 & 1
\end{bmatrix}
\begin{bmatrix}
1 & -\delta_z & \delta_y & 0 \\
\delta_z & 1 & -\delta_x & 0 \\
-\delta_y & \delta_x & 1 & 0 \\
0 & 0 & 0 & 1
\end{bmatrix}
\tag{B.9}
$$

or
$$
d\mathbf{T} = \mathbf{T}
\left[
\begin{bmatrix}
1 & 0 & 0 & d_x \\
0 & 1 & 0 & d_y \\
0 & 0 & 1 & d_z \\
0 & 0 & 0 & 1
\end{bmatrix}
\begin{bmatrix}
1 & -\delta_z & \delta_y & 0 \\
\delta_z & 1 & -\delta_x & 0 \\
-\delta_y & \delta_x & 1 & 0 \\
0 & 0 & 0 & 1
\end{bmatrix}
-
\begin{bmatrix}
1 & 0 & 0 & 0 \\
0 & 1 & 0 & 0 \\
0 & 0 & 1 & 0 \\
0 & 0 & 0 & 1
\end{bmatrix}
\right]
$$

$$
\tag{B.10}
$$

$$
\triangleq (\mathbf{T})(^T\mathbf{\Delta})
$$

where $^T\mathbf{\Delta}$ has the same structure as in Eq. (B.8), except that the definitions of δ and **d** are different. $\delta = (\delta_x, \delta_y, \delta_z)^T$ is the differential rotation about the principal axes of the **T** coordinate frame, and $\mathbf{d} = (d_x, d_y, d_z)^T$ is the differential translation along the principal axes of the **T** coordinate frame. From Eqs. (B.7) and (B.10), we obtain a relationship between $\mathbf{\Delta}$ and $^T\mathbf{\Delta}$:

$$
\mathbf{\Delta T} = (\mathbf{T})(^T\mathbf{\Delta})
$$

or
$$
^T\mathbf{\Delta} = \mathbf{T}^{-1}\mathbf{\Delta T} \tag{B.11}
$$

Using Eq. (2.2-27), Eq. (B.11) becomes

$$
^T\mathbf{\Delta} = \mathbf{T}^{-1}\mathbf{\Delta T} =
\begin{bmatrix}
\mathbf{n}\cdot(\delta\times\mathbf{n}) & \mathbf{n}\cdot(\delta\times\mathbf{s}) & \mathbf{n}\cdot(\delta\times\mathbf{a}) & \mathbf{n}\cdot(\delta\times\mathbf{p})+\mathbf{d} \\
\mathbf{s}\cdot(\delta\times\mathbf{n}) & \mathbf{s}\cdot(\delta\times\mathbf{s}) & \mathbf{s}\cdot(\delta\times\mathbf{a}) & \mathbf{s}\cdot(\delta\times\mathbf{p})+\mathbf{d} \\
\mathbf{a}\cdot(\delta\times\mathbf{n}) & \mathbf{a}\cdot(\delta\times\mathbf{s}) & \mathbf{a}\cdot(\delta\times\mathbf{a}) & \mathbf{a}\cdot(\delta\times\mathbf{p})+\mathbf{d} \\
0 & 0 & 0 & 0
\end{bmatrix}
$$

$$
\tag{B.12}
$$

where $\delta = (\delta_x, \delta_y, \delta_z)^T$ is the differential rotation about the principal axes of the base coordinate frame, and $\mathbf{d} = (d_x, d_y, d_z)^T$ is the differential translation along the principal axes of the base coordinate frame. Using the vector identities

$$
\mathbf{x}\cdot(\mathbf{y}\times\mathbf{z}) = -\mathbf{y}\cdot(\mathbf{x}\times\mathbf{z}) = \mathbf{y}\cdot(\mathbf{z}\times\mathbf{x})
$$

and
$$
\mathbf{x}\cdot(\mathbf{x}\times\mathbf{y}) = 0
$$

Eq. (B.12) becomes

$$
^T\Delta = \begin{bmatrix}
0 & -\delta \cdot (\mathbf{n} \times \mathbf{s}) & \delta \cdot (\mathbf{a} \times \mathbf{n}) & \delta \cdot (\mathbf{p} \times \mathbf{n}) + \mathbf{d} \cdot \mathbf{n} \\
\delta \cdot (\mathbf{n} \times \mathbf{s}) & 0 & -\delta \cdot (\mathbf{s} \times \mathbf{a}) & \delta \cdot (\mathbf{p} \times \mathbf{s}) + \mathbf{d} \cdot \mathbf{s} \\
-\delta \cdot (\mathbf{a} \times \mathbf{n}) & \delta \cdot (\mathbf{s} \times \mathbf{a}) & 0 & \delta \cdot (\mathbf{p} \times \mathbf{a}) + \mathbf{d} \cdot \mathbf{a} \\
0 & 0 & 0 & 0
\end{bmatrix}
$$

(B.13)

Since the coordinate axes **n**, **s**, **a** are orthogonal, we have

$$
\mathbf{n} \times \mathbf{s} = \mathbf{a} \qquad \mathbf{s} \times \mathbf{a} = \mathbf{n} \qquad \mathbf{a} \times \mathbf{n} = \mathbf{s}
$$

then Eq. (B.13) becomes

$$
^T\Delta = \begin{bmatrix}
0 & -\delta \cdot \mathbf{a} & \delta \cdot \mathbf{s} & \delta \cdot (\mathbf{p} \times \mathbf{n}) + \mathbf{d} \cdot \mathbf{n} \\
\delta \cdot \mathbf{a} & 0 & -\delta \cdot \mathbf{n} & \delta \cdot (\mathbf{p} \times \mathbf{s}) + \mathbf{d} \cdot \mathbf{s} \\
-\delta \cdot \mathbf{s} & \delta \cdot \mathbf{n} & 0 & \delta \cdot (\mathbf{p} \times \mathbf{a}) + \mathbf{d} \cdot \mathbf{a} \\
0 & 0 & 0 & 0
\end{bmatrix}
$$

(B.14)

If we let the elements of $^T\Delta$ be

$$
^T\Delta = \begin{bmatrix}
0 & -^T\delta_z & ^T\delta_y & ^Td_x \\
^T\delta_z & 0 & -^T\delta_x & ^Td_y \\
-^T\delta_y & ^T\delta_x & 0 & ^Td_z \\
0 & 0 & 0 & 0
\end{bmatrix}
$$

(B.15)

then equating the matrix elements of Eqs. (B.14) and (B.15), we have

$$
^Td_x = \delta \cdot (\mathbf{p} \times \mathbf{n}) + \mathbf{d} \cdot \mathbf{n} = \mathbf{n} \cdot [(\delta \times \mathbf{p}) + \mathbf{d}]
$$

$$
^Td_y = \delta \cdot (\mathbf{p} \times \mathbf{s}) + \mathbf{d} \cdot \mathbf{s} = \mathbf{s} \cdot [(\delta \times \mathbf{p}) + \mathbf{d}]
$$

$$
^Td_z = \delta \cdot (\mathbf{p} \times \mathbf{a}) + \mathbf{d} \cdot \mathbf{a} = \mathbf{a} \cdot [(\delta \times \mathbf{p}) + \mathbf{d}]
$$

(B.16)

$$
^T\delta_x = \delta \cdot \mathbf{n}
$$

$$
^T\delta_y = \delta \cdot \mathbf{s}
$$

$$
^T\delta_z = \delta \cdot \mathbf{a}
$$

Expressing the above equation in matrix form, we have

$$
\begin{bmatrix} {}^{\mathbf{T}}d_x \\ {}^{\mathbf{T}}d_y \\ {}^{\mathbf{T}}d_z \\ {}^{\mathbf{T}}\delta_x \\ {}^{\mathbf{T}}\delta_y \\ {}^{\mathbf{T}}\delta_z \end{bmatrix} = \begin{bmatrix} [\mathbf{n, s, a}]^T & [(\mathbf{p \times n}), (\mathbf{p \times s}), (\mathbf{p \times a})]^T \\ \mathbf{0} & [\mathbf{n, s, a}]^T \end{bmatrix} \begin{bmatrix} d_x \\ d_y \\ d_z \\ \delta_x \\ \delta_y \\ \delta_z \end{bmatrix} \quad \text{(B.17)}
$$

where $\mathbf{0}$ is a 3×3 zero submatrix. Equation (B.17) shows the relation of the differential translation and rotation in the base coordinate frame to the differential translation and rotation with respect to the $\mathbf{T}$ coordinate frame.

Applying Eq. (B.10) to the kinematic equation of a serial six-link manipulator, we have the differential of ${}^0\mathbf{T}_6$:

$$
d\,{}^0\mathbf{T}_6 = {}^0\mathbf{T}_6\,{}^{\mathbf{T}_6}\boldsymbol{\Delta} \quad \text{(B.18)}
$$

In the case of a six-link manipulator, a differential change in joint i motion will induce an equivalent change in ${}^0\mathbf{T}_6\,{}^{\mathbf{T}_6}\boldsymbol{\Delta}$

$$
d\,{}^0\mathbf{T}_6 = {}^0\mathbf{T}_6\,{}^{\mathbf{T}_6}\boldsymbol{\Delta} = {}^0\mathbf{A}_1\,{}^1\mathbf{A}_2 \cdots {}^{i-2}\mathbf{A}_{i-1}\,{}^{i-1}\boldsymbol{\Delta}_i\,{}^{i-1}\mathbf{A}_i \cdots {}^5\mathbf{A}_6 \quad \text{(B.19)}
$$

where ${}^{i-1}\boldsymbol{\Delta}_i$ is defined as the differential change transformation along/about the joint i axis of motion and is defined to be

$$
{}^{i-1}\boldsymbol{\Delta}_i = \begin{cases} \begin{bmatrix} 0 & -d\theta_i & 0 & 0 \\ d\theta_i & 0 & 0 & 0 \\ 0 & 0 & 0 & 0 \\ 0 & 0 & 0 & 0 \end{bmatrix} & \text{if link } i \text{ is rotational} \\[2em] \begin{bmatrix} 0 & 0 & 0 & 0 \\ 0 & 0 & 0 & 0 \\ 0 & 0 & 0 & dd_i \\ 0 & 0 & 0 & 0 \end{bmatrix} & \text{if link } i \text{ is translational} \end{cases} \quad \text{(B.20)}
$$

From Eq. (B.19), we obtain ${}^{\mathbf{T}_6}\boldsymbol{\Delta}$ due to the differential change in joint i motion

$$
{}^{\mathbf{T}_6}\boldsymbol{\Delta} = ({}^{i-1}\mathbf{A}_i\,{}^i\mathbf{A}_{i+1} \cdots {}^5\mathbf{A}_6)^{-1}\,{}^{i-1}\boldsymbol{\Delta}_i({}^{i-1}\mathbf{A}_i\,{}^i\mathbf{A}_{i+1} \cdots {}^5\mathbf{A}_6)
$$

$$
= \mathbf{U}_i^{-1}\,{}^{i-1}\boldsymbol{\Delta}_i\mathbf{U}_i \quad \text{(B.21)}
$$

where $\quad \mathbf{U}_i = {}^{i-1}\mathbf{A}_i\,{}^i\mathbf{A}_{i+1} \cdots {}^5\mathbf{A}_6$

Expressing $\mathbf{U}_i$ in the form of a general 4×4 homogeneous transformation matrix, we have

$$\mathbf{U}_i = \begin{bmatrix} n_x & s_x & a_x & p_x \\ n_y & s_y & a_y & p_y \\ n_z & s_z & a_z & p_z \\ 0 & 0 & 0 & 1 \end{bmatrix} \tag{B.22}$$

Using Eqs. (B.20) and (B.22) for the case of rotary joint i, Eq. (B.21) becomes

$$^{\mathbf{T}_6}\mathbf{\Delta} = \begin{bmatrix} 0 & -a_z & s_z & p_x n_y - p_y n_x \\ a_z & 0 & -n_z & p_x s_y - p_y s_x \\ -s_z & n_z & 0 & p_x a_y - p_y a_x \\ 0 & 0 & 0 & 0 \end{bmatrix} d\theta_i \tag{B.23}$$

For the case of a prismatic joint i, Eq. (B.21) becomes

$$^{\mathbf{T}_6}\mathbf{\Delta} = \begin{bmatrix} 0 & 0 & 0 & n_z \\ 0 & 0 & 0 & s_z \\ 0 & 0 & 0 & a_z \\ 0 & 0 & 0 & 0 \end{bmatrix} dd_i \tag{B.24}$$

From the elements of $^{\mathbf{T}_6}\mathbf{\Delta}$ defined in Eq. (B.15), equating the elements of the matrices in Eq. (B.15) and Eq. (B.23) [or Eq. (B.24)] yields

$$\begin{bmatrix} ^{\mathbf{T}_6}d_x \\ ^{\mathbf{T}_6}d_y \\ ^{\mathbf{T}_6}d_z \\ ^{\mathbf{T}_6}\delta_x \\ ^{\mathbf{T}_6}\delta_y \\ ^{\mathbf{T}_6}\delta_z \end{bmatrix} = \begin{cases} \begin{bmatrix} p_x n_y - p_y n_x \\ p_x s_y - p_y s_x \\ p_x a_y - p_y a_x \\ n_z \\ s_z \\ a_z \end{bmatrix} d\theta_i & \text{if link } i \text{ is rotational} \\[2em] \begin{bmatrix} n_z \\ s_z \\ a_z \\ 0 \\ 0 \\ 0 \end{bmatrix} dd_i & \text{if link } i \text{ is translational} \end{cases} \tag{B.25}$$

Thus, the jacobian of a manipulator can be obtained from Eq. (B.25) for $i = 1, 2, \ldots, 6$:

$$
\begin{bmatrix}
{}^{T_6}d_x \\
{}^{T_6}d_y \\
{}^{T_6}d_z \\
{}^{T_6}\delta_x \\
{}^{T_6}\delta_y \\
{}^{T_6}\delta_z
\end{bmatrix}
= \mathbf{J}(\mathbf{q})
\begin{bmatrix}
dq_1 \\
dq_2 \\
dq_3 \\
dq_4 \\
dq_5 \\
dq_6
\end{bmatrix}
\tag{B.26}
$$

where the columns of the jacobian matrix are obtained from Eq. (B.25). For the PUMA robot manipulator shown in Fig. 2.11 and its link coordinate transformation matrices in Fig. 2.13, the jacobian is found to be:

$$
\mathbf{J}_1(\theta) =
\begin{bmatrix}
J_{1x} \\
J_{1y} \\
J_{1z} \\
-[S_{23}(C_4 C_5 C_6 - S_4 S_6) + C_{23} S_5 C_6] \\
S_{23}(C_4 C_5 S_6 + S_4 C_6) + C_{23} S_5 S_6 \\
-S_{23} C_4 S_5 + C_{23} C_5
\end{bmatrix}
$$

where

$$
\begin{aligned}
J_{1x} = {} & [d_6(C_{23} C_4 S_5 + S_{23} C_5) + d_4 S_{23} + a_3 C_{23} + a_2 C_2](S_4 C_5 C_6 + C_4 S_6) \\
& - (d_6 S_4 S_5 + d_2)[C_{23}(C_4 C_5 C_6 - S_4 S_6) - S_{23} S_5 C_6]
\end{aligned}
$$

$$
\begin{aligned}
J_{1y} = {} & [d_6(C_{23} C_4 S_5 + S_{23} C_5) + d_4 S_{23} + a_3 C_{23} + a_2 C_2](-S_4 C_5 S_6 + C_4 C_6) \\
& - (d_6 S_4 S_5 + d_2)[-C_{23}(C_4 C_5 S_6 + S_4 C_6) + S_{23} S_5 S_6]
\end{aligned}
$$

$$
\begin{aligned}
J_{1z} = {} & [d_6(C_{23} C_4 S_5 + S_{23} C_5) + d_4 S_{23} + a_3 C_{23} + a_2 C_2](S_4 S_5) \\
& - (d_6 S_4 S_5 + d_2)(C_{23} C_4 S_5 + S_{23} C_5)
\end{aligned}
$$

$$
\mathbf{J}_2(\theta) =
\begin{bmatrix}
J_{2x} \\
J_{2y} \\
J_{2z} \\
S_4 C_5 C_6 + C_4 S_6 \\
-S_4 C_5 S_6 + C_4 S_6 \\
S_4 S_5
\end{bmatrix}
$$

where

$$J_{2x} = (d_6 S_3 C_5 + d_6 C_3 C_4 S_5 + d_4 S_3 + a_3 C_3 + a_2)(S_5 C_6)$$
$$- (-d_6 C_3 C_5 + d_6 S_3 C_4 S_5 - d_4 C_3 + a_3 S_3)(C_4 C_5 C_6 - S_4 S_6)$$

$$J_{2y} = -(d_6 S_3 C_5 + d_6 C_3 C_4 S_5 + d_4 S_3 + a_3 C_3 + a_2)(S_5 S_6)$$
$$+ (-d_6 C_3 C_5 + d_6 S_3 C_4 S_5 - d_4 C_3 + a_3 S_3)(C_4 C_5 S_6 + S_4 C_6)$$

$$J_{2z} = -(d_6 S_3 C_5 + d_6 C_3 C_4 S_5 + d_4 S_3 + a_3 C_3 + a_2)C_5$$
$$- (-d_6 C_3 C_5 + d_6 S_3 C_4 S_5 - d_4 C_3 + a_3 S_3)(C_4 S_5)$$

$$\mathbf{J}_3(\boldsymbol{\theta}) = \begin{bmatrix} (a_3 + d_6 C_4 S_5)(S_5 C_6) + (d_4 + d_6 C_5)(C_4 C_5 C_6 - S_4 S_6) \\ -(a_3 + d_6 C_4 S_5)(S_5 S_6) - (d_4 + d_6 C_5)(C_4 C_5 S_6 + S_4 C_6) \\ -(a_3 + d_6 C_4 S_5)C_5 + (d_4 + d_6 C_5)C_4 S_5 \\ S_4 C_5 C_6 + C_4 S_6 \\ -S_4 C_5 S_6 + C_4 S_6 \\ S_4 S_5 \end{bmatrix}$$

$$\mathbf{J}_4(\boldsymbol{\theta}) = \begin{bmatrix} d_6 S_5 S_6 \\ d_6 S_5 C_6 \\ 0 \\ -S_5 C_6 \\ S_5 S_6 \\ C_5 \end{bmatrix}$$

$$\mathbf{J}_5(\boldsymbol{\theta}) = \begin{bmatrix} d_6 C_6 \\ -d_6 S_6 \\ 0 \\ S_6 \\ C_6 \\ 0 \end{bmatrix}$$

$$\mathbf{J}_6(\boldsymbol{\theta}) = \begin{bmatrix} 0 \\ 0 \\ 0 \\ 0 \\ 0 \\ 1 \end{bmatrix}$$

B.3 STROBING FROM THE NEWTON-EULER EQUATIONS OF MOTION

The above two methods derive the jacobian in symbolic form. It is possible to numerically obtain the elements of the jacobian at time t explicitly from the Newton-Euler equations of motion. This is based on the observation that the ratios of infinitesimal hand accelerations to infinitesimal joint accelerations are the elements of the jacobian if the nonlinear components of the accelerations are deleted from the Newton-Euler equations of motion. From Eq. (B.2), the accelerations of the hand can be obtained by taking the time derivative of the velocity vector:

$$\begin{bmatrix} \dot{\mathbf{v}}(t) \\ \dot{\boldsymbol{\Omega}}(t) \end{bmatrix} = \mathbf{J}(\mathbf{q})\ddot{\mathbf{q}}(t) + \dot{\mathbf{J}}(\mathbf{q},\ \dot{\mathbf{q}})\dot{\mathbf{q}}(t) \tag{B.27}$$

where $\ddot{\mathbf{q}}(t) = [\ddot{q}_1(t),\ \ldots,\ \ddot{q}_6(t)]^T$ is the joint acceleration vector of the manipulator. The first term of Eq. (B.27) gives the linear relation between the hand and joint accelerations. The second term gives the nonlinear components of the accelerations and it is a function of joint velocity. Thus, a linear relation between the hand accelerations and the joint accelerations can be established from the Newton-Euler equations of motion, as indicated by the first term in Eq. (B.27). From Table 3.3 we have the following recursive kinematics (here, we shall only consider manipulators with rotary joints):

$$^i\mathbf{R}_0\boldsymbol{\omega}_i = {}^i\mathbf{R}_{i-1}({}^{i-1}\mathbf{R}_0\boldsymbol{\omega}_{i-1} + \mathbf{z}_0\dot{q}_i) \tag{B.28}$$

$$^i\mathbf{R}_0\dot{\boldsymbol{\omega}}_i = {}^i\mathbf{R}_{i-1}[{}^{i-1}\mathbf{R}_0\dot{\boldsymbol{\omega}}_{i-1} + \mathbf{z}_0\ddot{q}_i + ({}^{i-1}\mathbf{R}_0\boldsymbol{\omega}_{i-1}) \times \mathbf{z}_0\dot{q}_i] \tag{B.29}$$

$$^i\mathbf{R}_0\dot{\mathbf{v}}_i = ({}^i\mathbf{R}_0\dot{\boldsymbol{\omega}}_i) \times ({}^i\mathbf{R}_0\mathbf{p}_i^*) + ({}^i\mathbf{R}_0\boldsymbol{\omega}_i)$$
$$\times\ [({}^i\mathbf{R}_0\boldsymbol{\omega}_i) \times ({}^i\mathbf{R}_0\mathbf{p}_i^*)] + {}^i\mathbf{R}_{i-1}({}^{i-1}\mathbf{R}_0\dot{\mathbf{v}}_{i-1}) \tag{B.30}$$

The terms in Eqs. (B.29) and (B.30) involving ω_i represent nonlinear Coriolis and centrifugal accelerations as indicated by the third term in Eq. (B.29) and the second term in Eq. (B.30). Omitting these terms in Eqs. (B.29) and (B.30) give us the linear relation between the hand accelerations and the joint accelerations. Then if we successively apply an input unit joint acceleration vector $(\ddot{q}_1,\ \ddot{q}_2,\ \ldots,\ \ddot{q}_6)^T = (1, 0, 0, \ldots, 0)^T$, $(\ddot{q}_1,\ \ddot{q}_2,\ \ldots,\ \ddot{q}_6)^T = (0, 1, 0, \ldots, 0)^T$, $(\ddot{q}_1,\ \ddot{q}_2,\ \ldots,\ \ddot{q}_6)^T = (0, 0, 0, \ldots, 1)^T$, etc., the columns of the jacobian matrix can be "strobed" out because the first term in Eq. (B.27) is linear and the second (nonlinear) term is neglected. This numerical technique takes about $24n(n + 1)/2$ multiplications and $19n(n + 1)/2$ additions, where n is the number of degrees of freedom. In addition, we need $18n$ multiplications and $12n$ additions to convert the hand accelerations from referencing its own link coordinate frame to referencing the hand coordinate frame.

Although these three methods are "equivalent" for finding the jacobian, this "strobing" technique is well suited for a controller utilizing the Newton-Euler equations of motion. Since parallel computation schemes have been discussed and developed for computing the joint torques from the Newton-Euler equations of motion (Lee and Chang [1986b]), the jacobian can be computed from these schemes as a by-product. However, the method suffers from the fact that it only gives the numerical values of the jacobian and not its analytic form.

REFERENCES

Further reading for the material in this appendix may be found in Whitney [1972], Paul [1981], and Orin and Schrader [1984].

BIBLIOGRAPHY

Aggarwal, J. K., and Badler, N. I. (eds.) [1980]. "Motion and Time Varying Imagery," Special Issue, *IEEE Trans. Pattern Anal. Machine Intelligence*, vol. PAMI-2, no. 6, pp. 493–588.

Agin, G. J. [1972]. "Representation and Description of Curved Objects," Memo AIM-173, Artificial Intelligence Laboratory, Stanford University, Palo Alto, Calif.

Albus, J. S. [1975]. "A New Approach to Manipulator Control: The Cerebellar Model Articulation Controller," *Trans. ASME, J. Dynamic Systems, Measurement and Control*, pp. 220–227.

Ambler, A. P., et al. [1975]. "A Versatile System for Computer Controlled Assembly," *Artificial Intelligence*, vol. 6, no. 2, pp. 129–156.

Ambler, A. P., and Popplestone, R. J. [1975]. "Inferring the Positions of Bodies from Specified Spatial Relationships," *Artificial Intelligence*, vol. 6, no. 2, pp. 157–174.

Armstrong, W. M. [1979]. "Recursive Solution to the Equations of Motion of an N-link Manipulator," *Proc. 5th World Congr., Theory of Machines, Mechanisms*, vol. 2, pp. 1343–1346.

Astrom, K. J. and Eykhoff, P. [1971]. "System Identification—A Survey," *Automatica*, vol. 7, pp. 123–162.

Baer, A., Eastman, C., and Henrion, M. [1979]. "Geometric Modelling: A Survey," *Computer Aided Design*, vol. 11, no. 5, pp. 253–272.

Bajcsy, R., and Lieberman, L. [1976]. "Texture Gradient as a Depth Cue," *Comput. Graph. Image Proc.*, vol. 5, no. 1, pp. 52–67.

Ballard, D. H. [1981]. "Generalizing the Hough Transform to Detect Arbitrary Shapes," *Pattern Recog.*, vol. 13, no. 2, pp. 111–122.

Ballard, D. H., and Brown, C. M. [1982]. *Computer Vision*, Prentice-Hall, Englewood Cliffs, N.J.

Barnard, S. T., and Fischler, M. A. [1982]. "Computational Stereo," *Computing Surveys*, vol. 14, no. 4, pp. 553–572.

Barr, A., Cohen, P., and Feigenbaum, E. A. [1981–82]. *The Handbook of Artificial Intelligence*, vols. 1, 2, and 3, William Kaufmann, Inc., Los Altos, Calif.

Barrow, H. G., and Tenenbaum, J. M. [1977]. "Experiments in Model Driven Scene Segmentation," *Artificial Intelligence*, vol. 8, no. 3, pp. 241–274.

Barrow, H. G., and Tenenbaum, J. M. [1981]. "Interpreting Line Drawings as Three-Dimensional Surfaces," *Artificial Intelligence*, vol. 17, pp. 76–116.

Bejczy, A. K. [1974]. "Robot Arm Dynamics and Control," Technical Memo 33-669, Jet Propulsion Laboratory, Pasadena, Calif.

Bejczy, A. K. [1979]. "Dynamic Models and Control Equations for Manipulators," Technical Memo 715-19, Jet Propulsion Laboratory, Pasadena, Calif.

Bejczy, A. K. [1980]. "Sensors, Controls, and Man-Machine Interface for Advanced Teleoperation," *Science*, vol. 208, pp. 1327–1335.

Bejczy, A. K., and Lee, S. [1983]. "Robot Arm Dynamic Model Reduction for Control," *Proc. 22nd IEEE Conf. on Decision and Control*, San Antonio, Tex., pp. 1466–1476.

Bejczy, A. K., and Paul, R. P. [1981]. "Simplified Robot Arm Dynamics for Control," *Proc. 20th IEEE Conf. Decision and Control*, San Diego, Calif., pp. 261–262.

Bellman, R. [1970]. *Introduction to Matrix Analysis*, 2d edition, McGraw-Hill, New York.

Beni, G., et al. [1983]. "Dynamic Sensing for Robots: An Analysis and Implementation," *Intl. J. Robotics Res.*, vol. 2, no. 2, pp. 51–61.

Binford, T. O. [1979]. "The AL Language for Intelligent Robots," in *Proc. IRIA Sem. Languages and Methods of Programming Industrial Robots* (Rocquencourt, France), pp. 73–87.

Blum, H. [1967]. "A Transformation for Extracting New Descriptors of Shape," in *Models for the Perception of Speech and Visual Form* (W. Wathen-Dunn, ed.), MIT Press, Cambridge, Mass.

Bobrow, J. E., Dubowsky, S. and Gibson, J. S. [1983]. "On the Optimal Control of Robot Manipulators with Actuator Constraints," *Proc. 1983 American Control Conf.*, San Francisco, Calif., pp. 782–787.

Bolles, R., and Paul, R. [1973]. "An Experimental System for Computer Controlled Mechanical Assembly," Stanford Artificial Intelligence Laboratory Memo AIM-220, Stanford University, Palo Alto, Calif.

Bonner, S., and Shin, K. G. [1982]. "A Comparative Study of Robot Languages," *IEEE Computer*, vol. 15, no. 12, pp. 82–96.

Brady, J. M. (ed.) [1981]. *Computer Vision*, North-Holland Publishing Co., Amsterdam.

Brady, J. M., et al. (eds.) [1982]. *Robot Motion: Planning and Control*, MIT Press, Cambridge, Mass.

Bribiesca, E. [1981]. "Arithmetic Operations Among Shapes Using Shape Numbers," *Pattern Recog.*, vol. 13, no. 2, pp. 123–138.

Bribiesca, E., and Guzman, A. [1980]. "How to Describe Pure Form and How to Measure Differences in Shape Using Shape Numbers," *Pattern Recog.*, vol. 12, no. 2, pp. 101–112.

Brice, C., and Fennema, C. [1970]. "Scene Analysis Using Regions," *Artificial Intelligence*, vol. 1, no. 3, pp. 205–226.

Brooks, R. A., [1981]. "Symbolic Reasoning Among 3-D Models and 2-D Images," *Artificial Intelligence*, vol. 17, pp. 285–348.

Brooks, R. A. [1983*a*]. "Solving the Find-Path Problem by Good Representation of Free Space," *IEEE Trans. Systems, Man, Cybern.*, vol. SMC-13, pp. 190–197.

Brooks, R. A. [1983*b*]. "Planning Collision-Free Motion for Pick-and-Place Operations," *Intl. J. Robotics Res.*, vol. 2, no. 4, pp. 19–44,

Brooks, R. A., and Lozano-Perez, T. [1983]. "A Subdivision Algorithm in Configuration Space for Find-Path with Rotation," *Proc. Intl. Joint Conf. Artificial Intelligence* (Karlsuhe, W. Germany), pp. 799–808.

Bryson A. E. and Ho, Y. C. [1975]. *Applied Optimal Control*, John Wiley, New York.

Canali, C., et al. [1981*a*]. "Sensori di Prossimita Elettronici," *Fisica e Tecnologia*, vol. 4, no. 2, pp. 95–123 (in Italian).

Canali, C., et al. [1981*b*]. "An Ultrasonic Proximity Sensor Operating in Air," *Sensors and Actuators*, vol. 2, no. 1, pp. 97–103.

Catros, J. Y., and Espiau, B. [1980]. "Use of Optical Proximity Sensors in Robotics," *Nouvel Automatisme*, vol. 25, no. 14, pp. 47–53 (in French).

Chase, M. A. [1963]. "Vector Analysis of Linkages," *Trans. ASME, J. Engr. Industry*, Series B, vol. 85, pp 289–297.

Chase, M. A., and Bayazitoglu, Y. O. [1971]. "Development and Application of a Generalized d'Alembert Force for Multifreedom Mechanical Systems," *Trans. ASME, J. Engr. Industry*, Series B, vol. 93, pp. 317–327.

Chang, C. L., and Lee, R. C. T. [1973]. *Symbolic Logic and Mechanical Theorem Proving*, Academic Press, New York.

Chaudhuri, B. B. [1983]. "A Note on Fast Algorithms for Spatial Domain Techniques in Image Processing," *IEEE Trans. Systems, Man, Cybern.*, vol. SMC-13, no. 6, pp. 1166–1169.

Chow, C. K., and Kaneko, T. [1972]. "Automatic Boundary Detection of the Left Ventricle from Cineangiograms," *Comput. and Biomed. Res.*, vol. 5, pp. 388–410.

Chung, M. J. [1983]. "Adaptive Control Strategies for Computer-Controlled Manipulators," Ph.D. Dissertation, The Computer, Information, and Control Engineering Program, University of Michigan, Ann Arbor, Mich.

Cowart, A. E., Snyder, W. E., and Ruedger, W. H. [1983]. "The Detection of Unresolved Targets Using the Hough Transform," *Comput. Vision, Graphics, and Image Proc.*, vol. 21, pp. 222–238.

Craig, J. J. [1980]. *JARS: JPL Autonomous Robot System*, Robotics and Teleoperators Group, Jet Propulsion Laboratory, Pasadena, Calif.

Craig, J. J. [1986]. *Introduction to Robotics: Mechanics and Control*, Addison-Wesley, Reading, Mass.

Crandall, S. H., Karnopp, D. C., Kurtz, E. F., Jr., and Pridmore-Brown, D. C. [1968]. *Dynamics of Mechanical and Electromechanical Systems*, McGraw-Hill, New York.

Cross, G. R., and Jain, A. K. [1983]. "Markov Random Field Texture Models," *IEEE Trans. Pattern Anal. Mach. Intell.*, vol. PAMI-5, no. 1, pp. 25–39.

Darringer, J. A., and Blasgen, M. W. [1975]. "MAPLE: A High Level Language for Research," in *Mechanical Assembly*, IBM Research Report RC 5606, IBM T. J. Watson Research Center, Yorktown Heights, N.Y.

Davies, E. R., and Plummer, A. P. N. [1981]. "Thinning Algorithms: A Critique and a New Methodology," *Pattern Recog.*, vol. 14, pp. 53–63.

Davis, L. S. [1975]. "A Survey of Edge Detection Techniques," *Comput. Graphics Image Proc.*, vol. 4, pp. 248–270.

Davis, R. H., and Comacho, M. [1984]. "The Application of Logic Programming to the Generation of Plans for Robots," *Robotica*, vol. 2, pp. 137–146.

Denavit, J. [1956]. "Description and Displacement Analysis of Mechanisms Based on 2×2 Dual Matrices," Ph.D. Thesis, Mechanical Engineering, Northwestern U., Evanston, Ill.

Denavit, J., and Hartenberg, R. S. [1955]. "A Kinematic Notation for Lower-Pair Mechanisms Based on Matrices," *J. App. Mech.*, vol. 77, pp. 215–221.

Derksen, J., Rulifson, J. F., and Waldinger, R. J. [1972]. "The QA4 Language Applied to Robot Planning," Tech. Note 65, Stanford Research Institute, Menlo Park, Calif.

Dijkstra, E. [1959]. "A Note on Two Problems in Connection with Graphs," *Numerische Mathematik*, vol. 1, pp. 269–271.

Dodd, G. G., and Rossol, L. (eds.) [1979]. *Computer Vision and Sensor-Based Robots*, Plenum, New York.

Doran, J. E. [1970]. "Planning and Robots," in *Machine Intelligence*, vol. 5 (B. Meltzer and D. Michie, eds.), American Elsevier, New York, pp. 519–532.

Dorf, R. C. [1983]. *Robotics and Automated Manufacturing*, Reston Publishing Co., Reston, Va.

Drake, S. H. [1977]. "Using Compliance in Lieu of Sensory Feedback for Automatic Assembly," Report T-657, C. S. Draper Laboratory, Cambridge, Mass.

Dubowsky, S., and DesForges, D. T. [1979]. "The Application of Model Referenced Adaptive Control to Robotic Manipulators," *Trans. ASME, J. Dynamic Systems, Measurement and Control*, vol. 101, pp. 193–200.

Duda, R. O., and Hart, P. E. [1972]. "Use of the Hough Transformation to Detect Lines and Curves in Pictures," *Comm. ACM*, vol. 15, no. 1, pp. 11–15.

Duda, R. O., and Hart, P. E. [1973]. *Pattern Classification and Scene Analysis*, John Wiley, New York.

Duda, R. O., Nitzan, D., and Barrett, P. [1979]. "Use of Range and Reflectance Data to Find Planar Surface Regions," *IEEE Trans. Pattern Anal. Machine Intell.*, vol. PAMI-1, no. 3, pp. 259–271.

Duffy, J. [1980]. *Analysis of Mechanisms and Robot Manipulators*, John Wiley, New York.

Duffy, J., and Rooney, J. [1975]. "A Foundation for a Unified Theory of Analysis of Spatial Mechanisms," *Trans. ASME, J. Engr. Industry*, vol. 97, no. 4, Series B, pp. 1159–1164.

Dyer, C. R., and Rosenfeld, A. [1979]. "Thinning Algorithms for Grayscale Pictures," *IEEE Trans. Pattern Anal. Machine Intelligence*, vol. PAMI-1, no. 1, pp. 88–89.

Engelberger, J. F. [1980]. *Robotics in Practice*, AMACOM, New York.

Ernst, H. A. [1962]. "MH-1, A Computer-Oriented Mechanical Hand," *Proc. 1962 Spring Joint Computer Conf.*, San Francisco, Calif., pp. 39–51.

Fahlman, S. E. [1974]. "A Planning System for Robot Construction Tasks," *Artificial Intelligence*, vol. 5, no. 1, pp. 1–49.

Fairchild [1983]. *CCD Imaging Catalog*, Fairchild Corp., Palo Alto, Calif.

Falb, P. L., and Wolovich, W. A. [1967]. "Decoupling in the Design and Synthesis of Multivariable Control Systems," *IEEE Trans. Automatic Control*, vol. 12, no. 6, pp. 651–655.

Featherstone, R. [1983]. "The Calculation of Robot Dynamics Using Articulated-Body Inertia," *Intl. J. Robotics Res.*, vol. 2, no. 1, pp. 13–30.

Fikes, R. E., Hart, P. E., and Nilsson, N. J. [1972]. "Learning and Executing Generalized Robot Plans," *Artificial Intelligence*, vol. 3, no. 4, pp. 251–288.

Fikes, R. E., and Nilsson, N. J. [1971]. "STRIPS: A New Approach to the Application of Theorem Proving to Problem Solving," *Artificial Intelligence*, vol. 2, no. 3/4, pp. 189–208.

Fink, D. G. (ed.) [1957]. *Television Engineering Handbook*, McGraw-Hill, New York.

Finkel, R., et al., [1975]. "An Overview of AL, a Programming Language for Automation," *Proc. 4th Intl. Joint Conf. Artificial Intelligence*, pp. 758–765.

Franklin, J. W. and Vanderbrug, G. J. [1982]. "Programming Vision and Robotics Systems with RAIL," *SME Robots VI*, pp. 392–406.

Frazer, R. A., Duncan, W. T., and Collan, A. R. [1960]. *Elementary Matrices,* Cambridge University Press, Cambridge, England.

Freeman, H. [1961]. "On the Encoding of Arbitrary Geometric Configurations," *IEEE Trans. Elec. Computers*, vol. EC-10, pp. 260–268.

Freeman, H. [1974]. "Computer Processing of Line Drawings," *Comput. Surveys*, vol. 6, pp. 57–97.

Freeman, H., and Shapira, R. [1975]. "Determining the Minimum-Area Encasing Rectangle for an Arbitrary Closed Curve," *Comm. ACM,* vol. 18, no. 7, pp. 409–413.

Freund, E. [1982]. "Fast Nonlinear Control with Arbitrary Pole Placement for Industrial Robots and Manipulators," *Intl. J. Robotics Res.,* vol. 1, no. 1, pp. 65–78.

Fu, K. S. [1971]. "Learning Control Systems and Intelligent Control Systems: An Intersection of Artificial Intelligence and Automatic Control," *IEEE Trans. Automatic Control*, vol. AC-16, no. 2, pp. 70–72.

Fu, K. S. [1982a]. *Syntactic Pattern Recognition and Applications*, Prentice-Hall, Englewood Cliffs, N.J.

Fu, K. S. (ed.) [1982b]. Special Issue of *Computer* on Robotics and Automation, vol. 15, no. 12.

Fu, K. S., and Mui, J. K. [1981]. "A Survey of Image Segmentation," *Pattern Recog.,* vol. 13, no. 1, pp. 3–16.

Galey, B., and Hsia, P. [1980]. "A Survey of Robotics Sensor Technology," *Proc. 12th Annual Southeastern Symp. System Theory*, pp. 90–93.

Gantmacher, F. R. [1959]. *The Theory of Matrices,* 2 vols., Chelsea, New York.

Geschke, C. C. [1983]. "A System for Programming and Controlling Sensor-Based Robot Manipulators," *IEEE Trans. Pattern Anal. Machine Intell.,* vol. PAMI-5, no. 1, pp. 1–7.

Gilbert, E. G., and Ha, I. J. [1984]. "An Approach to Nonlinear Feedback Control with Application to Robotics," *IEEE Trans. Systems, Man, Cybern.,* vol. SMC-14, no. 2, pp. 101–109.

Gips, J. [1974]. "A Syntax-Directed Program that Performs a Three-Dimensional Perceptual Task," *Pattern Recog.,* vol. 6, pp. 189–200.

Goldstein, H. [1950]. *Classical Mechanics,* Addison-Wesley, Reading, Mass.

Gonzalez, R. C. [1983]. "How Vision Systems See," *Machine Design,* vol. 55, no. 10, pp. 91–96.

Gonzalez, R. C. [1985a]. "Industrial Computer Vision," in *Computer-Based Automation* (J. T. Tou, ed.), Plenum, New York, pp. 345–385.

Gonzalez, R. C. [1985b]. "Computer Vision," *McGraw-Hill Yearbook of Science and Technology*, McGraw-Hill, New York.

Gonzalez, R. C. [1986]. "Digital Image Enhacement and Restoration," in *Handbook of Pattern Recognition and Image Processing,* (T. Young and K.S. Fu, eds.), Academic Press, New York, pp. 191–213.

Gonzalez, R. C., and Fittes, B. A. [1977]. "Gray-Level Transformations for Interactive Image Enhancement," *Mechanism and Machine Theory*, vol. 12, pp. 111–122.

Gonzalez, R. C., and Safabakhsh, R. [1982]. "Computer Vision Techniques for Industrial Applications and Robot Control," *Computer*, vol. 15, no. 12, pp. 17–32.

Gonzalez, R. C., and Thomason, M. G. [1978]. *Syntactic Pattern Recognition: An Introduction*, Addison-Wesley, Reading, Mass.

Gonzalez, R. C., and Wintz, P. [1977]. *Digital Image Processing*, Addison-Wesley, Reading, Mass.

Goodman, J. W. [1968]. *Introduction to Fourier Optics*, McGraw-Hill, New York.

Green, C. [1969]. "Application of Theorem Proving to Problem Solving," *Proc. 1st Intl. Joint Conf. Artificial Intelligence*, Washington, D.C.

Grossman, D. D., and Taylor, R. H. [1978]. "Interactive Generation of Object Models with a Manipulator," *IEEE Trans. Systems, Man, Cybern.*, vol. SMC-8, no. 9, pp. 667-679.

Gruver, W. A., et al., [1984]. "Industrial Robot Programming Languages: A Comparative Evaluation," *IEEE Trans. Systems, Man, Cybern.*, vol. SMC-14, no. 4, pp. 321-333.

Guzman, A. [1969]. "Decomposition of a Visual Scene into Three-Dimensional Bodies," in *Automatic Interpretation and Classification of Images* (A. Grasseli, ed.), Academic Press, New York.

Hackwood, S., et al. [1983]. "A Torque-Sensitive Tactile Array for Robotics," *Intl.. J. Robotics Res.*, vol. 2, no. 2, pp. 46-50.

Haralick, R. M. [1979]. "Statistical and Structural Approaches to Texture," *Proc. 4th Intl. Joint Conf. Pattern Recog.*, pp. 45-60.

Haralick, R. M., Shanmugan, R., and Dinstein, I. [1973]. "Textural Features for Image Classification," *IEEE Trans. Systems, Man, Cybern.*, vol. SMC-3, no. 6, pp. 610-621.

Harmon, L. D. [1982]. "Automated Tactile Sensing," *Intl, J. Robotics Res.*, vol. 1, no. 2, pp. 3-32.

Harris, J. L. [1977]. "Constant Variance Enhancement—A Digital Processing Technique," *Appl. Optics*, vol. 16, pp. 1268-1271.

Hart, P. E., Nilsson, N. J., and Raphael, B. [1968]. "A Formal Basis for the Heuristic Determination of Minimum-Cost Paths," *IEEE Trans. Systems, Man, Cybern.*, vol. SMC-4, pp. 100-107.

Hartenberg, R. S., and Denavit, J. [1964]. *Kinematic Synthesis of Linkages*, McGraw-Hill, New York.

Hayer-Roth, R., Waterman, D., and Lenat, D. (eds.) [1983]. *Building Expert Systems*, Addison-Wesley, Reading, Mass.

Hemami, H. and Camana, P. C. [1976]. "Nonlinear Feedback in Simple Locomotion Systems," *IEEE Trans. Automatic Control*, vol. AC-19, pp. 855-860.

Herrick, C. N. [1976]. *Television Theory and Servicing*, 2d ed., Reston Publishers, Reston, Va.

Hillis, D. W. [1982]. "A High-Resolution Imaging Touch Sensor," *Intl. J. Robotics Res.*, vol. 1, no. 2, pp. 33-44.

Holland, S. W., Rossol, L., and Ward, M. R. [1979]. "CONSIGHT-I: A Vision-Controlled Robot System for Transferring Parts from Belt Conveyors," in *Computer Vision and Sensor-Based Robots* (G. G. Dodd and L. Rossol, eds.), Plenum, New York.

Hollerbach, J. M. [1980]. "A Recursive Lagrangian Formulation of Manipulator Dynamics and a Comparative Study of Dynamics Formulation Complexity," *IEEE Trans. Systems, Man, Cybern.*, vol. SMC-10, no. 11, pp. 730-736.

Hollerbach, J. M. [1984]. "Dynamic Scaling of Manipulator Trajectories," *Trans. ASME J. Dyn. Systems, Measurement and Control*, vol. 106, pp. 102-106.

Horn, B. K. P. [1977]. "Understanding Image Intensities," *Artificial Intelligence*, vol. 8, pp. 201-231.

Horowitz, S. L., and Pavlidis, T. [1974]. "Picture Segmentation by a Directed Split-and-Merge Procedure," *Proc. 2d Intl. Joint Conf. Pattern Recog.*, pp. 424-433.

Horowitz, R., and Tomizuka, R. [1980]. "An Adaptive Control Scheme for Mechanical Manipulators—Compensation of Nonlinearity and Decoupling Control," *Trans. ASME J. Dynamic Systems, Measurement, and Control*, to appear June 1986.

Hough, P. V. C. [1962]. "Methods and Means for Recognizing Complex Patterns," U. S. Patent 3,069,654.

Hu, M. K. [1962]. "Visual Pattern Recognition by Moment Invariants," *IEEE Trans. Inform. Theory*, vol. 8, pp. 179–187.

Huang, T. S., Yang, G. T., and Tang, G. Y. [1979]. "A Fast Two-Dimensional Median Filtering Algorithm," *IEEE Trans. Acoust., Speech, Signal Proc.*, vol. ASSP-27, pp. 13–18.

Huston, R. L., and Kelly, F. A. [1982]. "The Development of Equations of Motion of Single-Arm Robots," *IEEE Trans. Systems, Man, Cybern.*, vol. SMC-12, no. 3, pp. 259–266.

Huston, R. L., Passerello, C. E., and Harlow, M. W. [1978]. "Dynamics of Multirigid-Body Systems," *J. Appl. Mech.*, vol. 45, pp. 889–894.

Inoue, H. [1974]. "Force Feedback in Precise Assembly Tasks," *MIT Artificial Intelligence Laboratory Memo 308*, MIT, Cambridge, Mass.

Ishizuka, M., Fu, K. S., and Yao, J. T. P. [1983]. "A Rule-Based Damage Assessment System for Existing Structures," *SM Archives*, vol. 8, pp. 99–118.

Itkis, U. [1976]. *Control Systems of Variable Structure*, John Wiley, New York.

Jain, R. [1981]. "Dynamic Scene Analysis Using Pixel-Based Processes," *Computer*, vol. 14, no. 8, pp. 12–18.

Jain, R. [1983]. "Segmentation of Frame Sequences Obtained by a Moving Observer." Report GMR-4247, General Motors Research Laboratories, Warren, Mich.

Jarvis, R. A. [1983a]. "A Perspective on Range Finding Techniques for Computer Vision," *IEEE Trans. Pattern Anal. Machine Intell.*, vol. PAMI-5, no. 2, pp. 122–139.

Jarvis, R. A. [1983b]. "A Laser Time-of-Flight Range Scanner for Robotic Vision," *IEEE Trans. Pattern Anal. Machine Intell.*, vol. PAMI-5, no. 5, pp. 505–512.

Johnston, A. R. [1977]. "Proximity Sensor Technology for Manipulator End Effectors," *Mechanism and Machine Theory*, vol. 12, no. 1, pp. 95–108.

Kahn, M. E., and Roth, B. [1971]. "The Near-Minimum-Time Control of Open-Loop Articulated Kinematic Chains," *Trans. ASME, J. Dynamic Systems, Measurement and Control*, vol. 93, pp. 164–172.

Kane, T. R., and Levinson, D. A. [1983]. "The Use of Kane's Dynamical Equations in Robotics," *Intl. J. Robotics Res.*, vol. 2, no. 3, pp. 3–21.

Katushi, I., and Horn, B. K. P. [1981]. "Numerical Shape from Shading and Occluding Boundaries," *Artificial Intelligence*, vol. 17, pp. 141–184.

Ketcham, D. J. [1976]. "Real-Time Image Enhancement Techniques," *Proc. Soc. Photo-Optical Instrum. Engr.*, vol. 74, pp. 120–125.

Khatib, O. [1980]. "Commande Dynamique dans l'Espace Operationnel des Robots Manipulateurs en Presence d'Obstacles," Docteur Ingenieur Thesis, L'Ecole Nationale Superieure de l'Aeronautique et de l'Espace, Toulouse, France.

Kirk, D. E. [1970]. *Optimal Control Theory, An Introduction*, Prentice-Hall, Englewood Cliffs, N.J.

Klinger, A. [1972]. "Patterns and Search Statistics," in *Optimizing Methods in Statistics* (J. S. Rustagi, ed.), Academic Press, New York, pp. 303–339.

Klinger, A. [1976]. "Experiments in Picture Representation Using Regular Decomposition," *Comput. Graphics Image Proc.*, vol. 5, pp. 68–105.

Kohler, R. J., and Howell, H. K. [1963]. "Photographic Image Enhancement by Superposition of Multiple Images," *Phot. Sci. Engr.*, vol. 7, no. 4, pp. 241–245.

Kohli, D., and Soni, A. H. [1975]. "Kinematic Analysis of Spatial Mechanisms via Successive Screw Displacements," *J. Engr. for Industry, Trans. ASME*, vol. 2, series B, pp. 739–747.

Koivo, A. J., and Guo, T. H. [1983]. "Adaptive Linear Controller for Robotic Manipulators," *IEEE Trans. Automatic Control,* vol. AC-28, no. 1, pp. 162–171.

Landau, Y. D. [1979]. *Adaptive Control—The Model Reference Approach,* Marcel Dekker, New York.

Lee, B. H. [1985]. "An Approach to Motion Planning and Motion Control of Two Robots in a Common Workspace," Ph.D. Dissertation, Computer Information and Control Engineering Program, University of Michigan, Ann Arbor, Mich.

Lee, C. C. [1983]. "Elimination of Redundant Operations for a Fast Sobel Operator," *IEEE Trans. Systems, Man, Cybern.,* vol. SMC-13, no. 3, pp. 242–245.

Lee, C. S. G. [1982]. "Robot Arm Kinematics, Dynamics, and Control," *Computer,* vol. 15, no. 12, pp. 62–80.

Lee, C. S. G. [1983]. "On the Control of Robot Manipulators," *Proc. 27th Soc. Photo-optical Instrumentation Engineers,* vol. 442, San Diego, Calif., pp. 58–83.

Lee, C. S. G. [1985]. "Robot Arm Kinematics and Dynamics," in *Advances in Automation and Robotics: Theory and Applications* (G. N. Saridis, ed.), JAI Press, Conn., pp. 21–63.

Lee, C. S. G., and Chang, P. R. [1986a]. "A Maximum Piplined CORDIC Architecture for Robot Inverse Kinematics Computation," Technical Report TR-EE-86-5, School of Electrical Engineering, Purdue University, West Lafayette, Ind.

Lee, C. S. G., and Chang, P. R. [1986b]. "Efficient Parallel Algorithm for Robot Inverse Dynamics Computation," *IEEE Trans. Systems, Man, Cybern.,* vol. SMC-16, no. 4.

Lee, C. S. G., and Chung, M. J. [1984]. "An Adaptive Control Strategy for Mechanical Manipulators," *IEEE Trans. Automatic Control,* vol. AC-29, no. 9, pp. 837–840.

Lee, C. S. G., and Chung, M. J. [1985]. "Adaptive Perturbation Control with Feedforward Compensation for Robot Manipulators," *Simulation,* vol. 44, no. 3, pp. 127–136.

Lee, C. S. G., Chung, M. J., and Lee, B. H. [1984]. "An Approach of Adaptive Control for Robot Manipulators," *J. Robotic Systems,* vol. 1, no. 1, pp. 27–57.

Lee, C. S. G., Chung, M. J., Mudge, T. N., and Turney, J. L. [1982]. "On the Control of Mechanical Manipulators," *Proc. 6th IFAC Conf. Estimation and Parameter Identification,* Washington D.C., pp. 1454–1459.

Lee, C. S. G., Gonzalez, R. C., and Fu, K. S. [1986]. *Tutorial on Robotics,* 2d ed., IEEE Computer Press, Silver Spring, Md.

Lee, C. S. G., and Huang, D. [1985]. "A Geometric Approach to Deriving Position/Force Trajectory in Fine Motion," *Proc. 1985 IEEE Intl. Conf. Robotics and Automation,* St. Louis, Mo, pp. 691–697.

Lee, C. S. G., and Lee, B. H. [1984]. "Resolved Motion Adaptive Control for Mechanical Manipulators," *Trans. ASME, J. Dynamic Systems, Measurement and Control,* vol. 106, no. 2, pp. 134–142.

Lee, C. S. G., Lee, B. H., and Nigam, R. [1983]. "Development of the Generalized d'Alembert Equations of Motion for Mechanical Manipulators," *Proc. 22nd Conf. Decision and Control,* San Antonio, Tex., pp. 1205–1210.

Lee, C. S. G., Mudge, T. N., Turney, J. L. [1982]. "Hierarchical Control Structure Using Special Purpose Processors for the Control of Robot Arms," *Proc. 1982 Pattern Recognition and Image Processing Conf.,* Las Vegas, Nev., pp. 634–640.

Lee, C. S. G., and Ziegler, M. [1984]. "A Geometric Approach in Solving the Inverse Kinematics of PUMA Robots," *IEEE Trans. Aerospace and Electronic Systems,* vol. AES-20, no. 6, pp. 695–706.

Lewis, R. A. [1974]. "Autonomous Manipulation on a Robot: Summary of Manipulator Software Functions," Technical Memo 33-679, Jet Propulsion Laboratory, Pasadena, Calif.

Lewis, R. A., and Bejczy, A. K. [1973]. "Planning Considerations for a Roving Robot with Arm," *Proc. 3rd Intl. Joint Conf. Artificial Intelligence*, Stanford University, Palo Alto, Calif.

Lieberman, L. I., and Wesley, M. A. [1977]. "AUTOPASS: An Automatic Programming System for Computer Controlled Mechanical Assembly," *IBM J. Res. Devel.*, vol. 21, no. 4, pp. 321–333.

Lin, C. S., Chang, P. R., and Luh, J. Y. S. [1983]. "Formulation and Optimization of Cubic Polynomial Joint Trajectories for Industrial Robots," *IEEE Trans. Automatic Control*, vol. AC-28, no. 12, pp. 1066–1073.

Lozano-Perez, T. [1981]. "Automatic Planning of Manipulator Transfer Movements," *IEEE Trans. Systems, Man, Cybern.*, vol. SMC-11, no. 10, pp. 691–698.

Lozano-Perez, T. [1982]. "Spatial Planning, A Configuration Space Approach," *IEEE Trans. Comput.*, vol. C-32, no. 2, pp. 108–120.

Lozano-Perez, T. [1983a]. "Robot Programming," *Proc. IEEE*, vol. 71, no. 7, pp. 821–841.

Lozano-Perez, T. [1983b]. "Task Planning," in *Robot Motion: Planning and Control*, (M. Brady, et al., eds.), MIT Press, Cambridge, Mass.

Lozano-Perez, T., and Wesley, M. A. [1979]. "An Algorithm for Planning Collision-Free Paths Among Polyhedral Obstacles," *Comm. ACM*, vol. 22, no. 10, pp. 560–570.

Lu, S. Y., and Fu, K. S. [1978]. "A Syntactic Approach to Texture Analysis," *Comput. Graph. Image Proc.*, vol. 7, no. 3, pp. 303–330.

Luh, J. Y. S. [1983a]. "An Anatomy of Industrial Robots and their Controls," *IEEE Trans. Automatic Control*, vol. AC-28, no. 2, pp. 133–153.

Luh, J. Y. S. [1983b]. "Conventional Controller Design for Industrial Robots—A Tutorial," *IEEE Trans. Systems, Man, and Cybern.*, vol. SMC-13, no. 3, pp. 298–316.

Luh, J. Y. S., and Lin, C. S. [1981a]. "Optimum Path Planning for Mechanical Manipulators," *Trans. ASME, J. Dynamic Systems, Measurement and Control*, vol. 102, pp. 142–151.

Luh, J. Y. S., and Lin, C. S. [1981b]. "Automatic Generation of Dynamic Equations for Mechanical Manipulators," *Proc. Joint Automatic Control Conf.*, Charlottesville, Va., pp. TA-2D.

Luh, J. Y. S., and C. S. Lin. [1982]. "Scheduling of Parallel Computation for a Computer Controlled Mechanical Manipulator," *IEEE Trans. Systems, Man, Cybern.*, vol. SMC-12, pp. 214–234.

Luh, J. Y. S., and Lin, C. S. [1984]. "Approximate Joint Trajectories for Control of Industrial Robots Along Cartesian Path," *IEEE Trans. Systems, Man, Cybern.*, vol. SMC-14, no. 3, pp. 444–450.

Luh, J. Y. S., Walker, M. W., and Paul, R. P. [1980a]. "On-Line Computational Scheme for Mechanical Manipulators," *Trans. ASME, J. Dynamic Systems, Measurements and Control*, vol. 120, pp. 69–76.

Luh, J. Y. S., Walker, M. W., and Paul, R. P. [1980b]. "Resolved-Acceleration Control of Mechanical Manipulators," *IEEE Trans. Automatic Control*, vol. AC-25, no. 3, pp. 468–474.

Marck, V. [1981]. "Algorithms for Complex Tactile Information Processing," *Proc. Intl. Joint Conf. Artificial Intelligence*, pp. 773–774.

Markiewicz, B. R. [1973]. "Analysis of the Computed Torque Drive Method and Comparison with Conventional Position Servo for a Computer-Controlled Manipulator," Technical Memo 33-601, Jet Propulsion Laboratory, Pasadena, Calif.

Marr, D. [1979]. "Visual Information Processing: The Structure and Creation of Visual

Representations," *Proc. Intl. Joint Conf. Artificial Intelligence*, Tokyo, Japan, pp. 1108–1126.

Marr, D. [1982]. *Vision*, Freeman, San Francisco, Calif.

Martelli, A. [1972]. "Edge Detection Using Heuristic Search Methods," *Comput. Graphics Image Proc.*, vol. 1, pp. 169–182.

Martelli, A. [1976]. "An Application of Heuristic Search Methods to Edge and Contour Detection," *Comm. ACM*, vol. 19, no. 2, pp. 73–83.

Mason, M. T. [1981]. "Compliance and Force Control for Computer Controlled Manipulator," *IEEE Trans. Systems, Man, Cybern.*, vol. SMC-11, no. 6, pp. 418–432.

McCarthy, J., et al. [1968]. "A Computer with Hands, Eyes, and Ears," *1968 Fall Joint Computer Conf., AFIPS Proceedings*, pp. 329–338.

McDermott, J. [1980]. "Sensors and Transducers," *EDN*, vol. 25, no. 6, pp. 122–137.

Meindl, J. D., and Wise, K. D., (eds.) [1979]. "Special Issue on Solid-State Sensors, Actuators, and Interface Electronics," *IEEE Trans. Elect. Devices*, vol. 26, pp. 1861–1978.

Merritt, R. [1982]. "Industrial Robots: Getting Smarter All The Time," *Instruments and Control Systems*, vol. 55, no. 7, pp. 32–38.

Milenkovic, V., and Huang, B. [1983]. "Kinematics of Major Robot Linkages," *Proc. 13th Intl. Symp. Industrial Robots*, Chicago, Ill, pp. 16–31 to 16–47.

Mujtaba, S. M., and Goldman, R. [1981]. *AL User's Manual*, 3d ed., STAN-CS-81-889 CSD, Stanford University, Palo Alto, Calif.

Mujtaba, M. S., Goldman, R. A., and Binford, T. [1982]. "The AL Robot Programming Language," *Comput. Engr.*, vol. 2, pp. 77–86.

Mundy, J. L. [1977]. "Automatic Visual Inspection," *Proc. 1977 Conf. Decision and Control*, pp. 705–710.

Murray, J. J., and Neuman, C. P. [1984]. "ARM: An Algebraic Robot Dynamic Modeling Program," *Proc. Intl. Conf. Robotics*, Atlanta, Ga., pp. 103–113.

Myers, W. [1980]. "Industry Begins to Use Visual Pattern Recognition," *Computer*, vol. 13, no. 5, pp. 21–31.

Naccache, N. J. and Shinghal, R. [1984]. "SPTA: A Proposed Algorithm for Thinning Binary Patterns," *IEEE Trans. Systems, Man, Cybern.*, vol. SMC-14, no. 3, pp. 409–418.

Nagel, H. H. [1981]. "Representation of Moving Rigid Objects Based on Visual Observations," *Computer*, vol. 14, no. 8, pp. 29–39.

Nahim, P. J. [1974]. "The Theory of Measurement of a Silhouette Description for Image Processing and Recognition," *Pattern Recog.*, vol. 6, no. 2, pp. 85–95.

Narendra, P. M., and Fitch, R. C. [1981]. "Real-Time Adaptive Contrast Enhancement," *IEEE Trans. Pattern Anal. Machine Intell.*, vol. PAMI-3, no. 6, pp. 655–661.

Nau, D. S. [1983]. "Expert Computer Systems," *Computer*, vol. 16, pp. 63–85.

Nevatia, R., and Binford, T. O. [1977]. "Description and Recognition of Curved Objects," *Artificial Intelligence*, vol. 8, pp. 77–98.

Nevins, J. L., and Whitney, D. E. [1978]. "Computer-Controlled Assembly," *Sci. Am.*, vol. 238, no. 2, pp. 62–74.

Nevins, J. L., Whitney, D. E., et al. [1974–1976]. "Exploratory Research in Industrial Modular Assembly," NSF Project Reports 1 to 4; C. S. Draper Laboratory, Cambridge, Mass.

Newman, W. M., and Sproull, R. F. [1979]. *Principles of Interactive Computer Graphics*, McGraw-Hill, New York.

Neuman, C. P., and Tourassis, V. D. [1985]. "Discrete Dynamic Robot Modelling," *IEEE*

Trans. Systems, Man, Cybern, vol. SMC-15, pp. 193–204.

Neuman, C. P., and Tourassis, V. D. [1983]. "Robot Control: Issues and Insight," *Proc. Third Yale Workshop on Applications of Adaptive Systems Theory,* Yale University, New Haven, Conn., pp. 179–189.

Nigam, R., and Lee, C. S. G. [1985]. "A Multiprocessor-Based Controller for the Control of Mechanical Manipulators," *IEEE J. Robotics and Automation,* vol. RA-1, no. 4, pp. 173–182.

Nilsson, N. J. [1971]. *Problem-Solving Methods in Artificial Intelligence,* McGraw-Hill, New York.

Nilsson, N. J. [1980]. *Principles of Artificial Intelligence,* Tioga Pub., Palo Alto, Calif.

Noble, B. [1969]. *Applied Linear Algebra,* Prentice-Hall, Englewood Cliffs, N.J.

Oldroyd, A. [1981]. "MCL: An APT Approach to Robotic Manufacturing," presented at *SHARE 56,* Houston, Tex., March 9–13, 1981.

Ohlander, R., Price, K., and Reddy, D. R. [1979]. "Picture Segmentation Using a Recursive Region Splitting Method," *Comput. Graphics Image Proc.,* vol. 8, no. 3, pp. 313–333.

Orin, D. E. [1984]. "Pipelined Approach to Inverse Plant Plus Jacobian Control of Robot Manipulators," *Proc. Intl. Conf. Robotics,* Atlanta, Ga., pp. 169–175.

Orin, D. E., and Schrader, W. W. [1984]. "Efficient Computation of the Jacobian for Robot Manipulators," *Intl. J. Robotics Res.,* vol. 3, no. 4, pp. 66–75.

Orin, D. E., McGhee, R. B., Vukobratovic, M., and Hartoch, G. [1979]. "Kinematic and Kinetic Analysis of Open-Chain Linkages Utilizing Newton-Euler Methods," *Math. Biosci.,* vol. 43, pp. 107–130.

Papoulis, A. [1965]. *Probability, Random Variables, and Stochastic Processes,* McGraw-Hill, New York.

Park, W. T. [1981]. *The SRI Robot Programming System (RPS): An Executive Summary,* SRI International, Menlo Park, Calif.

Paul, R. P. [1972]. "Modeling, Trajectory Calculation, and Servoing of a Computer Controlled Arm," *Memo AIM-177;* Stanford Artificial Intelligence Laboratory, Palo Alto, Calif.

Paul, R. P. [1976]. "WAVE: A Model-Based Language for Manipulator Control," *Technical Paper MR76-615,* Society of Manufacturing Engineers, Dearborn, Mich. Also appears in *The Industrial Robot,* vol. 4, 1977, pp. 10–17.

Paul, R. P. [1979]. "Manipulator Cartesian Path Control," *IEEE Trans. Systems, Man, Cybern.,* vol. SMC-9, no. 11, pp. 702–711.

Paul, R. P. [1981]. *Robot Manipulator: Mathematics, Programming and Control,* MIT Press, Cambridge, Mass.

Paul, R. P. and Shimano, B. [1976]. "Compliance and Control" *Proc. Joint Automatic Control Conference,* Purdue University, West Lafayette, Ind.

Paul, R. P., Shimano, B. E., and Mayer, G. [1981]. "Kinematic Control Equations for Simple Manipulators," *IEEE Trans. Systems, Man, Cybern.,* vol. SMC-11, no. 6, pp. 449–455.

Pavlidis, T. [1977]. *Structural Pattern Recognition,* Springer-Verlag, New York.

Pavlidis, T. [1982]. *Algorithms for Graphics and Image Processing,* Computer Science Press, Rockville, Md.

Persoon, E., and Fu, K. S. [1977]. "Shape Discrimination Using Fourier Descriptors," *IEEE Trans. Systems, Man, Cybern.,* vol. SMC-7, no. 2, pp. 170–179.

Pieper, D. L. [1968]. "The Kinematics of Manipulators under Computer Control," Artificial Intelligence Project Memo No. 72., Computer Science Department, Stanford University, Palo Alto, Calif.

Pieper, D. L. and Roth, B. [1969]. "The Kinematics of Manipulators under Computer Control," *Proc. II Intl. Congr. Theory of Machines and Mechanisms,* vol. 2, pp. 159–168.

Pipes, L. A. [1963]. *Matrix Methods in Engineering,* Prentice-Hall, Englewood Cliffs, N.J.

Popplestone, R. J., Ambler, A. P., and Bellos, I. [1978]. "RAPT, A Language for Describing Assemblies," *Industrial Robot,* vol. 5, no. 3, pp. 131–137.

Popplestone, R. J., Ambler, A. P., and Bellos, I. [1980]. "An Interpreter for a Language Describing Assemblies," *Artificial Intelligence,* vol. 14, no. 1, pp. 79–107.

Raibert, M. H., and Craig, J. J. [1981]. "Hybrid Position/Force Control of Manipulators," *Trans. ASME, J. Dynamic Systems, Measurement, and Control,* vol. 102, pp. 126–133.

Raibert, M. H., and Tanner, J. E. [1982]. "Design and Implementation of a VLSI Tactile Sensing Computer," *Intl. J. Robotics Res.,* vol. 1, no. 3, pp. 3–18.

Rajala, S. A., Riddle, A. N., and Snyder, W. E. [1983]. "Application of the One-Dimensional Fourier Transform for Tracking Moving Objects in Noisy Environments," *Comput. Vision, Graphics, Image Proc.,* vol. 2, pp. 280–293.

Ramer, U. [1975]. "Extraction of Line Structures from Photographs of Curved Object," *Comput. Graphics Image Proc.,* vol. 4, pp. 81–103.

Reddy, D. R., and Hon, R. W. [1979]. "Computer Architectures for Vision," in *Computer Vision and Sensor-Based Robots* (G. G. Dodd and L. Rossol, eds.), Plenum, New York.

Requicha, A. [1980]. "Representation for Rigid Solids: Theory, Methods, and Systems," *Computing Surveys,* vol. 12, no. 4, pp. 437–464.

Requicha, A. [1983]. "Towards a Theory of Geometric Tolerancing," *Intl. J. Robotics Res.,* vol. 2, no. 4, pp. 45–60.

Requicha, A., and Voelcker, H. B. [1982]. "Solid Modeling: A Historical Summary and Contemporary Assessment," *IEEE Comput. Graphics and Applications,* vol. 2, no. 2, pp. 9–24.

Rich, E. [1983]. *Artificial Intelligence,* McGraw-Hill, New York.

Roberts, L. G. [1965]. "Machine Perception of Three-Dimensional Solids," in *Optical and Electro-Optical Information Processing,* (J. P. Tippett et al., eds.), MIT Press, Cambridge, Mass.

Rocher, F., and Keissling, A. [1975]. "Methods for Analyzing Three-Dimensional Scenes," *Proc. 4th Intl. Joint Conf. Artificial Intelligence,* pp. 669–673.

Rosen, C. A., and Nitzan, D. [1977]. "Use of Sensors in Programmable Automation," *Computer,* vol. 10, no. 12, pp. 12–23.

Rosenfeld, A., and Kak, A. C. [1982]. *Digital Picture Processing,* 2d ed., Academic Press, New York.

Roth, B., Rastegar, J., and Scheinman, V. [1973]. "On the Design of Computer Controlled Manipulators," *1st CISM-IFTMM Symp. Theory and Practice of Robots and Manipulators,* pp. 93–113.

Sacerdoti, E. D. [1977]. *A Structure for Plans and Behavior,* Elsevier, New York.

Sadjadi, F. A., and Hall, E. L. [1980]. "Three-Dimensional Moment Invariants," *IEEE Trans. Pattern Anal. Mach. Intell.,* vol. PAMI-2, no. 2, pp. 127–136.

Salari, E., and Siy, P. [1984]. "The Ridge-Seeking Method for Obtaining the Skeleton of Digital Images," *IEEE Trans. Systems Man, Cybern.,* vol. SMC-14, no. 3, pp. 524–528.

Saridis, G. N. [1983]. "Intelligent Robotic Control," *IEEE Trans. Automatic Control,* vol. AC-28, no. 5, pp. 547–557.

Saridis, G. N., and Lee, C. S. G. [1979]. "An Approximation Theory of Optimal Control for Trainable Manipulators," *IEEE Trans. Systems, Man, Cybern.,* vol. SMC-9, no. 3, pp. 152–159.

Saridis, G. N., and Lobbia, R. N. [1972]. "Parameter Identification and Control of Linear Discrete-Time Systems," *IEEE Trans. Automatic Control*, vol. AC-17, no. 1, pp. 52–60.

Saridis, G. N., and Stephanou, H. E. [1977]. "A Hierarchical Approach to the Control of a Prosthetic Arm," *IEEE Trans. Systems, Man, Cybern.*, vol. SMC-7, no. 6, pp. 407–420.

Scheinman, V. D. [1969]. "Design of a Computer Manipulator," *Artificial Intelligence Laboratory Memo AIM-92*, Stanford University, Palo Alto, Calif.

Shani, U. [1980]. "A 3-D Model-Driven System for the Recognition of Abdominal Anatomy from CT Scans," *Proc. 5th Intl. Joint Conf. Pattern Recog.*, pp. 585–591.

Shimano, B. [1979]. "VAL: A Versatile Robot Programming and Control System," *Proc. 3rd Intl. Computer Software Applications Conf.*, Chicago, Ill, pp. 878–883.

Shimano, B. E., and Roth, B. [1979]. "On Force Sensing Information and its Use in Controlling Manipulators," *Proc. 9th Intl. Symp. on Industrial Robots*, Washington, D.C., pp. 119–126.

Shirai, Y. [1979]. "Three-Dimensional Computer Vision," in *Computer Vision and Sensor-Based Robots* (G. G. Dodd and L. Rossol, eds.), Plenum, New York.

Siklossy, L. [1972]. "Modelled Exploration by Robot," Tech. Rept. 1, Computer Science Department, University of Texas, Austin, Tex.

Siklossy, L., and Dreussi, J. [1973]. "An Efficient Robot Planner which Generates its Own Procedures, *Proc. 3rd Intl. Joint Conf. Artificial Intelligence*, pp. 423, 430.

Silver, W. M. [1982]. "On the Equivalence of the Lagrangian and Newton-Euler Dynamics for Manipulators," *Intl. J. Robotics Res.*, vol. 1, no. 2, pp. 60–70.

Sklansky, J., Chazin, R. L., and Hansen, B. J. [1972]. "Minimum-Perimeter Polygons of Digitized Silhouettes," *IEEE Trans. Comput.*, vol. C-21, no. 3, pp. 260–268.

Snyder, W. E. [1985]. *Industrial Robots: Computer Interfacing and Control*, Prentice-Hall, Englewood Cliffs, N.J.

Spencer, J. D. [1980]. "Versatile Hall-Effect Devices Handle Movement-Control Tasks," *EDN*, vol. 25, no. 3, pp. 151–156.

Stepanenko Y., and Vukobratovic, M. [1976]. "Dynamics of Articulated Open-Chain Active Mechanisms," *Math. Biosci.*, vol. 28, pp. 137–170.

Suh, C. H., and Radcliffe, C. W. [1978]. *Kinematics and Mechanisms Design,* John Wiley, New York.

Sussman, G. J., Winograd, T., and Charniak, E. [1970]. "Micro-Planner Reference Manual," AI Memo 203, MIT Press, Cambridge, Mass.

Symon, K. R. [1971]. *Mechanics,* Addison-Wesley, Reading, Mass.

Sze, T. W., and Yang, Y. H. [1981]. "A Simple Contour Matching Algorithm," *IEEE Trans. Pattern Anal. Mach. Intell.*, vol. PAMI-3, no. 6, pp. 676–678.

Takase, K., Paul, R. P., and Berg, E. J. [1981]. "A Structural Approach to Robot Programming and Teaching," *IEEE Trans. Systems, Man, Cybern.*, vol. SMC-11, no. 4, pp. 274–289.

Takegaki, M., and Arimoto, S. [1981]. "A New Feedback Method for Dynamic Control of Manipulators," *Trans. ASME, J. Dynamic Systems, Measurement and Control*, vol. 102, pp. 119–125.

Tangwongsan, S., and Fu, K. S. [1979]. "An Application of Learning to Robotic Planning," *Intl. J. Computer and Information Sciences*, vol. 8, no. 4, pp. 303–333.

Tarn, T. J. et al. [1984]. "Nonlinear Feedback in Robot Arm Control," *Proc. 1984 Conf. Decision and Control,"* Las Vegas, Nev., pp. 736–751.

Taylor, R. H. [1976]. "The Synthesis of Manipulator Control Programs from Task-Level

Specifications," Report AIM-282, Artificial Intelligence Laboratory, Stanford University, Palo Alto, Calif.

Taylor, R. H. [1979]. "Planning and Execution of Straight Line Manipulator Trajectories," *IBM J. Res. Devel.*, vol. 23, no. 4, pp. 424–436.

Taylor, R. H., Summers, P. D., and Meyer, J. M. [1983]. "AML: A Manufacturing Language," *Intl. J. Robotics Res.*, vol. 1, no. 3, pp. 19–41.

Thompson, W. B. and Barnard, S. T. [1981]. "Lower-Level Estimation and Interpretation of Visual Motion," *Computer*, vol. 14, no. 8, pp. 20–28.

Thrall, R. M., and Tornheim, L. [1963]. *Vector Spaces and Matrices,* John Wiley, New York.

Tomita, F., Shirai, Y., and Tsuji, S. [1982]. "Description of Texture by a Structural Analysis," *IEEE Trans. Pattern Anal. Mach. Intell.*, vol. PAMI-4, no. 2, pp. 183–191.

Tomovic, R., and Boni, G. [1962]. "An Adaptive Artificial Hand," *IRE Trans. Automatic Control*, vol. AC-7, no. 3, pp. 3–10.

Toriwaki, J. I., Kato, N., and Fukumura, T. [1979]. "Parallel Local Operations for a New Distance Transformation of a Line Pattern and Their Applications," *IEEE Trans. Systems, Man, Cybern.*, vol. SMC-9, no. 10, pp. 628–643.

Tou, J. T. (ed.) [1985]. *Computer-Based Automation*, Plenum, New York.

Tou, J. T., and Gonzalez, R. C. [1974]. *Pattern Recogniton Principles*, Addison-Wesley, Reading, Mass.

Turney, J. L., Mudge, T. N., and Lee, C. S. G. [1980]. "Equivalence of Two Formulations for Robot Arm Dynamics," SEL Report 142, ECE Department, University of Michigan, Ann Arbor, Mich.

Turney, J. L., Mudge, T. N., and Lee, C. S. G. [1982]. "Connection Between Formulations of Robot Arm Dynamics with Applications to Simulation and Control," CRIM Technical Report No. RSD-TR-4-82, the University of Michigan, Ann Arbor, Mich.

Uicker, J. J. [1965]. "On the Dynamic Analysis of Spatial Linkages using 4×4 Matrices," Ph.D. dissertation, Northwestern University, Evanston, Ill.

Uicker, J. J., Jr., Denavit, J., and Hartenberg, R. S. [1964]. "An Iterative Method for the Displacement Analysis of Spatial Mechanisms," *Trans. ASME, J. Appl. Mech.*, vol. 31, Series E, pp. 309–314.

Unger, S. H. [1959]. "Pattern Detection and Recognition," *Proc. IRE*, vol. 47, no. 10, pp. 1737–1752.

User's Guide to VAL, [1979]. Version 11, 2d ed., Unimation, Inc., Danbury, Conn.

Utkin, V. I. [1977]. "Variable Structure Systems with Sliding Mode: A Survey," *IEEE Trans. Automatic Control,* vol. AC-22, pp. 212–222.

Vukobratovic, M., and Stokic, D. [1980]. "Contribution to the Decoupled Control of Large-Scale Mechanical Systems," *Automatica,* vol. 16, 1980, pp. 16–21.

Walker, M. W., and Orin, D. E. [1982]. "Efficient Dynamic Computer Simulation of Robotic Mechanisms," *Trans. ASME, J. Systems, Measurement and Control,* vol. 104, pp. 205–211.

Wallace, T. P., and Mitchell, O. R. [1980]. "Analysis of Three-Dimensional Movements Using Fourier Descriptors," *IEEE Trans. Pattern Anal. Machine Intell.*, vol. PAMI-2, no. 6, pp. 583–588.

Waltz, D. I. [1972]. "Generating Semantic Descriptions from Drawings of Scenes with Shadows," Ph.D. Dissertation, Artificial Intelligence Lab., MIT, Cambridge, Mass.

Waltz, D. I. [1976]. "Automata Theoretical Approach to Visual Information Processing," in *Applied Computation Theory* (R. T. Yeh, ed.), Prentice-Hall, Englewood Cliffs, N.J.

Webb, J. A., and Aggarwal, J. K. [1981]. "Visually Interpreting the Motion of Objects in

Space," *Computer*, vol. 14, no. 8, pp. 40–49.

Weiss, S. M., and Kulikowski, C. A. [1984]. *A Practical Guide to Designing Expert Systems*, Rowman and Allanheld, New Jersey.

Weska, J. S. [1978]. "A Survey of Threshold Selection Techniques," *Comput. Graphics Image Proc.*, vol. 7, pp. 259–265.

Wesley, M. A., et al., [1980]. "A Geometric Modeling System for Automated Mechanical Assembly," *IBM J. Res. Devel.*, vol. 24, no. 1, pp. 64–74.

White, J. M., and Rohrer, G. D. [1983]. "Image Thresholding for Optical Character Recognition and Other Applications Requiring Character Image Extraction," *IBM J. Res. Devel.*, vol. 27, no. 4, pp. 400–411.

Whitney, D. E. [1969a]. "Resolved Motion Rate Control of Manipulators and Human Prostheses," *IEEE Trans. Man-Machine Systems,* vol. MMS-10, no. 2, pp. 47–53.

Whitney, D. E. [1969b]. "State Space Models of Remote Manipulation Tasks," *Proc. Intl. Joint Conf. Artificial Intelligence*, Washington, D.C., pp. 495–508.

Whitney, D. E. [1972]. "The Mathematics of Coordinated Control of Prosthetic Arms and Manipulators," *Trans. ASME, J. Dynamic Systems, Measurement and Control,* vol. 122, pp. 303–309.

Will, P., and Grossman, D. [1975]. "An Experimental System for Computer Controlled Mechanical Assembly," *IEEE Trans. Comput.*, vol. C-24, no. 9, pp. 879–888.

Winston, P. H. [1984]. *Artificial Intelligence*, 2d ed., Addison-Wesley, Reading, Mass.

Wise, K. D. (ed.) [1982]. "Special Issue on Solid-State Sensors, Actuators, and Interface Electronics," *IEEE Trans. Elect. Devices*, vol. 29, pp. 42–48.

Wolfe, G. J., and Mannos, J. L [1979]. "Fast Median Filter Implementation," *Proc. Soc. Photo-Optical Inst. Engr.*, vol. 207, pp. 154–160.

Woods, R. E., and Gonzalez, R. C. [1981]. "Real-Time Digital Image Enhancement," *Proc. IEEE*, vol. 69, no. 5, pp. 643–654.

Wu, C. H., and Paul, R. P. [1982]. "Resolved Motion Force Control of Robot Manipulator," *IEEE Trans. Systems, Man, Cybern.*, vol. SMC-12, no. 3, pp. 266–275.

Yakimovsky, Y., and Cunningham, R. [1979]. "A System for Extracting Three-Dimensional Measurements from a Stereo Pair of TV Cameras," *Comput. Graphics Image Proc.*, vol. 7, pp. 195–210.

Yang, A. T. [1969]. "Displacement Analysis of Spatial Five-link Mechanisms Using 3×3 Matrices with Dual-Number Elements," *Trans. ASME, J. Engr. Industry,* vol. 91, no. 1, Series B, pp. 152–157.

Yang, A. T., and Freudenstein, R. [1964]. "Application of Dual Number Quaternian Algebra to the Analysis of Spatial Mechanisms," *Trans. ASME, J. Appl. Mech.*, vol. 31, series E, pp. 152–157.

Young, K. K. D. [1978]. "Controller Design for a Manipulator Using Theory of Variable Structure Systems," *IEEE Trans. Systems, Man, Cybern.,* vol. SMC-8, no. 2, pp. 101–109.

Yuan, M. S. C., and Freudenstein, R. [1971]. "Kinematic Analysis of Spatial Mechanisms by Means of Screw Coordinates," *Trans. ASME, J. Engr. Industry,* vol. 93, no. 1, pp. 61–73.

Zahn, C. T., and Roskies, R. Z. [1972]. "Fourier Descriptors for Plane Closed Curves," *IEEE Trans. Comput.*, vol. C-21, no. 3, pp. 269–281.

Zucker, S. W. [1976]. "Region Growing: Childhood and Adolescence," *Comput. Graphics Image Proc.*, vol. 5, pp. 382–399.

Zucker, S. W., and Hummel, R. A. [1981]. "A Three-Dimensional Edge Operator," *IEEE Trans. Pattern Anal. Mach. Intell.,* vol. PAMI-3, no. 3, pp. 324–331.

INDEX